International Conference on Advances in Materials Processing and Characterization (ICAMPC 2019)

IOP Conference Series: Materials Science and Engineering
Volume 764

Sathyamangalam, India
10 - 11 September 2019

Editors:

Prashanth Konda Gokuldoss
R Shanmuga Prakash
S Vignesh
V Hariprasad
S Arun Kumar

ISBN: 978-1-7138-3501-1
ISSN: 1757-8981

Printed from e-media with permission by:

Curran Associates, Inc.
57 Morehouse Lane
Red Hook, NY 12571

Some format issues inherent in the e-media version may also appear in this print version.

Additional copies of this publication are available from:

Curran Associates, Inc.
57 Morehouse Lane
Red Hook, NY 12571 USA
Phone: 845-758-0400
Fax: 845-758-2633
Email: curran@proceedings.com
Web: www.proceedings.com

TABLE OF CONTENTS

IOP Conf. Series: Materials Science and Engineering **764** (2020) 011001 doi:10.1088/1757-899X/764/1/011001

Stay Ahead

BANNARI AMMAN INSTITUTE OF TECHNOLOGY

Autonomous Institution, Accredited by NAAC with 'A' Grade

ICAMPC

International Conference on Advances in Materials Processing and Characterization (ICAMPC 2019)

10 – 11 September 2019

Organized by

Departments of Aeronautical, Automobile, Mechanical Engineering and Mechatronics

IOP Conf. Series: Materials Science and Engineering **764** (2020) 011001 doi:10.1088/1757-899X/764/1/011001

International Conference on Advances in Materials Processing and Characterization (ICAMPC 2019)

10 – 11 September 2019

Convenor

Dr K L Senthil Kumar

Chief Editor

Prof Prashanth Konda Gokuldoss

Editors

Prof R Shanmuga Prakash

Prof S Vignesh

Prof V Hariprasad

Prof S Arun Kumar

IOP Conf. Series: Materials Science and Engineering **764** (2020) 011001 doi:10.1088/1757-899X/764/1/011001

Index

IOP Conf. Series: Materials Science and Engineering **764** (2020) 011001 doi:10.1088/1757-899X/764/1/011001

Preface

The ***International Conference on Advances in Materials Processing and Characterization*** (ICAMPC 2019) was held on 10 and 11 September 2019 at Bannari Amman Institute of Technology (BIT), Sathyamangalam, India and was organized by the Departments of Aeronautical, Automobile, Mechanical and Mechatronics. The purpose of the conference was to collaborate among students, researchers and scientists around the world focusing on materials research involving topics such as additive manufacturing, powder metallurgy, composites, surface engineering, metal joining, materials testing, nanomaterials, numerical modeling and simulation, energy materials, smart materials etc. This conference provided an opportunity and platform for presenters to showcase their work and discuss them with like-minded researchers. We received over 150 research articles and short listed 64 articles based on peer review for presentation cum possible publication in IOP Conference Series: Materials Science and Engineering (p-ISSN:1757-8981, e-ISSN:1757-899X).

The invited keynote speakers were **Dr. N.L. Parthasarathi**, Scientist - E, Indira Gandhi Centre for Atomic Research, Kalpakkam, India and **Dr. K.G. Pradeep**, Assistant Professor, Department of Metallurgical and Materials Engineering, Indian Institute of Technology, Madras. The conference was partially sponsored by Indian Society for Technical Education (ISTE), New Delhi and Wollega University, Nekemte, Ethiopia.

The Editorial Board consisted of the Chief Editor, **Prof. Prashanth Konda Gokuldoss**, Department of Manufacturing and Industrial Engineering, Talinn University of Technology, Tallinn, Estonia and the Editors from BIT Sathyamnagalam were **Prof. R. Shanmuga Prakash**, Department of Mechanical Engineering, **Prof. S. Vignesh**, Department of Mechatronics, **Prof. V. Hariprasad**, Department of Aeronautical Engineering and **Prof. S. Arun Kumar**, Department of Mechatronics.

We are indebted to our reviewers, committee members and volunteers for their time, effort and suggestions. We are thankful to **Dr. S. V. Balasubramaniam**, Chairman, **Dr. M. P. Vijayakumar**, Trustee, **Dr. C. Palanisamy**, Principal, **Dr. K. Sivakumar**, Dean PDS, **Dr. P. Thangaraj**, Dean ATSD, **Dr. S. Jegadheeswaran**, Professor Incharge, R&D.

With Regards,
Dr. K.L. Senthil Kumar
Convenor, ICAMPC 2019
Professor & Head, Mechatronics
Bannari Amman Institute of Technology, Sathyamangalam, India.

IOP Conf. Series: Materials Science and Engineering **764** (2020) 011001 doi:10.1088/1757-899X/764/1/011001

About BIT

Bannari Amman Institute of Technology (BIT) established in 1996 by the Bannari Amman Group, serves as one of the citadels of higher education in the Southern part of India. BIT is an Autonomous Institution approved by AICTE, New Delhi and affiliated to Anna University, Chennai. The institute offers 19 undergraduate, 15 postgraduate programmes in Engineering, Technology and Management studies. All the eligible departments of Engineering, Technology and Management studies are recognised by Anna University, Chennai to offer Ph.D. programmes. The institution is re-accredited by National Assessment and Accreditation Council (NAAC) with a CGPA of 3.57 on a four point scale with A Grade. BIT is also certified for implementing 5S workplace management system and most of the eligible courses are accredited by National Board of Accreditation (NBA), New Delhi.

Nestled on the banks of the river Bhavani, BIT campus provides environment for learning in harmony with nature, away from the odds of the city life. The spacious and the earth hugging buildings punctuated with landscaped courtyards and pathways are designed to emphasise the business ethics and characteristics of an excellent centre for learning. The Campus hosts well planned academic blocks, computer centres, lecture halls, libraries, laboratories, conference halls, staff quarters, hostels, students' centres, gymnasium, ATM and general health clinic etc.

BIT aids to impart knowledge, teamwork, innovation, entrepreneurship, courage, sacrifice and duty which are innards of a meaningful life. BIT has transformed and ingrained various cutting edge administrative and academic reforms. The Institute caters to the needs of the students and members of faculty in realising their dreams and aspirations. BIT has been conferred with a number of accolades and recognitions from top-notch forums of high eminence.

Bannari Amman Institute of Technology
Sathyamangalam - TK
Erode District – 638 401
Tamilnadu, INDIA.
Ph: 04295-226000, Fax: 04295-226666
E-mail: stayahead@bitsathy.ac.in
Website: www.bitsathy.ac.in

IOP Conf. Series: Materials Science and Engineering **764** (2020) 011001 doi:10.1088/1757-899X/764/1/011001

Conference Message

IOP Conf. Series: Materials Science and Engineering **764** (2020) 011001 doi:10.1088/1757-899X/764/1/011001

BANNARI AMMAN INSTITUTE OF TECHNOLOGY
An Autonomous Institution Affiliated to Anna University Chennai - Approved by AICTE - Accredited by NAAC with "A" Grade
SATHYAMANGALAM - 638 401 ERODE DISTRICT TAMIL NADU INDIA
Stay Ahead
Ph: 04295-226000 / 221289 Fax: 04295-226666 E-mail: stayahead@bitsathy.ac.in Web: www.bitsathy.ac.in

Dr S V BALASUBRAMANIAM

Chairman, BIT

MESSAGE

I am delighted to note that the Departments of Mechatronics, Mechanical, Aeronautical and Automobile Engineering are jointly organizing an International Conference on Advances in Materials Processing and Characterization during 10-11 September 2019.

Technology is a necessity of the modern lifestyle. Every day there are new ideas and advancements in technology as the world becomes more involved in discovering and utilizing the potential of natural resources. Technology has improved human lives significantly by providing convenience and efficiency.

This conference is intended to bring together, a multi-disciplinary group of researchers and engineers from all over the world to present and exchange break-through ideas relating to the field of Materials Processing and Characterization. I hope the insightful paper presentations, panel discussions and workshops will provide ample opportunities for discussions, debate, and exchange of ideas and information among the conference participants. I appreciate the members of the organizing committee for their efforts and best wishes for the successful conduct of the conference.

S V BALASUBRAMANIAM

IOP Conf. Series: Materials Science and Engineering **764** (2020) 011001 doi:10.1088/1757-899X/764/1/011001

BANNARI AMMAN INSTITUTE OF TECHNOLOGY
An Autonomous Institution Affiliated to Anna University Chennai - Approved by AICTE - Accredited by NAAC with "A" Grade
SATHYAMANGALAM - 638 401 ERODE DISTRICT TAMIL NADU INDIA
Ph: 04295-226000 / 221289 Fax: 04295-226666 E-mail: stayahead@bitsathy.ac.in Web: www.bitsathy.ac.in
Stay Ahead

Dr M P VIJAYAKUMAR

Trustee, BIT

MESSAGE

It is a pleasure to know that the *International Conference on Advances in Materials Processing and Characterization* is being held on 10th and 11th September 2019 and is being jointly organized by the departments of Mechatronics, Mechanical, Aeronautical and Automobile Engineering.

Technology improvements can be done only by choosing better materials. It is imperative to select the right material for any given modern application. This is possible by researching the ways new materials can be made and how old materials can be synthesized better.

This conference helps young students, research scholars, academicians and industry personnel to learn interesting and new things. A conference, such as this one, acts as the place where bright minds around the world congregate to share their experiences and findings.

I wish that this conference is a success in sharing technology and ideas in processing materials and more importantly, finding uses for these materials.

M P VIJAYAKUMAR

IOP Conf. Series: Materials Science and Engineering **764** (2020) 011001 doi:10.1088/1757-899X/764/1/011001

BANNARI AMMAN INSTITUTE OF TECHNOLOGY
An Autonomous Institution Affiliated to Anna University Chennai - Approved by AICTE - Accredited by NAAC with "A" Grade
SATHYAMANGALAM - 638 401 ERODE DISTRICT TAMIL NADU INDIA
Stay Ahead
Ph: 04295-226000 / 221289 Fax: 04295-226666 E-mail: stayahead@bitsathy.ac.in Web: www.bitsathy.ac.in

Dr C PALANISAMY

Principal, BIT

MESSAGE

I have understood that an international conference on 'Advances in Materials Processing and Characterization' is jointly organized by four of our departments. Materials are so important that human civilization has been demarcated by the materials used during those times - from the Stone Age and Bronze Age to the now Silicon Age.

Materials Engineering and research is the study of matter and how it is made. I hope this conference will bring out the latest research on materials.

Work on materials science and engineering will yield better materials for the future and will provide sustainable alternatives. This will impact every part of society in a positive manner, by having the best possible material for everything we can invent or imagine.

In the current scenario, emerging materials will find use in solving technological and environmental problems in society. I am certain that this conference aims at disseminating information and technological advancements on materials for the benefit of humankind and society. I wish everyone at the conference the best.

C PALANISAMY

BANNARI AMMAN INSTITUTE OF TECHNOLOGY
An Autonomous Institution Affiliated to Anna University Chennai - Approved by AICTE - Accredited by NAAC with "A" Grade
SATHYAMANGALAM - 638 401 ERODE DISTRICT TAMIL NADU INDIA
Stay Ahead
Ph: 04295-226000 / 221289 Fax: 04295-226666 E-mail: stayahead@bitsathy.ac.in Web: www.bitsathy.ac.in

Dr K SIVAKUMAR

Dean PDS, BIT

MESSAGE

I am delighted that the *International Conference on Advances in Materials Processing and Characterization* is being organized.

Modern materials include a wide range of materials, not only metals, polymers and ceramics but also semiconductors, biological materials, optical and magnetic materials, smart materials and nanostructured materials.

Everything is composed of materials, from computer chips to artificial hips. The future will exist only by discovering and designing new materials.

The ways by which materials are processed dictate how they can be put to use. Different processing routes allow for different structures and therefore different properties. These differences in properties result in different applications.

I am confident that these discussions will be held at our conference. Every attendee will be benefited out of this conference.

K SIVAKUMAR

IOP Conf. Series: Materials Science and Engineering **764** (2020) 011001 doi:10.1088/1757-899X/764/1/011001

BANNARI AMMAN INSTITUTE OF TECHNOLOGY
An Autonomous Institution Affiliated to Anna University Chennai - Approved by AICTE - Accredited by NAAC with "A" Grade
SATHYAMANGALAM - 638 401 ERODE DISTRICT TAMIL NADU INDIA
Stay Ahead Ph: 04295-226000 / 221289 Fax: 04295-226666 E-mail: stayahead@bitsathy.ac.in Web: www.bitsathy.ac.in

Dr P THANGARAJ

Dean TSD, BIT

MESSAGE

I am very happy to be a part of the Advisory Committee of this International Conference. The theme of this conference, Advances in Materials Processing and Characterization, has a paramount importance in core disciplines of Mechanical Engineering, Automobile Engineering, Aeronautical Engineering, and Mechatronics. Bannari Amman Institute of Technology (BIT) is a place where skills are nurtured organically. I believe that the campus provides a vibrant ambience to the external delegates for knowledge sharing forums.

I am certain that the deliberations at the highest level in this conference will gain benefit to all the aspiring fraternity of researchers and faculty members, particularly in the field of materials science and engineering.

I wish success of this international conference on advances in Materials Processing and Characterization!

P THANGARAJ

IOP Conf. Series: Materials Science and Engineering **764** (2020) 011001 doi:10.1088/1757-899X/764/1/011001

BANNARI AMMAN INSTITUTE OF TECHNOLOGY
An Autonomous Institution Affiliated to Anna University Chennai - Approved by AICTE - Accredited by NAAC with "A" Grade
SATHYAMANGALAM - 638 401 ERODE DISTRICT TAMIL NADU INDIA
Ph: 04295-226000 / 221289 Fax: 04295-226666 E-mail: stayahead@bitsathy.ac.in Web: www.bitsathy.ac.in
Stay Ahead

Dr S JEGADHEESWARAN

Professor I/C R&D, BIT

MESSAGE

The research and development in materials and manufacturing has long been a significant impact on the country's economy as well as on the societal development. We, as research and development (R&D) team at Bannari Amman Institute of Technology (BIT), are committed to encourage the fellow colleagues, scholars, and students to involve in various activities that include sharing of acquired knowledge. We are glad to host this International Conference on Advances in Materials Processing and Characterization (ICAMPC) 2019 at BIT, Sathyamangalam.

This conference will deepen and broaden the knowledge on materials and contribute innovation and responsiveness to ever-increasing international needs. I am sure this conference will add enough thrust on the emerging technologies and provide a platform for the researchers and industry personnel to share their findings and cutting-edge research.

I wish ICAMPC 2019 a grand success!

S JEGADHEESWARAN

IOP Conf. Series: Materials Science and Engineering **764** (2020) 011001 doi:10.1088/1757-899X/764/1/011001

Wollega University

Dr HASSEN YUSUF

The President, Wollega University

MESSAGE

I am happy that an ***International Conference on Advances in Materials Processing and Characterization*** is being held in India. It is a pleasure to see an Ethiopian university and an Indian institution come together at this conference. I am glad that Wollega University is pioneering such efforts in Ethiopia. Education and the advancement of knowledge are important for the development of any nation and its people. I am sure that knowledge transfer at this conference will benefit the teachers and academicians of both institutions and countries. This conference will surely bring to the fore some of the excellent research happening around the world.

I wish the organizers all the best in conducting an international conference such as this one.

Hassen Yusuf

IOP Conf. Series: Materials Science and Engineering **764** (2020) 011001 doi:10.1088/1757-899X/764/1/011001

Keynote Address

Dr. N.L. Parthasarathi
Scientist - E
Indira Gandhi Centre for Atomic Research (IGCAR)
Kalpakkam, India

The main objective of materials processing is to enhance the structural features (e.g., crystal structure, microstructure, size, and shape) required for the product to perform well in its intended application. Materials processing is central to the field of materials science and engineering, and is a vital step in manufacturing. Plasma spraying process is a widely used surface modification method employed to achieve different types of coatings. Thanks to its high-temperature heat source, thermal plasma, this makes it possible to melt practically all possible material including all ceramic materials and refractory metals. Plasma Transferred Arc (PTA) Welding is a thermal process for applying wear and corrosion resistant layers on surfaces of metallic materials. The welding process employs a constricted high-energy plasma arc between a non-consumable electrode and the base material. The prime difference between spraying and coating is spraying does not establish a metallurgical bonding similar to the plasma transferred arc welding (hardfacing). Hardfacing is a metalworking process where harder or tougher material is applied to a base metal. It is welded to the base material, and generally takes the form of specialized electrodes for arc welding or filler rod for oxyacetylene and gas tungsten arc welding (TIG) welding. For high temperature applications, Nickel or Cobalt based powder metal alloys are used in (PTA) also called powder plasma welding and thermal spray processes like high-velocity oxygen fuel coating (HVOF), plasma spray, spray and fuse, etc. X-ray powder diffraction (XRD) is a rapid analytical technique primarily used for phase identification of a crystalline material and can provide information on unit cell dimensions. The analyzed material is finely ground, homogenized, and average bulk composition is determined. Optical metallography entails examination of materials using visible light to provide a magnified image of the micro- and macrostructure. In scanning electron microscopy (SEM), the surface of the specimen is bombarded with a beam of electrons to provide information for producing an image.

IOP Conf. Series: Materials Science and Engineering **764** (2020) 011001 doi:10.1088/1757-899X/764/1/011001

Snapshots of ICAMPC 2019

ICAMPC 2019 inauguration.

Welcome address by Dr. K.L. Senthil Kumar, Convenor, ICAMPC 2019.

IOP Conf. Series: Materials Science and Engineering **764** (2020) 011001 doi:10.1088/1757-899X/764/1/011001

Presidential address by Dr. C. Palanisamy, Principal, BIT.

Keynote address by Dr. N.L. Parthasarathi, Scientist - E, IGCAR.

IOP Conf. Series: Materials Science and Engineering **764** (2020) 011001 doi:10.1088/1757-899X/764/1/011001

Delegates gathering for keynote address.

A delegate presenting his research work in a session.

IOP Conf. Series: Materials Science and Engineering **764** (2020) 011002 doi:10.1088/1757-899X/764/1/011002

List of Committees

Advisory Committee

Dr C Palanisamy, Principal, BIT Sathy

Dr K Sivakumar, Dean PDS, BIT Sathy

Dr P Thangaraj, Dean ATSD, BIT Sathy

Dr S Jegadheeswaran, Prof I/C, R&D, BIT Sathy

Dr M Ravi Kumar, HoD Mechanical Engineering, BIT Sathy

Prof K Rajasekar, HoD Aeronautical Engineering, BIT Sathy

Dr N Natarajan, HoD Automobile Engineering, BIT Sathy

Organizing Committee

Convenor

Dr K L Senthil Kumar, Professor, Mechatronics, BIT Sathy

Organizing Secretaries

Dr V C Uvaraja, Mechanical Engineering, BIT Sathy

Dr T Ramesh Kumar, Mechanical Engineering, BIT Sathy

Dr V Srisaran, Mechatronics, BIT Sathy

Coordinators

Mr S Arun Kumar, Mechatronics, BIT Sathy

Mr V Hariprasad, Aeronautical Engineering, BIT Sathy

Mr S Dhayaneethi, Automobile Engineering, BIT Sathy

Sponsors

Indian Society for Technical Education (ISTE), Tamil Nadu Section.

Wollega University, Ethiopia.

IOP Conf. Series: Materials Science and Engineering **764** (2020) 011002 doi:10.1088/1757-899X/764/1/011002

Sub Committees

Committee	Faculty In Charge
Keynote Speakers	Dr Ragupathy Karu (Auto)
Paper Followup + Book of Abstracts	Mr V Hari Prasad (Aero)
Certificate / Banner	Mr L S Gokul (Mech) Mr D Yuvaraj (Mech)
Refreshment	Dr R Rajiev (Mech) Mr D Lakshmanan (Aero)
Accommodation	Mr V D Tamilarasan (Mech)
Reception, Inauguration, Help Desk	Ms K K Megavarthini (Mechatronics) Ms K Poomani (Mechatronics)
Registration	Mr G Sundarraju (Mech)
Finance	Mr S K Dhinesh (Mechatronics) Mr P Nagarajan (Mechatronics) Ms T Jayapriya (Mechatronics)
Publication	Mr R Shanmuga Prakash (Mech) Mr S Vignesh (Mechatronics)
Transport	Mr S Raghavendra Prabhu (Mechatronics)
Programme Schedule	Mr A Tajdeen (Mech) Mr N Bhuvanesh (Mech)

We also thank the contribution by other Teaching and Non-Teaching staff of BIT, Sathyamangalam.

IOP Conf. Series: Materials Science and Engineering **764** (2020) 011003 doi:10.1088/1757-899X/764/1/011003

Peer review statement

All papers published in this volume of *IOP Conference Series: Materials Science and Engineering* have been peer reviewed through processes administered by the proceedings Editors. Reviews were conducted by expert referees to the professional and scientific standards expected of a proceedings journal published by IOP Publishing.

IOP Conf. Series: Materials Science and Engineering **764** (2020) 012001 doi:10.1088/1757-899X/764/1/012001

A comparative study of solid lubricant types on the microstructure and mechanical behaviour of AA7075-zirconium boride aerospace composites

P Loganathan[1], A Gnanavelbabu*[1], K Rajkumar[2], S Ayyanar[1]
[1]Department of Industrial Engineering, CEG Campus, Anna University, Chennai-600025, India.
[2]Department of Mechanical Engineering, SSN College of Engineering, Chennai-600110, India.

*Corresponding author email: agbabu@annauniv.edu

Abstract. Aluminium alloy is the most favourable material for industrial usage because of outstanding low density, specific strength and resistance to oxidation. In this study, zirconium diboride (5 wt. % ZrB_2) as hard ceramic particles and solid lubricants as hexagonal boron nitride (2.5 wt. % hBN) and tungsten disulfide (2.5 wt. % WS_2) particles were used for fabrication of AA 7075 hybrid composite. The fabrication of hybrid composites is done through liquid metallurgy technique. The uniform distribution particles on matrix material were analysed through Optical Microscope (OM) and X-Ray Diffraction (XRD) was used to identify the phase of alloy and hybrid composite materials. The Micro Vickers hardness and tensile strength were used to evaluate the mechanical strength of alloy and composites. The microstructure of composites reveals fairly distributed ceramic and solid lubricant particles over the base matrix. The XRD peaks reveals presence of AA7075, ZrB2, hBN and WS2 with low intensity peaks. AA7075/ZrB_2/WS_2 composite shows an improved hardness and tensile strength as compared to AA7075/ZrB_2/hBN composite. It was observed that the influence of WS2 on AA7075/ZrB_2 composites have a higher bending strength than hBN composite and hBN particles offers easy grain movement during plastic deformation. Tensile fractured surfaces of alloy and hybrid composites were analysed through Scanning Electron Microscope (SEM). It reveals a ductile fracture in AA7075 alloy whereas ductile-brittle is observed in hybrid composites.

1. Introduction

Zinc based aluminium alloy is used for energy saving applications, due to its greater durability to undergo fatigue loading. AA7075 is mainly used in aerospace structural material as a load bearer [1-3]. Improvement in mechanical properties of an alloy is attained through hard ceramic reinforcement particles [4]. Optimal material properties are attained through hybrid composite for required applications and it's achieved through the selection of primary and secondary particles which are not affecting base matrix properties [5]. The strengthening of aluminium alloy is normally done through even distribution of ceramic particles in the matrix. Compared to other ceramic particles, boride based particles are stronger, and its refractory nature is able to withstand elevated temperatures [6]. Major

IOP Conf. Series: Materials Science and Engineering **764** (2020) 012001 doi:10.1088/1757-899X/764/1/012001

applications include hypersonic rocket nozzle inserts and re-entry vehicles systems in the aerospace industries [7].

Good machinability characteristics and improvement in wear resistance of composites are attained through the addition of a solid lubricant. As compare to graphite, hBN have better frictional heating properties with thermal shock resistance and density of hBN is (2.25 g cm-3) which is lower than that of pure aluminium [8]. Tungsten disulphide (WS_2) are well known solid lubricant with lamellar structure with extremely low frictional properties [9-10].

Ashwath and Anthony Xavior [11] studied the mechanical behaviour of SiC and Al_2O_3 hard ceramic particles reinforced with AA2900 and AA2024 alloy. Al_2O_3 played a crucial role in improving the strength of the composite. Hamid Reza Ezatpour et al. [12] concluded that the mechanical and microstructural effect on AA6061/ Nano Al_2O_3 particles was developed through stir casting. The strength and ductility of the composite reduced as increasing in A Al_2O_3 nano particles due to agglomeration of particles and induced porosity.

Syed Nasimul Alam and Lailesh Kumar [13] carried out detailed analysis on Al matrix reinforced with Natural Flake Graphite (NFG) fabricated through powder metallurgy route. The reduction in tensile strength of composite due to brittle nature of Al_4C_3 precipitate at the interface of NFG and Al matrix. Agglomeration of graphite particles leads to poor densification of composite, leading to low level of hardness for composite. Ravi Kumar et al. [14] fabricated AA6063/TiC composite with varying wt. %. The addition of TiC particles reduces ductility of the composite, but other way it increases resistance to the plastic deformation.

Ravi Kumar et al. [15] analysed the microstructural behaviour and mechanical property of AA6082/ZrO_2/Coconut Shell Ash (CSA) hybrid composite. The addition of CSA acts as a soft lubrication medium. VinodKumar et al. [16] studied the microstructure, hardness and tensile behaviour of AA7075/TiB_2 composite processed through ultrasonic assisted stir casting. Agglomerates of TiB_2 particles were broken-down into smaller particles through ultrasonic effect, this provides a fine distribution with a good interfacial bonding between matrix and reinforcement particles which improves the better load bearing capacity. Dinaharan et al. [17] developed AA6061/Al_2Cu composites through friction stir Welding Process (FSP). Large size Al_2Cu particles were broken into smaller particles upto 15 times less than the original size due to severe deformation.

Mathan Kumar and Kumaraswamidhas [18] attempted an analysis on the mechanical and wear behaviour of AA6061/ZrB_2/AlN composite fabricated through stir casting. The mechanical and wear resistance of the composites increases with increasing reinforcement due to strong bonding nature. Kenneth Kanayo Alaneme and Bethel Jeremiah Bamike [19] concluded the hardness and tensile strength decrease with addition of quarry dust (QD) on AA6063/quarry dust (QD)/SiC hybrid composite.

Literatures reveal the application of hBN and WS_2 on AA7075/ZrB_2 composite provides a great opportunity to improve its mechanical and tribological properties. Therefore, this research work investigate the effects of hBN/WS_2 solid lubricant particles on the microstructural characterization and mechanical behaviour of AA7075/ZrB_2 composite.

2. Experimental Setup

The stir casting method was adopted for fabrication of hybrid composites. AA7075 alloy was base matrix with hard ceramic reinforcement particle as ZrB_2 with average particle size is 5.80 μm. WS_2 and hBN were solid lubricants with average particle size of less than 5.00 μm. Base matrix was melted at 800 °C in an electrical resistance furnace. Argon gas was continuously supplied to furnace at a constant flow rate of 2 lpm to avoid any undesirable reactions. Before mixing of reinforcement particles into melt, they were pre-heated at 250°C for improving wettability nature and remove moisture content. The stirring process carried out through coated steel blades to create vortex formation, the pre-heated particles were transferred to molten melt. The stirring speed was maintained at 450 rpm for 5-10 min. Due to smaller size of reinforcement particles, a good bonding and fine distribution was achieved at lower wt.% which was selected for fabrication of composites. The molten melt was poured into a pre-heated steel die with diameter 20 mm and length of 250 mm. Also composite was fabricated in plate form with dimension of 100 X 100 X 10 mm. The T6 heat treatment was carried out for alloy and fabricated hybrid composites. The fabricated samples are shown in figure 1 and its designation are displayed in table 1.

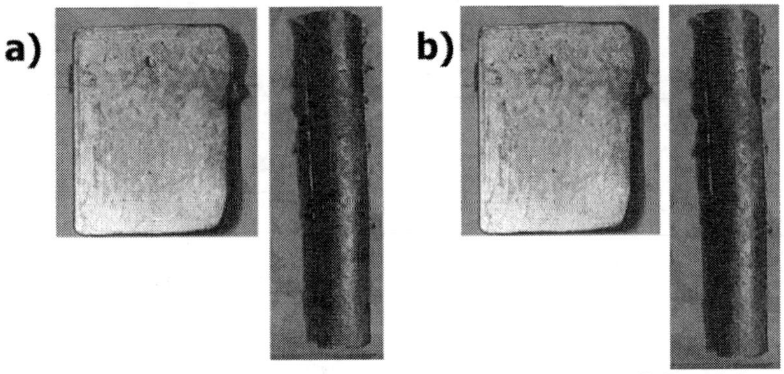

Figure 1. Fabricated AA7075 hybrid composites (a) AA7075+ 5 wt. % ZrB_2+ 2.5 wt. % hBN, (b) AA7075+ 5 wt. % ZrB_2+ 2.5 wt. % WS_2.

Table 1. Fabricated AA7075 hybrid composites.

S.No	Composition	Material
1	Ascast	A0
2	AA7075+ ZrB_2 (5wt%) + 2.5 wt.% hBN	A1
3	AA7075+ ZrB_2 (5wt%) + 2.5 wt.% WS_2	A2

Heat treated samples were subjected to microstructural analysis through an optical microscope. The different grades of SiC abrasive papers were used for polishing sample surface and Keller's reagent were used as etchant medium to enhance the visibility of the grain boundaries. XRD is used for phase analysis of fabricated hybrid composites and X'Pert high score plus software used to identify the phases in the composites. The ASTM E384-16 standard as used for Micro Vickers hardness of the composites with applied load of 500 gf with a dwell time for 15 second. Seven readings were taken at various locations of composite and average value is calculated. The ASTM E8 standard obtained for tensile strength of composite in a computerized universal testing machine (Make: MTS, 100 KN capacity). Initial strain of 0.01 mm/s is fixed for all experiments. Three replicates were carried out and the average values were recorded. Fracture surface of the tested hybrid composites was analysed through SEM (Scanning Electron Microscope).

3. Results and discussions

3.1. Optical Microstructure and XRD Analysis

Figure 2 reveals the optical micrographs and XRD analysis of alloy and hybrid composites. Figure 2a shows the micrograph of A0 alloy. It reveals formation of coarse and fine precipitates of MgZn, AlZn, and Al2MgCu are observed in and around the grain boundaries due to the effect of T6 treatment [20-21].

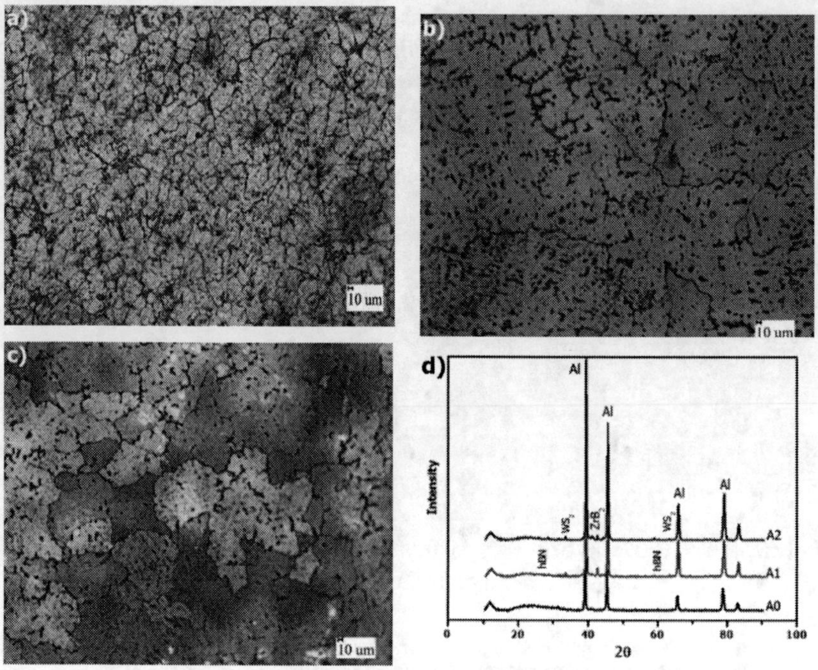

Figure 2. Optical microstructure and XRD of hybrid composites. **2 (a)** A0; **2 (b)** A1; **2 (c)** A2; **2 (d)** XRD analysis of hybrid composites.

The optical micrograph of fabricated hybrid composites is displayed in figure 2(b-c). The micrographs reveals fine distribution of ZrB_2/WS_2/hBN particles over the base matrix and it was attained through dispersion strengthening mechanism and also suspending of particles for longer time period in the molten melt based on difference in wettability factor among the base matrix and reinforcements, this fine distribution of particles tends to increase the mechanical strength of AA7075 hybrid composites [22]. The reinforcement particles obstructed the grain growth in matrix material and also it refines the grain structure.

The X-ray diffraction for the alloy and hybrid composites are presented in figure 2(d). The analysis confirmed no interfacial reaction among the fabricated composites. The diffraction pattern of the Al and reinforcement particles were observed. The presence of WS_2 and hBN is characterized as a solid lubricant showing a lower intensity of peaks exhibiting a binary relationship [23] due to film substrate over the AA7075/ZrB_2 composite materials.

3.2. Mechanical behaviour of AA7075/ZrB2 hybrid Composites

The variation in hardness of AA7075 and hybrid composites is presented in figure 3. The presence of hard ceramic ZrB_2 particle which act as load bearing and resist the dislocations within the matrix which leads to improve hardness of the composites than base matrix. The effect on hardness of composite materials were associated with fine distribution of reinforcement particles and their bonding interface between the AA7075 [24-25]. The interfacial bonding strength of WS_2 was relatively higher than that of hBN, hence hardness of A2 hybrid composite (135.25 VHM) is higher than A1 hybrid composite (127.12 VHM). The Orowan and Hall-Petch strengthening mechanisms is tends to increase the hardness of the composite due to strong interface between the reinforcement and matrix material.

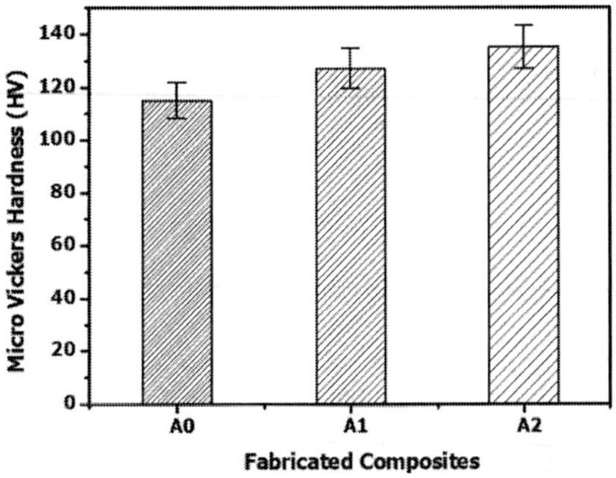

Figure 3. Variation of micro vickers hardness in AA7075 and AA7075 hybrid composites.

3.3. Tensile Strength

Figure 4 shows the tensile strength of alloy and composites. The tensile strength of composite is increased with addition of reinforcement particles as compare to A0 due to the effective load transferring between the matrix and reinforcement particles through better interfacial bonding and effect of strain gradient strengthening in the composites. The addition of hBN particles increase the brittleness of the composite, hence it decrease the tensile strength as compared to WS_2 composite (228 MPa).

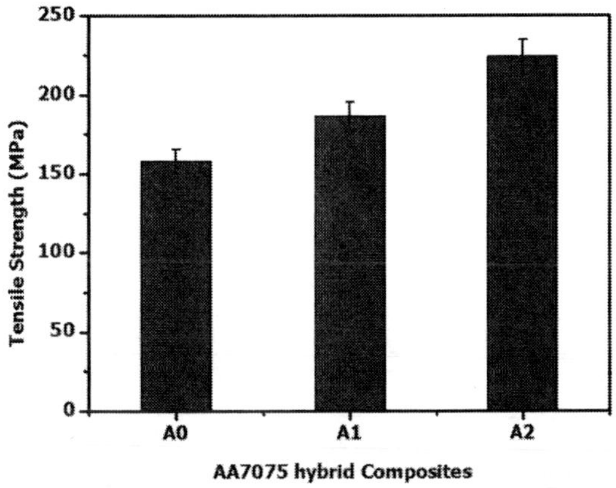

Figure 4. Variation of tensile strength in AA7075 and AA7075 hybrid composites.

Figure 5 (a) shows a fracture surface morphology of A0 alloy which reveals large amount of plastic flow prior to failure and indicating ductile mode of fracture. Figure 5 (b-c) shows the fracture analysis of hybrid composites, which reveals both ductile and brittle fracture mode. ZrB_2 particles act as grain refinement and it reduce the ductility of the composite with smaller size of dimples. The reduction of tensile strength of A1 composite is due to low density of hBN particles occupying high weight fraction in the composite and have their low strength when compared to WS_2 composite. The uniform distribution of particles reduce the crack propagation and improve the tensile strength.

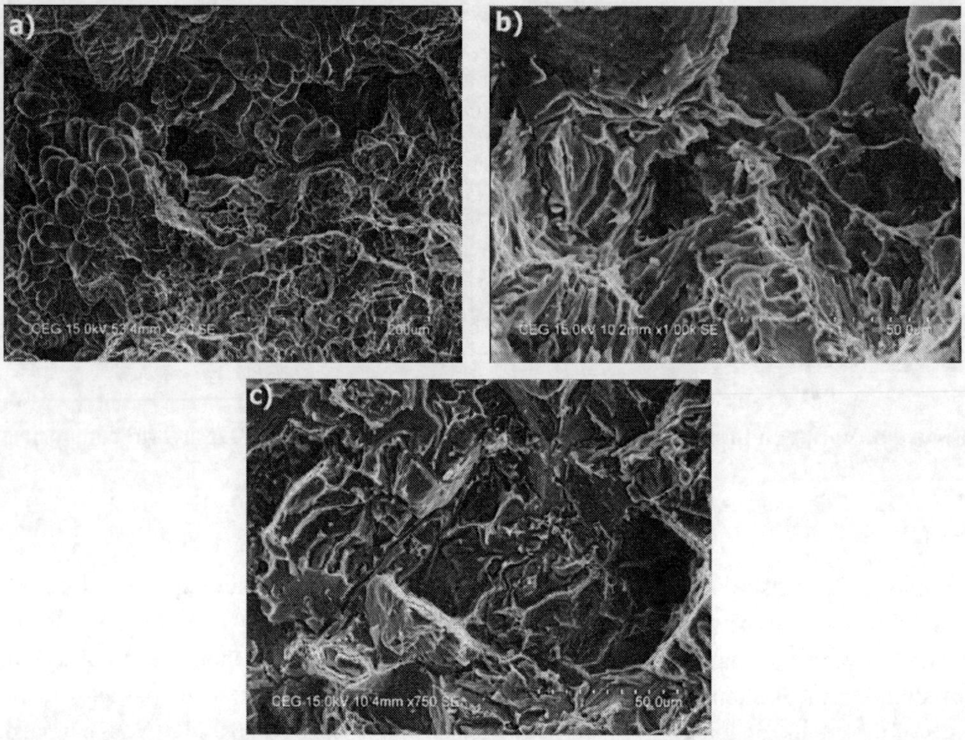

Figure 5. Fracture morphology of AA7075/ZrB2 composites. **5 (a)** A0; **5 (b)** A1; **5 (c)** A2.

4. Conclusions

AA7075-ZrB2 hybrid composites were successfully fabricated using the stir casting method with different solid lubricant particles.

- The fine distribution of particles on the matrix material are ensured through optical microscope and phase formation by XRD analysis.
- Hybrid composites shows higher hardness as compared to base alloy. A2 composite shows a higher hardness than A1 due slight improvement in wettability of WS2 with the matrix.
- A2 composite shows higher tensile than A1 due to the smaller grain size of composite which improved strength of the composite.
- Fractography of composites showed both ductile and brittle mode of failure,

Acknowledgments

Authors gratefully acknowledge the support of Department of Science and Technology–Science and Engineering Board (DST-SERB), (Grant No: EEQ/2017/000382) New Delhi, India.

References

[1] Rajmohan T, Palanikumar K, Ranganathan S, "Evaluation of mechanical and wear properties of hybrid aluminium matrix composites", Transactions of nonferrous metals society of China, Vol. 23(9), pp. 2509-2517, 2013.

[2] Xiu Z, Yang W, Dong R, Hussain M, Jiang L, Liu Y and Wu G, "Microstructure and mechanical properties of 45 vol.% SiCp/7075Al composite", Journal of Materials Science & Technology, Vol. 31(9), pp. 930-934, 2015.

[3] Saravanan L, Senthilvelan T, "Investigations on the hot workability characteristics and deformation mechanisms of aluminium alloy-Al2O3 nanocomposite", Materials & Design, Vol. 79, pp. 6-14, 2015.

[4] Sharma P, Sharma S, Khanduja D, "A study on microstructure of aluminium matrix composites", Journal of Asian Ceramic Societies, Vol. 3(3), pp. 240-244, 2015.

[5] Singh J, Chauhan A, "Characterization of hybrid aluminum matrix composites for advanced applications–A review", Journal of Materials Research and Technology, Vol. 5(2), pp. 159-169, 2016.

[6] Deaquino-Lara R, Gutiérrez-Castañeda, E, Estrada-Guel I, Hinojosa-Ruiz G, García-Sánchez E, Herrera-Ramírez JM, Pérez-Bustamante R, Martínez-Sánchez R, "Structural characterization of aluminium alloy 7075–graphite composites fabricated by mechanical alloying and hot extrusion", Materials & Design, Vol. 53, pp. 1104-1111, 2014.

[7] Meléndez-Martınez JJ, Domınguez-Rodrıguez A, Monteverde F, Melandri C, De Portu G, "Characterisation and high temperature mechanical properties of zirconium boride-based materials", Journal of the European Ceramic Society, Vol. 22(14-15), pp. 2543-2549, 2002.

[8] Chen C, Guo L, Luo J, Hao J, Guo Z, Volinsky AA, "Aluminium powder size and microstructure effects on properties of boron nitride reinforced aluminium matrix composites fabricated by semi-solid powder metallurgy", Materials Science and Engineering: A, Vol. 646, pp. 306-314, 2015.

[9] Huang SJ., Lin CC., Huang JY, Tenne R, "Mechanical behaviour enhancement of AZ31/WS_2 and AZ61/WS_2 magnesium metal matrix nanocomposites", Advances in Mechanical Engineering, Vol. 10(2), pp. 16878140-17753442, 2018.

[10] Xie ZW, Wang LP, Wang XF, Huang L, Yang LU, Yan JC, "Mechanical performance and corrosion behaviour of TiAlSiN/WS_2 multilayer deposited by multi-plasma immersion ion implantation and deposition and magnetron sputtering", Transactions of Nonferrous Metals Society of China, Vol. 21, pp. s470-s475, 2011.

[11] Ashwath P, Xavior MA, "Effect of ceramic reinforcements on microwave sintered metal matrix composites", Materials and Manufacturing Processes, Vol. 33(1), pp. 7-12, 2018.

[12] Ezatpour HR, Sajjadi SA, Sabzevar MH, Huang Y, "Investigation of microstructure and mechanical properties of Al6061-nanocomposite fabricated by stir casting", Materials & Design, Vol. 55, pp. 921-928, 2014.

[13] Alam SN, Kumar L, "Mechanical properties of aluminium based metal matrix composites reinforced with graphite nano platelets", Materials Science and Engineering: A, Vol. 667, pp.16-32, 2016.

[14] Kumar KR, Kiran K, Sreebalaji VS, "Micro structural characteristics and mechanical behaviour of aluminium matrix composites reinforced with titanium carbide", Journal of Alloys and Compounds, Vol. 723, pp. 795-801, 2017.

[15] Kumar KR, Pridhar T, Balaji VS, "Mechanical properties and characterization of zirconium oxide (ZrO2) and coconut shell ash (CSA) reinforced aluminium (Al 6082) matrix hybrid composite", Journal of Alloys and Compounds, Vol. 765, pp.171-179, 2018.

[16] Meti VKV, Shirur S, Nampoothiri J, Ravi, K.R., Siddhalingeshwar, I.G., "Synthesis, Characterization and Mechanical Properties of AA7075 Based MMCs Reinforced With TiB_2 Particles Processed Through Ultrasound Assisted In-Situ Casting Technique", Transactions of the Indian Institute of Metals, Vol. 71(4), pp.841-848, 2018.

[17] Dinaharan I., Balakrishnan M, Selvam JDR, Akinlabi ET, "Microstructural characterization and tensile behavior of friction stir processed AA6061/Al_2Cu cast aluminum matrix composites", Journal of Alloys and Compounds, Vol. 781, pp.270-279, 2019.

[18] Kumar NM, Kumaraswamidhas LA, "Characterization and tribological analysis on AA 6061 reinforced with AlN and ZrB_2 in-situ composites", Journal of Materials Research and Technology, Vol. 8(1), pp.969-980, 2019.

[19] Kenneth KA, Bethel JB, "Characterization of mechanical and wear properties of aluminium based composites reinforced with quarry dust and silicon carbide", Ain Shams Engineering Journal, Vol. 9, pp.2815–2821, 2018.

[20] Kumar PV, Reddy GM, Rao KS, "Microstructure, mechanical and corrosion behaviour of high strength AA7075 aluminium alloy friction stir welds–Effect of post weld heat treatment". Defence Technology, Vol. 11(4), pp. 362-9, 2015.

[21] Dasgupta R, Meenai H, "SiC particulate dispersed composites of an Al–Zn–Mg–Cu alloy: Property comparison with parent alloy", Materials Characterization, Vol. 54(4-5), pp.438-445, 2005.

[22] Rajkumar K, Thushal NA, Gnanavelbabu A, Sabarinathan P, "Experimental investigations on the Wire Electrochemical Micro Machining (WECM) integrity of AA6061-TiB_2 composite". Materials Today: Proceedings, Vol. 5(2), pp. 6990-6998, 2018.

[23] Liu, S., Wang, Y., Muthuramalingam, T. and Anbuchezhiyan, G, "Effect of B_4C and MOS_2 reinforcement on micro structure and wear properties of aluminum hybrid composite for automotive applications", Composites Part B: Engineering, Vol. 176, pp.107329, 2019.

[24] Manickam D, Santhanam SKV, Sivagnanam K, "Experimental investigation of LM25 alloy reinforced with Sic, Gr and Moa particles", Materiali in Tehnologije, Vol. 53(3), pp. 395-398, 2019.

[25] Gnanavelbabu A, Muthazhagan C, Rajkumar K, Ayyanar S, Loganathan P, "Experimental Investigation and Reliability Analysis on Wear Performance of AA6061-B_4C-Nanographite Hybrid Composites" Materials Today: Proceedings, Vol. 5(2), pp. 8436-8445, 2018.

IOP Conf. Series: Materials Science and Engineering **764** (2020) 012002 doi:10.1088/1757-899X/764/1/012002

Design and analysis of hybrid reconfigurable wheel for armoured vehicles

D Elil Raja*, **S Prathap Singh and T Prabhuram**
Department of Mechanical Engineering, St. Joseph's Institute of Technology, Chennai, India.

*Author for correspondence email: elilraja76@gmail.com

Abstract. The armoured vehicles are designed to carry heavy loads and move along a variety of terrains. The mobility is the major factor that makes the armoured vehicle move freely and rapidly over the terrains. Currently wheeled vehicles and tracked vehicles are used for off-road conditions. This paper a new hybrid reconfigurable wheel has been designed and analyzed that combines the benefits of the wheels and tracks. The reconfigurable wheel contains triangular configuration and circular configuration which automatically changed with respect to the speed of the vehicle. The triangular configuration will give more traction and stability while the circular configuration will provide easy mobility. In this paper, the various components required for the proposed mechanism are designed using CATIA V5 R6 and detailed stress and displacement analysis are carried out using ANSYS R18.2. The optimum design parameters are obtained and simulation of the reconfigurable wheel is performed. The reconfigurable wheel takes a triangular configuration when the speed is below 20kmph and transformed into circular configuration while the speed increases beyond 20kmph which assures stability and mobility of the armoured vehicle.

1. Introduction

Over the past decade the revolution and innovations has been evolved to gain the traction in the armoured vehicles but still the speed will not attained for the armoured vehicle. This ideology of the hybrid reconfigurable track-wheel mechanism in armoured vehicle ensures the traction and speed of the wheel. This configurable wheel proves more control and speed in off-road and mud tracks. This mechanism shows and proves that the wheel acts both track and wheel with respect to input speed. In this mechanism at low speed and high torque the wheel works as in the form of track at the same time for the high speed it works normally as wheel. This mechanism confirms the stability and high durability of the vehicle over off-road and irregular terrains. This principle and mechanism can be widely used in military wheeled vehicles, machine carrying wagons, gun carrying machines, military off-road vehicles, and heavy defensive machines for quick action. Reconfiguration of the locomotion is materialized by a different reconfiguration system, which consists of an actuating system, a pair of symmetric reconfigurable frames that are ingrained in three geared track wheels. The mechanical and electronics principles are used in the proposed configurable wheel. Y.G. Kim et al. [1] developed a driving control system for a mobile platform with wheel-track hybrid system for rough terrain for the civil environment. J. S. Dai et al. [2] used the reconfigurable mechanism for the robot control system. J. S. Dai et al. [3] introduced Mobility in metamorphic mechanisms for foldable kinds. S. K. Lim et al. [4] tried a new system for varying geometry with single-tracked mechanism for robot. K. Srinivasa Rao et al. [5] developed a new alloy wheels and carried analysis using analysis software. Miroslav Demić et al. [6] designed the steering system to minimize the kinematic errors. In order to eliminate the kinematic error, they proposed one new method with reduced wear. Andrzej Wiśniewski et al. [7]

collected the information about wheeled armoured vehicles and compared to get better future development. N.Rishi Kanth et al. [8] designed a crawler robot to climb, transportation, rescue and suit for deferent working places. This reduced the requirement of more information to run the robot. Aliakbar Alamdari et al. [9] introduced an analytical approach to modelling, analysis and reconfiguration of articulated leg-wheel systems. This approach increased the traction and stability margin. T. Maneewarn et al. [10] conducted a kinematic analysis for task-based reconfiguration of wheel-arm robots.

2. Methodology for hybrid wheel

The methodologies of the hybrid wheel-track mechanism used in the armoured vehicles and heavy vehicles explain that the proposed reconfigurable wheel will act as a track at low speed and wheel at high speed. The speed of the wheel has been sensed by sensors and if the input speed is less than 20Km/hr the hybrid wheel will act as a track and if the input speed is more than 20km/hr the track has transformed to a wheel. This mechanism works with respect to the speed and torque produced by the gearbox's output shaft. The hybrid wheel ensures the vehicle's durability and stability of the wheel over off-road and unpredictable terrains. These hybrid wheels consist of an actuating system and two symmetrical reconfigurable frames. To make a triangular shape for track mode three gears are used and also ten different gear shaped teeth that are meshed with rollers during track mode. To form circular shape these three gears are meshes with internal gears and the shaft meshes with the mainframe during the circular mode to rotate the whole wheel.

3. Design specifications

The proposed hybrid reconfigurable wheel consists of different parts to form the triangular shape and circular shape. Mainframe-in, Mainframe-out, Piston, Wheel, Roller, Drive shaft and Toggle gear are the parts used in the proposed hybrid wheel. The mainframe-in is the main component of the hybrid wheel that acts as a chassis in the configured wheel. Mainframe-out is another major component in the hybrid wheel that receives power through the driveshaft from the gearbox and transmits output power when the wheel is in a circular configuration. The mainframe's inner gears are used to convert the wheel from track mode to wheel mode. Steel 1020 are the material used the mainframe in and mainframe out.

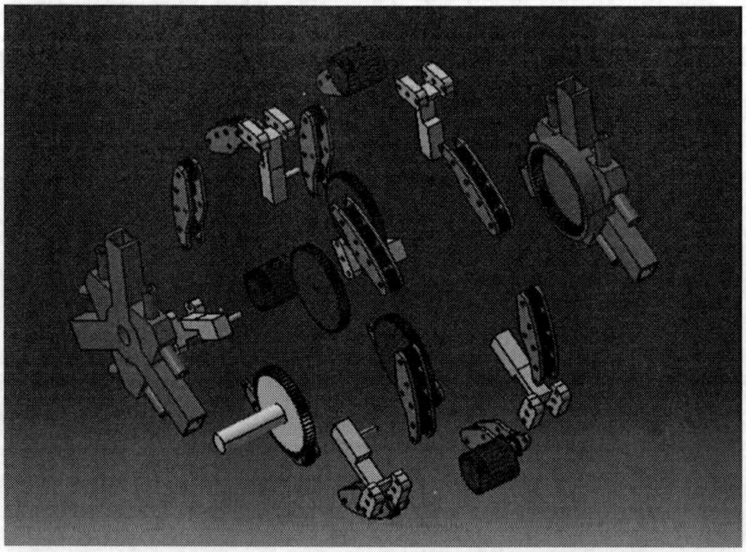

Figure 1. Exploded view of the parts used in reconfigured wheel.

The Piston plays a major role while converting from triangle shape into a circular shape with a help of hydraulic actuating system. This actuating system drives the mainframe up and down and it is made up of Al 2025.The piston is used in six different places with same design and material. The arm is the component used to modify the tire's shape. This arm is coupled with the mainframe-mounted piston and spring and this system facilitates the rotation of the tires. Totally there are 12 arms used with the same material and design. While the shape is triangular, the roller is used to transmit power tangentially to the tire. The toggle gear is used to transfer power with free rotation from the drive shaft to the roller. The roller has no constrain about direction of rotation. Six toggle gears are used with same specification and material. This shaft is used to distribute power to the wheel from the gearbox. All the parts are designed based on the engine standards. Figure 1 demonstrates all the components related to hybrid wheel.

All designed parts are assembled together that will show as a front and isometric view of the configured wheel for circular and Triangular position in figure 2a, figure 2b and figure 3a and figure 3b respectively.

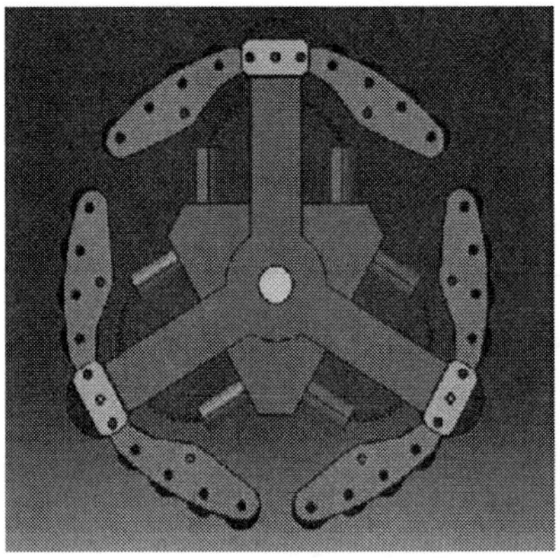

Figure 2a. Front view of the circular configuration.

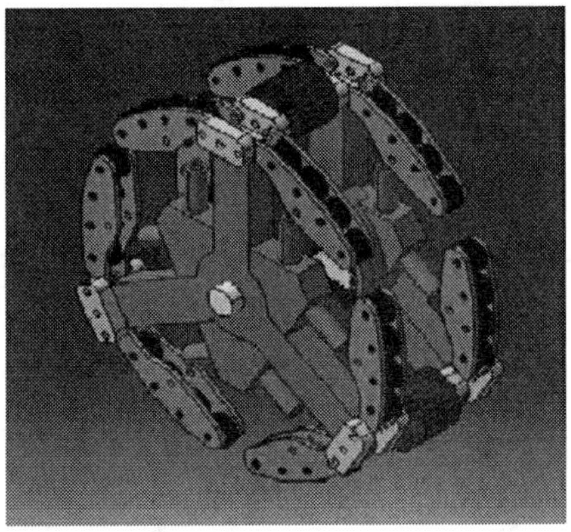

Figure 2b. Isometric view of the circular configuration.

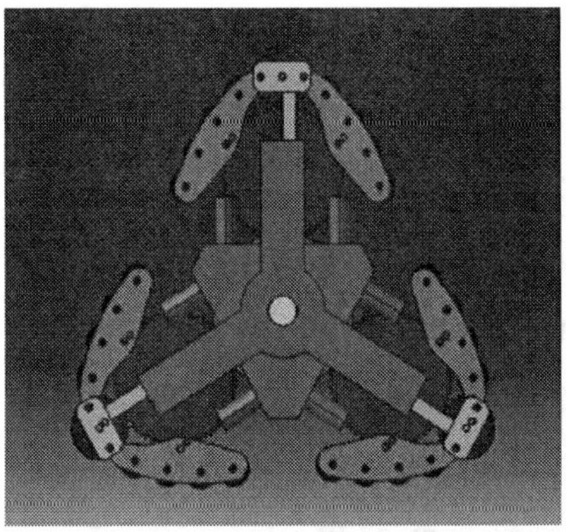

Figure 3a. Front view of the Triangular shape.

Figure 3b. Isometric view of the Triangular shape.

IOP Conf. Series: Materials Science and Engineering **764** (2020) 012002 doi:10.1088/1757-899X/764/1/012002

4. Working principle of hybrid Wheel

Figure 4 shows the track configuration of wheel for the speed less than 20 kmph. During the working, the Piston will elongate towards the outside and arm will contract towards the inside to form a triangular shape. This Configuration of the wheel can sustains speed up to 20 kmph. The output of the transmission system is connected to the wheel through the driveshaft to drive the vehicle. The output of the driveshaft is connected to the mainframe-in. The mainframes were connected with a hydraulic piston arm to move up and down while the leg is constrained to the piston and mainframes. The up and down motion of the piston will shift the wheel shape from circular to triangular. The position of the arm adjusts automatically when the piston in the mainframe is pushed up using the hydraulic system. The roller is used for moving the layer in triangular locomotion. The roller is meshed for rotation in the same direction using the toggle gear connected between the driveshaft and the roller. The entire frame is rotated in circular locomotion by keeping the layer constrained. The piston in the mainframe will be pulled back and the arm will change the shape of the wheel to circular in shape. The output of the drive shaft is connected to the wheel and the wheel will rotate continuously. The speed of the rotating wheel was calculated using a sensor, in order to monitor the motion of the wheel. This sensor plays an important role in shifting the configuration from one shape to another shape. The shape of the wheel configuration was changed according to the speed of the wheel. If the speed of the wheel is less than 20 kmph, the configuration is track mode to initially gain traction, and if the speed is above 20 kmph, the configuration changes the shape in to wheel mode.

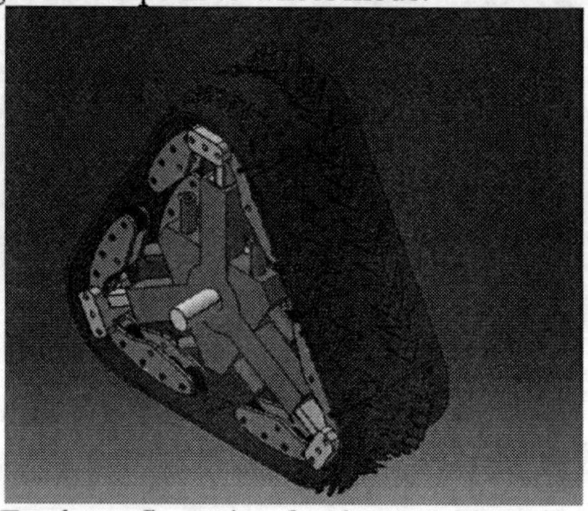

Figure 4. Track configuration for the speed less than 20 kmph.

Figure 5 shows the track configuration of wheel for the speed above 20 kmph. If the speed increased above 20 kmph, the shape of the wheel was changed to circular from triangular. This circular design is obtained after action of piston contracting and arm extracting as same like the action happened in the triangular shape.

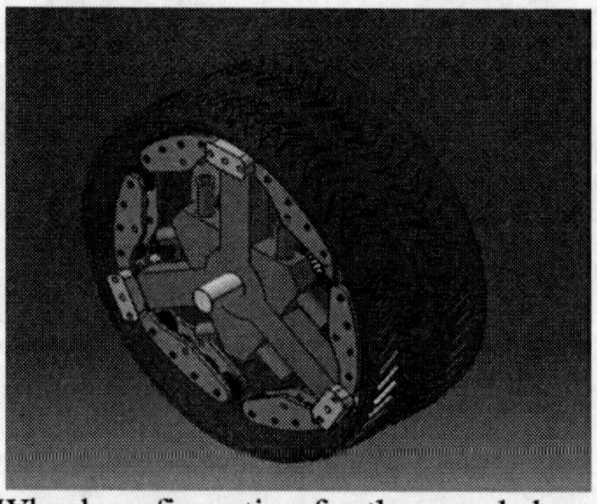

Figure 5. Wheel configuration for the speed above 20 kmph.

IOP Conf. Series: Materials Science and Engineering **764** (2020) 012002 doi:10.1088/1757-899X/764/1/012002

5. Result and Discussion

The parts are modelled using CATIA V5 software, and the analysis was carried out using ANSYS software (R18.2). For all parts associated with the configured wheel the structural analysis was performed.

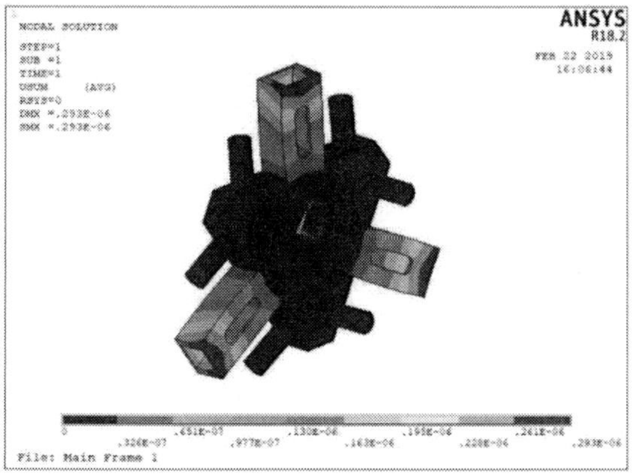

Figure 6a. Displacement analysis on mainframe.

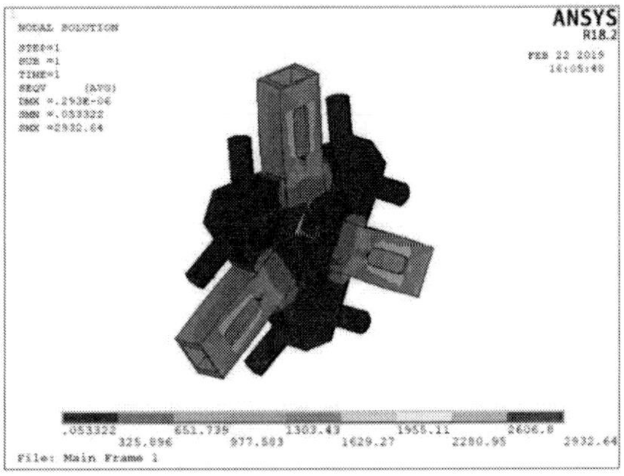

Figure 6b. Stress analysis on mainframe.

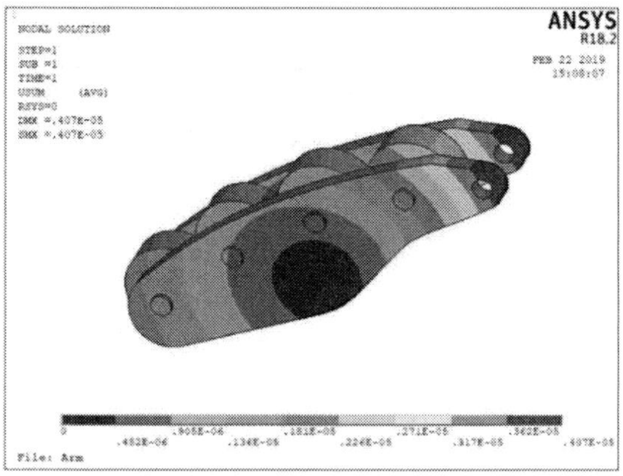

Figure 7a. Displacement analysis on arm.

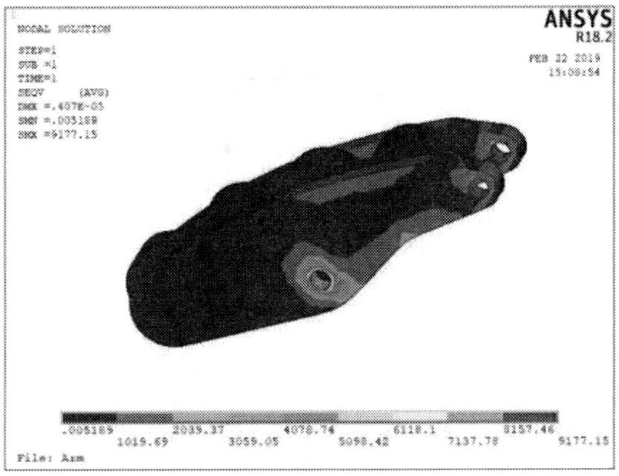

Figure 7b. Stress analysis on arm.

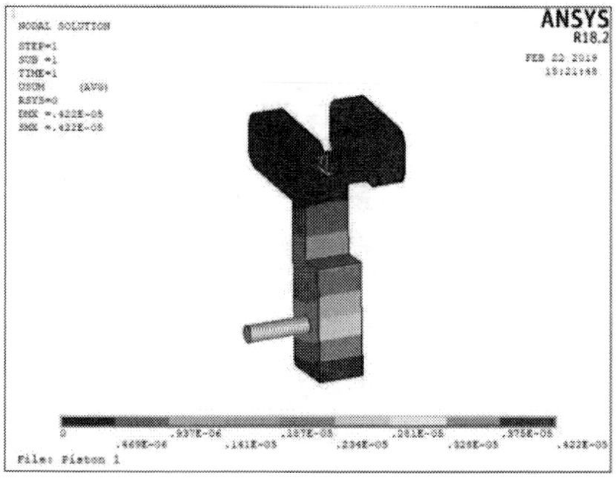

Figure 8a. Displacement analysis on piston.

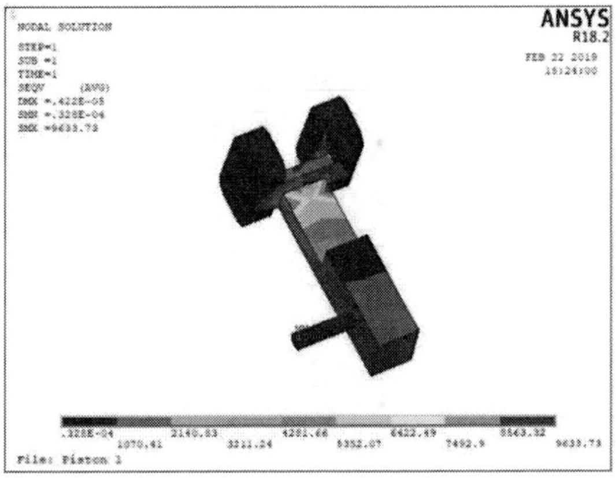

Figure 8b. Stress analysis on piston.

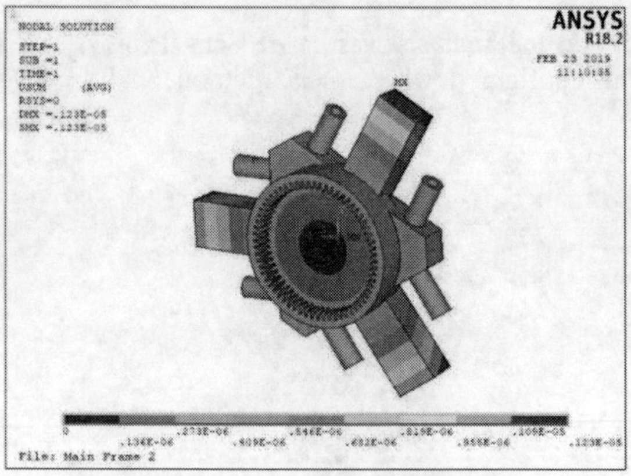

Figure 9a. Displacement analysis on mainframe out.

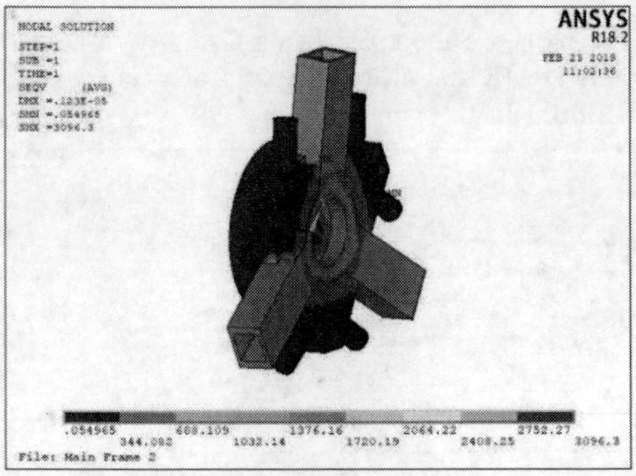

Figure 9b. Stress analysis on mainframe out.

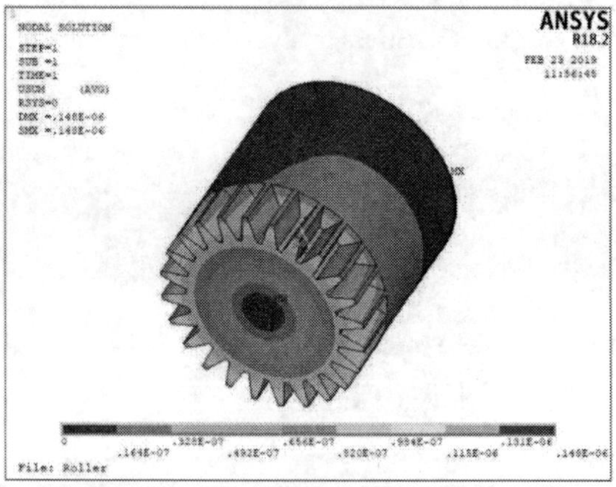

Figure 10a. Displacement analysis on roller.

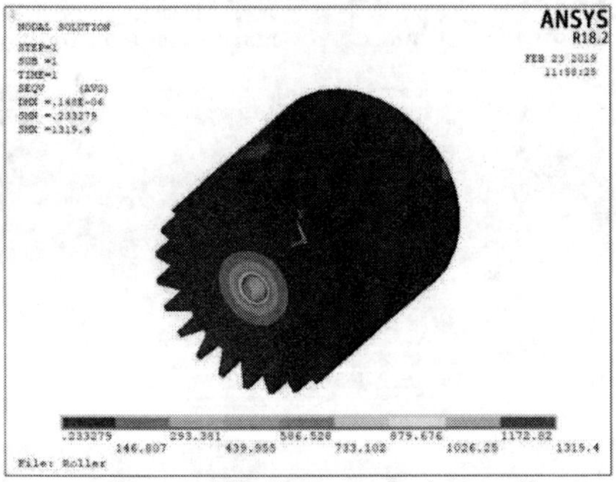

Figure 10b. Stress analysis on roller.

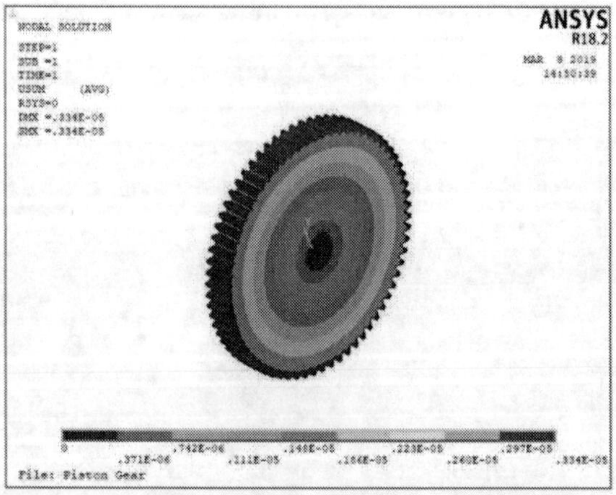

Figure 11a. Displacement analysis on toggle gear.

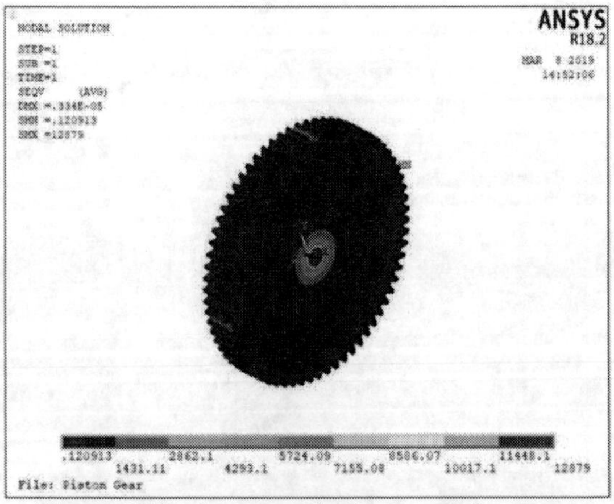

Figure 11b. Stress analysis on toggle gear.

IOP Conf. Series: Materials Science and Engineering **764** (2020) 012002 doi:10.1088/1757-899X/764/1/012002

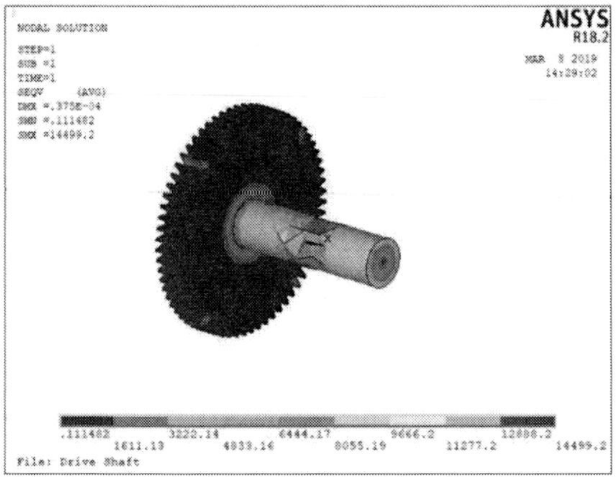

Figure 12a. Displacement analysis on drive shaft.

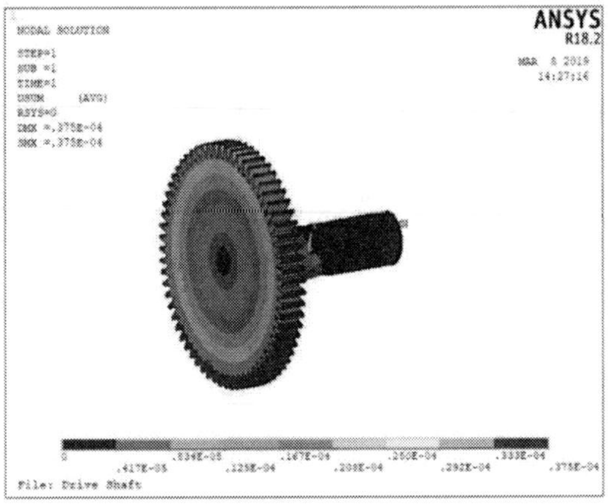

Figure 12b. Stress analysis on drive shaft.

Figure 6a and figure 6b illustrate the Mainframe displacement and stress analysis. In the mainframe, the maximum displacement and stress are 0.293 x 10^{-6} mm and 2932.64 N/mm^2. Mainframe failure begins when the load increases from the outer leg portion. The displacement and stress analysis of the arm is shown in figure 7a and figure 7b. 0.407 x 10^{-5}mm and 9177.15 N/mm^2 are the maximum displacement and stress induced in the arm. The arm failure will occur due to the maximum displacement on the arm's outer edge. Figure 8a and figure 8b show the piston's displacement and stress analysis. The piston's maximum displacement and stress are 0.422 x 10^{-5} mm and 9633.73 N/mm^2. Displacement is the major factor in determining the piston's strength. Figure 9a and figure 9b demonstrate the mainframe displacement and stress analysis. In the mainframe out, the maximum displacement and stress are 0.123x 10^{-5} mm and 3096.3 N/mm^2. The stress in the mainframe affects the mainframe out directly. The stress in the mainframe out is also greater than the stress in the mainframe. Figure 10a and figure 10b demonstrates the roller's displacement and stress analysis. The 0.148x 10^{-6} mm and 1319.4N/mm^2 are the maximum displacement and stress induced in the roller. On the circular portion of the roller, the maximum displacement was achieved if the load increased from the material's allowable limit. The displacement and stress analysis of toggle gear are shown in Figure 11a and figure 11b. The toggle gear's total displacement and pressure are 0.334 x 10^{-5} mm and 12879N/mm^2. Gear is the highly affected part due to load. The major failure on the outer teeth was seen. The displacement and stress analysis on the drive shaft is shown in Figure 12a and figure 12b. The drive shaft's maximum displacement and stress are 0.334 x 10^{-5} mm and 12879 N/mm^2.

All the parts associated with the reconfiguration wheel, the displacement analysis were performed; the values obtained from the analysis are within the allowable value. The stress analysis also conducted for all the parts used in the triangular and circular configuration the stress values obtained from the analysis are within the allowable limits of the selected materials. The displacement values also within the allowable limits of the proposed materials.

6. Conclusion

The components used in the hybrid reconfigurable wheel for heavy-armed vehicle was modelled using CATIA V6 R9 and the parts are analysed using ANSYS R18.2 software. The components involved in the reconfigured wheel for both traction and wheel are subjected to displacement and stress analysis the values obtained from the analysis reveals that all the deflection values and stress values are within the allowable limits that will not affect the working condition of the system. This hybrid wheel concept is designed for heavy-armoured vehicles that carry artillery, armed weapons, missile carriers, combat vehicles, and other military-operated vehicles. This hybrid wheel ensures the vehicle's durability and stability of the wheel over mud and off-roads. The proposed wheel design offers both traction and speed of the wheel. The design and the working principle of the hybrid reconfigurable wheel are verified by running over a simulation using computer software.

IOP Conf. Series: Materials Science and Engineering **764** (2020) 012002 doi:10.1088/1757-899X/764/1/012002

References

[1] Y.G. Kim, J.-H. Kwak, J. Kim, J. An, K.D. Lee, "Adaptive driving mode control of mobile platform with wheel-track hybrid type for rough terrain in the civil environment" International Conference 390 on Control Automation and Systems, pp. 86–90, 2010.

[2] J. S. Dai, M. Zoppi, X. W. Kong (Eds.), "Advances in reconfigurable mechanisms and robots I", Springer Verlag, London, 2012.

[3] J. S. Dai, J. Rees Jones, "Mobility in metamorphic mechanisms of foldable/erectable kinds", Transactions of the ASEM, Journal of Mechanical Design, Vol. 121 (3) pp.375–382, 1999.

[4] S. K. Lim, D. I. Park, Y. K. Kwak, B.-S. Kim, S.-W. Jeon, "Variable geometry single-tracked mechanism for a rescue robot", IEEE International Workshop on Safety, Security and Rescue Robotics, pp.111–115, 2005.

[5] K. Srinivasa Rao, M .Rajesh ,G.Sreedhara Babu, Design And Analysis of Alloy Wheels, International Research Journal of Engineering and Technology, Vol. 4 (6), pp.2036-2042, 2017.

[6] Miroslav Demić, Miroslav Demid, "Analysis of Influence of Design Parameters on Steered Wheels Shimmy of Heavy Vehicles", Vehicle System Dynamics: International Journal of Vehicle Mechanics and Mobility, Vol. 26(5), pp. 343 – 379, 2007.

[7] Andrzej Wiśniewski, Piotr Rybak, "Chosen development problems of wheeled armoured transporters", Journal of KONES Powertrain and Transport, Vol. 25(3), pp. 506 - 510, 2018.

[8] N.Rishi Kanth, A.Srinath, J. Suresh Kumar, "Design and analysis of intelligent ground vehicle", International Journal of Scientific & Engineering Research, Vol. 8(3), pp. 1963 - 1970, 2017.

[9] Aliakbar Alamdari, Venkat Krovi, "Active reconfiguration for performance enhancement in articulated wheeled vehicles", Proceedings of Proceedings of the ASME 2014 Dynamic Systems and Control Conference, pp. 1 - 10, 2014.

[10] T. Maneewarn, S. Suthaweesub, "Kinematic analysis for task-based reconfiguration of wheel-arm robots", ICAR '05. Proceedings. 12th International Conference on Advanced Robotics, 2005.

IOP Conf. Series: Materials Science and Engineering **764** (2020) 012003 doi:10.1088/1757-899X/764/1/012003

Modelling and analysis of femur bone from CT scan

P Kishore[1, a], Anjan Kumar Dash[1, b], K Dhilip Kumar[2, c]
[1]School of Mechanical Engineering, SASTRA Deemed to be University, Thanjavur, India.
[2]Physiotherapist, Sanjay Physiotherapy Centre, Thanjavur, India.

Email: [a]kishorepandiyanmech@gmail.com, [b]anjandash@mech.sastra.edu, [c]dilipphysio70@gmail.com

Abstract. About 200 million peoples around the world are affected by different types of bone diseases. Osteoporosis is one among them which is a low bone density disease which increases the porosity of the bone and leads to severe pain and fracture of the human bone. Women's are most affected by this disease due to various reasons. It can be only predicted after the failure or fracture of the bone when the sudden impact of the load. The femur is the longest bone in the human body which is mainly affected by means of fracture due to osteoporosis. The strength of the bone mainly depends on its density and can be determined by the Bone Mineral Density (BMD) Test. Hence it is necessary to numerically predict the stress and deformation of bone for the osteoporosis patient in their daily activities. Nowadays many researches have been undergone in Finite Element Analysis (FEA) of human bone in order to predict the stress and deformation when subjected to various loads in daily life. In this work, Structural analysis is carried out on femur bone which is modelled from CT scan report and material properties are assigned from osteoporosis patient. The loading conditions are based on the force which exerts on the femur bone during routine activities. The analysis is carried out with two different boundary conditions and it is seen that the femur head and distal end of the femur is subjected towards maximum normal stress and deformation.

1. Introduction

Osteoporosis is a state which devastates strength of the bone which makes the bone become fragile especially for the women after the postmenopausal period where the weakness of the bone progress over a certain year and can be identified only when the unexpected impact or slight fall which causes the bone to fracture. About 200 million people around the world are affected by this disease and in India 46 million people are affected by osteoporosis. It is mainly occurred after postmenopausal due to low calcium density and vitamin D deficiency which result in extensive soreness, pain, disability, and perhaps results in the death of patients. The different aspects in osteoporosis are Nutritional factors, Nutritional status, Lifestyle, drug use and preceding history of fractures. The rich calcium and vitamin foods, exposure of sunlight in the evening which gives vitamin D, Changing lifestyle and rest are some of the treatment for the patient suffering from osteoporosis [1]. The risk of fracture is reduced by hormone replacement therapy in supplement with calcium and Vitamin D complements [2]. Now a day's various researches have been undergoing on the FEA of the human bone in order to determine the stresses induced in the human bone on the application of practical force which people encounters in their daily life. FEA is also useful in studying the fracture mechanics of the human bone on the external application of the force. The femur bone is the longest bone in the human body which is stressed more during walking and step climbing. These bones are subjected to maximum load during normal human activities. The patient with osteoporosis has mostly subjected towards fracture of the femur bone which is seen from the medical history. Researchers have been carried out to determine the fractural load distribution of stress and fracture morphology of the human femur bone. The fracture load and fracture morphology of the femur bone was studied practically and by means of the numerical method where both showed that good agreement with each other and also the internal bone mineral density significantly influences the fracture morphology high value of bone mineral density gives improved stiffness and strength [3]. The vibration and stress characteristics of human femur bone have been analysed and determined the deformation of the femur on the application of the load based on the human weight and the corresponding stress distribution are evaluated [4]. The human bones are imported into a solid model by the CT scan and MRI scan which is useful for FEA. The material properties and boundary condition for analysis of the femur bone are derived from the CT scan report and the linear relation between apparent bone density and Hounsfield unit (HU) can be derived from the numerical equation where the strength of the human bone depends on the density and porosity [5]. The femur is designed in CAD software and the FEA is carried out. The material properties of the human bone like given density poisons ratio and Young's modulus are imported to the designed femur model and the stress analysis were carried out with various load which acts on the femur bone during walking, standing, jumping and running [6]. The numerical analysis also carried out for finding the various stress and deformation when subjected towards various load on the healthy femur and the broken femur with the prosthesis. The analysis showed that the titanium prosthesis gives less deformation when compared with other materials [7]. The failure of the femur bone is mainly due to the sudden accident and continuous vibration which acts on the proximal area of the femur. The neck and bone shaft of the femur bone are the areas which are mainly affected by the vibration on femur which is determined from the nodal analysis with ten resonance frequency [8]. FEA is carried out for finding the stress and fracture morphology of the human femur when subjected towards the load [9]. Since osteoporosis patients are majorly affected by fracture of femur bone due to varying stress and deformations. FEA is one of the powerful models for predicting stress and deformation when subjected by the external load. In this work, a step to predict the stress and deformation of the femur bone for osteoporosis patients during their normal daily activities is taken. The material properties of the femur bone of osteoporosis patient are taken from the literatures [9], [10], [11]. The FEA are carried out in ANSYS 19 software and the results are evaluated using various boundary conditions.

2. Modelling of the Femur Bone

The 3-D model of the femur bone was constructed using the CT scan report which is in DICOM format from a patient of 52 years old female. This CT scan report is imported into the 3-D slicer software. The 3-D slicer is open-source and free software for visualizing the medical images and which is very much useful in analyzing the CT scan and MRI scan report of the patients. From the software 3-D model of the lower leg is extracted and the required femur bone is cropped from the whole leg segment by the region of interest option as shown in Figure 1. Then in editor option, the complete 3-D model was obtained with the threshold frequency of 756. Finally, the 3-D model meshes in editor mode and complete triangular meshed solid geometry is obtained from the software. Then this femur model is saved in .stl format and it is imported in the blender software.

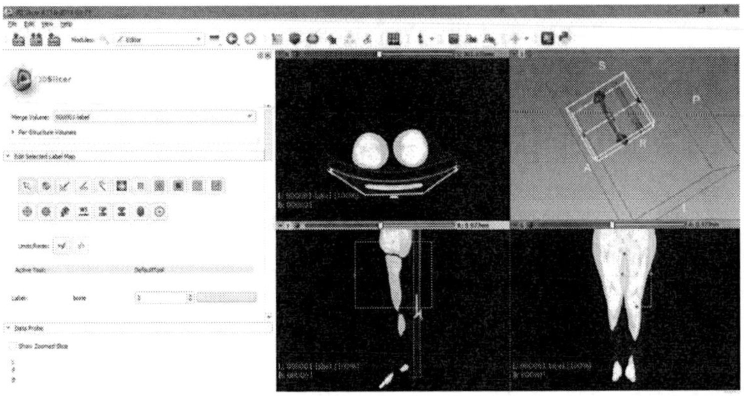

Figure 1. CT scan report of a patient imported from the 3-D slicer.

Blender is open-source 3D computer graphics software. The femur model extracted from the 3D slicer has to be cleaned further for removing the unwanted objects apart from bone and the surface of the bone are made smoother as shown in Figure 2. This cleaned 3D model is stored in .stl format and this 3D model is taken for further analysis.

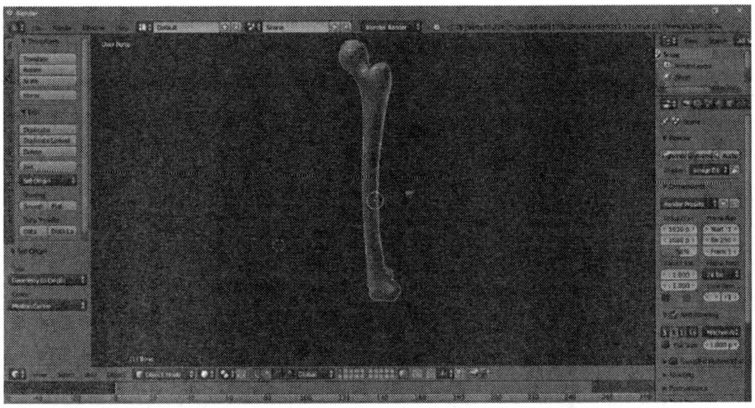

Figure 2. The Surface cleaned model from Blender software.

The Computer-Aided Design (CAD) software CATIA V5 R20 is used for converting the .stl file into a .stp format for importing the femur model in ANSYS as shown in Figure 3.

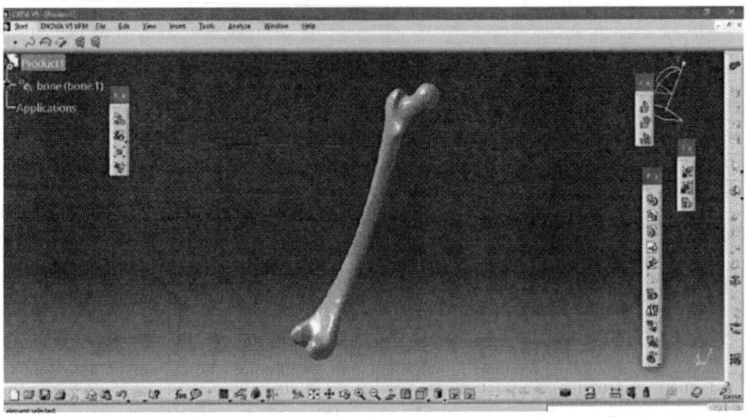

Figure 3. Conversions of the .stl file into ANSYS format in CATIA V5.

3. Assigning Material Property to the solid model

The Biomechanical capability of human bone strength can be determined based on bone strength. The essential constraint for structural analysis is assigning the material property to the solid model. The density of the bone plays an important role in determining the strength of osteoporosis patients. Bone Mineral Density (BMD) is measured from the central DXA test. The test measures the BMD and compares the score with the well-known standard and gives the T value. The T value of the normal healthy person varies from +1 to -1. If the T score value is less than -1 to -2.5 then the patients are affected with low bone density disease and are subjected to a high risk of fracture [10]. From his research, the Bone mineral density of osteoporosis patient is 0.750 g/cm3. The material property of the osteoporosis patient varies based on the porosity of the bone [11]. If the porosity increases the density and yield stress of the bone decreases linearly and ultimately decreases the strength of the bone and increases the probability of fracture. From the experimental studies, the density of the bone varies from 0.9 to 1.3 g/cm3. Hence from the literature, the material properties of the bone taken for structural analysis are shown in table 1.

Table 1. Bone Material Properties.

S. No	Parameter	Value
1	Density (g/cm^3) [10], [11]	0.75
2	Young's Modulus (MPa) [9]	10500
3	Poisson's ratio	0.3
4	Shear Modulus (Pa)	4038

4. Loading Condition

The tetrahedral meshing is done under patch conforming algorithm for the 3 Dimensional models with the slow transition which creates 250307 nodes and 173870 elements. Normally tetrahedral meshing is used for complex geometry for good surface finish. The final geometry of the meshed femur model is shown in Figure 4.

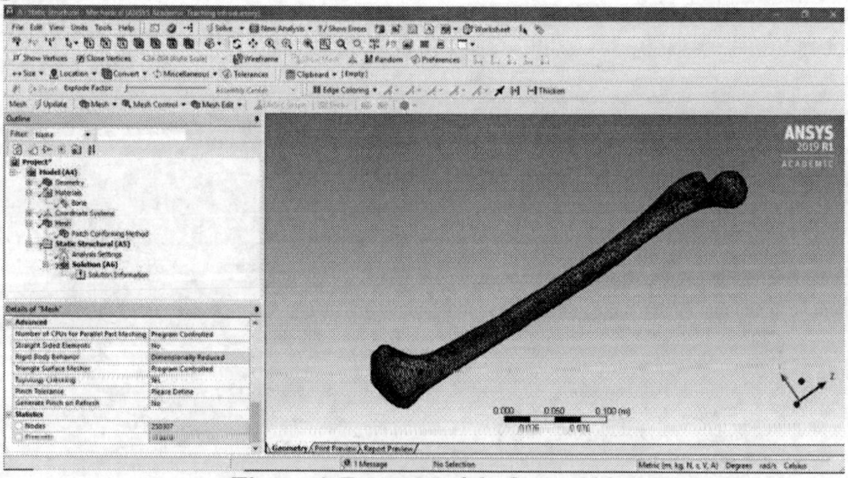

Figure 4. Femur Model after meshing.

IOP Conf. Series: Materials Science and Engineering **764** (2020) 012003 doi:10.1088/1757-899X/764/1/012003

The femur bone is subjected towards load during walking, jumping and running. During our normal daily activities, the loads exerted on the femur varies from 266N to 690N [12]. The patients with osteoporosis feel pain on their daily activities in the hip joint due to the weak bone mineral density and also even it results in failure of the femur bone when the sudden impact of forces acts on the femur. This causes the patients to the breakdown of their mobility due to severe pain. In general femur head and the distal end of the femur is liable towards more stress compared to another region because it is the place where it comes to meet the hip socket and knee joint. Hence the femur model is analysed with three different loads (270N, 400N, and 650N) under the subsequent condition.

- Case 1: Head of the femur is assumed to be fixed support and the varying loads have been applied on the distal end of the femur in order to calculate the normal stress and total deformation of femur bone at this type boundary condition.
- Case 2: Distal end of the femur is assumed to be fixed support and varying loads were applied on the head of the femur for evaluating the normal stress and total deformation of the femur bone.

5. Result and Discussion

Normal stress and Total deformation of the femur bone were calculated under two different cases and are illustrated above. First the Case 1 is discussed with the 3 different loading conditions (270N, 400N, and 650N). The normal stress induced in the femur is shown in Figure 5 (a) and (b) and Table 2 exemplifies the maximum, minimum and average normal stress induced in the femur bone under boundary condition.

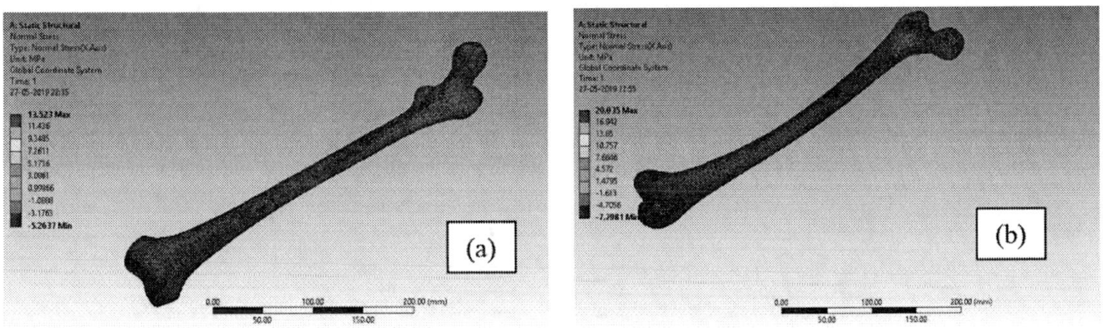

Figure 5. Normal stress (case 1) at **(a)** 270N and **(b)** 400N.

From Table 2, it is seen that Normal stress value increases gradually with the increase of the load. The femur neck exhibits maximum stress with the given two loads. When the load of 650N applied the Normal stress increases abruptly and the structural analysis for the 650N load cannot predict by the software as the loading condition exceeds the maximum value.

IOP Conf. Series: Materials Science and Engineering **764** (2020) 012003 doi:10.1088/1757-899X/764/1/012003

Table 2. Normal stress at different loads (case 1).

Load (N)	Minimum (Mpa)	Maximum (Mpa)	Average (Mpa)
270	-5.2637	13.523	-0.12819
400	-7.7981	20.035	-0.18991
650	-	-	-

Hence for the patients with these material properties, if the sudden impact of 650N acts on the femur head leads to fracture of the neck and leads to unpredictable pain for the patients. With the load of 270N and 400N, the stress-induced the femur neck is found to be maximum and leads to the stepping stone of fracture. Similarly, the deformation of femur bone under the same boundary condition is carried out are shown in Figure 6 (a) and(b) and the average of deformation are shown in Table 3.

Table 3. Deformation (case 1) at different loads.

Load (N)	Minimum (mm)	Maximum (mm)	Average (mm)
270	0	27.601	9.2896
400	0	40.89	13.762
650	-	-	-

It is also seen that the deformation for the minimum load of 270N is found to be higher and increases rapidly for the higher loads. Hence from the analysis under bounder condition case 1, the patients of low bone density when they experience a load of 270N it leads to failure of the femur bone, especially at the neck. Hence intense care and treatment must be an encounter for enhancing the bone density.

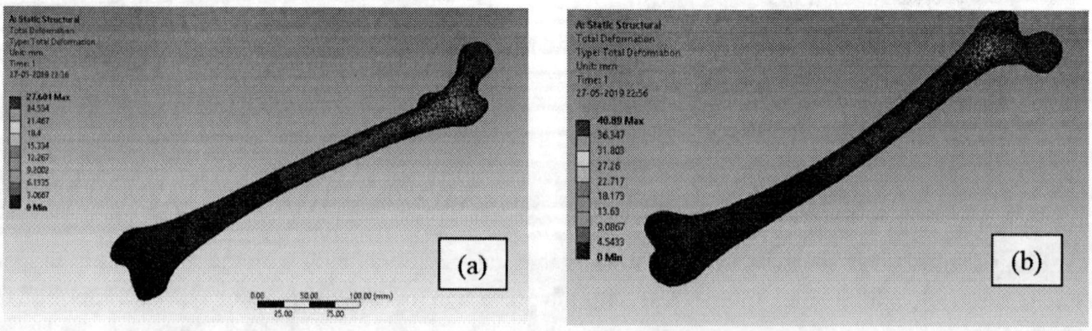

Figure 6. Normal stress (case 2) at **(a)** 270N and **(b)** 400N.

Now in Case 2, the distal end of the femur is assumed to be force acting area and the head of the femur is considered as fixed support. For the same boundary condition the structural analysis in the distal end of the femur was carried out and the result of normal stress is shown in Figure 7 (a), (b) and (c) and total deformation are shown in Figure 8 (a), (b) and (c). Table 4 and 5 denotes the normal stress and the deformation values of minimum, maximum and average.

IOP Conf. Series: Materials Science and Engineering **764** (2020) 012003 doi:10.1088/1757-899X/764/1/012003

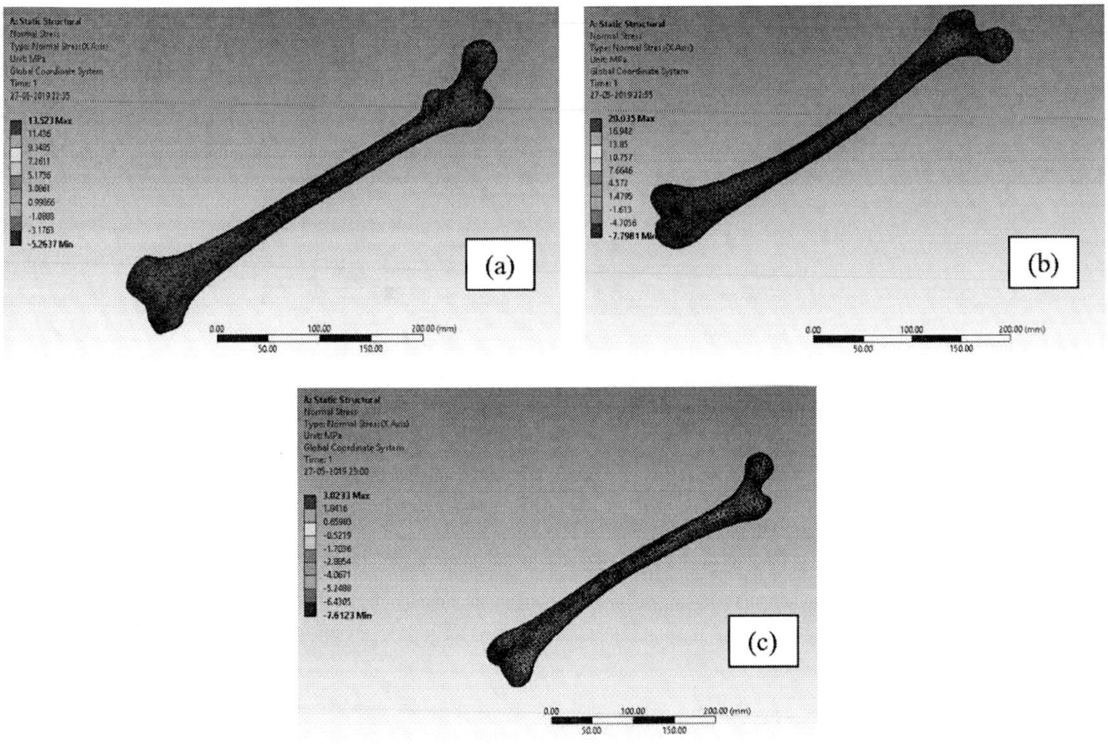

Figure 7. Normal stress (case 2) at **(a)** 270N, **(b)** 400N and **(c)** 650N.

Table 4. Normal stress at different loads (case 2).

Load (N)	Minimum (Mpa)	Maximum (Mpa)	Average (Mpa)
270	-4.6451	10.442	1.3011e-003
400	-6.8816	15.469	1.9275e-003
650	-11.527	25.911	3.2285e-003

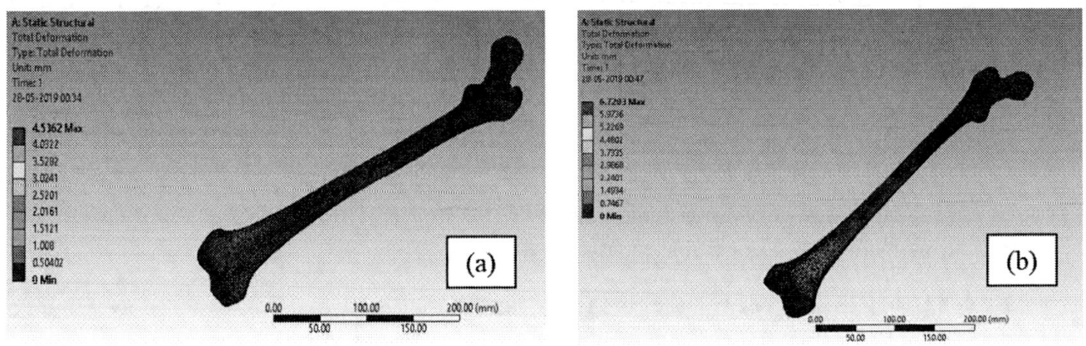

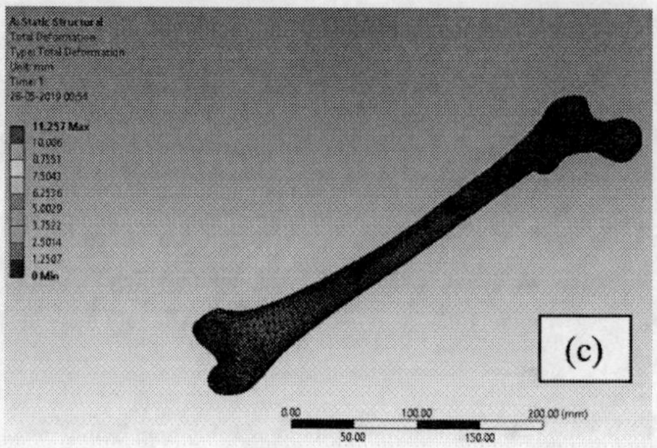

Figure 8. Total deformation (case 2) at **(a)** 270N, **(b)** 400N and **(c)** 650N.

From the Case 2 analysis, it is seen that Normal stress increases by 5% of the incremental load values taken for the study. The distal end of the femur is affected extremely at the incremental of loads. At maximum load of 650N the entire femur bone is stressed at maximum level hence there exists a higher risk of failure especially at the distal end of the femur. When the distal end is loaded by fixing femur head as fixed support it is seen that maximum deformation occurs at the distal end than the surface of the femur. Even in the minimum load of 270N, it is clearly seen that the distal end of the femur has a high risk of failure.

Table 5. Total deformation at different loads (case 2).

Load (N)	Minimum (mm)	Maximum (mm)	Average (mm)
270	0	4.5362	1.6479
400	0.	6.7203	2.4414
650	0.	11.257	4.0893

6. Conclusion

The 3-D model of the femur bone obtained from the CT scan data of 52 years old female is taken for the study. From the static structural analysis, the following conclusions are made.

- Structural analysis of the femur bone model is carried out with two different boundary conditions and the normal stress and deformation studies are undergone to find the probability of risk of failure of a femur bone for the patient affected by osteoporosis.
- In case 1 boundary condition the maximum deformation is seen in the area of the head of the femur and the Maximum normal stress also acts at the femur head.
- In case 2 boundary conditions under the same loading condition the maximum deformation and normal stress are observed at the distal end of the femur.
- By comparing the above two cases the femur head is the area where it is exposed to a high risk of fracture and even unpredictable pain may induce for the patient with osteoporosis.

IOP Conf. Series: Materials Science and Engineering **764** (2020) 012003 doi:10.1088/1757-899X/764/1/012003

- Hence Finite Element Analysis is helpful to visualize the stress and deformation of the femur bone under various boundary conditions. The people having low bone density have to take proper treatment and exercise in order to keep away from the risk of fracture.

References

[1] Bassiouny I Anuradha, V Khadilkar and Rubina M Mandlik, "Epidemiology and treatment of osteoporosis in women: an Indian perspective", International Journal of women's Health Care, Vol. 7 (3), pp. 841–850, 2015.

[2] Al-Azzawi F, Barlow D, Hillard T, Studd J, Williamson J and Rees, "Prevention and treatment of osteoporosis in women", Writing Group on Osteoporosis for the British Menopause Society Council, Vol. 13 (4), pp. 178–181, 2007.

[3] M. Marco, E. Giner, J. R. Caeiro-Rey, M. H. Miguélez, and R. Larraínzar-Garijo, "Numerical modeling of hip fracture patterns in human femur", Computer Methods and Program in Biomedicine, Vol. 173, pp. 67–75, 2019.

[4] P. B. Kumar and D. R. Parhi, "Vibrational Characteristics and Stress Analysis in a Human Femur Bone", Materials today Proceedings, Vol. 4 (9), pp. 10084–10087, 2017.

[5] X. Artru and K. A. Ispirian, "Bunch Coherence in Parametric X-Ray Radiation", Electron-photon Interaction in Dense Media, Vol. 17, pp. 191–196, 2012.

[6] K. C. N. Kumar, T. Tandon, P. Silori, and A. Shaikh, "Biomechanical Stress Analysis of a Human Femur Bone Using ANSYS", Materials today Proceedings, Vol. 2(4–5), pp. 2115–2120, 2015.

[7] P. S. R. Senthil Maharaj, R. Maheswaran, and A. Vasanthanathan, "Numerical analysis of fractured femur bone with prosthetic bone plates", Procedia Engineering, Vol. 64, pp. 1242–1251, 2013.

[8] A. Kumar, H. jaiswal, T. Garg, and P. P. Patil, "Free Vibration Modes Analysis of Femur Bone Fracture Using Varying Boundary Conditions based on FEA", Procedia Materials Science, Vol. 6, pp. 1593–1599, 2014.

[9] Yeokyeong Lee, Naomichi Ogihara, and Taeyong Lee, "Assessment of finite element models for prediction of osteoporotic fracture", Journal of the Mechanical Behaviour of Biomedical Materials, Vol. 97, pp. 312-320 2019.

[10] J. Bover et al., "Osteoporosis, bone mineral density and CKD–MBD complex (I): Diagnostic considerations", Nefrologia, Vol. 38(5), pp. 476–490, 2018.

[11] G. Osterhoff, E. F. Morgan, S. J. Shefelbine, L. Karim, L. M. McNamara, and P. Augat, "Bone mechanical properties and changes with osteoporosis", Injury, Vol. 47, pp. S11–S20, 2016.

[12] John P. Paul, "Load actions on the human femur in walking and some resultant stresses", Experimental Mechanics, Vol. 11(3), pp. 121–125, 2006.

IOP Conf. Series: Materials Science and Engineering **764** (2020) 012004 doi:10.1088/1757-899X/764/1/012004

Investigating the fracture behaviour of carburized and carbonitrided 16MnCr5

Cm Vivek
Department of Mechanical Engineering, Periyar Maniammai Institute of Science and Technology, Vallam, India.

Corresponding author email: vivekintense@gmail.com

Abstract. Low carbon alloys are low in cost and easily shapeable. Chemical composition of a material determines the mechanical properties. In this paper 16MnCr5, low alloy steel is surface treated via carburizing and carbonitriding in sealed quench furnace and quenched with oil. The effects of tensile deformation, strain rate behaviour and fracture characteristics were investigated by scanning electron microscope (SEM). A detailed investigation was done on case hardness how the surface treatment alters the hardness of the material from surface to core. The fracture of the specimen after surface treatment shows results with an intergranular fracture with projected burs on the fractured specimen. This paper investigates the fracture mechanism of surface treated material.

1. Introduction

Gas carburizing and carbonitriding is the widely used process for hardening the surface of the steel. The process is usually carried out in low carbon steels of 0.14-0.25%. Properties of materials changes with the change in metallurgical structure and the presence of inclusions in it. The surface treatment is mainly based on the process parameters; the atmosphere of material treated, quenching medium and materials characteristics. These factors influence the carbon diffusion rate or mass coefficient transfer. The main aim of this experimental study is to experimentally develop and understand the carburizing and carbonitriding process and study their influence in fracture mechanism. Optimizing the process parameter will provide better efficiency. Hence parameter selection plays a pivotal role in the case hardening process. A number of investigations are studied on case hardening of various steels.

Non-metallic inclusions play a pivotal in determining the fracture, corrosion and cyclic load bearing capacity. These properties determine the performance of steel, which aids in choosing for application. There are more reservations about the word inclusion, where various researchers comprise carbides and carbonitrides with other particles. [1] 13Cr-4Ni martensitic stainless steel fractograph images reveal much cuboidal shaped titanium carbonitride particle. [2] 17CrNiMo6 reheated, forged and annealed were austenitized by cyclic heat treatment in a salt bath furnace. Fractograph image shows quasi cleavage in as received 17CrNiMo6.The austenitized 17CrNiMo6 with fine microstructure with quasi cleavage and complex river patterns of small cleavages. [3] Ingot melted in a vacuum and categorized into four categories based on their treatments. The fracture stress shows that ultrafine ferrite/cementite (Uf-F/C) has the maximum fracture stress among other parameters taken. Uf-F/C was heat treated and warm rolled which exhibits fine grain [4].

Optimizing the parameters by saturation–value method in low-pressure carburization of 16Cr3NiWMoVNbE reveals a reduction in diffusion coefficient [5]. Using Helium as a quenching medium provides improved mechanical properties of the core of case hardened steels. No negative effects were observed in the microstructure of 16MnCr5 while quenching with helium [6]. Cryogenic treatment does not bring significant improvement to mechanical properties in 16MnCr5. Cryogenically treated 16Mncr5 shows hardness and wear resistance increased; however, toughness decreases with soaking time [7].

2. Materials and Methods

The 16MnCr5 is a Fe-alloy highly used for automotive applications because of its machinability and shape ability. However, the low-alloy steels have a low hardness value surface treatment is used to enhance the case hardness of the material. Gas Carburizing and Carbonitriding are done in the specimen. The process cycle for the gas carburizing and carbonitriding are shown in Figure 1 and Figure 2 respectively. The material is loaded in sealed quenched furnace surface treated according to the process cycle. The carbon boost potential of 0.95% and carbon + urea boost potential of 1.25 is employed in the gas carburizing and carbonitriding process. The material is quenched using mineral oil and annealing is carried out to relieve internal stress [8].

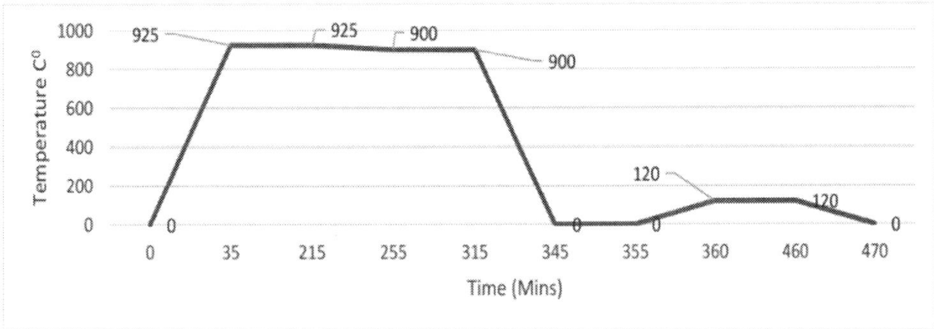

Figure 1. Carburizing cycle.

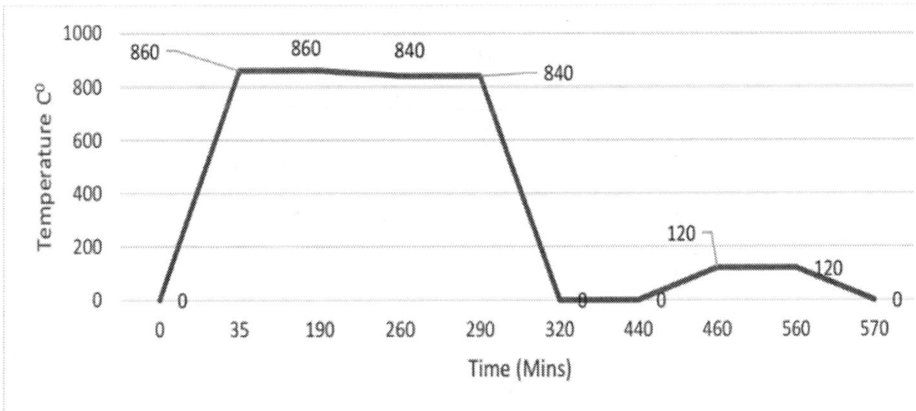

Figure 2. Carbonitrided cycle.

3. Properties Observation

3.1. Chemical Composition Testing

To analyze the composition of the material, the chemical composition test was carried out using OES (Optical Emission Spectroscopy). The chemical composition of the specimens is shown in Table 1.

Table 1. Chemical composition of the specimens.

Composition for	C	Si	Mn	P	S	Cr
As received condition	0.159	0.2	1.14	0.009	0.028	1.02
Carburized	0.6	0.203	1.19	0.013	0.024	1.11
Carbonitrided	0.494	0.216	0.77	0.023	0.043	0.0015

3.2. Tensile Test

Material surface treated both Carburized and Carbonitrided are tested for tensile strength and Young's modulus. The specimen is machined as per the ASTM standard. Length of rod =300mm, Length of gauge =200mm, Diameter of gauge=12mm. Figure 4 and Figure 5 shows the load vs. displacement graph for 16MnCr5, Carburized specimen, and Carbonitrided specimen respectively. It is found that the tensile strength of 387.077 N/mm^2 is obtained in the Carburized sample. However, the Percentage of elongation in Carbonitrided specimen is 5% where the specimen material has about 20%. Fracture with deformation forming cup and cone structure is observed in as-received condition, where the surface treated specimen was observed with brittle fractures with projected burs.

Because of strain hardening and necking the specimen elongates up to 20% of its size cause ductile pull. The cup and cone structure is obtained shown in figure 3. From figure 4, the load vs. displacement curves shows the conventional ductile fracture with strain hardening and necking.

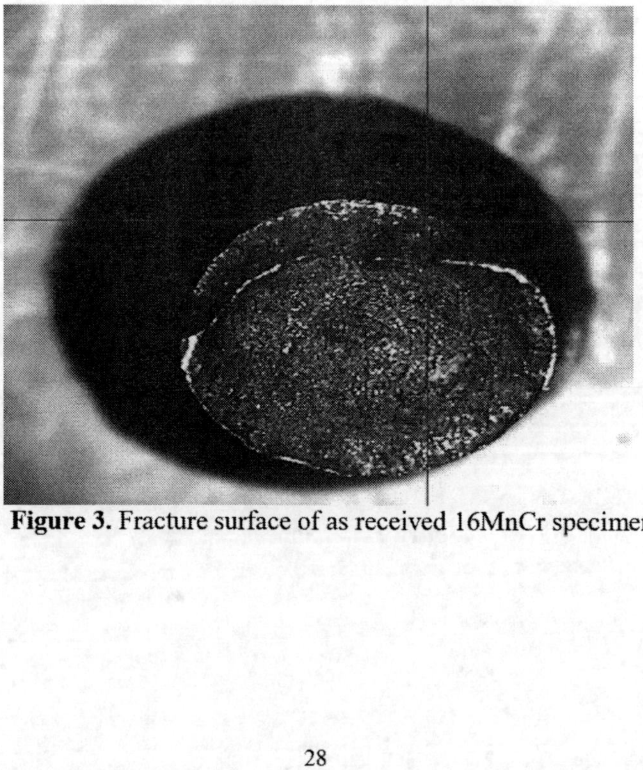

Figure 3. Fracture surface of as received 16MnCr specimen.

In Figure 5 (a) and 5 (b), load vs. displacement graph shows the material failed under brittle, necking zone is not visible and stain hardening is at minimal. Hence the elongation of Carburized and Carbonitrided specimen is 5% and 6% only.

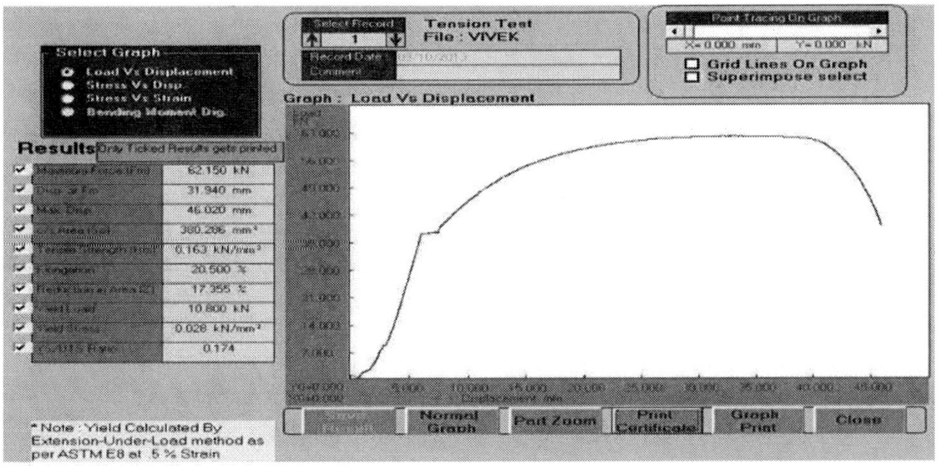

Figure 4. Load vs. Displacement – as received 16MnCr specimen.

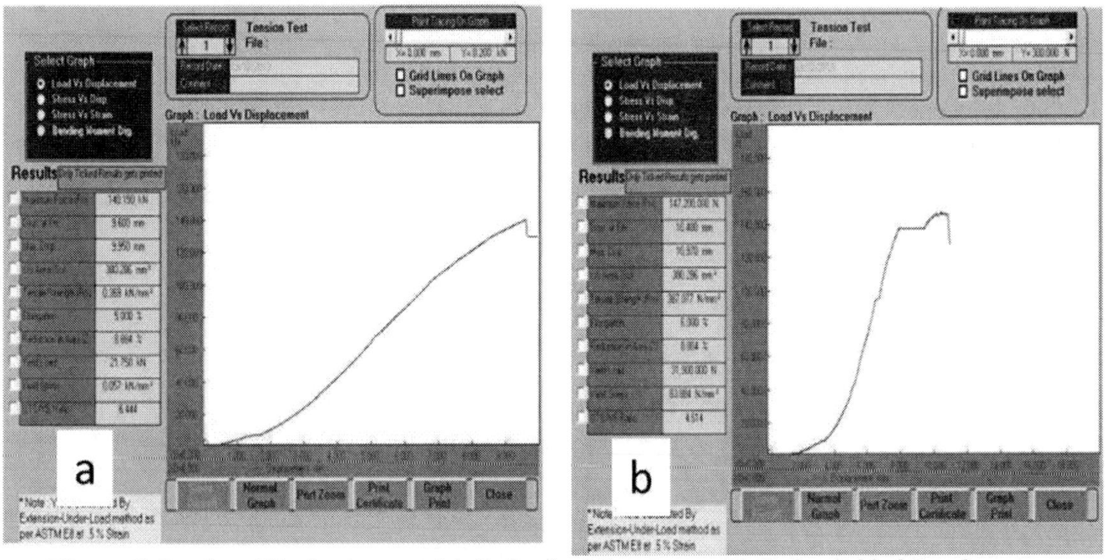

Figure 5. Load vs. Displacement – **(a)** Carburized Specimen; **(b)** Carbonitrided Specimen.

3.3. Case Depth

By finding the effective case depth of material after successful surface treatment, the hardness of the specimen decrease with distance penetrated from the surface. Figure.6 shows the case depth of surface-treated specimens. It is found that Carburized specimen with 720 HV in surfaces, when penetrated through the case has HV of 491 at 0.6 mm. However in Carbonitrided specimen surface has a relatively low hardness of 705.6 HV, when penetrated through the case has HV of 528.8. The Carbonitrided specimen exhibits bur formation in the fracture zone. Carbon influences the crack initiation and crack propagation by which yield strength of the material is increased up to 83.8 N/mm2 and 57.8 N/mm2 in Carburized and Carbonitrided specimen, respectively. The Carburized specimen's core has 370HV and Carbonitrided specimen has 330HV. The as-received specimen's hardness is 230HV; hence an increase in hardness, carbon content prolongs the crack initiation and propagation and increases the maximum load-bearing capacity of the material.

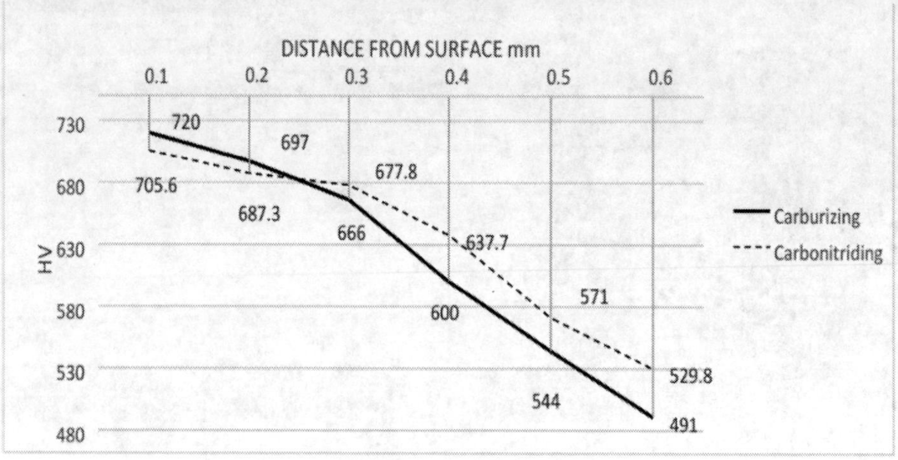

Figure 6. Case depth.

3.4. Hardness Test

Resistance for intending is measured by using the Rockwell hardness test. Carburized specimen has better hardness in both case and core. For consideration of 16MnCr5, diamond indenter is chosen as per Rockwell scale testing standards and hardness of the specimen is measured. The measurement is done on the C scale. The hardness values of as received, Carburized and Carbonitrided specimen are shown in Figure 7.

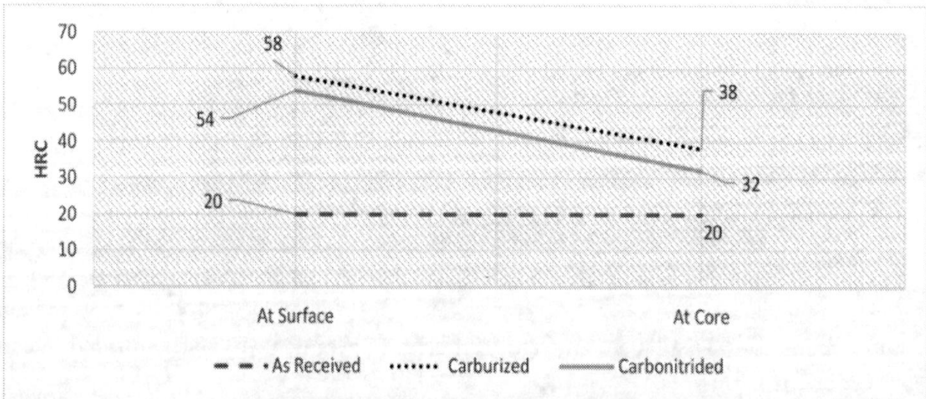

Figure 7. Rockwell hardness.

4. Microstructural Observations

4.1. Microstructure

Metallographic Test is used to analyze the microstructure of the specimen. The specimen is of length 20 mm and diameter 22 mm. The face of the material is polished using four grades of emery sheets to get finely polished outlook without a scratch. The specimen polished in a grinding machine with the addition of alumina. Then Nital, a mixture of 4% nitrogen and 96% of ethanol is applied partially to face of the specimen and dried for 5 minutes. Then the specimen is placed under the optical microscope to obtain a clear microstructure.

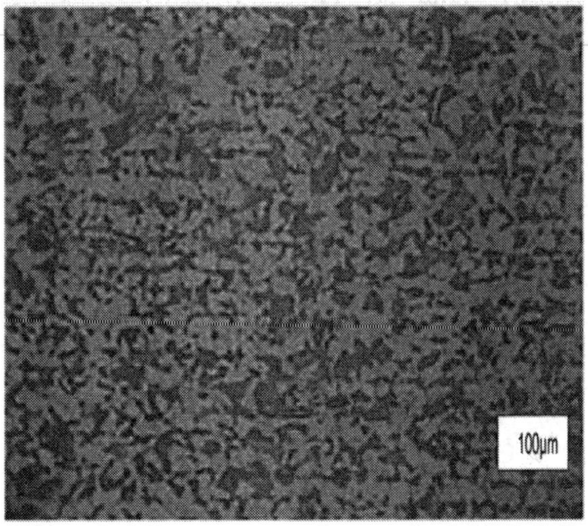

Figure 8. Microstructure of as received 16MnCr5.

The optical microscopic study of as received 16MnCr5 steel at 100X reveals ferrite and pearlite structures shown in figure 8, the proportion being found to be 83 % and 17 % respectively by applying lever rule method. The studies on after heat treatment also done at same 100X. As depicted in figure 9 (a), the Carburized specimen shows two different structures, a fine martensite on its case and a low carbon martensite on its core. The Carbonitrided specimen on the other hand, is with martensite structure at its surface, retaining austenite in traces and the core having same as that of Carburized specimen as shown in figure 9 (b).

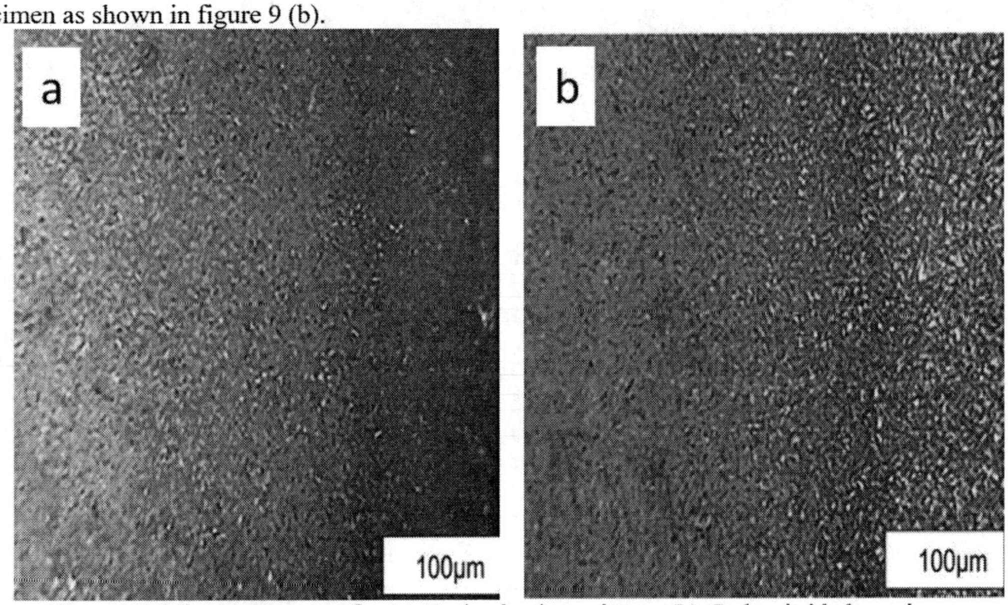

Figure 9. Microstructures of – **(a)** Carburized specimen; **(b)** Carbonitrided specimen.

4.2. Fractography

Observation on tensile fractured specimens characterizes the crack initiation and propagation of the Carburized and Carbonitrided specimen shown in Figure 10 (a) and 10 (b).

IOP Conf. Series: Materials Science and Engineering **764** (2020) 012004 doi:10.1088/1757-899X/764/1/012004

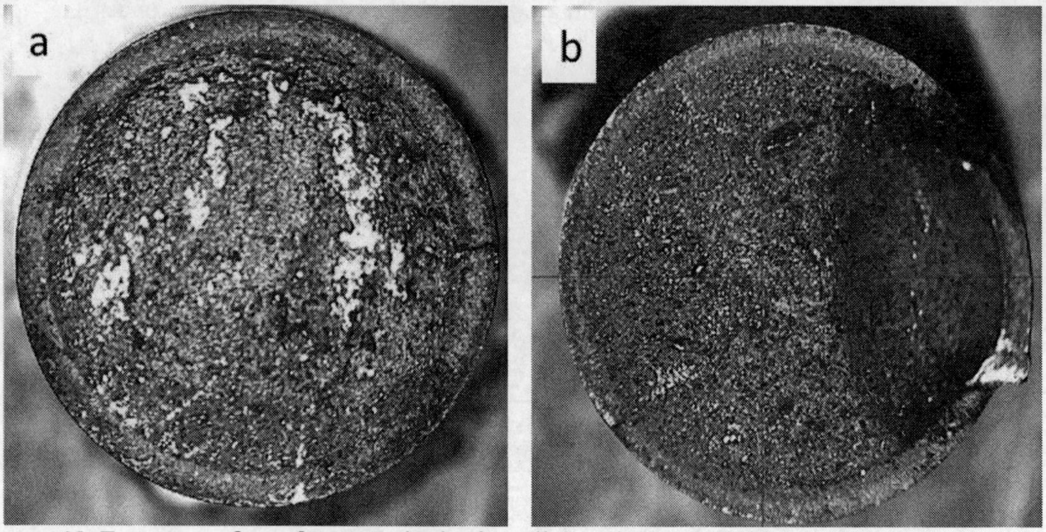

Figure 10. Fracture surface of – **(a)** Carburized tensile specimen; **(b)** Carbonitrided tensile specimen.

Scanning electron microscope images of fracture area of specimens are shown in Figures, 11 (a), 12 (a), 13 (a) and 14 (a) taken from Carburized specimens while figures 11 (b), 12 (b), 13 (b) and 14 (b) are taken from Carbonitrided specimen.

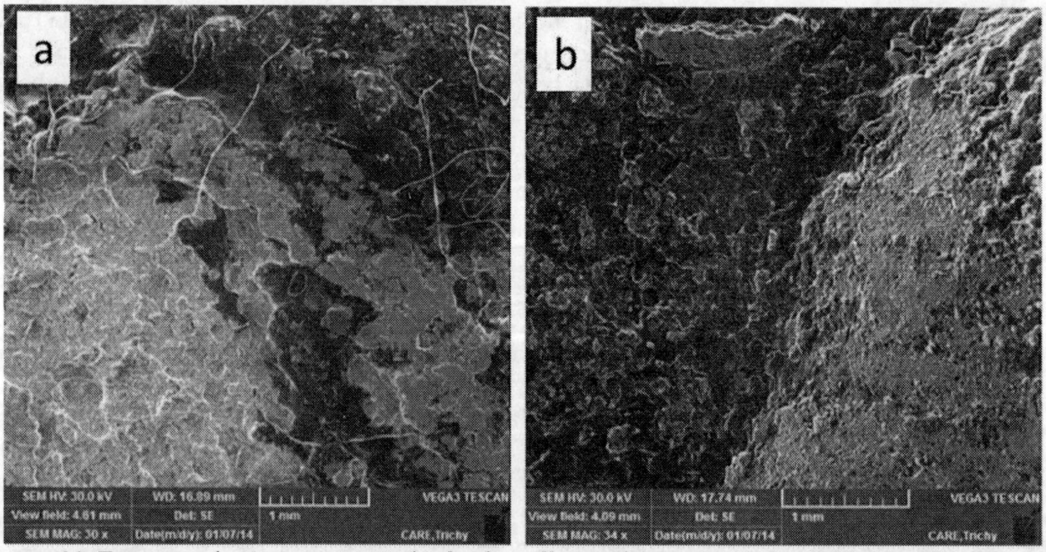

Figure 11. Fractography 1 mm – **(a)** Carburized tensile specimen; **(b)** Carbonitrided tensile specimen.

Figure 11 (a) and 11 (b) reveals the presence of inclusion-forming the case of the material. The fracture zone is shown in figure 12 (a) and 12 (b) shows visible fine faceted cleavages, the new crack nucleus is initiating from neared grains and propagate. Figure 13 (a), 13 (b) and Figure 14 (a), 14 (b) reveal the grain boundaries with dimple formation. Some regions are structure less because of slip band rupture and due to pulling force of the specimen underwent during tensile testing. The fracture found to be intermediate between ductile and brittle.

IOP Conf. Series: Materials Science and Engineering **764** (2020) 012004 doi:10.1088/1757-899X/764/1/012004

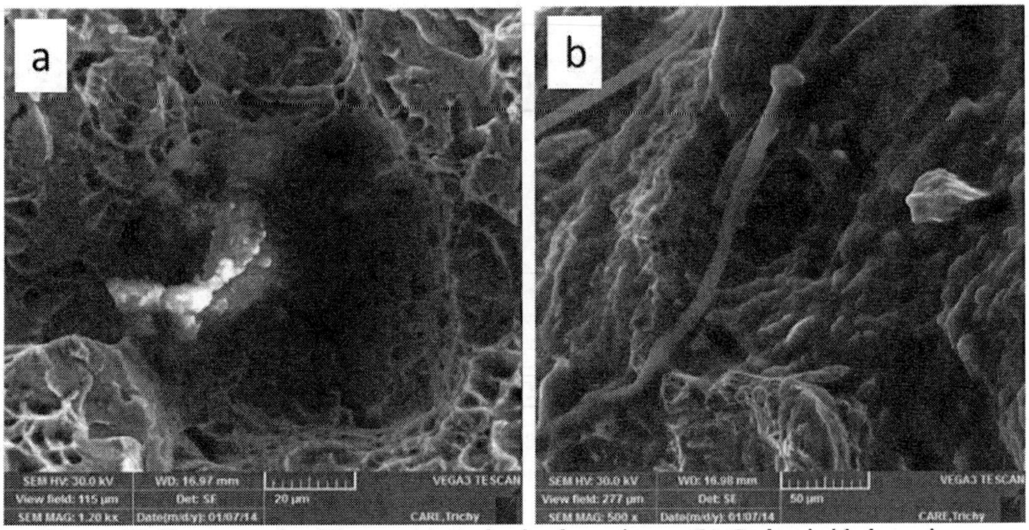

Figure 12. Fractography 50 µm – **(a)** Carburized specimen; **(b)** Carbonitrided specimen.

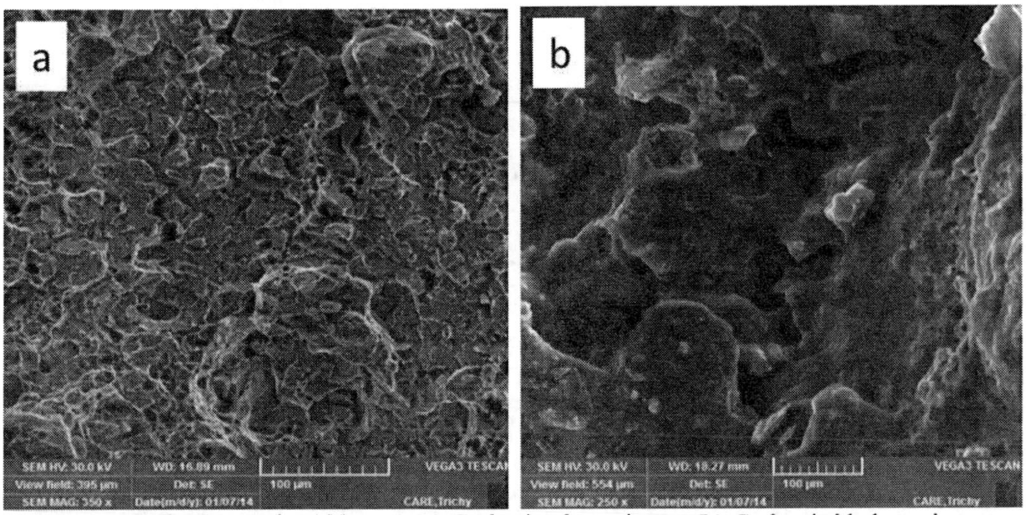

Figure 13. Fractography 100µm – **(a)** Carburized specimen; **(b)** Carbonitrided specimen.

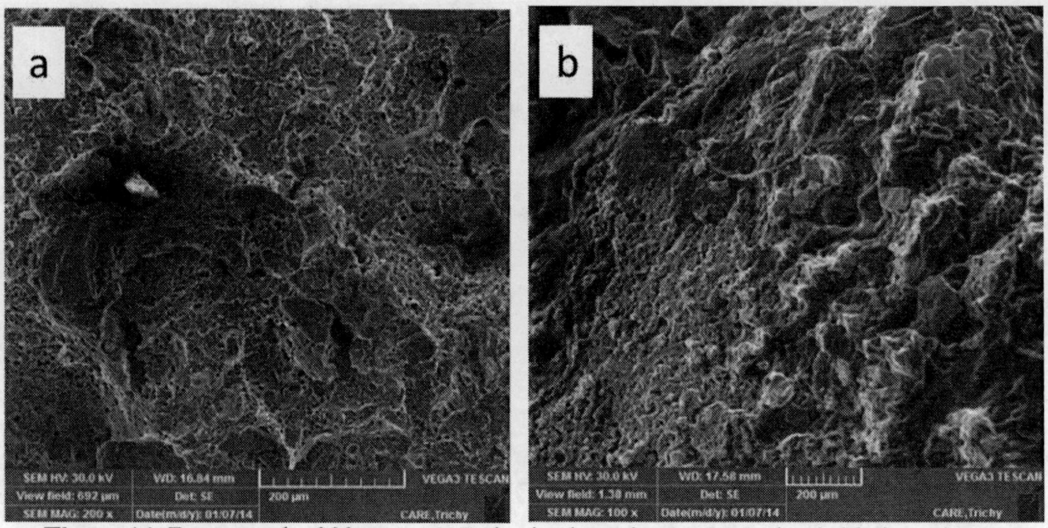

Figure 14. Fractography 200µm – **(a)** Carburized specimen; **(b)** Carbonitrided specimen.

However macro observations suggest that the case-hardened specimens are brittle fractured. When crack propagates from case to core reaching the grain boundary, the crack doesn't propagate. It is because the core is retained with austenite where the case of the specimen is martensite because of surface treatment. But the fractograph images reveals cleavage and dimples in core spots of specimen results in an intergranular fracture.

5. Conclusion

16Mncr5 low carbon steels was surface treated by gas carburizing and gas carbonitriding process. Carbon enriches up to 0.600% of weight in the carburizing process and 0.494 % in the carbonitriding process. The Surface hardness of 59HRC is found in Carburized and 54 HRC in Carbonitrided specimen. The transformation from austenite to martensite in Micro structural changes were observed in both surface treated specimens. Reduction associated with volume expansion results when austenite transforms to martensite. The fracture mechanism in Carburized and Carbonitrided specimen study reveals that case exhibiting brittleness and core with ductility.

References

[1] André Luiz Vasconcellos daCosta e Silva, "The effects of non-metallic inclusions on properties relevant to the performance of steel in structural and mechanical applications", Journal of Materials Research and Technology, Vol.8(2), pp.2408-2422, 2019, doi:10.1016/j.jmrt.2019.01.009

[2] P.Wang, S.P.Lu, N.M.Xiao, D.Z.Li, Y.Y.Li, "Effect of delta ferrite on impact properties of low carbon 13Cr–4Ni martensitic stainless steel", Materials Science and Engineering: A, Vol.527(13–14), pp.3210-3216, 2010, doi:10.1016/j.msea.2010.01.085

[3] Chunfang Wang, Maoqiu Wang, Jie Shi, Weijun Hui, Han Dong, "Effect of microstructural refinement on the toughness of low carbon martensitic steel", Scripta Materialia, Vol.58(6), pp.492-495, 2008, doi:10.1016/j.scriptamat.2007.10.053

[4] Hanamura, Yin Fuxing, Nagai Kotobu, "Ductile-brittle transition temperature of ultrafine ferrite/cementite microstructure in a low carbon steel controlled by effective grain size", ISIJ international, Vol.44(3), pp.610-617, 2004.

[5] Wang Haojie, Bin Wang, Zhaodong Wang, Yong Tian, R.D.K. Misra, "Optimizing the low-pressure carburizing process of 16Cr3NiWMoVNbE gear steel", Journal of Materials Science & Technology, Vol.35(7), pp.1218-1227, 2019, doi:10.1016/j.jmst.2019.02.001

[6] Atraszkiewicz R, B. Januszewicz, W. Stachurski, K. Dybowski, A. Rzepkowski, "High pressure gas quenching: Distortion analysis in gears after heat treatment", Materials Science and Engineering: A, Vol.558, pp.550-557, 2012, doi:10.1016/j.msea.2012.08.047

[7] Nandu Mohan, Sanjivi Arul, "Effect of Cryogenic Treatment on the Mechanical Properties of Alloy Steel 16MnCr5", Materials Today: Proceedings, Vol.5(11), Part 3, pp.25265-25275, 2018, doi:10.1016/j.matpr.2018.10.329

[8] Cm.Vivek, "Influence Of Carburizing and Carbonitriding in 16mncr5 to Enhance MechanicalProperties", International Journal of Innovations in Engineering and Technology, Vol.7(3), pp.261-266, 2016.

IOP Conf. Series: Materials Science and Engineering **764** (2020) 012005 doi:10.1088/1757-899X/764/1/012005

Solid particle erosion studies on SiC reinforced functionally graded aluminium matrix composites

S Prathap Singh[1]*, T Prabhuram[1], K Vinoth Babu[2], D Elilraja[1], M Uthayakumar[3], T P D Rajan[4]
[1]Department of Mechanical Engineering, St. Joseph's Institute of Technology, Chennai, India.
[2]Department of Mechanical Engineering, Sri Indu College of Engineering and Technology, Telangana, India.
[3]School of Automotive and Mechanical Engineering, Kalasalingam Academy of Research and Education, Krishnankoil, India.
[4]Material Science and Technology Division, CSIR-NIIST, Trivandrum, Kerala, India.

*Corresponding author email: prathap.singh50@gmail.com

Abstract. The main objective of this present study is to fabricate functionally graded Aluminium alloy reinforced with silicon carbide composite by horizontal centrifugal casting process and investigate its erosion wear behaviour using Taguchi technique. Microstructural examination confirmed that the gradient structure across the radial direction and maximum particle reinforcement was found at the outer zone of the fabricated functionally graded metal matrix composite. Solid particle erosion wear experiments were carried out on the fabricated functionally graded metal matrix composite using abrasive air jet erosion test rig. Erosion process parameters such as zone distance from outer surface, Erodent velocity and Exposure time were examined using L9 orthogonal array table. The effect of erosion parameters on the erosion rate was investigated by signal-to-noise ratio and Analysis of Variance. Analysis of Variance result concluded that zone distance from the outer surface had 91.46% influence on erosion rate followed by exposure time (2.54%) and erodent Velocity (0.54%). Finally, regression equation was formed to find the optimal erosion rate and that has been validated by confirmatory experiment.

1. Introduction

Functionally graded materials are the newly developing materials and it has been attracted by the researcher because of its better gradient structure. It has two or more material added together with different volume fraction to get continuously varying properties. A component with better wear resistant on the outer surface when compared with the inner surface can be achieved through functionally graded materials [1]. Now a day, solid particle erosion is one of the most considerable obstacles that affect the tribological behaviour of the metal matrix composites. Here the solid abrasive particles move with high impact velocity and strike on the surface with different directions and immediately take away some materials from the surface of the composite. Because of this action, the normal properties of the composite can be altered due to the localized impact of high velocity solid particles [2]. W. Wu et al compared the effect of reinforcement on the mechanical and wear properties of the heat treated aluminium and pure aluminium. They conducted an analysis on the aluminium with as-cast, annealed, as-quenched, and T6 conditions. They concluded that lack of ductility was the major problem for erosion rate. Moreover, heat treatment had a little considerable effect on the erosion [3]. Mengyao Dong et al studied the effect of solid particle erosion on the thermoplastic polyurethane nanocomposite reinforced with a different weight percentage of carbon nanotubes fabricated through hot pressing technique. The influence of impact parameters such as impingement angle, impact velocity and size of impact particles on the surface roughness of the fabricated composites was studied. They concluded that the addition of carbon nanotubes improved the erosion resistance and also the increase in impact velocity increased the erosion rate [4]. Mehmet Bagci et al had done a erosion study on the epoxy and glass fibers matrix composite reinforced with 0, 15 and 30% boric acid particles fabricated through hand lay-up technique. The impact of different erosion parameters such as fiber direction, erodent size, percentage boric acid particles, impingement angle and impact velocity were studied. Moreover, Taguchi technique was used to found the combination of erosion parameters which affect the erosion rate of the composite. They concluded that the addition of boric acid particles increased the strong bonding between epoxy composite and also contributed 52.51% on the erosion rate of the composite [5]. M. Pethuraj et al conducted erosion studies on the Aluminum metal matrix composite reinforced with sillimanite fabricated by vacuum assisted stir casting method. They found the effect of different varying operation parameters such as erodent velocity, angle of impingement and weight percentage of the reinforcement on the erosion rate of the composite. The results were revealed that the addition of reinforcement decreased the erosion rate. Moreover, the high reinforced composite had a high erosion rate when compared with pure Aluminum [6]. K. Balamurugan et al carried erosion study on the silicon carbide reinforced magnesium composite. They found the influence of erosion governing parameters such as erodent particle. impact angle, erosion velocity, and discharge rate on the erosion rate and Taguchi technique were used to found the optimal parameters that are affected the erosion rate of the composite. The conclusion stated that erosion velocity has a significant effect on the composite compared with other parameters [7]. S. Vigneshwaran et al had done erosion experimental study on the red mud filled and unfilled sisal fiber reinforced hybrid composites. Taguchi analysis was used to found the factor which is highly affected the erosion rate among various factors such as filler material, fiber concentration, erodent velocity, impact angle and erodent discharge rate. The significant finding shows that erodent velocity had the greatest effect on the erosion rate of the composite [8]. Many researchers have conducted erosion study on the different composite but no one reported the erosion study on the Aluminium functionally graded metal matrix composite. This is the seed for the present study.

The main objective of the present study is to fabricate Aluminum Functionally Graded Metal Matrix Composite reinforced with SiC particles (FGMMC) through stir casting process followed by a vertical centrifugal casting process. Then the erosion wear behavior of FGMMC was found by using air jet erosion tester apparatus using L9 orthogonal array table. Finally, the optimal factors that are influencing the erosion wear are calculated.

2. Experimental Details

2.1. Materials and Methods

The major base material preferred for this study was Aluminium alloy Al6061 as a result of its wide range of automotive applications and reinforcement elected for this study was Silicon Carbide(SiC) with a standard size of 15μm (20 wt%). The Aluminium alloy Al6061 had a lower density of 2.7 g/cm^3 compared to SiC particle which had a density of 3.21 g/cm^3.

Functionally graded metal matrix composites were fabricated through two steps namely stir casting process and vertical centrifugal casting process. Initially, Aluminium alloy was melted in the electrical resistance furnace to 740°C and then SiC particles were added to the molten metal slowly. Mechanical stirring was done on the molten metal to get uniform distribution of SiC. Then the melt was poured into a vertical centrifugal casting machine which was rotating with a speed of 1200 rpm. Because of the centrifugal force, high density SiC particles are thrown to the outer surface of the fabricated disc during centrifugal casting. Finally, a disc with a diameter of 250 mm and thickness 25 mm was fabricated and is shown in figure 1.

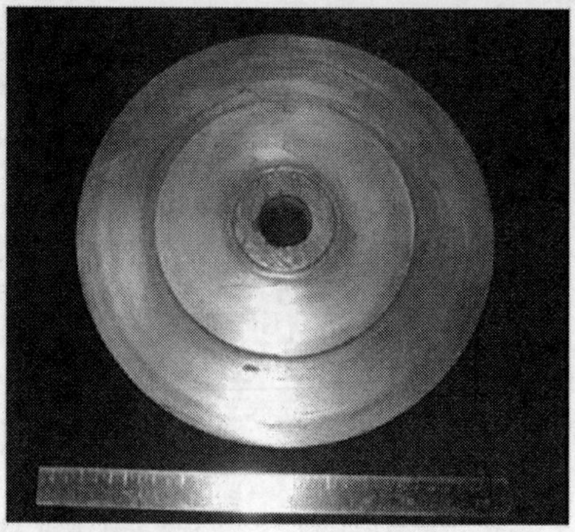

Figure 1. Al6061-20%SiC functionally graded metal matrix composite disc.

The microstructural examination was performed on the fabricated disc to ensure the distribution of SiC particles towards the inner surface. Based on the microstructural study three regions were selected to erosion study based on the distribution of SiC particles. A cut portion from the fabricated disc with three selected zone is shown in figure 2. A specimen with a dimension of 10 x 10 x 4 mm was cut from the three regions of the disc to erosion study.

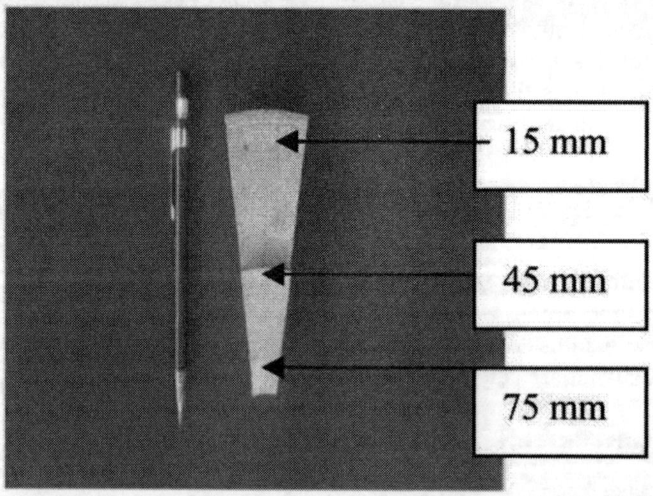

Figure 2. Cut portion from the fabricated disc with three identified zones.

2.2. Solid Particle Erosion

Abrasive air jet erosion tester was used to study the erosion wear performance and the apparatus is shown in figure 3. During erosion, high velocity air mixed with uniform size abrasive erodent. This mixture was made to impact on the target material through a nozzle. Alumina particles with an average size of 50 microns are selected as an erodent. Impingement angle of 90° was maintained throughout the experiment. The erosion study was conducted on the specimen with three erosion control process parameters such as zone distance from the outer surface, erodent velocity and exposure time with different levels and is consolidated in table 1. L9 orthogonal array table was elected to conduct the erosion experiment.

Table 1. Erosion process parameters and their levels.

S. No	Zone distance from the outer surface (mm)	Erodent velocity (m/s)	Exposure time (min)
1	15	100	10
2	45	150	15
3	75	200	20

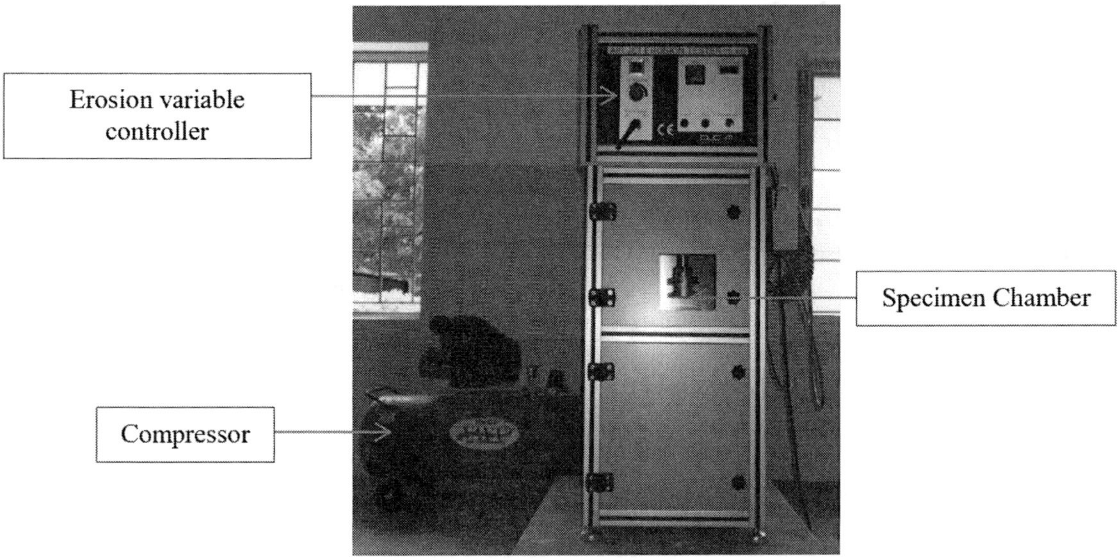

Figure 3. Air jet erosion tester apparatus.

Finally, Erosion wear rate was calculated for each combination of erosion parameters using the following formula [9].

$$\text{Erosion rate} = \frac{W_m}{W_p}$$

Where W_m is the weight loss in the FGMMC specimen and it was found by using a precision weighing machine and W_p is the weight of the impact particles (testing time × feed rate).

3. Results and discussions

3.1. Microstructural Examination

Figure 4 shows the microstructure of the specimen cut from the three regions of the fabricated disc. Figure 4(a) shows the microstructure observed on the specimen cut from the 15 mm distance from the outer surface. This shows a high concentration of SiC particles. Because the high density SiC particles are propelled to the outer surface of the composite during centrifugal casting and this portion is called SiC particle rich zone. Figure 4(b) shows the microstructure observed on the specimen cut from the 45 mm distance from the outer surface. This shows the intermediate concentration of SiC particles distribution. This portion is called SiC particle intermediate zone. Figure 4(c) shows the microstructure observed on the specimen cut from the 75 mm distance from the outer surface. This shows a lower concentration of SiC particles. This portion is called SiC particle depletion zone.

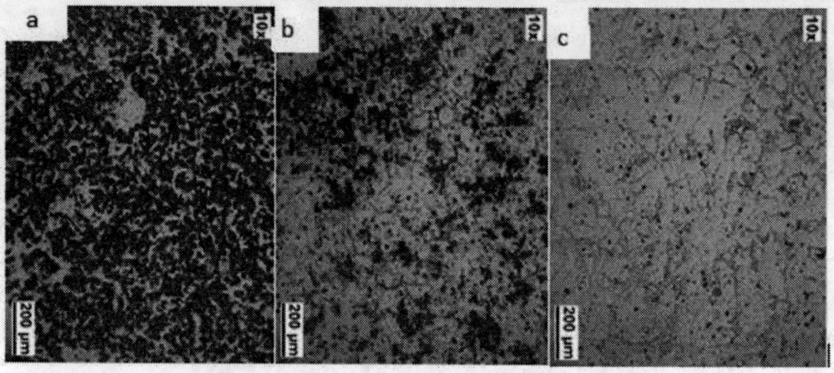

Figure 4(a). Microstructure of SiC particle rich zone; **Figure 4(b).** Microstructure of SiC particle intermediate zone; **Figure 4(c).** Microstructure of SiC particle depletion zone.

3.2. Erosion Test

The major intention of the experiment is to determine the optimum combination of erosion parameters to get a lower erosion rate. Table 2 shows the experimental results of erosion rate with a different combination of erosion parameters and corresponding S/N ratios.

Table 2. Erosion experimental results and their corresponding S/N ratio.

S. No	Zone distance from the outer surface (mm)	Erodent velocity (m/s)	Exposure time (min)	Erosion rate 10^{-5} (g/m^2)	S/N ratio (dB)
1	15	100	10	1.25	7.604
2	15	150	15	1.16667	8.203
3	15	200	20	1.225	7.779
4	45	100	15	1.46667	6.215
5	45	150	20	1.375	6.776
6	45	200	10	1.3	7.263
7	75	100	20	1.785	4.509
8	75	150	10	2.1	3.098
9	75	200	15	2.16667	2.826

IOP Conf. Series: Materials Science and Engineering **764** (2020) 012005 doi:10.1088/1757-899X/764/1/012005

3.3. Taguchi Analysis

Taguchi method was utilized to determine the effect of each erosion parameters on the response. Signal to Noise ratio for the erosion parameters namely zone distance from the outer surface, erodent velocity and exposure time were calculated by feeding the calculated erosion rate to the MINITAB 17 software. "Smaller the Better" quality characteristic was selected to find minimum wear rate. The response table for the signal to noise ratio for the calculated erosion rate is shown in table 3.

Table 3. S/N ratio response table for erosion rate.

Erosion parameters	Level-1	Level-2	Level-3	Delta	Rank
Zone distance from the outer surface (mm)	-1.680	-2.791	-6.064	4.384	1
Erodent velocity (m/s)	-3.433	-3.516	-3.586	0.153	3
Exposure time (min)	-3.554	-3.794	-3.187	0.607	2

From table 3, it was observed that zone distance from the outer surface was found to be the most significant process parameter that is affecting the erosion rate. Moreover, erodent velocity had only a small amount of impact on the erosion rate when compared with zone distance from the outer surface and exposure time. Figure 5 shows the main effect plot that shows the impact of all erosion parameters with different levels. The optimum erosion parameters that are affected the erosion rate can be found from this main effect plot. So the optimum erosion process parameters needed for getting lower erosion rate are 15 mm zone distance from the outer surface, 100 m/s erodent velocity and 20 min exposure time. Moreover, ANOVA analysis was done to confirm the above said optimum erosion parameters.

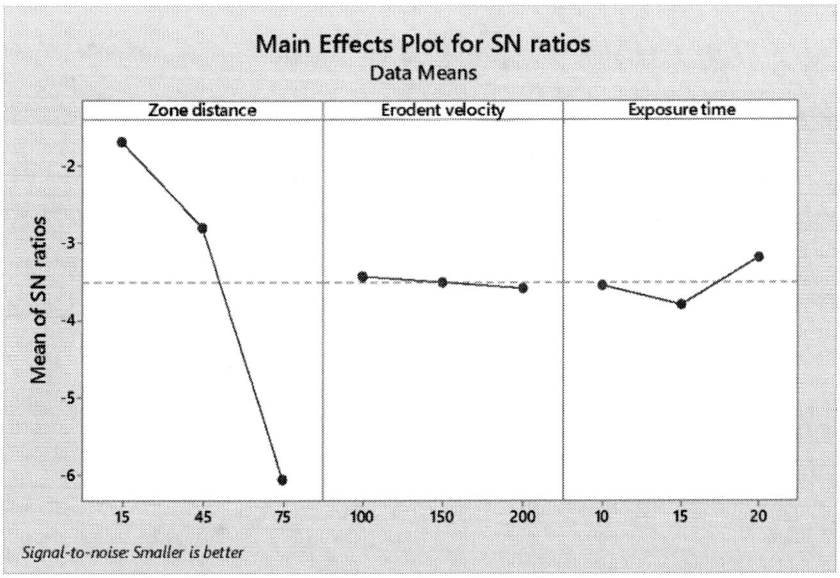

Figure 5. Main effect plot for erosion parameters on erosion rate.

IOP Conf. Series: Materials Science and Engineering **764** (2020) 012005 doi:10.1088/1757-899X/764/1/012005

Analysis of Variance was done by using MINITAB 17 software. The results obtained in the ANOVA analysis are shown in table 4. The ANOVA results reviled that zone distance from the outer surface had a greater impact on the erosion rate when comparing with erodent velocity and exposure time. The zone distance from the outer surface had 91.46% contribution on the erosion rate whereas the exposure time had 2.49 % contribution and erodent velocity had only 0.54 % contribution on the erosion rate. The error had only a small significant role on the erosion rate which shows the experiment was conducted without missing any erosion parameters.

Table 4. Results of ANOVA and the percentage of the significance of each parameter.

	Degree of freedom	Sum of Squares	Mean Sum of Squares	Contribution %	Rank
Zone distance from the outer surface (mm)	2	1.07847	0.539233	91.46	1
Erodent velocity (m/s)	2	0.00647	0.003233	0.54	3
Exposure time (min)	2	0.02944	0.014720	2.49	2
Error	2	0.06480	0.032400	5.49	
Total	8	1.17917		100	

3.4. Confirmation Test

Confirmation test has been conducted with optimum erosion parameters. After finding the optimum erosion parameters from taguchi analysis, it is necessary to find the optimum erosion rate with optimum erosion parameters. A regression equation was derived from MINITAB 17 software to find the predicted optimum erosion rate and the equation is shown below.

Erosion rate = 0.972 + 0.01339 Zone distance + 0.00063 Erodent velocity - 0.0088 Exposure time

The predicted value calculated was 1.05985. The experimental erosion wear was calculated using the abrasive air jet tester and the experimental erosion value was 0.9985. The theoretical and experimental erosion values are less than the values which are already calculated in the erosion experiment. The results of the confirmation erosion test with optimum erosion rate are tabulated in table 5.

Table 5. Results of the confirmation experiment with optimum erosion rate.

	Optimal machining Parameter	
	Predicted	Experimental
Erosion rate	1.05985	0.9985

4. Conclusion

This study presented the fabrication of Aluminium functionally graded metal matrix composite reinforced with SiC particles through stir casting process followed by vertical centrifugal casting process successfully and the following conclusions are drawn,

Microstructure examination confirmed the gradient distribution of SiC particles along the radial direction from outer surface and erosion test was conducted by using taguchi L9 orthogonal array table. The optimum erosion parameters calculated using the taguchi method to get lower erosion rate was zone distance from the outer surface of 15 mm, erodent velocity of 100 m/s and exposure time of 20 minute.

ANOVA result revealed that the zone distance from the outer surface had the highest amount of contribution on the erosion rate followed by erodent velocity and exposure time.

The confirmation experiment was conducted for the optimum erosion parameters and the confirmation test result showed a lesser amount of erosion rate compared with all the combinations.

References

[1] Bassiouny I Saleh, Mahmoud H. Ahmed, "Development of Functionally Graded Tubes Based on Pure Al/Al_2O_3 Metal Matrix Composites Manufactured by Centrifugal Casting for Automotive Applications", Metals and Materials International, Vol. 25(4), pp.1 - 11, 2019, doi:10.1007/s12540-019-00391-3

[2] N M Barkoulaa, G C Papanicolaoub, J Karger-Kocsisa, "Prediction of the residual tensile strengths of carbon-fiber/epoxy laminates with and without interleaves after solid particle erosion", Composites Science and Technology, Vol. 62, pp. 121 - 130, 2002.

[3] W Wu, K C Goretta, J L Routbort, "Erosion of 2014 A1 reinforced with SiC or A1203 particles", Materials Science and Engineering, Vol. 151, pp. 85 - 95, 1992.

[4] Mengyao Dong, Chuan Wang, Hu Liu, Chuntai Liu, Changyu Shen, Jiaoxia Zhang, hengxinzhuo Jia, Tao Ding, Zhanhu Guo, "Enhanced Solid Particle Erosion Properties of Thermoplastic Polyurethane-Carbon Nanotube Nanocomposites", Macromolecular Materials and Engineering, Vol. 304(5), pp. 1 - 11, 2019, doi: 10.1002/mame.201900010

[5] Mehmet Bagci, Huseyin Imrek, "Application of Taguchi method on optimization of testing parameters for erosion of glass fiber reinforced epoxy composite materials", Materials and Design, Vol. 46, pp. 706 - 712, 2013, doi:10.1016/j.matdes.2012.11.024

[6] M Pethuraj, S Rajakarunakaran, S Rajesh, "Solid Particle Erosive Behaviour of Sillimanite reinforced Aluminum metal matrix composites", Materials Research Express, Vol. 5(6), pp. 1 - 28, 2018, doi:10.1088/2053-1591/aac8ba

[7] K Balamurugan, M Uthayakumar, M Ramakrishna, U T S Pillai, "Air Jet Erosion Studies on Mg/SiC Composite", Silicon, pp. 1 - 11, 2019, doi:10.1007/s12633-019-00148-y

[8] S Vigneshwaran, M Uthayakumar, V Arumugaprabu, "Solid particle erosion study on redmud - an industrial waste reinforced sisal/polyester hybrid composite", Materials Research Express, Vol. 6(6), pp. 1- 22, 2019, doi:10.1088/2053-1591/ab0a44

[9] Mengyao Dong, Qiang Li, Hu Liu, Chuntai Liu, Evan K. Wujcik, Qian Shao, Tao Ding, Xianmin Ma, Changyu Shen, Zhanhu Guo, "Thermoplastic polyurethane-carbon black nanocomposite coating: Fabrication and solid particle erosion resistance", Polymer, Vol. 158, pp. 381 - 390, 2018, doi:10.1016/j.polymer.2018.11.003

IOP Conf. Series: Materials Science and Engineering **764** (2020) 012006 doi:10.1088/1757-899X/764/1/012006

Influence on performance and emission characteristics of diesel engine by introducing medium strength magnetic field in fuel and air lines

Chandrasekaran M*, Prakash KB, Prakash S, Ravikumar M

Department of Mechanical Engineering, Bannari Amman Institute of Technology, Coimbatore-638 401, India.

*Corresponding author E-mail: mchand.87@gmail.com

Abstract. Manufacturing In this work, the energy of electromagnet is used to treat the fuel and airline for reducing the fuel consumption and emissions in four-stroke diesel engine. The experiment comprises Internal Combustion (IC) Engine test rig and electromagnets with different intensities (4500, 6000, 8500 and 10000 gauss), are installed along the honeycomb structured mini channel fuel line and air suction line. Experiments are carried out in four-stroke diesel engine test rig for variable engine load by applying medium strength magnetic field on air and fuel supply line. The medium strength magnetic field is used to improve the atomization process effectively, due to that increment in the rate of disintegration of the droplets, as result the viscosity of the fuel decreases. Because of cohesive force between the fuel molecules are affected by the magnetic field and it leads to less surface tension force effect, thus helps to improve atomization process. It is observed that, the impact of magnetic field treatment considerably increased the brake thermal efficiency about 13% and maximum fuel reduction of 11% was obtained for field intensity 10000G and also found that the HC, CO, PM, and NOx emissions are decreased about 10.4%-14.2%, 24–25%, 13%–15%, and 8.4%-11% respectively, but CO2 is increased up to 12.2% because of association of oxygen molecules with CO which is critical point and it can be treated in the catalytic converter.

1. Introduction

As of late, there are such a significant number of endeavours pertaining of better performance and emission limit of engines per unit fuel supply. Bharath Stage (Indian Emission Standards) emission standards are more stringent towards emission norms that demands exhaust treatment and improved performance modification in all relevant aspects in engines. Generally, diesel engine is one of serious plant for generating the pollution over gasoline engine and diesel fuel has much energy density with complex molecular structure [1].

Magnetic field is introduced as external stimuli, which ionizes the fuel as per the theory of magnetic field mutual action. Complex hydrocarbon chain have different critical physical behaviour during atomization process because it has thickly stuffed structures called pseudo compounds, which can additionally compose into groups [2]. Stereochemistry is explained the isomerism of that normal structure para state of hydrogen and useful ortho state by magnetic strength. At the time of combustion para arrangement is not right orient to combine oxygen with hydrogen due to poor penetration of oxygen in air/fuel mixture. Magnetic field strength helps to polarize the hydrocarbons and it converts critical para state into ortho state of hydrogen, due to that favourable isomers was obtained, which increase space between hydrogen in molecular chain and set platform for interlocking oxygen with hydrogen more[3]. Also it has been likewise noticed that when the fuel goes through a magnetic field, there atomization process improved by affecting surface tension of the fuel why because fluidity ($1/\mu$) improved for mixture formation. Thus, it leads to complete combustion in engines [4].

Some of investigations have assessed the effect of magnetic field on diesel fuels, gasoline for reducing emissions as well as fuel consumption [5] and also trying to predict right optimized operating parameter at higher energy performance. Previous work [6] is reported 22%, 12% and 7% reduction in CO, HC and fuel consumption by pre combustion treatment (5000 gauss magnetic field is applied) and post treatment in gasoline engine. Impact on CO2 emission by introducing magnetic field in four stroke single cylinder engine was captured with catalytic converter which makes clear that after treatment is also needed [6]. Everyone focused on introduction of magnetic field in fuel line pre combustion treatment, however limited work was carried out on airline. Applying effective magnetic field on both fuel and airline are very critical and challenging, because it was seriously affected by air gap in interface.

The objectives of this work are to improve performance and reduce emissions using medium strength magnetic field in a diesel engine by introducing magnetic flux in fuel and air supply line.

2. Experimental setup

2.1. Magnetic Inducer

A magnetic field is introduced in fuel line which has honeycomb structure made by multi mini channel (dia>3mm). The challenge of introducing mini channels is to maintain pressure drop and it is helpful to disperse magnetic field in the fuel line. Airline also connected with magnetic coil which is faced to North Pole, due to that state of orientation is changed from Para state to Ortho state which open to adhere more oxygen with fuel hydrocarbon. So that proper combustion takes place, the surface tension force (cohesive of fuel bunch) also was affected by magnetic field and it leads to better atomization for attain air/fuel mixture. The coil is connected with controlled magnetic field system, for that required power is trapped from the battery and it was connected to regulator which controls the applied magnetic field as per need flux strength (4000~10000 Gauss) [9]. The specification of the diesel engine is shown in the below table 1.

Table 1. Specification of diesel engine.

Specification	Details of specification
Model	DAF8
Make	Kirloskar
Type	Four stroke, Water cooled, Diesel
No. of cylinder	1
Bore	110 mm
Rated speed	1800 rpm
Stroke length	85 mm
Combustion principle	Compression Ignition Engine
Maximum Load	11 kg
Compression ratio 3 port	18:1
Lubrication system	Forced lubrication

IOP Conf. Series: Materials Science and Engineering **764** (2020) 012006 doi:10.1088/1757-899X/764/1/012006

2.2. Experimental Procedure

The test runs are conducted in mono cylinder, four stroke water cooled diesel engine arrangements shown in the figure 1 to find the performances and exhaust gas qualities. Engine is coupled with eddy current dynamometer to apply electrical load on engine effectively. Pollutants from exhaust stream is measured by using five gas analyzer and results observed are CO, NOx in ppm scale but all others in percentage (%). The experiment loads are swinging between 25% ~100% at rated speed of the above said engine (1800 rpm). Before taking observation results, the engine is warmed up for 30 minutes that was continued with 4 minutes for further every sampling cycle test. At any time, three Particulate Matter (PM) and Polycyclic Aromatic Hydrocarbons (PAHs) tests were gathered during each and every run, it acquires significant result data. The samples are gathered from exhaust for PAH examinations by utilizing isokinetic samples and the values noted cautiously with appropriate trial techniques that are adhered with standard norms. The data acquisition system is collecting all the data's from various locations in engine and also integrated with gas analyzer. It measures the presence of such gases like HC, CO, NOX and CO2, at every above said load with and without magnetic field.

Figure 1. Photographic view of experimental setup.

3. Results and discussion

The results were observed by normal condition as well as magnetic field applied condition that are interpreted by performance characteristics.

3.1. Influencing of magnetic field on engine performance

The performance of engine is denoted by two parameters i.e. Brake Specific Fuel Consumption (BSFC) and Brake Thermal Efficiency (BTE). The BSFC (g kW^{-1} h^{-1}) is considered as the fuel burning up per unit mechanical energy generation. In general, when engine load increases BSFC decreases but its value remains same and there are no notable changes in the specific fuel consumption. The BSFC and BTE are calculated based on the equation (1) and equation (2), in which, V indicates the mass of the fuel used in each run (g), P is mechanical work generation in (kW), and t represents the time taken for volume fuel usage, and H indicates the calorific value (kJ g^{-1}) of the diesel fuel.

$$BSFC = V/Pt \quad (g\ kw^{-1}\ h^{-1}) \tag{1}$$

$$BTE = Pt/VH \quad (\%) \tag{2}$$

Equation 3, is used to calculate the total fuel consumption, where Vr is the CC of fuel utilized to produce power, (L) and D is the distance travelled by the vehicle.

$$\text{Total Fuel consumption} = \text{Vr/D (lt. km}^{-1}) \quad (3)$$

Previous report [10] produced similar result for both BSFC and fuel consumption at 50% loading condition in their work. BSFC is decreased significantly by 11%, while put variation in engine load from 25% to 50%. In case of brake thermal efficiency at superior loads the combustion efficiency will be better in diesel engine due to rise in cylinder temperature, which improves the atomization and evaporation process. Likewise similar experiment study [11] towards BTE was obtained that says BTE increased due to ionization process while increasing the voltage applied on magnetic inducer. Applied magnetic intensity on the fuel and airline is influencing the engine performance and emission characteristics and it is proportional to the rate of atomization. From the figure 2, BTE is increased approximately 13%, when compared to the conditions without and with magnetic fields (8500G) respectively. Hence the mechanical efficiency increased by 14% under part load operation at magnetic strength 10000G.

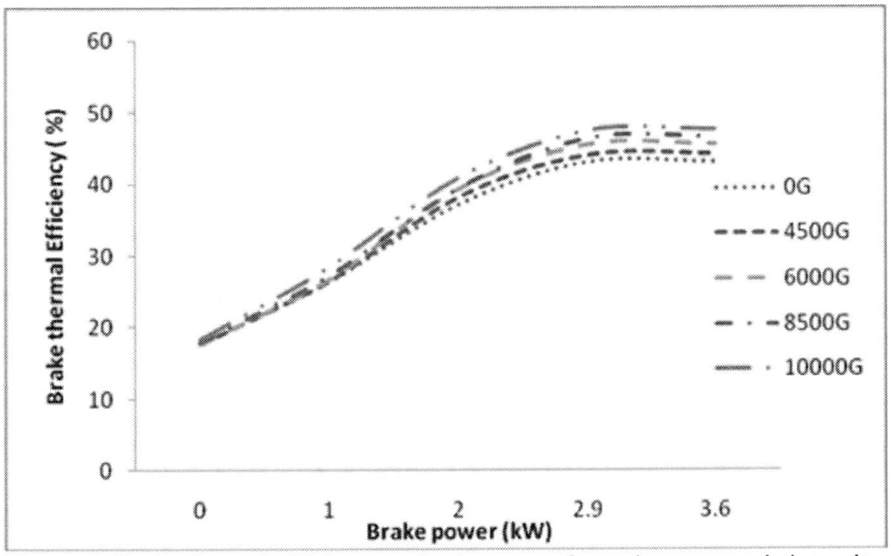

Figure 2. Brake thermal efficiency vs brake power for various magnetic intensity.

3.2. Magnetic field impact on Mechanical efficiency

From the below figure 3, it is clear that the mechanical efficiency increases with increase in magnetic field (10000 gauss) applied to the fuel and air line by magnet inducer. In part load operation the mechanical efficiency increased by 6% to 8%, when compare with and without magnetic field, respectively. Previous work [11] reported similar result that mechanical efficiency increases with increase in applied magnetic field due to higher heat release rate by ionization of fuel, it was open to attain effective increase in mechanical efficiency of engine at part load as well as peak load operations, when the intensity of magnetic field is 10000 gauss.

IOP Conf. Series: Materials Science and Engineering **764** (2020) 012006 doi:10.1088/1757-899X/764/1/012006

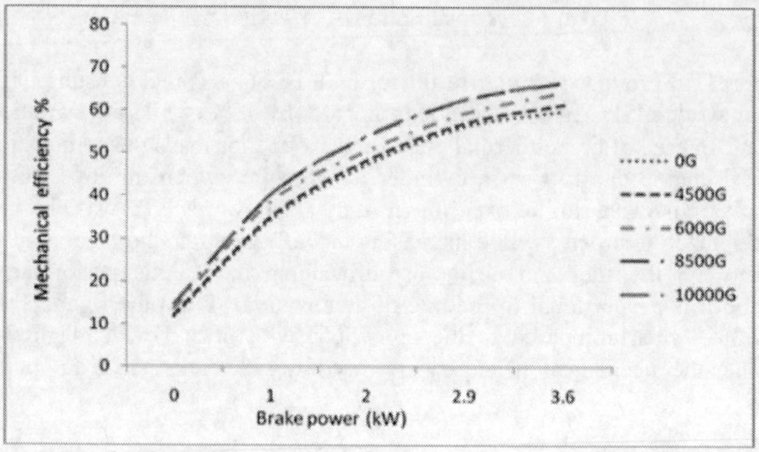

Figure 3. Mechanical efficiency vs brake power for various magnetic intensity.

3.3. Magnetic field influence on Pollutants formation

The emission prediction have done at part load conditions for normal diesel and it was treated by four different magnetic fields of 4500G, 6000G, 8500G and 10000G, and these are tabulated in Table 2.

Table 2. Emission levels at 50% loading condition with different magnetic field.

Magnetic field (Gauss)	CO (ppm) ±14	HC (%)	CO_2 (%)	PM (mg Nm^{-3})	NO_x (ppm)
0	2030	16.9	3.2	44	350
4500	2019	16.7	3.4	42.4	342
6000	2001	15.8	3.65	41.7	338
8500	1859	14.5	3.8	39.9	335
10000	1865	12.7	4.2	39	330

Thus it is clear (from the Table 2) that the effect of magnetic field in the fuel line is used to reduce emission, especially CO is more in engine exhaust due to incomplete combustion [11]. The medium magnetic strength is taking care of significant decrease in the pollutant formation, diesel was treated with 10000 gauss magnetic condition gives a better result in CO emission is 1865 ppm which is reduced by 25% (from the figure 4) at part load condition with respect to applied magnetic strength.

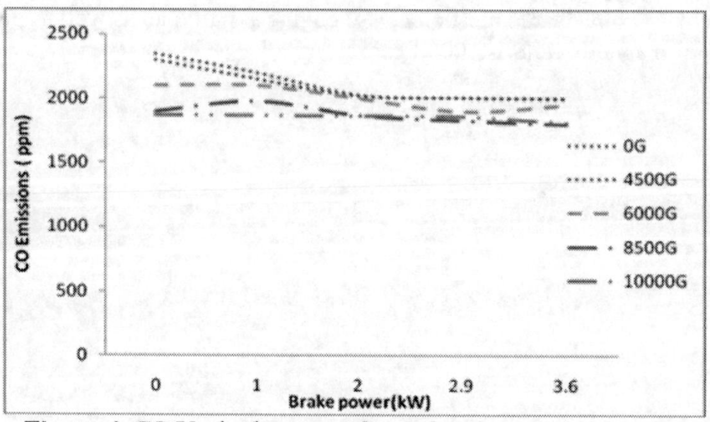

Figure 4. CO Vs. brake power for various magnetic intensity.

HC emission at part load condition is noted for diesel engine with and without magnetic field.it was emitted due to incomplete combustion of fuel and causes of line fault or storage tank ,so minor escape through leak and evaporation [12]. At full load condition, the amount of HC emissions are found to be 16.7% for diesel under normal conditions, when magnetic field is introduced in fuel and air line corresponding HC values are approximately decreased by 10.4%~14.2% at 10000 gauss. The experiment clearly shows that the increase in magnetic field reduces the HC emission significantly when compared to normal operating condition.

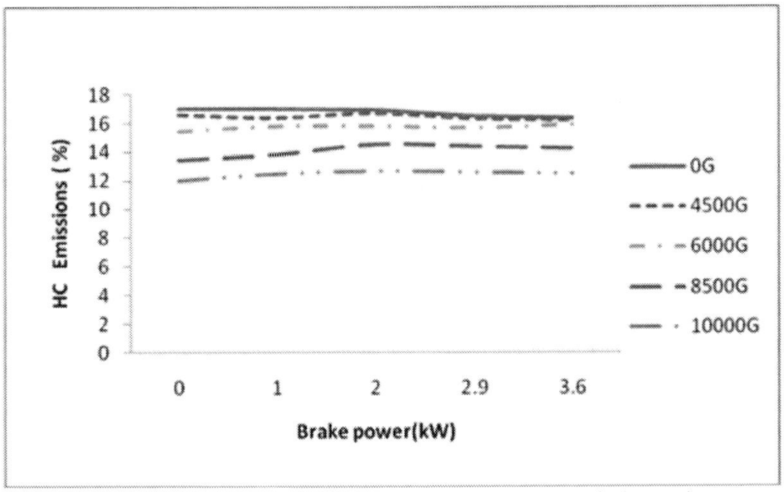

Figure 5. HC vs brake power for various magnetic intensity.

The figure 6 shows the variation of CO_2 emission with respect to different engine power is critical which not affected by magnetic field. The CO_2 emissions are directly proportional to the quantity of fuel consumed by the engine and it is increased under all operating conditions [13]. At full load condition it was reached maximum 12.2% which is treated by post treatment process (by catalytic converter).

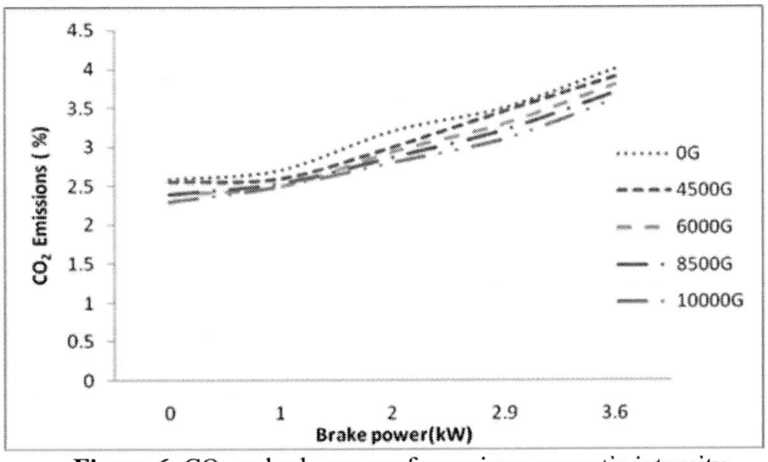

Figure 6. CO_2 vs brake power for various magnetic intensity.

The mechanism of NOx formation demands higher cylinder temperature, at no load and part loads condition, the NOx emission was less due to low consumption of fuel and slightly reduced in cylinder temperature [14-15]. From the figure 7, it is getting increased at peak loads due to higher combustion temperature, the amount of NOx emissions are 350 ppm, 330 ppm for diesel with and without magnetic field, respectively. NOx emission is reduced based on over period of running time are decreased by 8.4%, 11.6% for 8500 gauss and 10000 gauss respectively.

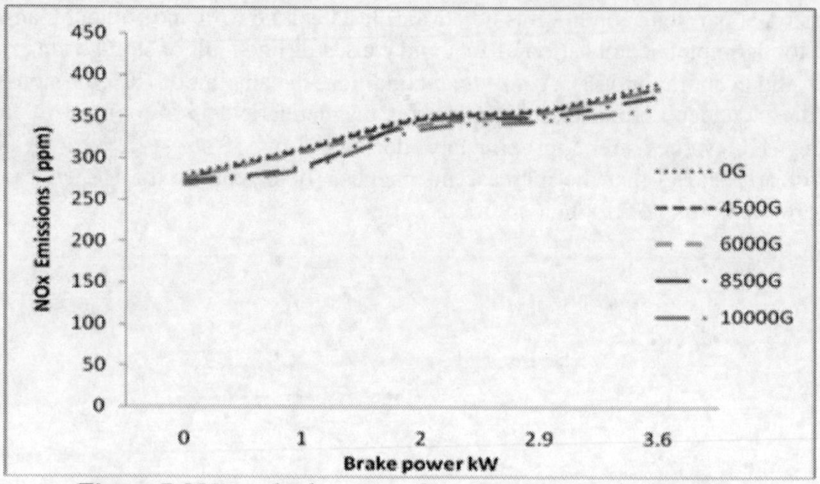

Figure 7. NOx vs brake power for various magnetic intensity.

Particulate matters (PM) are formed due to incomplete combustion of fuel in combustion chamber and presents of impurities in both fuel and air, as result shoot formation takes place. It was noted in the below figure 8, that significant reduction in PM shows efficient combustion achieved in combustion chamber. PM reduction (13% ~ 15%) was found in medium strength magnetic field 8500 gauss and maximum reduction in 10000gauss strength.

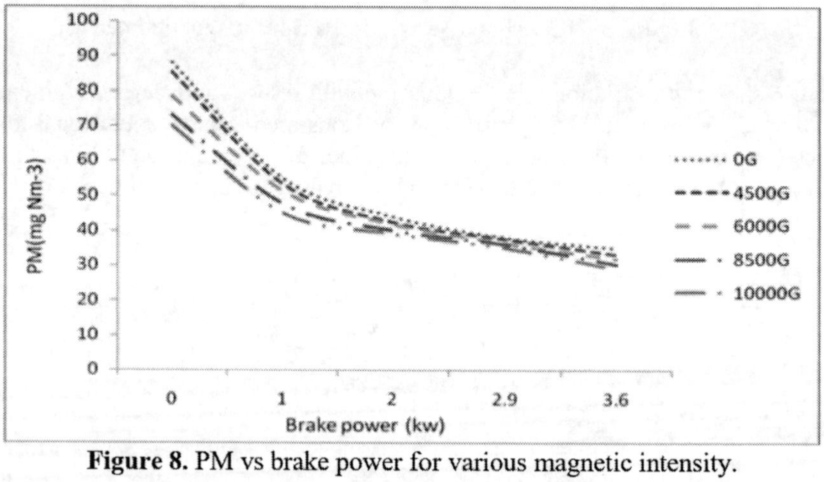

Figure 8. PM vs brake power for various magnetic intensity.

4. Conclusions

Introducing a magnetic field in the fuel and airline brought about a decrease in BSFC by 11%, while the BTE is enhanced by approximately 13%. In view of emitted pollutant reduction in percentage, the HC, CO, PM, and NOx emissions are decreased about 10.4%-14.2%, 24–25%, 13%–15%, and 8.4%-11% respectively. CO2 emission is increased by 12.2% due to association of oxygen with CO and forms CO2. Particulate matters are encounter by filter and the presents were reduced by 15% at part load with the magnetic field inducer.

References

[1] A. S. Faris, S. K. Al-Naseri, N. Jamal, R. Isse, M. Abed, Z.Fouad, A. Kazim, N. Reheem, A. Chaloob, H. Mohammad, H. Jasim, J. Sadeq, A. Salim, A. Abas, “Effects of magnetic field on fuel consumption and exhaust emissions in two stroke engine”, Energy Procedia, Vol.18, pp.

327-338, 2012.
[2] Rongjia Tao, “Investigate effects of magnetic fields on fuels”, Department of Physics,Temple University, Philadelphia, PA 19122, March 15, 2004.
[3] Hejun Guo, Z. Liu, Y.Chen and R. Yao, “A study of magnetic effects on the physicochemical properties of individual hydrocarbons”, Logistical Engineering College, Chongqing 400042, P.R China, pp.16-220, 1994.
[4] Farrag A. El Fatih, G. M. Saber, “Effect of fuel magnetism on engine performance and emissions”, Australian Journal of Basic and Applied Sciences, Vol.4 (12), pp.6354-6358, 2010.
[5] Habbo A. R. A., Khalil R. A. and Hammoodi H. S., “Effect of magnetizing the fuel on the performance of S.I.engine”, Al-Rafidain Engineering, Vol.19 (6),pp.84-90, 2011.
[6] V. Ugare, N. Bhave and S. Lutade, “Performance of spark ignition engine under the influence of magnetic field”, International Journal of Research in Aeronautical and Mechanical Engineering, Vol.1(3), pp.36-43, 2013.
[7] P. Govindasamy, S. Dhandapani, "Performance and Emissions Achievements by Magnetic Energizer with a Single Cylinder Two Stroke Catalytic Spark Ignition Engine", Journal of Scientific and Industrial Research, vol. 66, pp. 457-463,2007.
[8] P. Govindasamy, S. Dhandapani, “Experimental investigation of cyclic variation of combustion parameters in catalytically activated and magnetically energized two-stroke si engine”, Journal of Energy & Environment, Vol. 6, pp. 45-59, 2007.
[9] P. Patel, G. Rathod and T. Patel, “Effect of magnetic field on performance and emission of single cylinder four stroke diesel engine”, IOSR Journal of Engineering, Vol.4 , pp. 28-34, 2014.
[10] S. Jain and S. Deshmukh, “Experimental investigation of magnetic fuel conditioner (M.F.C.) in I.C. engine”, IOSR Journal of Engineering, Vol. 2 (7), pp. 27-31, 2017.
[11] C. A. Okoronkwo, C. C. Dr. Nwachukwu, L.C. Dr. Ngozi and J.O. Igbokwe, “The effect of electromagnetic flux density on the ionization and the combustion of fuel (an economy design project)”, American Journal of Scientific and Industrial Research, Vol.1, pp. 527-531, 2010.
[12] P. Govindasamy, S. Dhandapani, “Experimental investigation of the effect of magnetic flux to reduce emissions and improve combustion performance in a two stroke, catalytic-coated spark-ignition engine”, International Journal of Automotive Technology, Vol.8(5), pp. 533-542, 2007.
[13] A. R. Attar, P. Tipole, V. Bhojwan and S. Deshmukh, “Effect of magnetic field strength on hydrocarbon fuel viscosity and engine performance”, International Journal of Mechanical Engineering and Computer Applications, vol.1 (7), pp. 94-98, 2013.
[14] Tretyakov, I.G., Rybak, M.A. and Stepanenko, E.Yu., “Method of magnetic treatment for liquid hydrocarbons”, In the proceedings of the International Conference on Surface Engineering Application, Washington, USA, vol.6, pp. 80-83,1985.
[15] Al Dossary, Rashid. M .A., “The effect of magnetic field on combustion and emissions”, Master’s thesis, King Fahd University of Petroleum and Minerals, 2009.

IOP Conf. Series: Materials Science and Engineering **764** (2020) 012007 doi:10.1088/1757-899X/764/1/012007

ARM based electronic load controller for micro-hydropower stations

Sritha.P[*] and Preethi.G
Department of Electrical and Electronics Engineering, Bannari Amman Institute of Technology, Coimbatore-638 401, India.

[*]Author for correspondence E-mail: srithap@bitsathy.ac.in

Abstract. This paper presents a method for maintaining constant voltage and frequency of synchronous generator using Advanced RISC Machine (ARM) controller based Electronic Load Controller. Constant frequency and voltage can be generated by a Synchronous Generator only when a constant electrical load is maintained. It is achieved by using Electronic Load Controller (ELC).When synchronous generator is connected to a turbine or any sudden changing loads its voltage and frequency also varies. This Electronic Load Controller consist of ARM LPC 1768, ACS712 current sensor, Insulated-Gate Bipolar Transistor (IGBT), an auxiliary load and consumer load and is implemented using Pulse Width Modulation (PWM) technique. This Electronic Load Controller provides effective and easy way of controlling voltage and frequency for linear loads which obeys the ohm's law. PWM technique is the fast response method compared to other methods having smooth speed control. So the load will be switched on/off properly and output voltage and frequency will be maintained.

1. Introduction

Due to low power ratings of micro hydro power stations, synchronous and asynchronous generators are mostly used to maintain the input hydro power constant due to fixed speed turbine. This will maintain the output of the synchronous generator to be kept constant. The major disadvantage of standalone micro hydro power stations is the variation of consumer load with the speed of the water turbine. Thus variation in speed of water turbine will give rise to changes in voltage and frequency of the generator output. The invention of ELC decreases the risk and complexity in maintaining the output of the synchronous generator used in micro hydropower stations. An ELC continuously measures and controls the generated power.

The basic concept of ELC is that it allows the generator to run at their full load power and electronic load is used just to observe the excess power generated by the synchronous generator. ELC measures the generator output speed and compares with the reference speed to control the power absorbed by the dump load to make the speed constant.An ELC is designed to maintain the output of the generators which are used in micro hydropower system. When the load connected increases above the rating of the generator, the excitation capacitors will supply the demand and when the consumer load decreases, then the excess power will be consumed by the dump load.Voltage and frequency can be kept constant by maintaining constant load. The controller adjusts for changes in the fundamental load via consequently varying the measure of amount of power delivered to the resistive load. This resistive load is also known as ballast load.

Neural-Network (NN) based ELC is implemented to control the isolated asynchronous generator which is connected to the uncontrolled turbine [1]. Sometimes the excess power to dump load is greater than the consumer load, hence it is reduced by using Distributed Electronic Load Controller [2]. The sensor circuits are used along with DSP processor based distribution static compensator used as an electronic load controller (ELC) to maintain constant load for self-excited induction generator (SEIG) [3]. ELC offers the elemental reactive power and conjointly compensates harmonics of currents that could be a connection of voltage source converter and dc-link capacitance with a chopper which gives the control signal to chopper for turn on the auxiliary load and therefore improves the power quality of the overall system [4]. Controller maintains the constant load by varying loads of dump load and also frequency is made constant by giving frequency as feedback variable [5]. An analysis and design of ELC for 3-phase synchronous generator suitable for micro-hydropower stations with constant-input power is delivered. It also provides mathematical modelling of synchronous generator [6]. ELC was designed using uncontrolled rectifier, chopper switch and ballast load [7]. Decoupled electronic load controller along with voltage regulator is designed to control the isolated asynchronous generator [8]. Current controlled voltage source inverter based ELC is proposed. Fuzzy logic is implemented instead of PI controller [9]. Fuzzy logic based electronic load controller also developed. Fuzzy logic gives the output control pulse in simple ways [10].

This paper proposed an Electronic Load Controller which is simpler to design; operation is easier, cost efficient and easy to control that uses Pulse Width Modulation technique to generate pulses to smooth turn on of IGBT. LPC1768 is used to generate Pulse Width Modulation (PWM) pulse.

2. Block diagram of proposed Electronic load controller

Electronic Load Controller (ELC) used in power plants to maintain constant frequency and voltage of generator by diverting excessive energy to dummy loads. Power diversion is done by Insulated Gate Bipolar Transistors (IGBT). By using ELC, power input of generator will be equal to power output of generator. The block diagram of the proposed Electronic Load Controller is shown in figure 1.

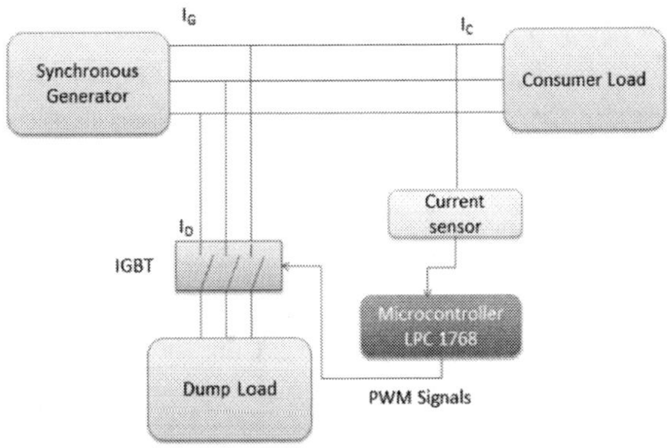

Figure 1. Block diagram of proposed Electronic load controller.

Synchronous generator is connected to the consumer load and dump load/ballast load. Current sensor is connected to any one of the phases of consumer load to measure the phase current and sensed value is given to the microcontroller LPC1768. PWM pulses are given to the IGBT via IGBT driver circuit. IGBT is then connected to the dump load. Optocoupler is connected between microcontroller and the IGBT driver.

Synchronous generator has rated voltage of 440V and rated current of 3A. The output voltage of Synchronous generator is kept constant. For this a constant current was fixed (2A) and the current to the consumer load is sensed continuously. The sensed current was read by the ARM LPC1768. If the sensed current is lower than the fixed current, then the microcontroller produces the PWM pulses according to the difference in the current. The PWM pulses is given to the optocoupler which is used to separate the low power and the high power circuit. The same pulses is also applied to the IGBT gate which connects the dump load to the generator output and the same current is maintained in the generator and the output voltage is maintained at constant. The microcontroller is programing using these equations (1), (2), (3) and (4).

Let,
P_G = Power generated
P_C = Power consumed by consumer load
P_D = Power consumed by Dump load
I_D = Dump load current
I_G =Generator output current
I_C = Consumer load current
We know that,

$$P_D = P_C + P_D \quad \text{Eqn. (1)}$$

$$I_G = I_C + I_D \quad \text{Eqn. (2)}$$

Generator current is kept constant at 2A (IG).

Current sensor output = consumer load current.

$$I_D = I_G - I_C \quad \text{Eqn. (3)}$$

$$I_D = 2 - I_C \quad \text{Eqn. (4)}$$

2.1. Dump load/ballast load

Dump load/Ballast load is fundamentally resistive load (obstruction) used to dump the extra power generated by the generator in the form of heat. Usually air heaters, water heaters will guarantee long life, maintenance free and economically useful. When the dump load is turned on, it consumes the full excess power generated irrespective of the rating of the dump load hence dump loads must be turned on/ off by PWM pulses to avoid excess power to the loads otherwise loads will damage.

2.2. Insulated Gate Bipolar Transistor (IGBT)

IGBT is a 3-terminal power semiconductor switch which is used to control the high current circuit. They are similar to working of the MOSFET. In spite of the fact that they display quick turn-on, their turn-off is slower than a MOSFET in terms of current fall time. Trench IGBT FGA25N120ANTD is used which is 1200-V non-punch through (NPT). To withstand the voltage of 400V, FGA25N120ANTD IGBT is used. The IGBTs are connected in series inverted mode as it conducts for the positive cycle only. So there are two IGBTs used for each phase. The IGBT FGA25N120ANTD is shown in the Figure.2. For each IGBT, heat sink is attached to cool the IGBT as it conducts high current.

2.2.1. IGBT configuration per phase

The IGBT configuration per phase is shown in the figure 2.

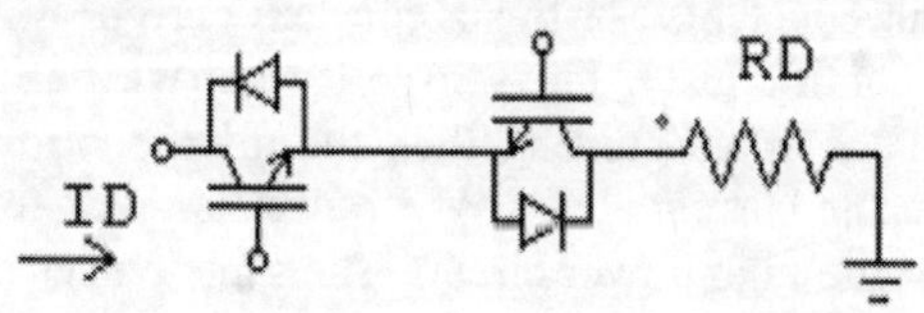

Figure 2. IGBT configuration.

IOP Conf. Series: Materials Science and Engineering **764** (2020) 012007 doi:10.1088/1757-899X/764/1/012007

IGBT conducts for only positive cycle and series inverted IGBT topology is implemented to control the turn on and turn off of the ballast load because one IGBT conducts in positive cycle whereas other conducts in negative cycle. The series inverted IGBT is connected to each phase. The PWM pulses from the LPC 1768(PIN 21, 22, 23) for each phase R, Y, B are connected to the IGBT's in every phase. The same PWM pulse is applied to two IGBT's connected in the same phase.

2.3. ACS current sensor

It is used to measure either AC or DC current up to 20A and it is an accurate sensor. This sensor can be used to measure high AC current. The pin VCC and ground of current sensor is connected to the LPC1768.VOUT is used to give power supply to sensor and the output voltage is proportional to the connected/consumer load. The current is taken at pin OUT. The sensor is connected in between the generator and the consumer load. The features of ACS current sensor are,

- Output sensitivity - 100 mV/A.
- Extremely stable output offset voltage.
- Factory trimmed for accuracy.
- Magnetic hysteresis is nearly zero.

2.3.1. Interfacing current sensor with microcontroller

The purpose of using microcontroller is to generate PWM pulses for switching of dump loads via IGBT. The PWM pulse must be generated as per the current to the consumer load. This process is done by microcontroller LPC 1768.

Out of 40 pins, pin 15 is used to sense current from the current sensor, pin 21,22,23 are used to take out generated PWM pulses to turn on the dump loads. Reasons for using this LPC 1768 are,

- Low power consumption.
- Easy to program.
- Online compiler is readily available to check the code.
- Easy to generate the PWM pulses.

The current to the consumer load is sensed by current sensor ACS 712 and given to microcontroller through pin 15. From that current to the dump load is calculated and the current through three phases are obtained. The sinusoidal current is converted into PWM pulse by comparison with constant DC supply. Then the generated PWM pulse from pin 21, 22, 23 is given to IGBT in each phase respectively through optocoupler and gate driver for turning dump load ON and OFF.

2.4. Optocoupler

Optocoupler is placed between the microcontroller and driver circuit to prevent microcontroller LPC1768 from high voltages. In this, MOC3021 optocoupler is used. It has 6 pins. Out of 6 pins, four pins are used. PIN1: Anode is connected to the LPC1768 and receives the PWM pulse from the microcontroller.PIN2: Cathode is connected to the ground.PIN4 and PIN6: Main terminal is connected to the IGBT driver to give PWM pulse.

2.5. IGBT driver

The magnitude of PWM pulses from the microcontroller is not enough to drive the IGBT, hence IGBT driver is used in between the microcontroller and the IGBT. IRS 2110 driver is used to drive the IGBT. The IRS2110 is given with PWM pulses from the optocoupler which is then amplified by the driver and then given to the IGBT gate where it is connected in series inverted mode.

3. Pulse Width Modulation

In Pulse Width Modulation, the analog signal is encoded in to series of signal pulses. The PWM pulse is produced by comparing the reference signal with the carrier signal. The frequency of carrier signal should be higher than the reference signal so that waveform perceived by the load must be smooth as possible. The figure 3 represents the sinusoidal PWM technique.
PWM Sampling Theorem states that, "a continuous time signal can be represented in its samples and can be reconstructed back when sampling frequency f_s is higher than or equal to the two times the highest frequency of signal".

$$f_s = 2f_h$$

The PWM technique is based on the two signals: carrier signal and modulation signal. The modulation signal should be generated from the magnitude of current flow to the consumer load. The consumer load current can be varied from 0A to 2A. The carrier signal is generated as triangular wave which is of very higher frequency (switching frequency of IGBT). The two signals are generated correctly by programming the ARM microcontroller. Whenever the magnitude of modulation signal is greater than the magnitude of carrier signal, the output switching pulse is presented. Then the switching pulse is applied to the IGBT to control the load connected to the synchronous generator.

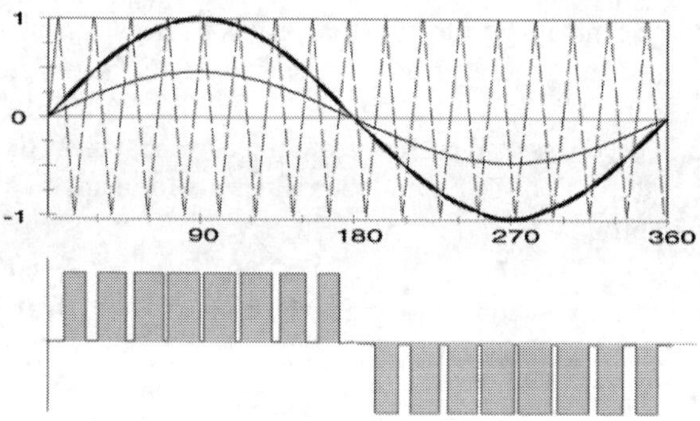

Figure 3. PWM pulses generation.

3.1. Advantages of proposed system

The advantages of the proposed method are,
- No hammer effect from load changes.
- High reliability and simple circuit.
- Low maintenance.
- ELC have low wear and tear on mechanical parts.
- It can be easily fitted in electrical system and easy to operate.
- Less expensive.

4. Simulation and results

4.1. Simulation circuit
The simulation for the proposed ELC is carried out in the MATLAB software. The Simulink model of the proposed system is shown in the figure 4. It consists of synchronous machine, consumer load, ballast load, PWM pulse generation circuit and excitation system for the synchronous machine. The ballast and consumer load is connected in parallel to the synchronous generator. Whenever there is sudden change in the load occurs, the voltage across the generator varies. The variation in the voltage causes drastic effect to the other loads connected to the machine. The current in any of the phase is

measured and is compared with the reference current. The error is measured and is given to the PID controller. The control signal is generated by PID controller and is compared with carrier signal. In MATLAB carrier signal is generated by the Repeating sequence block. The output of the comparator gives PWM pulses. The PWM pulses controls the IGBT switching and the constant load is maintained. Hence, the output voltage and frequency of the synchronous generator is kept constant.

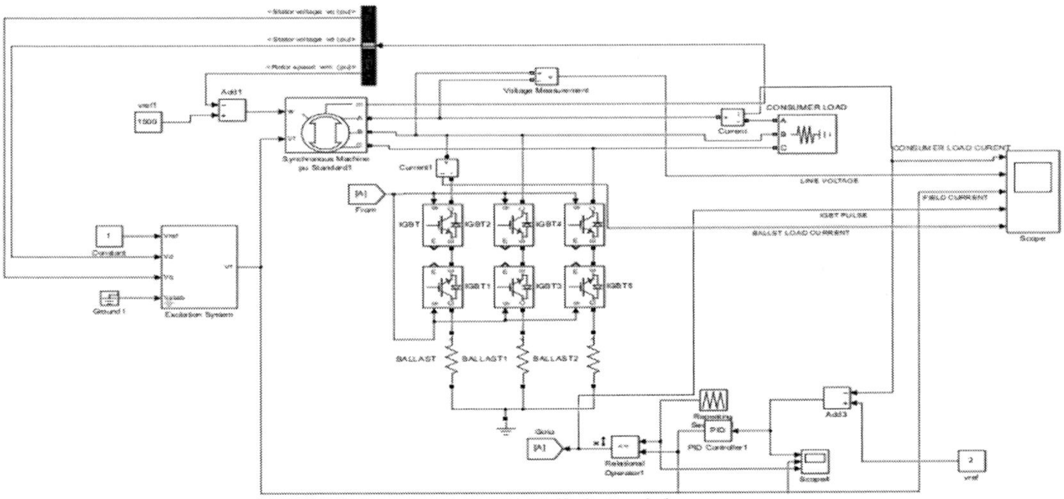

Figure 4. Simulink model.

4.1.1. Results of the simulation

The voltage level of the synchronous generator without ELC and with ELC is shown in the figure 5 and figure 6.Without ELC whenever load changes, the output voltage level changes and cannot be kept constant. With ELC is connected in parallel with consumer load to the synchronous generator, then the voltage is maintained at constant. There are three different regions A, B, C shown in the simulation output. Whenever voltage level is reduced by 10V, then ELC will enter makes the voltage constant.

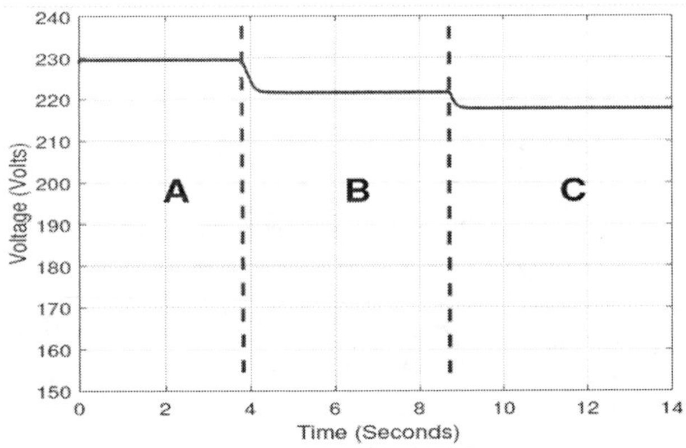

Figure 5. Voltage control without ELC.

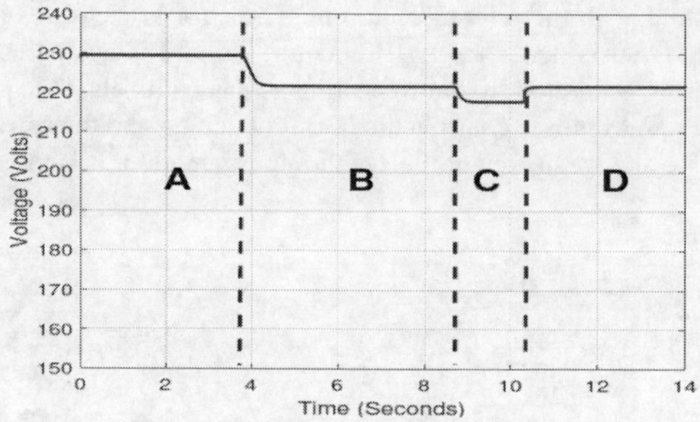

Figure 6. Voltage control with ELC.

The proposed system is implemented using the Arduino UNO board. The relay and solenoid valve are controlled automatically and cab be controlled manually using the Blynk Application on an android mobile phone. The android phone is connected to the board using the ESP8266 Wi-Fi module. The four parameters namely moisture, temperature, humidity and intensity is monitored and controlled using the ON/OFF switch button displayed on the Blynk Application. The sensor results viz., humidity, temperature, moisture, sunlight intensity and the status of the motor is linked with the Blynk app shown in figure 3 is the reading taken from the field model created in an area of about 30cm×30cm. When the moisture is above 250% the motor is turned OFF automatically or manually through the Blynk Application. Also the device is programmed such that if the light intensity is above 25% the motor will be still ON until the moisture reaches 300% to compensate the effect of sunlight.

5. Results and discussion

The below figures show the PWM pulses corresponding to the dump load. When consumer load current is IC =2A, then the current absorbed by the dump load is 0A. The PWM pulse generated from LPC1768 has no pulses and IGBT will not turn on hence the ballast load is not connected to the generator. The corresponding pulse waveform is shown in figure 7.

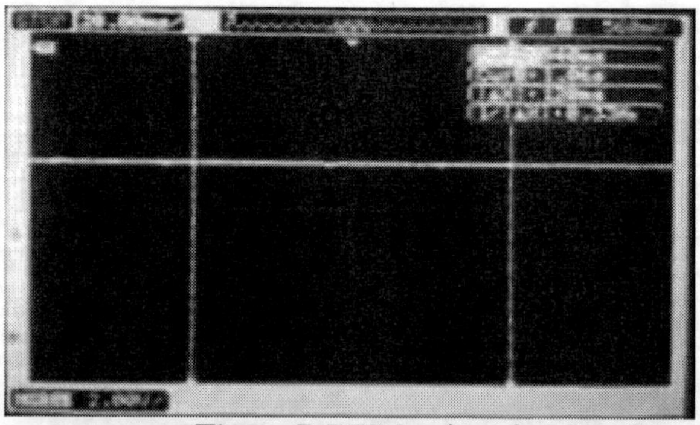

Figure 7. PWM pulse when $I_C = 2A$.

When consumer load current is IC =1A, then the current absorbed by the dump load is ID=1A. The PWM pulse generated from LPC1768 has pulses according to the current magnitude and so IGBT will turn on and the dump load is connected to the generator. The corresponding pulse waveform is shown in figure 8.

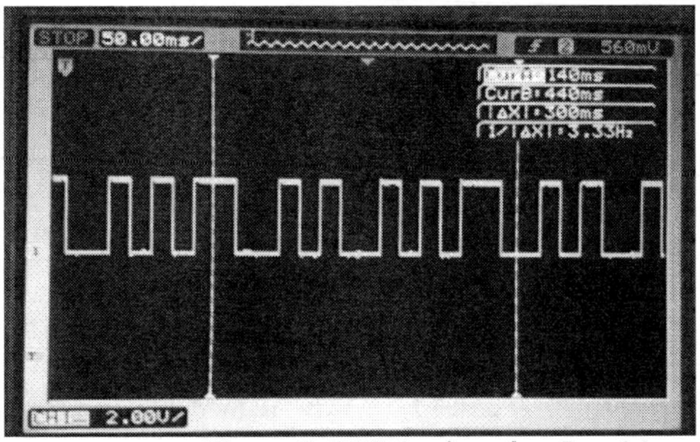

Figure 8. PWM pulse when $I_C = 1A$.

When consumer load current IC =0A, then the current absorbed by the dump load is ID=2A. The PWM pulse generated from LPC1768 has pulses according to the current magnitude and so IGBT will turn on and the ballast load is connected to the generator. The corresponding pulse waveform when IC = 0A is shown in figure 9.

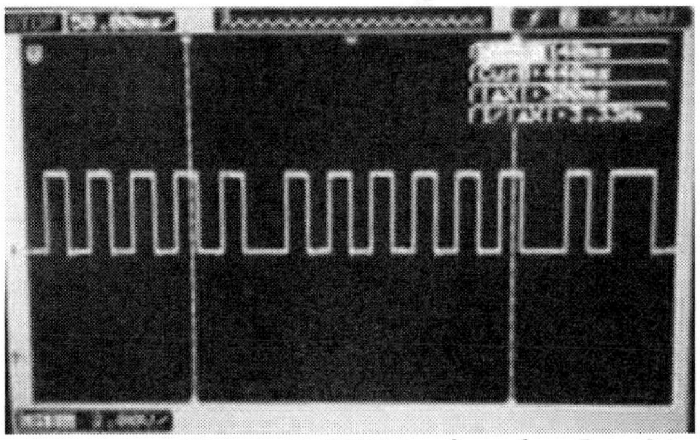

Figure 9. PWM pulse when $I_C = 0A$.

The hardware output and model of electronic load controller is shown in figure 10 and figure 11.

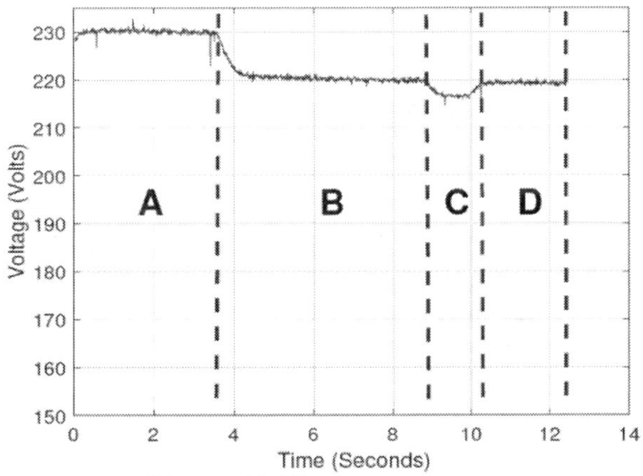

Figure 10. Hardware output.

IOP Conf. Series: Materials Science and Engineering **764** (2020) 012007 doi:10.1088/1757-899X/764/1/012007

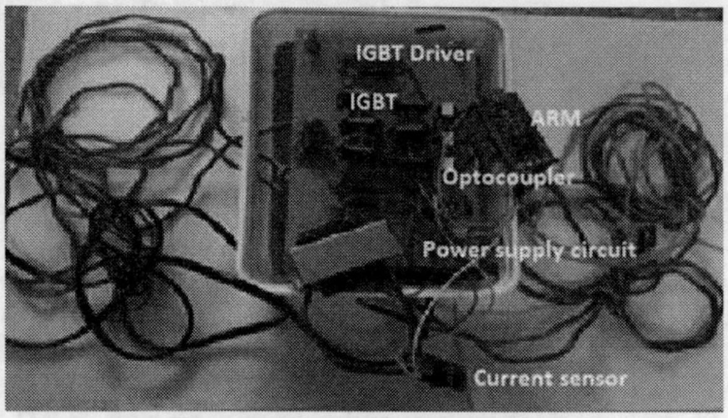

Figure 11. Hardware model.

The hardware setup of the arm based electronic load controller is shown in figure 12.

Figure 12. Hardware setup.

6. Conclusion

The arm based electronic load controller maintains voltage and frequency for linear loads by switching the loads at constant time intervals. The software simulation for various load values were stimulated and the model is also implemented in hardware. The socio-economic impacts of the proposed system are,

- This can help in getting electricity to rural parts and hill areas of the world which is not directly connected to the grid.
- Charging or water heating as dump loads for surplus power.

References

[1] Bhim Singh and V. Rajagopal, "Neural-Network-Based Integrated Electronic Load Controller for Isolated Asynchronous Generators in Small Hydro Generation", IEEE Transactions on Industrial Electronics, Vol.58(9), 2011.

[2] Babak Nia Roodsari and Edwin Peter Nowicki, "Analysis and experimental Investigation of the Improved Distributed Electronic Load Controller", IEEE Transactions on Energy Conversion, Vol. 33(3), 2018.

[3] Rajasekhara Reddy Chilipi, Bhim Singh, and S. S. Murthy, "Performance of a Self-Excited Induction Generator with DSTATCOM-DTC Drive-Based Voltage and Frequency Controller", IEEE Transactions On Energy Conversion, Vol. 29(3), 2014.

[4] Rajagopal B. Singh G.K. Kasal, "Electronic Load Controller with Power Quality Improvement of Isolated Induction Generator For Small Hydro Power Generation", ET Renewable Power Generation, Vol.5(2), pp.202–213, 2011.

[5] Ujjwal Kumar Kalla, Bhim Singh, and S.Sreenivasa Murthy, "Modified Electronic Load Controller for Constant Frequency Operation With Voltage Regulation of Small Hydro-Driven Single-Phase SEIG", IEEE Transactions On Industry Applications, Vol.52(4), 2016.

[6] Nan Win Aung, AungZeYa, "Design of Electronic Load Controller By Using Combination Method For Micro-Hydro Power Plant And Its Control And Monitoring Program Simulation", International Journal of Electrical, Electronics and Data Communication, Vol.3(6), 2015.

[7] GuilliamNel, and Wesley Doorsamy, "Development of an Intelligent Electronic Load Controller for Stand-Alone Micro-Hydropower Systems", 2018 IEEE PES/IAS Power Africa.

[8] Bhim Singh, V Rajagopal, Ambrish Chandra and Kamal-Al-Haddad, "Decoupled Electronic Load Controller for Asynchronous Generator in Pico-Hydro Power Generation," 5th International Conference on Industrial and Information Systems, ICIIS 2010, Jul 29 - Aug 01, 2010.

[9] Gaound Chetana, S K Shah, S J Patel, "Analysis and design of electronic load controller using FLC for self-excited induction generator", 1st International Conference on Emerging Technology Trends in Electronics, Communication and Networking 978-1-4673-1627-9/12, 2012.

[10] Eshani Mishra, Sachin Tiwari,"Fuzzy Logic Control based Electronic Load Controller for Self Excited Induction Generator", International Conference on Electrical Power and Energy Systems (ICEPES), Maulana Azad National Institute of Technology, Bhopal, India. Dec 14-16, 2016.

IOP Conf. Series: Materials Science and Engineering **764** (2020) 012008 doi:10.1088/1757-899X/764/1/012008

Thermo-Mechanical analysis of fused filament fabrication process

N Manikandan[1*], T Vignesh[1], C Prasath[1] and Mohamed Ismail[2]

[1]Department of Mechatronics Engineering, Sri Krishna College of Engineering and Technology, Coimbatore-641 008, India.
[2]Department of Mechatronics Engineering, Bannari Amman Institute of Technology, Sathyamangalam-638 401, India.
*Author for correspondence E-mail: manikandan@skcet.ac.in

Abstract. Manufacturing industry has been revolutionized by 3D printing or Additive Manufacturing (AM) processes introduced in the late 1980s. Subsequent to intensive research, the technology is even accessed by hobbyists. A layer-by-layer deposition of a material produces a prototype in the commercialized 3-dimensional (3D) printing techniques. Among the 3D printing processes, Fused Filament Fabrication (FFF) or Fused Deposition Modelling (FDM) process is the commonly used process for prototyping, manufacturing functional parts, etc. The process involves the extrusion of a molten polymer filament sequentially through a heated nozzle in the required pattern on to a platform. The FDM process primarily depends on the thermal gradient as the deposited molten material cools, solidifies and bonds with the previous layer. This temperature gradient greatly affects the produced parts microstructure and in turn the macrostructure of the final part. Numerous attempts to experimentally characterize the thermal and mechanical properties of structures fabricated with FDM have been reported in the literature. However, few attempts have been made with 2D models to predict the thermal and mechanical behaviour of the FDM printed components. In this study, a 3D thermo-mechanical model has been developed using ANSYS® workbench using the element birth and death effect. The temperature evolution, maximum principal stress and the deformation due to the thermal gradient were calculated using the developed model. The developed Finite Element Analysis (FEA) model was tested with properties of Poly-Lactic Acid (PLA) material. Through the analysis, it was shown that the model can be used as a predictive tool to understand the thermo-mechanical behaviour of FDM printed parts and also the modelling method used has a great influence on the thermal gradient of the printed parts.

1. Introduction

The Additive Manufacturing (AM) technology has rapidly gained popularity since its inception in the late 1980's. One of the key advantages of the AM process is the zero-waste generation during the fabrication process. The various categories of AM include Stereolithography (SLA), Laser Engineered Net Shaping (LENS), Fused Deposition Modelling (FDM) or Fused Filament Fabrication (FFF), Direct Metal Laser Sintering (DMLS), Laminated Object Manufacturing (LOM), Electron Beam Melting (EBM) and Digital Light Processing (DLP) [1]. Among the above-mentioned AM processes, the FDM process is the most widely used process due to its ease of use, material availability and affordability. In the FDM process, filaments made of a polymer material usually Acrylonitrile butadiene styrene (ABS) or Poly-Lactic Acid (PLA) is partially melt and deposited on a heated bed. Even though the process is even used by hobbyists, the deposition process is still not completely explored [2]. As the process is primarily driven by the thermal behaviour, the thermal gradient of the filaments during and after deposition is crucial for determining the dimensional accuracy and quality of the fabricated product. Other parameters such as scanning speed and orientation of material deposition also plays an important role in the final quality of the product [3-5].

As the temperature is crucial in the FDM process, several monitoring processes were developed. A non-intrusive method for FDM machine monitoring was developed by Wu. et al. which detailed the use of acoustic emissions to monitor the FDM machine conditions [6]. Another method for online monitoring of the layered manufacturing processes using an optical camera was also implemented to detect process defects [7]. Experimental measurements using embedded thermocouples were also demonstrated to determine the quality of layer bonding in the FDM processes [8]. However, limitations such as scan quality, S/N ratio and location of the sensors exist in these systems. Meanwhile the use of Infra-Red (IR) thermal imaging systems has gained popularity in comparing with the measurement techniques. But the cost of the IR system is too high which is sometimes even twice the cost of a desktop 3D printer [9].

Considering the drawbacks in the experimental methods, it is essential to develop predictive models to characterize and understand the thermal behaviour of the FDM process. Few numerical simulations using 2D models were established to predict the thermal gradient of the deposited filament in the FDM process. A 2D heat transfer analysis was performed by Rodriguez and Thomas by assuming rectangular cross sections and neglecting the effects of conduction between the filament and the heated bed [10]. Indirect results such as the dimension accuracy of the thermal behaviour was also demonstrated by Zhang et al [11]. Several theoretical models without the validation of the thermal behaviour with the experimental results were also found in literature [12–17]. Primarily, the simulations were carried out in ANSYS APDL and APDL coding was used to solve the model. The theoretical models developed also did not address the phase transition heat transfer and the heat loss/generation into the governing equation. FEA simulation using commercially available software packages such as Abaqus and MATLAB were also used to model the FDM process [18-19]. These simulations also lacked the verification of the model with experimental results. For addressing these drawbacks, this research focusses on developing a 3D finite element model considering the effects of conduction, convection and radiation on the thermal behaviour of the FDM deposition process. A commercial software ANSYS® Workbench is used to simulate the material deposition process. The developed thermal and mechanical model is simulated with PLA as print material. Finally, the FEA results is validated with the experimental results of the thermal gradient measured using an IR thermal camera in the previous research works.

2. Thermo-Mechanical modelling

2.1. Finite Element Model

A three-dimensional coupled thermo-mechanical model was created to study the FDM process. The commercially available ANSYS R19.2 was used for the simulation. Initially, the thermal model was solved, and the results of the thermal model were given as input to the structural analysis to determine the effect due to thermal stress. The geometry of the model was generated with a dimension of 15 (length) X 0.4 (width) X 10 (height) mm. SOLID90 element type with 20 nodes was used to generate the mesh of 0.4mm element size. The size of the model was selected to study the results using a single line generation of the part without using infill. The filament part was printed on a bed of size 15 (length) X 10 (width) X 4 (thickness) mm. The bed size was chosen very small to understand the heat transfer behavior between the deposited filament and the bed. The model has 25 layers of material deposition that equals to 10 mm of the part height. Figure 1 shows the meshed solid model of the fabricated part and the heated bed.

IOP Conf. Series: Materials Science and Engineering **764** (2020) 012008 doi:10.1088/1757-899X/764/1/012008

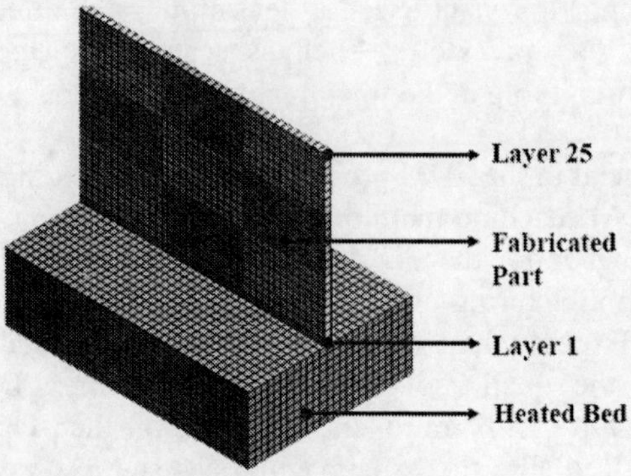

Figure 1. Finite Element Mesh of the FDM Model.

The element birth and death concept in ANSYS R19.2 Workbench was used to activate and deactivate a particular element at a given time and load condition. The birth and death concept has been successfully implemented in simulating temperature gradients in welding and melting of material [20–22]. The element birth and death method is applied to the FDM process to better replicate the material deposition process. Figure 2 (a) and (b) show the element birth and death effects utilized in the developed model.

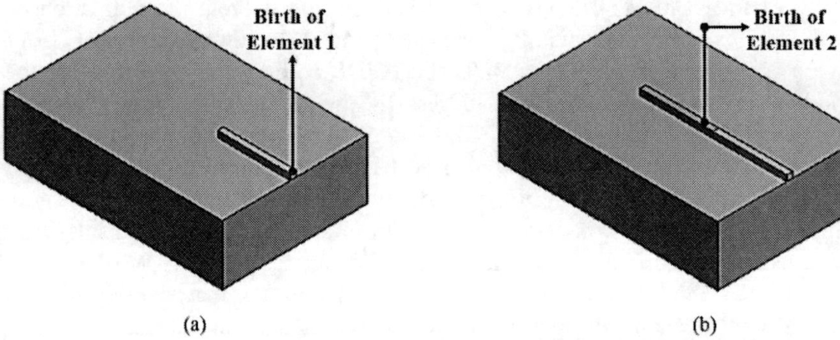

Figure 2. Birth and Death of elements during the FDM simulation. **2 (a)** Birth of element 1 during load case 1; **2 (b)** Birth of element 2 during load case 2.

2.2. Thermal boundary conditions and assumptions

The study basically involves two steps – Thermal analysis and a structural analysis. The following boundary conditions were applied to solve the model.

- The initial temperature for each element was maintained at 190°C for the PLA filament printing.
- The first three elements will be in contact with the heated bed. The thermal contact resistance was calculated using the relation explained in [23]. The value was found to be 77.33 $Wm^{-2}\,°C^{-1}$.
- The value of convective heat transfer between the fabricated part surface to the ambient air was set to 100 $Wm^{-2}\,°C^{-1}$. The value was calculated using the function for forced convection over a rectangular body [23]. The forced convective value was calculated considering the cooling fan speed over the fabricated part at approximately 2.5-3 m/s [24].
- The effect of radiation was neglected in the simulation as the effect of convective heat transfer is considerable in the FDM process [13]. But it was proven that the effect of radiation should be considered for the FDM process by Natalie Rudolph et al. [15]. So, an emissivity value of 0.95 was considered as suggested in their research work.
- The structural analysis was performed with all the degrees of freedom of the bottom surface of the first layer arrested.

The following assumptions were also made based on the till date research findings in the simulation of material deposition in the FDM process.

- The initial temperature is maintained constant throughout the simulation process.
- The speed of material deposition is equal to the travel speed of the nozzle.
- The extruded material can be modeled as a finite volume with imposed heat energy.

2.3. Material properties of PLA

The simulation was carried out with commercially available WOL3D PLA filament. The material was chosen as it is commercially available and is one of the most commonly used polymer filaments for prototype part fabrication. The material properties for PLA was adapted from [25] and [18]. The values of thermal conductivity and specific heat are temperature dependent and the density values are temperature independent. Table 1 shows the physical, thermal properties and the operating conditions of PLA used for the simulation.

Table 1. Properties of PLA used for simulation.

Characteristics	Unit	Value
Density	kg m^{-3}	1252
Thermal Conductivity	W m^{-1} °C^{-1}	
190 °C		0.195
109 °C		0.197
48 °C	°C	0.111
Specific Heat (C_p)	J kg^{-1} °C^{-1}	
190 °C		2060
100 °C		1955
55 °C		1590
Print Temperature	°C	190
Build Plate Temperature	°C	60
Ambient Temperature	°C	22
Glass Transition Temperature (T_g)	°C	65

3. Simulation results and validation

3.1. Thermal simulation

The overall thermal simulation was carried out for a total time period of 22.5 seconds. The time period was calculated based on the time required to print the part at a nozzle speed of 50 mm/s. A time step of 0.3 seconds for each load step was simulated. Figure 3 (a) and (b) shows the front view temperature distribution of the fabricated PLA part and the heated bed at the first layer and the final layer (10 mm) respectively. The temperature distribution is observed to replicate the actual material deposition process. It is also observed that the heated bed has a minimum effect on the temperature distribution of the fabricated part. This condition is observed since the heated bed is of high density and mass compared to the printed part. The heated bed almost maintains the room temperature during the simulation process. This effect is in agreement with the similar phenomenon observed by Zhou et al. [16]. The final time step shows three distinct layers: the heated bed at 60°C, the bottom layers cooled down to room temperature of 22°C and the last few layers above the glass transition temperature. The temperature distribution is such that the temperature reduces from the bottom layer towards the middle layers of the part and the last layer will have the deposition temperature.

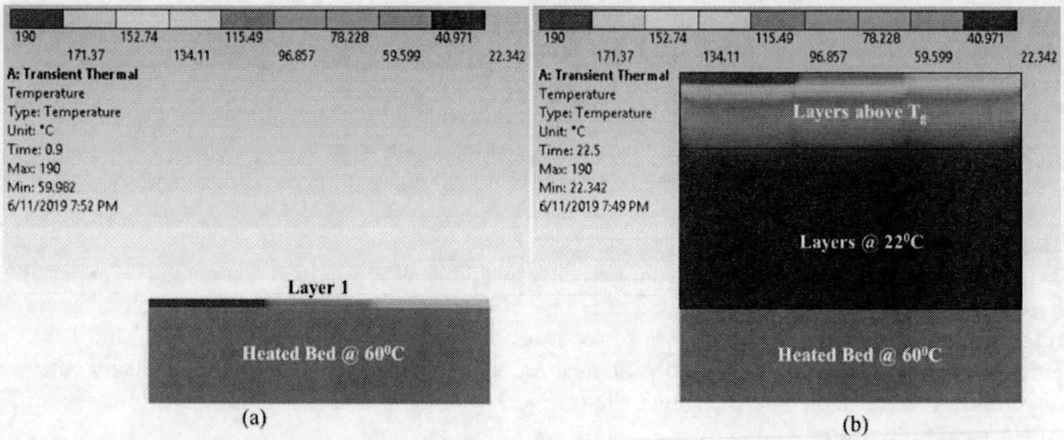

Figure 3. Temperature distribution during the FDM simulation of PLA (Front View). **3 (a)** at 0.9 seconds after the deposition of first layer; **3 (b)** at 22.5 seconds after the deposition of last layer.

The key parameter observed in the simulation is the Diffusion Time (t_{diff}). Diffusion time is the time taken for the deposited filament to reach the glass transition temperature from the deposited temperature. This factor determines the bonding quality between the layers and in turn the part strength. Figure 4 shows the temperature distribution comparison curves for the first ten layers for the PLA part simulation. From the chart, the average diffusion time was found to be 4.89 seconds. The calculated value is compared with the analytical t_{diff} value of 5.27 calculated by Li et al. for the parts fabricated with different process parameters [26]. The values agree with variation of 7.2 %. The comparison shows that the developed thermal model can be used for the simulation of 3D FDM part fabrication process.

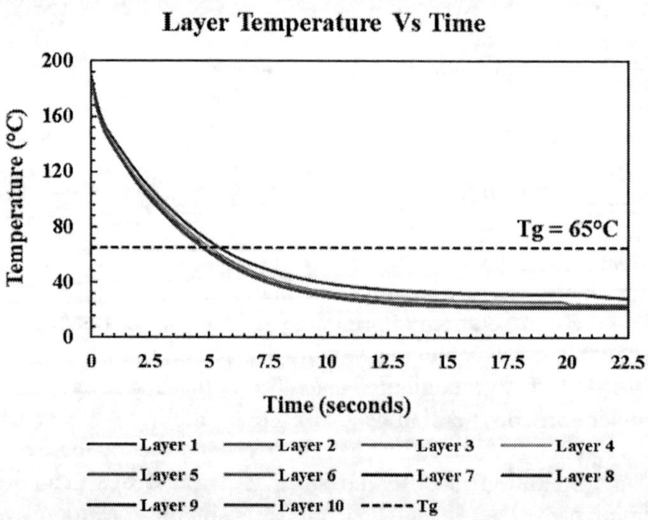

Figure 4. Temperature distribution comparison curves for the first ten layers of PLA.

3.2. Structural simulation

The structural simulation was carried out with the base of first layer fixed to the heated bed. The thermal load from the thermal simulation was used as input to the structural simulation. The maximum principal stress and the deformation in X and Y directions were measured from the simulation and compared with the samples printed. Three samples of PLA slabs of same dimensions as the simulation model were printed using INSTA3D FDM printer. The nozzle temperature was maintained at 190°C. The bed temperature was maintained at the recommended temperature of 60°. All the three samples

IOP Conf. Series: Materials Science and Engineering **764** (2020) 012008 doi:10.1088/1757-899X/764/1/012008

are printed with same process parameters and filament. Figure 5 shows the printed samples in comparison with a millimeter measurement scale.

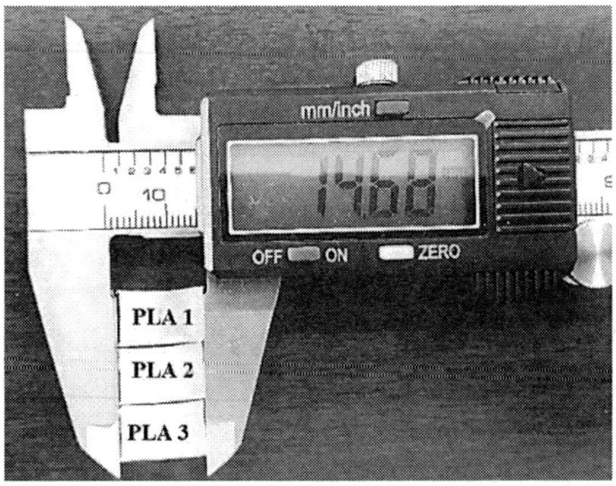

Figure 5. Samples of PLA slabs printed using INSTA3D FDM Printer.

Figure 6 (a) and (b) show the deformation comparison in X-direction and Y-direction for the PLA samples. The deformation in X-direction is maximum at the top edge showing similarity with the simulation results. Sagging effect is observed in the Y-direction as the part is built layer by layer. The simulation shows similar sagging effect in the Y-direction. The sagging may be attributed to the cooling effect of the bottom layers causing the middle layers to fall in the Y-direction during layer deposition.

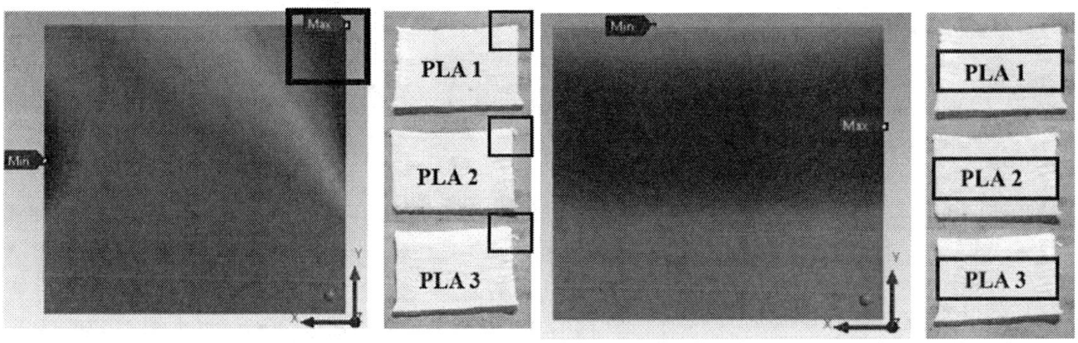

Figure 6. Deformation comparison between simulation and printed parts, **6 (a)** in X- Direction; **6 (b)** in Y-Direction.

4. Future outlook

The thermo-mechanical model was developed considering the effects of temperature, convection and radiation during the FDM process. The developed framework was proved to be applicable for a simple slab geometry. The model can be further extended for full complex part simulations which has support structures with the understanding of temperature distribution. The current work did not involve the study of adhesion effects between the printed layers, factors such as change in nozzle temperature, infill pattern and the number of walls in the fabricated part. The parametrical study still needs addressing with intensive theoretical and experimental investigations. Further the model can be improved with the inclusion of voids in the deposited filament material.

5. Conclusions

In this research, the thermal behavior of the FDM process was modelled and simulated numerically using ANSYS R19.2; diffusion time, maximum thermal stress and distortion parameters were evaluated. The model demonstrated the applicability of the element birth and death effects to simulate the FDM material deposition process. The developed FDM model can predict the diffusion time with a bias of 7.2% which is comparable with the results available in literature. The simulation results revealed the following characteristics regarding the simulation of FDM process:

- The element birth and death technique can be used for the simulation of FDM process.
- The effects of radiation should be considered during the thermal simulation.
- The temperature comparison of the deposited layers show that the diffusion time plays an important role in the thermal behavior of the material deposition.
- Sagging effect is observed when the part is printed as single layer thin walls.

References

[1] S. A. M. Tofail, E. P. Koumoulos, A. Bandyopadhyay, S. Bose, L. O'Donoghue, and C. Charitidis, "Additive manufacturing: scientific and technological challenges, market uptake and opportunities", Materials Today, vol. 21(1), pp. 22–37, 2018.

[2] B. N. Turner, R. Strong, and S. A. Gold, "A review of melt extrusion additive manufacturing processes: I. Process design and modeling", Rapid Prototyping Journal, vol. 20(3), pp. 192–204, 2014.

[3] K. Thrimurthulu, P. M. Pandey, and N. V. Reddy, "Optimum part deposition orientation in fused deposition modeling", International Journal of Machine Tools Manufacturing, vol. 44 (6), pp. 585–594, 2004.

[4] P. Alexander, S. Allen, and D. Dutta, "Part orientation and build cost determination in layered manufacturing", CAD Computer Aided Design, vol. 30(5), pp. 343–356, 1998.

[5] Y. Li, H. Yang, X. Lin, W. Huang, J. Li, and Y. Zhou, "The influences of processing parameters on forming characterizations during laser rapid forming", Materials Science and Engineering, vol. 360 (1–2), pp. 18–25, 2003.

[6] H. Wu, Y. Wang, and Z. Yu, "In situ monitoring of FDM machine condition via acoustic emission", International Journal of Advanced Manufacturing Technology, vol. 84 (5–8), pp. 1483–1495, 2016.

[7] T. Fang, M. A. Jafari, I. Bakhadyrov, A. Safari, S. Danforth, and N. Langrana, "Online defect detection in layered manufacturing using process signature", Institute of Electrical and Electronics Engineers, pp. 4373–4378, 2002.

[8] Q. Sun, G. M. Rizvi, C. T. Bellehumeur, and P. Gu, "Effect of processing conditions on the bonding quality of FDM polymer filaments", Rapid Prototyping Journal, vol. 14(2), pp. 72–80, 2008.

[9] R. B. Dinwiddie *et al.*, "Infrared imaging of the polymer 3D-printing process", Proceedings of SPIE, vol. 9105, pp. 1–12, 2014.

[10] Jose F. Rodriguez, James P. Thomas, and John E. Renaud, "Characterization of the mesostructure of fused deposition acrylonitrile butadiene styrene materials", Rapid Prototyping Journal, vol. 6 (3), pp. 175–186, 2000.

[11] Y. Zhang and K. Chou, "A parametric study of part distortions in fused deposition modelling using three-dimensional finite element analysis", Proceedings of the Institution of Mechanical Engineers, Part B: Journal of Engineering Manufacture, vol. 222(8), pp. 959–967, 2008.

[12] S. Dabiri, S. Schmid, and G. Tryggvason, "Fully Resolved Numerical Simulations of Fused Deposition Modeling", Proceedings of the ASME 2014 International Manufacturing Science and Engineering Conference, 2014.

[13] S. F. Costa, F. M. Duarte, and J. A. Covas, "Thermal conditions affecting heat transfer in FDM/FFE: a contribution towards the numerical modelling of the process", Virtual and

Physical Prototyping, vol. 10(1), pp. 35–46, 2015.

[14] C. H. Blake Courter, Vishal Savane, Jing Bi, Siddarth Dev, "Finite Element Simulation of the Fused Deposition Modelling Process", NAFEMS World Congress, 2017.

[15] N. Rudolph, J. Chen, and T. Dick, "Understanding the temperature field in fused filament fabrication for enhanced mechanical part performance", AIP Conference Prooceedings, pp. 1–5, 2019.

[16] Y. Zhou, T. Nyberg, G. Xiong, and D. Liu, "Temperature Analysis in the Fused Deposition Modeling Process", Proceedings -rd International Conference on Information Science and Control Engineering, pp. 678–682, 2016.

[17] K. Chockalingam, M. Karthic, and N. Jawahar, "Development of Thermal and Structural Deformation Model to Predict the Part Build Dimensional Error in Fused Deposition Modeling", International Journal of Advanced Design and Manufacturing Technology, vol. 11(2), pp. 27–37, 2018.

[18] A. Cattenone, S. Morganti, G. Alaimo, and F. Auricchio, "Finite Element Analysis of Additive Manufacturing Based on Fused Deposition Modeling: Distortions Prediction and Comparison With Experimental Data", Journal of Manufacturing Science and Engineering , vol. 141 (1), 2018.

[19] S. Costa, F. Duarte, and J. A., "Using MATLAB to Compute Heat Transfer in Free Form Extrusion," in MATLAB - A Ubiquitous Tool for the Practical Engineer, InTech Open, 2011.

[20] N. Kathari, K. R. Reddy, and J. B. Kalyan, "A Critical Comparative Study of Welding Simulations Using FEA", International Journal of Innovative Research in Science, Engineering and Technology, pp. 11170–11178, 2016.

[21] A. Damale and K. Nandurrkar, "Thermo-Mechanical Finite Element Analysis and Experimental Validation of Weld Induced AngularDistortion in MMAW Butt Welded Plates", International Journal of Engineering Research & Technology, vol. 1(6), pp. 1–7, 2012.

[22] P. Sagar, D. Kumar Gope, and S. Chattopadhyaya, "Thermal analysis of TIG welded Ti-6Al-4V plates using ANSYS", IOP Conference Series: Materials Science and Engineering, vol. 377(1), pp. 4–10, 2018.

[23] J. Gibbins, "Thermal Contact Resistance of Polymer Interfaces", University of Waterloo, 2006.

[24] X. Zhou, S. J. Hsieh, and Y. Sun, "Experimental and numerical investigation of the thermal behaviour of polylactic acid during the fused deposition process", Virtual and Physical Prootyping, vol. 12, no. 3, pp. 221–233, 2017.

[25] R. L. Shady Farah , Daniel G. Anderson, "Physical and mechanical properties of PLA, and their functions in widespread applications-A comprehensive review", Advanced Drug Delivery Reviews, p. 26, 2016.

[26] L. Li, C. Bellehumeur, and P. Gu, "Investigation of Bond Formation in FDM Process", 13th Solid Freeform Fabrication Symposium, pp. 400–407, 2002.

IOP Conf. Series: Materials Science and Engineering **764** (2020) 012009 doi:10.1088/1757-899X/764/1/012009

A review on friction stir welding of aluminium alloys and the effects on tool geometry

Santhoshkumar S*, Senthil Kumar K L, Kalil Rahiman M and Mathankumar P
Department of Mechanical Engineering, Bannari Amman Institute of Technology, Sathyamangalam - 638401 India.

*Corresponding author email: santhoshbe2@gmail.com

Abstract. Friction stir welding (FSW) is a novel technique used to join similar and dissimilar alloys. Aluminium alloys of 2XXX, 6XXX and 7XXX are reviewed and presented in this paper. In FSW welding process, heat is generated during the process and workpieces are joined without melting the aluminium alloys. The tool moves along the soften surface for joining aluminium alloys. Tool materials, tool geometry, welding speed, rotational speed, axial forces are examined and the effects on welding surfaces after welding process has been analysed. Friction stir welding can be widely used in aviation, marine and vehicle body construction industries because of its mechanical properties where fusion welding process revealed deformities on the welding materials to be joined. In this review article, the recent research of friction stir joining of aluminium alloys and its applications in various industries are explicated, including the quality of weld of aluminium alloys, the evaluation on microstructure in weld nuggets, and tool wear. The conclusion of this review is unequivocally recommended that aluminium alloys are suggested for future research with numerous industrial applications..

1. Introduction

Friction stir welding is a technique used to join Aluminium alloys, invented by The Welding Institute (TWI), UK in 1991. In FSW technique, a rotating non consumable welding tool is used for heat generation between workpiece due to friction and causes deformation at welding nugget. FSW has vast applications in aircraft, ship building, automobile body construction, and railway industries. Thus fundamental studies on rotational speed, weld nugget, microstructure, mechanical properties and process parameters have recently been studied.

The formation of the oxide layers in conventional welding is significant due to coefficient of thermal expansion, high thermal conductivity, and shrinkages during solidification process. After welding aluminium alloys using fusion welding technique, oxide formation at weld nugget cannot be avoided and mechanical properties are also not improved on the surface of workpiece. To overcome this problem, solid state welding technique such as FSW is used to join Al alloys. Further, FSW has a certain advantages over conventional welding methods because it reduces distortions, residual stresses in weld nugget.

2. Principle of FSW Technique

Friction Stir Welding is a solid state welding process in which a cylindrical shape shouldered tool rotates on the surface of the workpiece to be joined. The workpiece is which firmly clamped in the fixture of FSW machine and during the rotation of the tool, the axial force is applied on the welding surface. A non-consumable tool is used to join workpieces without melting it during the welding process and the workpiece is plasticized due to the generation of heat and friction.

3. Microstructure and Mechanical Properties of FSW Joints

In Automobile, Aerospace and Railway industries, reduction of weight is important to fabricate heavy machine parts because of limitation in the strength to weight ratio. Due to this property, materials such as copper alloys steels, wrought iron, titanium and magnesium alloys are replaced by aluminium and its alloys. Thomas et al. [1] investigated the microstructure, hardness and tensile strength of aluminium alloys Al Mg4.5Mn0.7 (AA5083) and AlZn6Mg2Cu (AA7075). During hot working process microstructures of aluminium alloys are similar and the hardness is varied in alloy AA5083. The mechanical properties of welded AA5083 and AA7075 6.0 mm alloys are as high as 100% and 72% respectively compared with base material.

After welding process, grain structure of aluminium alloys are improved drastically due to recrystallization of microstructure. Liu et al [2] investigated the microstructure of AA6061-T6 alloy by varying the welding speeds from 300 to 1000 rpm and traverse speeds of 0.15 to 0.25 cm/s. TEM (Transmission Electron Microscope) and light microscopy are used for microstructure analysis and they have divided the result in two phases. The first phase and the second phase results the variation in residual hardness between 55 WHN to 65 WHN. In weld nuggets, the grain size is 10μm and the hardness of work piece is 100μm approximately. Residual hardness is improved in the first and second phase after the solidification process but the effects of microstructure at low temperature is not investigated. Z.H. Fu et al. [3] studied the dissolution in weld nugget and the effect on microstructure during the joining of aluminium alloys at low temperature which eliminates the major problem of conventional welding processes and it is performed in the presence of inert gas. During the welding process, oxidation is drastically reduced due to the presence of inert gas.

By varying rotational speed of welding in Friction Stir Welding, harness of weld nugget has been improved by Micro hardness test. P. Cavaliere et al. [4] investigated the mechanical properties and micro structure of AA 6056 by varying rotational speeds of 500, 800 and 1000 rpm and welding speeds of 40, 56 and 88 mm/min. The joints after welding process are characterized by using micro hardness test. Residual micro hardness is also important parameters during the welding process and grain structure is also improved after solidification process. Li et al. [5] studied the residual microstructures of AA2024 and AA6061. The super plastic flow is found by residual, equiaxed grain ranging from 1 to 15 mm and reveals 40% reduction in residual micro hardness and 50% residual micro hardness reduction in the AA2024.

However, post weld treatment has not been absorbed by many researchers after the welding process. Rhodes et al. [6] investigated the mechanical properties and the behaviour of micro structure after post treatment of AA7075 alloy with the speed 5m/ min and they observed the low dense dislocation and also nugget is recrystallized and the strength of weld is enhanced after welding process. After solidification process, hardness and microstructure has been improved but dispersion of other elements over the weld nugget is not observed after post weld treatment.

Dispersion of alloying elements such as Si and Cu, Mg are used to improve the hardness during the welding process. Ghosh et al. [7] studied the phase transformation in weld zone and grain refinement of A356 and A6061-T6 aluminium alloys and studied the dispersion of Si particles on weld nugget.

IOP Conf. Series: Materials Science and Engineering **764** (2020) 012009 doi:10.1088/1757-899X/764/1/012009

Phase transformation from liquid state to solid state is played a vital role during the welding process but mechanical properties by post weld aging weld for base metal during the initial stage are not observed during the process. Yan et al. [8] attempted the effect of base metals T7451, T62 and also studied the post weld heat treatment of AA7050 aluminium alloy. The mechanical properties of base 7075 aluminium increases during initial temper the join strength is also increased by post weld aging method. However, ultimate tensile strength improves the similar and dissimilar metals.

During welding process, ultimate strength has been improved at low welding speed and high welding speed. Sakthivel et al. [9] attempted the metallurgical and mechanical properties of the similar and dissimilar aluminium alloys by varying speed from 50 mm/min to 175 mm/min and they have found that parent material has ultimate tensile strength of 84 MPA at low welding speed of 50 mm/min and 80 at higher welding speed 175mm/min. By varying welding speed, strength of parent metal has been improved after welding process.

Super plastic impairment of Aluminium alloy improves the grain structure after welding process. S. Benavides et al. [10] investigated a study on low temperature FSW of aluminium 2024 and studied super plastic impairment using active reformation of superfine equiaxed grains. At low temperature, the hardness and the equiaxed grains are developed after the solidification process. The effect of heat affected zone is recorded during the welding process. Priya et al. [11] studied the microstructure and metallurgical properties of AA6061 and it is observed that hardness is increased in weld nugget and the hardness is not improved in HAZ (Heat Affected Zone).

Formation of onion ring and the microstructure are observed on the surface of weld nugget by many researches and similarly hardness has been improved after the welding process. Da Silva et al. [12] investigated the features of microstructure and mechanical properties of dissimilar AA2024-T3 and AA7075-T6 joints and performed hardness and tensile test. The SEM result reveals the formation of onion ring is negligible.

Aluminium alloys and its compositions are used by many researchers but they use only one or two type of similar and dissimilar aluminium alloys to analyse the microstructure and metallurgical properties. Rajakumar et al. [13] investigated six types of aluminium alloys including 7xxx series alloy such as AA7075 and AA7039. The micro structural analysis is performed in weld nugget. Ductility of aluminium alloys and tensile strength are analysed after the welding process. Leitao et al. [14] found the mechanical and metallurgical behaviour of AA5182-H111 and AA60616-T4 and the size of grains of AA5182-H111 are based on tensile strength and ductility of aluminium alloys.

Tensile strength and mechanical properties are varied at low welding speeds and higher welding speeds. O F Flores [15] attempted to find the tensile characteristics in weld nugget. It is observed that at the lowest spindle speed tensile strength is decreased. Palanivel et al. [16] investigated AA5083-H111 and AA6351-T6 using tensile strength aluminum alloys. The tensile strength and welding surface without any defects are developed during the FSW process. Sato Y.S. et al. [17] studied on aluminium alloy 1100 using stored roll bonding and the grain structure is relatively improved after FSW process. Shujun Chen et al. [18] absorbed the effect of tensile strength in weld zone and HAZ. The tensile strength of the welded joint of 100–400 A has improved by 2.74%–7.38%, and for 500A and 600A has increased by 17.11%. In HAZ and weld zone, tensile strength has not been improved after the welding process.

Researcher are not only analysed the mechanical and metallurgical properties and they are also observed concentric formations, cavities and smooth crown during the welding process. Lombard et al. [19] investigated the properties of friction stir welded AA5083-H321 aluminum alloys parameter and observed concentric ring in the weld zone and the width of the nugget was of the order of the pin diameter. Scialpi, et al. [20] investigated on the mechanical and microstructures properties of Friction Stir welded joints based on shoulder geometries. It is recorded that smooth crown with fillet and cavity is produced after the FSW process.

Thermo-mechanically affected zone (TMAZ) is affected by varying welding speeds and mechanical properties are also affected after post weld treatment process. Singh et al. [21] studied the mechanical properties and microstructure of 7039 Al alloys after the joining process. The rotary speed of 635 rpm and the welding speed of 8 and 12 mm/min are used to weld aluminium alloys. It is recorded that the stir zone has maximum coarse grain than the thermo-mechanically affected zone (TMAZ) and the strength based on yield and tensile strength are relatively improved in the friction stir welding joints. Rajamanickam et al. [22] attempted to study the mechanical properties of aluminum alloy AA2014 based on tool rotation. From the analysis of tensile property data of joints, it is observed that tensile properties have been influenced based on parameters such as rotational and welding speed. From the above literature surveys, researchers have been attempted to analyse mechanical, metallurgical properties, concentric rings, cavities, HAZ and TMAZ by Friction Stir Welding process. From the above literatures surveys, researches are used only similar metals for their research.

4. Analysis on Tool Geometry

Microstructure, generation of heat, flow of material affects the tool geometry and the life of tool is an important feature which depends on parameter during the welding process. The parameters such as axial force, rotation, tool tilt angle, shoulder and pin diameter, welding speed have been studied and mechanical characteristics of Friction Stir Welded joints are also studied including tool life. Important parameters such as axial tool pressure (F), rotational speed (N) and traverse speed (S), on weld properties have been studied by researchers. Acerra.F et al. [23] investigated the aluminium alloy of AA7075-AA2024 and observed that heat generation is higher when the diameter of shoulder in tool is increased after the welding process and the major defects are analysed in welding zone and the elements of coating blank.

Tool geometry, shoulder diameter and pin are also an important parameters which affect the microstructure and mechanical properties. Fonda R.W. et al. [24] studied the morphology of aluminium alloys based on grain structure in FSW Process and observed that FSW tool has free from damages after the post welding process and grain structure is relatively improved near the welding tool. Finer type of grains is developed and texture follows FCC structure during the welding process. Texture of FSW welded aluminium alloys have been analysed and FCC structure is formed after the post welding process. Zaho et al. [25] the pin and shoulder plays a vital role in material flow which controls the characteristics of welding during the joining process. FSW has a weld defects and contours belong to the tool design and geometry. Material flow during the welding process is done based on pin and shoulder diameter.

Researchers are also used simulations such as CFD (Computational fluid dynamics), Analysis of variance (ANOVA) based on FEM (finite element model) to analyse the FSW aluminium alloys. Buffa.G et al. [26] made a study on the simulation based on model has been developed in which studies the geometry of tool in weld nugget. The three dimensional finite element model (FEM) based on cylindrical geometry has been investigated using FSW tool and behaviour of tool over the weld, microstructure of tool have been analysed. Mechanical properties and tensile strength has been analysed by Analysis of variance (ANOVA). Mohanty et al. [27] investigated the parameters such as speed of tool and other geometries of AA1100 aluminium alloys and Analysis of variance (ANOVA) are utilized to find the mechanical properties and tensile strength is relatively increased by increasing rotational and welding speeds.

IOP Conf. Series: Materials Science and Engineering **764** (2020) 012009 doi:10.1088/1757-899X/764/1/012009

Trivex tool and Triflute tools are compared by varying the transverse and axial forces and the mechanical properties and metallurgical properties of aluminium alloys have been compared after the welding process. Colegrove [28] investigated the material flow of FSW process using the CFD. Trivex as well as MX are used to weld alloys and the result is compared with Triflute tools. The transverse and axial force are relatively decreased by using Trivex tool and increased by using Triflute tool. CFD analysis is also used for the comparisons of aluminium alloys.

The morphology of aluminium alloys have been varied by Tool geometry, Transverse, longitudinal motions. Mahoney et al. [29] investigated the properties based on transverse and longitudinal motions. Tool geometry and microstructure have been analysed after the aging of 7075T65 Al alloy. It is observed that the yield strength, tensile strength is poor in HAZ after the post weld treatment process. Fujii Hidetoshi et al. [30] studied the microstructure and metallurgical properties after the post welding treatment and triangular type of profile is used to weld aluminium alloys. The tool has less impact at low welding speed. The strength is relatively decreased in HAZ by researchers and they are also made an attempt to modify the shape of the tool. It results that the strength in weld nugget is improved after the post treatment process.

The microstructure and mechanical properties are relatively improved by varying the tool diameter, shoulder diameter and pin. Elangovan et al. [31] studied the pin and shoulder diameter of tool on FSW of AA6061 aluminum alloy and shoulder diameters with different profiles of tool pin are used to join the welds. Their study is shown observations such as macrostructure analysis and transverse tensile properties. Micro structure and mechanical properties are relatively improved based on tool shoulder and pin profile.

Optimization Techniques such as Taguchi method is used to analyse the factors on tool geometry. Lakshminarayanan et al. [32] studied Taguchi method to find the factors which affect the tensile strength of the welded of FSW RDE-40 aluminium alloy and it observed that the influence of tensile strength of the weld is based on tool geometry. Mechanical properties such as tensile strength have been improved for FSW welded aluminium alloys.

Researchers are also used threaded and unthreaded tools to analyse the material flow in the weld nugget after the welding process. Oliver Lorrain et al. [33] investigated threaded pins and its application in industries. In the first phase, it is not unthreaded and this is because of the tool wear. The material flow has been analysed both threaded and unthreaded pin profiles. From above literature surveys, it has been found that they are varied the tool shape, shoulder diameter, pin to improve the weld nugget and analysed the material flow after FSW Process.

5. Conclusion

In this review, the friction stir welded Aluminium alloys have been presented and the simulation techniques are also validated with its process parameters. It is observed that similar and dissimilar FSW welded aluminium alloys have been utilized for industrial applications owing to its mechanical properties. For the past decades, few researchers are examined dissimilar aluminium alloys and analysed mechanical and metallurgical properties. In this paper, the grain structure of FSW joints, cavities, formation of onion rings in weld nugget zone of 2XXX, 6XXX, 7XXX series aluminium alloys are examined and SEM analysis reveals the deformities in aluminium alloys. In future, dissimilar aluminium alloys are recommended to enhance the quality of weld in industries and commercial applications.

IOP Conf. Series: Materials Science and Engineering **764** (2020) 012009 doi:10.1088/1757-899X/764/1/012009

References

[1] W.M. Thomas, C.J.Dawes, "Friction stir process welds aluminium alloys", Welding Journal. Vol.75, pp 41-45, 1996.

[2] Liu, G, Murr, LE, Nion, CS, Mcclure, JC & Vega, FR, "Micro structural aspects of the friction stir welding of 6061-T6 aluminum", Scripta Materialia, vol.37, pp.355-361, 1997.

[3] Z. H. Fu, D. Q. He, and Wang H, "Friction stir welding of aluminium alloys", Journal of Wuhan University of Technology- Materials Science Edition, Vol.19, pp 61-64,2004.

[4] P. Cavalier, G. Campanile. and F. Panella, "Effect of welding parameters on mechanical and microstructure properties of AA6056 joints produced by friction stir welding", Journal of materials processing technology, Vol.180, 2006, pp. 263 270.

[5] Li Y, Murr, LE, Brown, D & McClure, JC, "Flow visualization and residual micro structures associated with the friction stir welding of 2024 aluminum to 6061 aluminum", Materials Science and Engineering , Vol.A271, pp.213-223, 1998.

[6] Rhodes, C. G., Mahoney, M. W., Bingel, W. H., Spurling, R. A. And Bampton, C. C, "Effects of friction stir welding on microstructure of 7075 aluminium", Scripta Materialia, Vol.36(1), pp.69-75, 1997.

[7] Ghosh, M, Kumar, K, Kailas, SV & Ray, AK, "Optimization of friction stir welding parameters for dissimilar aluminum alloys", Materials and Design, Vol.31, pp.3033-3037, 2010.

[8] Yan, J. and Reynolds, A. P., "Effect of initial base metal temper on mechanical properties in AA7050 friction stir welds", Science and Technology of Welding and Joining, Vol.14(4), pp.282-287, 2009.

[9] Sakthivel T, Sengar G.S, Mukhopadhyay J, "Effect of welding speed on micro structure and mechanical properties of friction stir welded aluminium", International journal of Advanced Manufacturing Technology, Vol.43(5-6), 468-473, 2009.

[10] S. Benavides, Y. LI, L. E. Murr, D. Brown and J.C. McClure, "Low temperature friction stir welding of 2024 Aluminum", Scripta Materialia, Vol.41(8), pp. 809-815, 1999.

[11] Priya, R, Sarma, VS & Rao, KP, "Effect of post weld heat treatment on the microstructure and tensile properties of dissimilar friction stir welded AA 2219 and AA 6061 alloys", Transactions of the Indian Institute of Metals, Vol.62(1), pp.11-19, 2009.

[12] Da Silva, AM, Arruti, E, Janeiro, G, Aldanondo, E, Alvarez, P & Echeverria, A, "Material flow and mechanical behaviour of dissimilar AA2024-T3 and AA7075-T6 aluminium alloys friction stir welds", Materials and Design, Vol.32(4), pp.2021-2027, 2011.

[13] Rajakumar, S. Balasubramanian, V, "Establishing relationships between mechanical properties of aluminium alloys and optimized friction stir welding process parameters", Materials and Design, Vol.40(10), pp.17-35, 2012.

[14] Leitao, C, Leal, RM, Rodrigues, DM, Loureiro, A & Vilaca, P, "Mechanical Behaviour of Similar and Dissimilar AA5182-H111 and AA6016-T4 Thin Friction Stir Welds", Materials and Design, Vol.30(1), pp.101-108, 2009.

[15] O V Flores, "Micro Structural Issues in friction Stir Welded aluminum alloy", Scripta Materialia, Vol.38(5), pp. 703-708, 1998.

[16] Palanivel, R, Koshy Mathews, P & Murugan, N, "Development of mathematical model to predict the mechanical properties of friction stir welded AA6351 aluminum alloy", Journal of Engineering Science and Technology Review, Vol.4(1), pp. 25-31, 2011.

IOP Conf. Series: Materials Science and Engineering **764** (2020) 012009 doi:10.1088/1757-899X/764/1/012009

[17] Y. S Sato, Y. Kurihara, S.H.C Park, H. Kokawa and N. Tsuji, "Friction stir welding of ultrafine grained Al Alloy 1100 produced by Accumulative Roll-Bonding", Scripta Materialia, Vol.50(1), pp. 57-60, 2004.

[18] S Chen, H Zhang, X Jiang, T Yuan ,Y Han, X Li, "Mechanical properties of electricassisted friction stir welded 2219 aluminum alloy", Journal of Manufacturing Processes, Vol.44, pp. 197-206, 2019.

[19] H. Lombard, D. G. Hattingh, A. Steuwer, and M. N. James, "Effect of process parameters on the residual stresses in AA5083-H321 friction stir welds", Materials Science and Engineering A, vol.501, pp.119-124, 2009.

[20] Scialpi, L.A.C. De Fillippis and P. Cavaliere, "Influence of shoulder geometry on microstructure and mechanical properties of friction stir welded 6082 aluminum alloy", Materials and Design, Vol.28, pp.1124-1129, April 2006.

[21] Singh, RKR, Sharma, C, Dwivedi, DK, Mehta, NK & Kumar, P, "The microstructure and mechanical properties of friction stir welded Al-Zn-Mg alloy in as welded and heat treated conditions", Materials and Design, Vol.32, pp.682-687, 2011.

[22] N. Rajamanickam, V. Balusamy, M. G. Reddy, and K. Natarajan, "Effect of process parameters on thermal history and mechanical properties of friction stir welds", Materials and Design, Vol.30, pp. 2726-2731, 2009.

[23] Acerra.F, Buffa.G, Fratini.L and Troiano.G, "On the FSW of AA2024- T4 and AA7075-T6 T-joints: an industrial case study", International Journal of Advanced Manufacturing Technology, Vol.48, pp.1149–1157, 2010.

[24] R. W Fonda, J. F. Bingert and K. J Colligan, "Development of grain structure During stir welding", Scripta Materialia ,Vol.51, pp. 243-248, 2004.

[25] Y.H. Zhao, S. B Lin and F. X. Qu, "The influence of pin geometry on bonding and mechanical properties in friction stir welded 2014 alloy", Materials Letters, Vol.59(23), pp. 2948-2952, 2005.

[26] Buffa, G., Hua, J., Shivpuri, R. and Fratini, L., "Design of the friction stir welding tool using the continuum based FEM model", Materials Science and Engineering: A, Vol.419(1–2), pp.381-388, 2006.

[27] Mohanty, HK, Venkateswarlu, D, Mahapatra, MM, Pradeep Kumar, N & Mandal, R, "Modeling the Effects of Tool Probe Geometriesand Process Parameters on Friction Stirred Aluminium Welds", Journal of Mechanical Engineering and Automation, Vol.2(4), pp.74-79, 2012.

[28] Colegrove, P. A. and Shercliff.H.R, "3-Dimensional CFD modelling of flow round a threaded friction stir welding tool profile", Journal of Materials Processing Technology, Vol.169, pp.320–327, 2005.

[29] Mahoney, M., Rhodes, C., Flintoff, J., Bingel, W. and Spurling, R, "Properties of friction-stir welded 7075 T651 aluminium", Metallurgical and Materials Transactions A, Vol.29(7), pp.1955- 1964, 1998.

[30] Fujji Hidetoshi and Cri Ling., "Effect of tool shape on mechanical properties and microstructure of friction stir welded aluminum alloys", Journal of Materials Science and Engineering, Vol.54, pp. 25-31, 2006.

IOP Conf. Series: Materials Science and Engineering **764** (2020) 012009 doi:10.1088/1757-899X/764/1/012009

[31] K. Elangovan, and V. Balasubramaniam, "Influences of tool pin profile and tool shoulder diameter on the formation of friction stir processing zone in AA6061 aluminium alloy", Materials and Design, Vol.29, pp.362-373, 2008.

[32] A K Lakshminarayanan, V Balasubramanian and K Elangovan, "Effect of Welding Processes on Tensile Properties of AA6061 Aluminium Alloy Joints", International Journal of Advanced Manufacturing Technology, Vol.40, 286-296, 2009.

[33] Olivier Lorrain, Veronique Favier, Hamid Zahrouni, Didier Lawrjaniec, "Understanding the material flow path of friction stir welding process using unthreaded tools", Journal of Materials Processing Technology, Vol.210(4), pp.603-609, 2010.

Optimization of machining parameters in CNC turning of AA6061-B4C-CNT hybrid composites using Grey-fuzzy method

A Gnanavelbabu[1*] , V Arunachalam[1], K T Sunu Surendran[1] and K Rajkumar[2]
[1]Department of Industrial Engineering, CEG Campus, Anna University, Chennai - 600025, India.

[2]Department of Mechanical Engineering, SSN College of Engineering, Chennai - 603110, India.

[*]Corresponding author email: agbabu@annauniv.edu

Abstract. Aluminium hybrid composites and its machining receives a great attention due to their superior properties like good strength to weight ratio, good castability, machinability, and low cost, suitable for various applications in automotive and aerospace industries. CNC turning is considered as an ideal method for machining of aluminium hybrid composites. Nevertheless, the parameters of turning of aluminium hybrid composites need to be optimized for better and precise machining. Also, the presence of solid lubricants such as Carbon nanotube, Graphite, hexagonal Boron Nitride etc. promotes the machinability of aluminium hybrid composites. In this study, CNC turning on AA6061 matrix based hybrid composites reinforced with Boron carbide (5-15 vol.%) and Carbon nanotube (15 vol.%) was performed. Cutting Speed (rpm):60-180, feed (mm/rev): 0.10-0.20, depth of cut (mm): 0.50-1.00 and percentage of B4C (vol.%) 5-15 were the controlling parameters and responses namely tangential force (N), cutting power (W) and tool wear (mm3/min) were considered for the optimization. Experimental design was made as per Taguchi's L27 orthogonal array and Grey-fuzzy analytical tool was applied for the optimization. The optimal setting for turning operation on this hybrid composite using Grey-fuzzy analysis were 120 rpm, 0.20 mm/rev, 0.5 mm and 15 vo.% for cutting speed, feed, depth of cut and reinforcement respectively.

1. Introduction

Aluminium alloy matrix ceramic particulates reinforced composites possess excellent combination of mechanical and physical properties; promising for structural applications. Aluminium Matrix Composites (AMCs) shows better quality characteristics due to the combined effect of properties of the matrix and ceramic reinforcements. Matrix possesses high toughness and good ductility; whereas the ceramics shows high modulus and high strength [1]. Aluminium Hybrid Composites (AHCs) are the advanced composites and are mostly preferred due to their superior properties for automotive and aerospace applications [2]. AHCs containing ceramic particles such as B4C, SiC, TiC and soft lubricants such as Carbon nanotube (CNT), hexagonal Boron Nitride (hBN) were shown much improvements in machinability, wear resistance and strength [3-4]. Stir casting is the most popular and effective method for fabricating hybrid composites as it ensures a uniform distribution of different reinforcements [5]. Turning is considered as the feasible method for machining of cylindrical components from hybrid composites. Even though, the optimization of turning parameters for the particular hybrid composite is necessary for obtaining good quality characteristics.

Grey Relational Analysis (GRA) is a versatile method used for optimization and that analyses uncertain relations among the factors in a given system. In addition, Grey-fuzzy approach is popular tool for optimization as it has the combined advantages of both grey system and fuzzy logic approach. Sankar et al. applied Taguchi-Grey technique for optimizing non-conventional machining parameters for graphite solid lubricant reinforced AA6061-B4C based hybrid composites [6]. Ravinder Kumar et al. studied the feasibility on turning of AA7075/ SiC and AA7075/SiC/Graphite hybrid composite by varying the input parameters such as cutting speed, feed rate and approach angle. The machinability of the hybrid composite was better due to the presence of graphite solid lubricant [7]. Kalpesh tank et al. applied ANOVA, fuzzy and desirability function analysis in order to find the optimal combination of input parameters for finest surface finish characteristics in turning process [8]. Surendra Kumar et al. investigated the CNC turning characteristics of aluminium alloy 8011 and optimized the turning parameters by Taguchi-Fuzzy method. They ranked the parameters which significantly influenced the quality characteristics as feed, depth of cut and cutting speed [9]. Senthilkumar et al. presented an approach for optimizing the turning parameters of AISI 1045 steel using grey fuzzy logic. They found an improvement in performance of the turning process by applying hybrid techniques [10]. Palanisamy et al. applied Grey-Fuzzy approach and successfully determined the optimal parameter setting for CNC turning process [11]. Biswajit Das et al. optimized the CNC milling parameters for Al–4.5%Cu–TiC composites using grey- fuzzy logic approach and reported that cutting speed has a major role in finding the Grey Fuzzy Reasoning Grade [12].

The literature survey reveals that, no research works has been reported particularly on the optimization of CNC turning parameters for AA6061-B4C-CNT hybrid composites using Grey-Fuzzy logic method. Therefore, the main research objective of this work is to find an optimal combination of turning parameters over the responses such as tangential force, cutting power and tool wear by applying Grey-fuzzy tool.

2. Materials and Methodology

2.1. Materials

AA6061 was chosen as the matrix material as it is the most commonly used precipitation- hardened aluminium alloy and due to its favorable properties like good mechanical properties and wettability. Boron Carbide (B4C) is a hard ceramic commonly used as particulate reinforcements in aluminium to improve the wear resistance and strength. Carbon nanotube (CNT) is well known for its excellent lubricating properties, uniform dispersion with AMMC and improved mechanical properties. B4C and CNT particles with sizes 35-40μm and 30nm respectively were used as the reinforcements. The details of composition are shown in table 1. The volume of B_4C is varied at 5, 10 and 15% and the volume of CNT kept constant at 15%. Thus, three different compositions were considered in this study so as to examine the cutting performance at different volume of reinforcements. The hybrid composites were fabricated through liquid metallurgy route (stir casting).

Table 1. Volumetric composition of materials.

S. No.	AA6061 (Vol.%)	B_4C (Vol.%)	CNT (Vol.%)
1	80	5	15
2	75	10	15
3	70	15	15

2.2. Parameters and levels
The fabricated composites were machined using CNC turning center (Make: Jobber XL). Cutting speed, feed and depth of cut were considered as the parameters of the CNC machine. The details of factors and levels considered for the optimization are given in Table 2. The design and conduct of experiments were based on Taguchi's L_{27} orthogonal array.

Table 2. Various levels of input parameters.

S. No.	Parameters	Low	Medium	High
1	Cutting speed (rpm)	60	120	180
2	Feed (mm/rev)	0.1	0.15	0.2
3	Depth of cut (mm)	0.5	0.75	1
4	Reinforcement (Vol. %)	5	10	15

Tangential force (R_1), tool wear (R_2) and cutting power (R_3) were considered as the responses. Tangential force and tool wear values were directly obtained from the CNC turning machine. The cutting power was calculated using the equation 1.

$$Pc = \frac{D * F * S * Kc}{60 * 1000 * \eta} \tag{1}$$

Where, Pc-Cutting power; D-Depth of cut; F-feed; S-Cutting speed; Kc -Specific cutting force and η-Machine Coefficient.

2.3. *Grey Relational Analysis (GRA)*
In GRA, the S/N ratio is calculated initially to reduce the variability in the process by minimizing the effects of uncontrollable factors. As our objective is minimizing all the three responses, the "smaller the better" was implied. In GRA, the normalizing of values was done using equation 2. Further, equation 3 was used to determine the Grey Relational Coefficient (GRC).

$$xi * (k) = \frac{max.\ xi\ (k) - xi\ (k)}{max.\ xi\ (k) - min.\ xi\ (k)} \tag{2}$$

Where, xi*(k): normalized data and xi (k): observed data for the ith experiment by using k^{th} response.

$$\xi i(k) = \frac{\Delta min. + \varsigma \Delta max.}{\Delta i(k) + \varsigma \Delta max.} \tag{3}$$

Where, Δi (k): absolute value of the difference between xi* (k) and x0(k), Δmax: global maximum values, Δmin: global minimum values.

To expand or compress the range of GRC, the distinguishing coefficient (ζ) was taken as 0.5. Finally, Grey Relational Grade (GRG) values were calculated by taking the average of GRCs. The G-fuzzy is the combination of Grey relational approach and fuzzy logic theory. In this approach, GRA technique effectively utilized to convert the multi-objective problem in to a single-objective. The highest GRG represents the minimum value for input parameters. However, fuzzy logic theory effectively reduces the degree of uncertainty in the obtained GRG values. Each input variable has low, medium, high subsets in its membership function as shown in table 3. The range of output has Extremely Large (EL), Extremely Small (ES), Very Large (VL), Very Small (VS), Large (L), Medium (M), Small (S), Large Medium (LM) and Small- Medium (SM) in its membership function as show in table 4.

IOP Conf. Series: Materials Science and Engineering **764** (2020) 012010 doi:10.1088/1757-899X/764/1/012010

Table 3. Range of fuzzy subsets for input membership functions.

Condition	Range	Membership function
Low	-0.5 – 0.5	
Medium	0 – 1	Triangular
High	0.5 – 1.5	

Table 4. Range of fuzzy subsets for output membership functions.

Condition	Range	Membership function
Extremely large	0.8672 – 1.126	
Extremely small	-0.125 – 0.125	
Very large	0.75 – 1	
Very small	0 – 0.125	
Large	0.625 – 0.875	Triangular
Medium	0.375 – 0.625	
Small	0.125 – 0.375	
Large medium	0.5012 – 0.7512	
Small-medium	0.25 – 0.5	

Figures 1 (a) and (b) indicate the fuzzy rule editor and membership functions for output GFRG (Grey Fuzzy Reasoning Grade) respectively. For fuzzification of the GRC, the input and output variables are created in the Fuzzy Inference System (FIS). Fuzzy rule was constructed by simple IF-THEN rule with condition and conclusion. The centroid method was chosen to calculate the GFRG from the fuzzy multi-response output.

Grey relational coefficients are $\eta 1$, $\eta 2$, $\eta 3$. GRC of tangential force, cutting power, tool wear are taken as $\eta 1$, $\eta 2$, $\eta 3$ respectively. Multi-response output is taken as ξ.

Rule 1: if $\eta 1$ is $A11$, $\eta 2$ is $A12$ and $\eta 3$ is $A13$, then ξ is $D1$,
else
Rule 2: if $\eta 1$ is $A21$ $\eta 2$ is $A22$ and $\eta 3$ is $A23$, then ξ is $D2$
else
....
....
Rule n: if $\eta 1$ is $An1$, $\eta 2$ is $An2$ and $\eta 3$ is $An3$, then ξ is $D1n$.

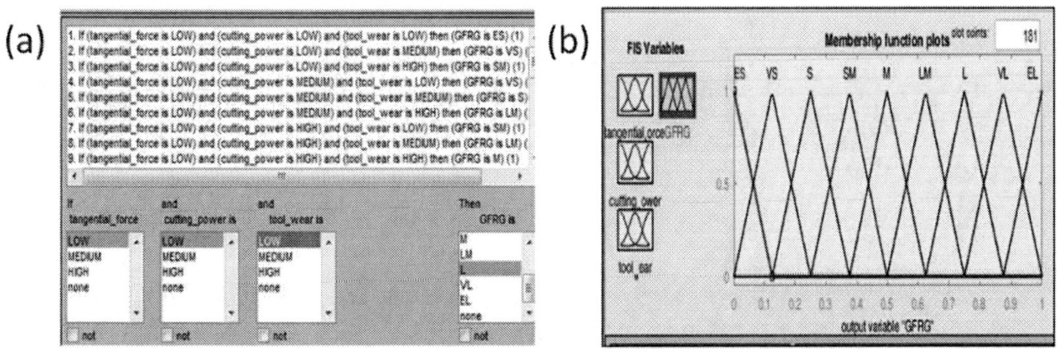

Figure 1. **(a)** Fuzzy rule editor; **(b)** Typical membership functions for output GFRG.

3. Result and Discussion

The composites as per the volumetric composition were successfully fabricated through stir -bottom pouring casting process. Three specimens of each composition were prepared as shown in figure 2(a). CNC turning was performed on the composite specimens as per the structured L 27 orthogonal array. Sample specimens after CNC turning process are shown in figure 2(b). The experimental results were shown in the below table 5.

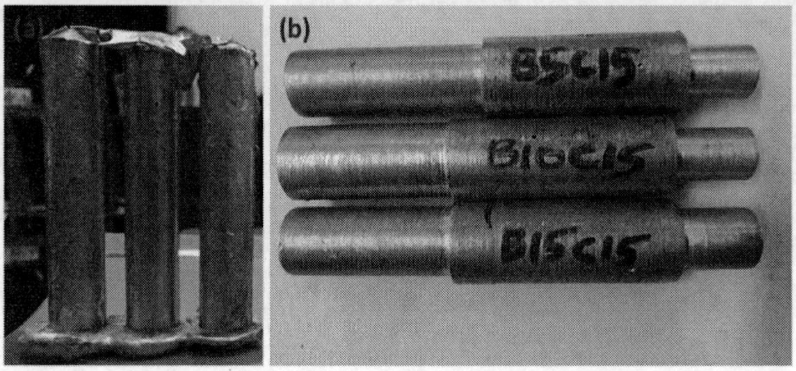

Figure 2. (a) Sample specimens (as-cast), **(b)** Samples after CNC turning process.

The S/N ratio, normalized S/N ratio, corresponding GRC values and GRG of all 3 responses were calculated for the different experimental trials and are shown in table 6. The single representation of all the three responses corresponding to each trial is shown by the GRG value and a higher value of GRG is always desirable. The peak value of GRG is observed for the 23rd trial, (0.851). This demonstrates the closeness of optimum parameter setting.

Table 5. Experimental results of turning on hybrid composites.

Trial No.	Cutting speed (rpm)	Feed (mm/rev)	Depth of cut (mm)	Reinfor-cement (Vol. %)	Tang. force (N)	Cutting power (W)	Tool wear (mm^3/min)
1	60	0.1	0.5	5	12	3.72	0.14
2	60	0.1	0.5	10	13	4.03	0.23
3	60	0.1	0.5	15	15	4.65	0.21
4	60	0.15	0.75	5	6	1.86	0.16
5	60	0.15	0.75	10	10	3.1	0.26
6	60	0.15	0.75	15	17	5.27	0.16
7	60	0.2	1	5	16	4.96	0.15
8	60	0.2	1	10	12	3.72	0.19
9	60	0.2	1	15	20	6.2	0.19
10	120	0.1	0.75	5	13	8.06	0.2
11	120	0.1	0.75	10	11	6.82	0.31
12	120	0.1	0.75	15	11	6.82	0.27
13	120	0.15	1	5	12	7.44	0.23
14	120	0.15	1	10	13	8.06	0.27
15	120	0.15	1	15	20	12.4	0.25
16	120	0.2	0.5	5	11	6.82	0.21
17	120	0.2	0.5	10	12	2.48	0.19
18	120	0.2	0.5	15	23	14.26	0.18
19	180	0.1	1	5	10	9.3	0.23

Trial No.	Cutting speed (rpm)	Feed (mm/rev)	Depth of cut (mm)	Reinfor-cement (Vol. %)	Tang. force (N)	Cutting power (W)	Tool wear (mm³/min)
21	180	0.1	1	15	10	9.3	0.25
22	180	0.15	0.5	5	8	7.44	0.19
23	180	0.15	0.5	10	15	13.95	0.32
24	180	0.15	0.5	15	16	14.88	0.19
25	180	0.2	0.75	5	7	6.51	0.25
26	180	0.2	0.75	10	13	12.09	0.29
27	180	0.2	0.75	15	16	14.88	0.27

MATLAB R2016a software was used for fuzzy Logic analysis to find the GFRG values. Three triangular membership functions were applied to the GRC of Tangential force, cutting power and tool wear. Figure 3 shows triangular membership functions that are applied to the responses. For FIS, rules are framed and applied for the prediction of GFRG's for all 27 experiments. A comparison of GFRG values with GRG values is shown in figure 4. While comparing the GRG with GFRG values, the GFRG values are improved.

Table 6. S/N ratio, GRC and GRG values.

Trial No.	S/N ratio			Normalised S/N ratio			GRC			GRG
	R1	R2	R3	R1	R2	R3	R1	R2	R3	
1	-21.58	-11.41	17.08	0.52	0.33	0.00	0.508	0.429	0.333	0.423
2	-22.28	-12.11	12.77	0.58	0.37	0.60	0.541	0.443	0.556	0.513
3	-23.52	-13.35	13.56	0.68	0.44	0.49	0.611	0.472	0.495	0.526
4	-15.56	-5.39	15.92	0.00	0.00	0.16	0.333	0.333	0.374	0.347
5	-20.00	-9.83	11.70	0.38	0.25	0.75	0.447	0.399	0.666	0.504
6	-24.61	-14.44	15.92	0.78	0.50	0.16	0.690	0.500	0.374	0.521
7	-24.08	-13.91	16.48	0.73	0.47	0.08	0.649	0.486	0.353	0.496
8	-21.58	-11.41	14.42	0.52	0.33	0.37	0.508	0.429	0.442	0.460
9	-26.02	-15.85	14.42	0.90	0.58	0.37	0.828	0.543	0.442	0.604
10	-22.28	-18.13	13.98	0.58	0.71	0.43	0.541	0.629	0.468	0.546
11	-20.83	-16.68	10.17	0.45	0.62	0.96	0.477	0.571	0.929	0.659
12	-20.83	-16.68	11.37	0.45	0.62	0.79	0.477	0.571	0.709	0.586
13	-21.58	-17.43	12.77	0.52	0.67	0.60	0.508	0.600	0.556	0.555
14	-22.28	-18.13	11.37	0.58	0.71	0.79	0.541	0.629	0.709	0.626
15	-26.02	-21.87	12.04	0.90	0.91	0.70	0.828	0.851	0.626	0.768
16	-20.83	-16.68	13.56	0.45	0.62	0.49	0.477	0.571	0.495	0.514
17	-21.58	-7.89	14.42	0.52	0.14	0.37	0.508	0.367	0.442	0.439
18	-27.23	-23.08	14.89	1.00	0.98	0.30	1	0.961	0.418	0.793
19	-20.00	-19.37	12.77	0.38	0.77	0.60	0.447	0.689	0.556	0.564
20	-20.00	-19.37	12.04	0.38	0.77	0.70	0.447	0.689	0.626	0.587
21	-20.00	-19.37	12.04	0.38	0.77	0.70	0.447	0.689	0.626	0.587
22	-18.06	-17.43	14.42	0.21	0.67	0.37	0.389	0.600	0.442	0.477
23	-23.52	-22.89	9.90	0.68	0.97	1.00	0.611	0.942	1	0.851
24	-24.08	-23.45	14.42	0.73	1.00	0.37	0.649	1	0.442	0.697
25	-16.90	-16.27	12.04	0.11	0.60	0.70	0.361	0.557	0.626	0.515
26	-22.28	-21.65	10.75	0.58	0.90	0.88	0.541	0.834	0.808	0.727
27	-24.08	-23.45	11.37	0.73	1.00	0.79	0.649	1	0.709	0.786

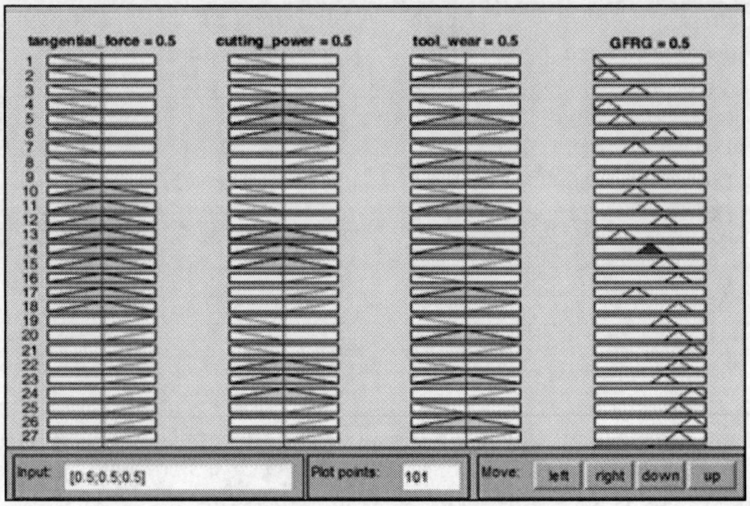

Figure 3. Fuzzy logic rules view.

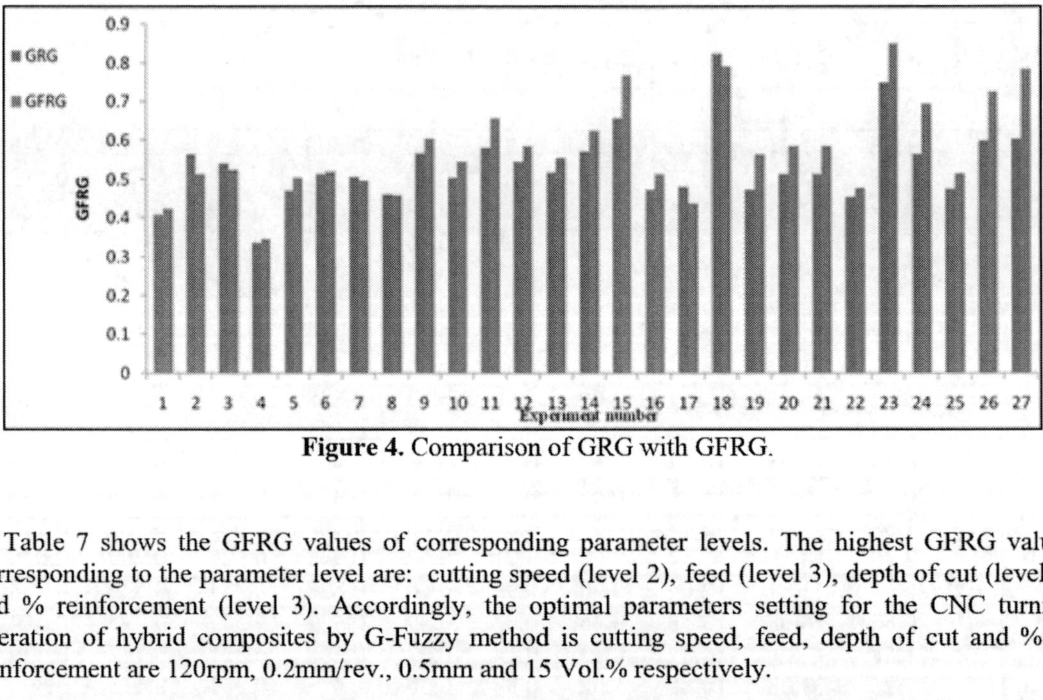

Figure 4. Comparison of GRG with GFRG.

Table 7 shows the GFRG values of corresponding parameter levels. The highest GFRG values corresponding to the parameter level are: cutting speed (level 2), feed (level 3), depth of cut (level 1) and % reinforcement (level 3). Accordingly, the optimal parameters setting for the CNC turning operation of hybrid composites by G-Fuzzy method is cutting speed, feed, depth of cut and % of reinforcement are 120rpm, 0.2mm/rev., 0.5mm and 15 Vol.% respectively.

Table 7. GFRG values and optimum parameter levels.

Parameters	GFRG			Corresponding
	Level 1	Level 2	Level 3	Optimum value
Cutting speed	0.4860	**0.5973**	0.5497	120 rpm
Feed	0.5162	0.5380	**0.5546**	0.2 mm/rev
Depth of cut	**0.5623**	0.5147	0.5317	0.5 mm
Reinforcement	0.4615	0.5547	**0.5926**	15 Vol. %

4. Conclusion

The fabrication of AA6061-B4C-CNT hybrid composites and CNC turning operation as per Taguchi's L27 orthogonal array were done successfully. Grey Relational Analysis and Fuzzy logic were combined to find out the optimum setting parameters. Grey relational coefficient, Grey relational grade and Grey fuzzy reasoning grade were calculated accordingly for G-Fuzzy analysis. The optimum combination of turning parameters for AA6061-B4C-CNT hybrid composites was cutting speed - 120 rpm, Feed - 0.20mm/rev, Depth of cut - 0.5mm and Reinforcement - 15Vol. %.

Acknowledgement

The authors wish to acknowledge the financial support (Grant No.: EEQ/2017/000382) by Science and Engineering Research Board (SERB), Govt. of India.

References

[1] Jaswinder Singh, Amit Chauhan, "Characterization of hybrid aluminum matrix composites for advanced applications – A review", Journal of Materials Research and Technology, vol. 5(2), pp. 159-169, 2016.

[2] M O Bodunrin, K K Alanemea, L H Chown, "Aluminium matrix hybrid composites: A review of reinforcement philosophies; mechanical, corrosion and tribological characteristics", Journal of Materials Research and Technology, vol. 4, pp. 434-445, 2015.

[3] Barada Prasanna Sahoo, Diptikanta Das, "Tribological behaviour of ceramic and carbon nano-tube reinforced metal matrix composites - a review", Materials Today: Proceedings, Vol. 5(9), pp. 20549-20559, 2018.

[4] V M Ravindranath, G S Shiva Shankar, S Basavarajappa, N G Siddesh Kumar, "Dry sliding wear behavior of hybrid aluminum metal matrix composite reinforced with boron carbide and graphite particles", Materials Today: Proceedings, vol. 4, pp. 11163–11167, 2017.

[5] K Anand Babu, P Venkataramaiah, K Dharma Reddy, "Mechanical characterization of aluminium hybrid metal matrix composites synthesized by using stir casting process", Materials today proceedings, vol. 5(14), pp. 28155-28163, 2018.

[6] M Sankar, A Gnanavelbabu, K Rajkumar, "Parametric Optimisation of Abrasive Assisted Electro Chemical Machining of Al6061-B4C-Gr Using Taguchi-Grey Technique", Journal of the Balkan Tribological Association, vol. 23(1), pp. 67-87, 2017.

[7] Ravinder Kumar, Santram Chauhan, "Study on surface roughness measurement for turning of Al 7075/10/SiCp and Al 7075 hybrid composites by using response surface methodology (RSM) and artificial neural networking (ANN)", Measurement, vol. 65, pp. 166-180, 2015.

[8] Kalpesh Tank, Nishith Shetty, Gaurang Panchal, Aman tukrel, "optimization of turning parameters for the finest surface characteristics using desirability function analysis coupled with fuzzy methodology and ANOVA", Materials today proceedings, vol. 5(5), pp. 13015-13024, 2018.

[9] Surendra Kumar Saini, Sharad Kumar Pradhan, "Optimization of Multi-Objective Response during CNC Turning using Taguchi-Fuzzy Application", Procedia Engineering, vol. 97, pp. 141-49, 2014.

[10] N Senthilkumar, J Sudha, V Muthukumar, "A grey-fuzzy approach for optimizing machining parameters and the approach angle in turning AISI 1045 steel", Advances in Production Engineering & Management, vol. 10, pp. 195-208, 2015.

[11] D Palanisamy, P Senthil, "Application of Grey-Fuzzy Approach for Optimization of CNC Turning Process", Materials today proceedings, vol. 5(2), pp. 6645–6654, 2018.
[12] Biswajit Das, S Roy, R N Rai, S C Saha, "Application of grey fuzzy logic for the optimization of CNC milling parameters for Al–4.5%Cu–TiC MMCs with multi-performance characteristics", International Journal of Engineering Science and Technology, vol. 19(2), pp. 857-865, 2016.

IOP Conf. Series: Materials Science and Engineering **764** (2020) 012011 doi:10.1088/1757-899X/764/1/012011

Optimization of Abrasive Water Jet Machining Parameters on AA6061/B_4C/hBN Hybrid Composites using Grey-RSM

A Gnanavelbabu*, V Arunachalam, K T Sunu Surendran and P Saravanan
Department of Industrial Engineering, CEG Campus, Anna University, Chennai - 600025, India.

*Corresponding author email: agbabu@annauniv.edu

Abstract. Aluminium Matrix Hybrid Composites (AMHCs) possess an excellent combination of physical and mechanical properties suitable for a variety of applications in automobile and aerospace industries. Addition of solid lubricant hexagonal Boron Nitride (hBN) reduces the complexity in machining of aluminium hybrid composites. Abrasive Water Jet Machining (AWJM) is a non-traditional technique applied for the machining of alloys and composites, due to its better cutting performances. In this work, AA6061 matrix hybrid composites reinforced with B_4C (5, 10 and 15vol.%) and hBN (15vol.%) were developed through liquid metallurgy route. The composite were then machined using AWJM. Process parameters considered were abrasive flow rate, water pressure, traverse speed, percentage of reinforcement and mesh size over the responses Kerf Taper Angle (KTA) and Surface roughness (Ra). The experimental design was based on L_{27} orthogonal array and Grey-RSM analytical tool was applied for optimization. The optimal values of parameter setting for AWJM machining process using Grey RSM methodology were 340 g/min, 200 MPa, 60.06 mm/min, 5.01% and 80 for abrasive flow rate, water pressure, traverse speed, percentage of reinforcement and mesh size were respectively.

1. Introduction

A wide range of engineering applications are imposed on Metal Matrix Composites (MMCs) due to their high specific strength, stiffness, wear resistance and dimensional stability [1]. Due to the advanced physical and mechanical properties, Aluminium Matrix Hybrid Composites (AMHCs) are majorly used for automobile and aerospace applications [2]. Aluminium alloy 6061 is the most commonly used matrix material. When it reinforced with hard Boron Carbide (B_4C) ceramic particles, the strength, hardness, wear resistance, thermal and chemical stability have been improved [3]. In addition, the solid lubricant namely hexagonal Boron Nitride (hBN), is added for improving the machinability, wear and thermal resistance of AMHCs [4]. One of the feasible and economical ways for fabricating hybrid composites is the mechanical stirring assisted casting which ensures a fine distribution of reinforcements in the matrix [5].Machining of hybrid composites in conventional methods is a great challenge. To overcome this difficulty, many newer techniques are employed nowadays. Abrasive Water Jet Machining (AWJM) process provides a high-quality cutting for composites with a great surface finish suitable for producing various components for automotive and aerospace applications. The basic principle behind AWJM is to increase the material removal rate (MRR) from the surface by utilizing the abrasiveness of fine-grained abrasive particles flowing in a high velocity water jet. Almost any type of material can be cut by AWJM which includes hard brittle materials, soft materials etc. [6-8].

The quality of cutting on usually depends on abrasive material, work piece material, thickness of the specimen etc. and other machine controlling parameters [9]. Hence, optimization of AWJM process parameters is important for obtaining better cutting quality characteristics. Response Surface Methodology (RSM) is the popular technique for determining the optimal parameter settings. Further, Integrated Grey-Taguchi approach coupled with RSM (GT-RSM) is used for the optimization of many industrial processes. Jagadish et al. applied ANOVA and RSM for optimizing the AWJM parameters for surface roughness (Ra) and Process Time (PT) on green composites. Stand-off Distance (SoD), Nozzle Speed (NS) and Pressure within the Pumping System (PwPS) were the parameters taken for optimizing [10]. Manoj M. et al. applied Taguchi-DEAR methodology for optimizing the AWJM parameters for AA 7075 matrix based composites by considering the input parameters standoff distance, transverse speed and Water jet pressure over MRR, Ra and taper angle [11]. Ravi Kumar et al. in their optimization study using RSM concluded that transverse speed, percentage of reinforcement and standoff distance were the significant parameters which influenced the MRR and Ra of aluminium composites [12]. Santhanakumar et al. applied Grey theory based RSM methodology for optimizing the AWJM process parameters for aluminium hybrid composites. Parameters such as SoD, traverse speed, abrasive flow rate and water pressure were considered for optimizing the responses striation zone length, angle and Ra [13].

To minimize the surface roughness (Ra) and Kerf Taper Angle (KTA) of B_4C and hBN reinforced AA6061 hybrid composites, an optimal process parameter setting is necessary for AWJM process. From the literature review, it is clearly understandable that application of Grey-RSM for acquiring optimal setting for AWJM parameters for AA6061-B_4C-hBN hybrid composites has not much been attempted. Therefore, the prime aim of this work is to apply Grey-RSM methodology in order to optimize the AWJM process parameters on the fabricated hybrid composites.

2. Materials and Methodology

AA6061 is the widely used aluminium alloy which has good corrosion resistances, high strength to weight ratio, dimensional stability and is chosen as the matrix material. Boron Carbide (B_4C) with an APS (Average Particle Size) of 10-25 μm and hexagonal Boron Nitride (hBN) with an APS of 5μm were used as the reinforcements. The volumetric composition of the materials for fabricating the hybrid composites through stir casting is given in table 1.

Table 1. Volumetric composition.

S. No.	AA6061 (Vol.%)	B_4C (Vol.%)	hBN (Vol.%)
1	80	5	15
2	75	10	15
3	70	15	15

The various input parameters of the machine along with their levels considered for optimization is given in Table 2. Standoff distance (1.5mm), impinging angle (90^0) and focusing tube length (75mm) were kept constant throughout the study. Responses considered for this study were Surface roughness (Ra) and Kerf Taper Angle (KTA).

Table 2. Levels of AWJM parameters.

S. No.	Parameters	Low	Medium	High
1	A - Mesh size (size)	80	100	120
2	B - Abrasive Flow Rate (g/min)	240	340	440
3	C - Water Pressure (MPa)	125	200	275
4	D -Traverse Speed (mm/min)	60	90	120
5	E - % of reinforcement (Vol.%)	5	10	15

The cutting of hybrid composites of size 100 mm x 100 mm and thickness 10 mm was carried out in an AWJM machine (Make: OMAX- 2626). The image of the AWJM machine and specimen after AWJM process are shown in figure 1a and b. The cutting quality of the specimens was analysed by measuring the kerf width, kerf taper angle and surface roughness. Video measurement system (VMS2010 F) was utilized for the measurement of kerf width. The samples after AWJM process placed in VMS in order to find the bottom and top kerf width. Using the kerf width and sample thickness, KTA is calculated by equation 1. The effect on various AWJM parameters on the surface finish was studied. A non-contact profilometer was utilized for measuring surface roughness value (Ra).

$$\theta = \tan -1 \frac{(w_t - w_b)}{2t} \quad (1)$$

w_t = Top kerf width , w_b = Bottom kerf width , t = Thickness of the work piece.

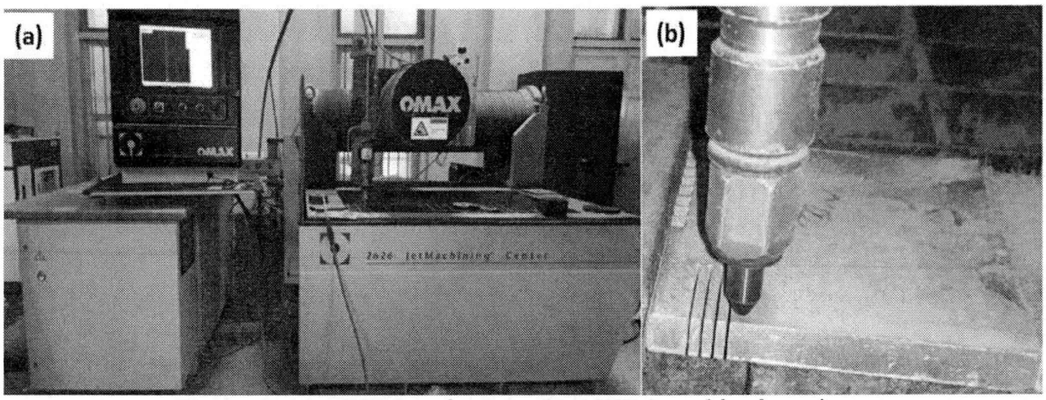

Figure 1. (a) AWJM equipment; **(b)** AWJM machined specimen.

3. Result and Discussion

AWJM was performed on the hybrid composites as per the L_{27} orthogonal array and the experimental results obtained are shown in table 3. The optimization of parameters was done by Grey-Response Surface Methodology (G-RSM) using design expert software. In grey relational analysis the optimization was performed until the Grey Relational Grade (GRG) obtained. When GRG value is higher, it indicates that the parameter is nearest to the optimal value. In G-RSM, the GRG values representing the multiple responses were taken as single performance and are used in the generation of responses surface graphs.

Table 3. Experimental results.

S.No	A	B	C	D	E	KTA (°)	Ra (μm)
1	80	240	125	60	5	0.400667	3.6032
2	80	240	125	60	10	0.381367	3.6185
3	80	240	125	60	15	0.358276	3.6248
4	80	340	200	90	5	0.390369	3.6189
5	80	340	200	90	10	0.373955	3.6288
6	80	340	200	90	15	0.372869	3.6392
7	80	440	275	120	5	0.502271	3.6299
8	80	440	275	120	10	0.438276	3.6388
9	80	440	275	120	15	0.309356	3.6507
10	100	240	125	120	5	0.608369	3.6353
11	100	240	125	120	10	0.469256	3.6448
12	100	240	125	120	15	0.400896	3.6532
13	100	340	200	60	5	0.454166	3.5948

IOP Conf. Series: Materials Science and Engineering **764** (2020) 012011 doi:10.1088/1757-899X/764/1/012011

S.No	A	B	C	D	E	KTA (°)	Ra (µm)
14	100	340	200	60	10	0.444256	3.6085
15	100	340	200	60	15	0.410929	3.6185
16	100	440	275	90	5	0.637416	3.6295
17	100	440	275	90	10	0.464267	3.6332
18	100	440	275	90	15	0.315247	3.6408
19	120	240	125	90	5	0.659476	3.6188
20	120	240	125	90	10	0.626374	3.6241
21	120	240	125	90	15	0.585339	3.6332
22	120	340	200	120	5	0.61828	3.6288
23	120	340	200	120	10	0.435167	3.6353
24	120	340	200	120	15	0.396609	3.6408
25	120	440	275	60	5	0.5463	3.6241
26	120	440	275	60	10	0.470676	3.6302
27	120	440	275	60	15	0.369276	3.6383

3.1. Grey Relational Grade (GRG)

The S/N ratio, Normalized S/N ratio, Grey Relational Coefficient (GRC) and GRG for each trial were calculated and the values obtained are shown in table 4. A large GRG among the trials is considered as optimal value. The highest GRG value obtained was 0.735 on 16th trial.

Table 4. Grey Relational Grades (GRGs).

Trial No.	S/N		S/N (Normalised)		GRC		GRG
	R1	R2	R1	R2	GRC (R1)	GRC (R2)	
1	7.94	-11.13	0.34	0.14	0.43	0.37	0.400
2	8.37	-11.17	0.28	0.41	0.41	0.46	0.433
3	8.92	-11.19	0.19	0.52	0.38	0.51	0.445
4	8.17	-11.17	0.31	0.41	0.42	0.46	0.440
5	8.54	-11.20	0.25	0.58	0.40	0.55	0.473
6	8.57	-11.22	0.25	0.76	0.40	0.68	0.538
7	5.98	-11.20	0.64	0.60	0.58	0.56	0.569
8	7.17	-11.22	0.46	0.75	0.48	0.67	0.576
9	10.19	-11.25	0.00	0.96	0.33	0.92	0.628
10	4.32	-11.21	0.89	0.70	0.82	0.62	0.723
11	6.57	-11.23	0.55	0.86	0.53	0.78	0.652
12	7.94	-11.25	0.34	1.00	0.43	1.00	0.716
13	6.86	-11.11	0.51	0.00	0.50	0.33	0.418
14	7.05	-11.15	0.48	0.24	0.49	0.40	0.442
15	7.72	-11.17	0.38	0.41	0.44	0.46	0.451
16	3.91	-11.20	0.96	0.60	0.92	0.55	0.735
17	6.66	-11.21	0.54	0.66	0.52	0.59	0.557
18	10.03	-11.22	0.02	0.79	0.34	0.70	0.521
19	3.62	-11.17	1.00	0.41	1.00	0.46	0.730
20	4.06	-11.18	0.93	0.50	0.88	0.50	0.691
21	4.65	-11.21	0.84	0.66	0.76	0.59	0.678
22	4.18	-11.20	0.91	0.58	0.85	0.55	0.700
23	7.23	-11.21	0.45	0.70	0.48	0.62	0.549
24	8.03	-11.22	0.33	0.79	0.43	0.70	0.565

Trial No.	S/N		S/N (Normalised)		GRC		GRG
	R1	R2	R1	R2	GRC (R1)	GRC (R2)	
25	5.25	-11.18	0.75	0.50	0.67	0.50	0.585
26	6.55	-11.20	0.55	0.61	0.53	0.56	0.545
27	8.65	-11.22	0.23	0.75	0.39	0.66	0.529

3.2. Optimization

A statistical technique used widely for empirical model building is Response Surface Methodology (RSM). It provides the optimized results for different combinations of input parameters. In this study, RSM was applied on GRG values in order to find the optimum combination of parameters. The results of Analysis Of Variance (ANOVA) are shown in figure 2. The model "F value" 10.50 indicate the model significance. The p-value should be less than 0.05 shows significant. ANOVA shows that combination of parameters AB (Mesh size and Abrasive flow rate), AE (Mesh Size and Reinforcement %), BC (Abrasive Flow Rate and Water Pressure) are significant. R-sq. value 0.9131 closes to one show that the model is fit.

Source	Sum of Squares	df	Mean Square	F Value	p-value Prob > F	
Model	0.27	13	0.021	10.50	< 0.0001	significant
A-Mesh size	7.111E-004	1	7.111E-004	0.35	0.5617	
B-Abrasive flow	9.825E-004	1	9.825E-004	0.49	0.4962	
C-Water pressu	0.000	0				
D-Traverse spe	5.512E-003	1	5.512E-003	2.75	0.1212	
E-% reinforcem	2.948E-003	1	2.948E-003	1.47	0.2469	
AB	0.012	1	0.012	5.89	0.0305	
AC	0.000	0				
AD	3.769E-003	1	3.769E-003	1.88	0.1936	
AE	0.016	1	0.016	8.21	0.0132	
BC	0.034	1	0.034	17.12	0.0012	
BD	9.235E-003	1	9.235E-003	4.61	0.0513	
BE	3.258E-003	1	3.258E-003	1.63	0.2247	
CD	0.000	0				
CE	0.000	0				
DE	9.375E-004	1	9.375E-004	0.47	0.5061	
A^2	1.659E-003	1	1.659E-003	0.83	0.3795	
B^2	0.000	0				
C^2	0.000	0				
D^2	0.000	0				
E^2	5.315E-003	1	5.315E-003	2.65	0.1275	
Residual	0.026	13	2.005E-003			
Cor Total	0.30	26				

Figure 2. ANOVA for GRG.

Summary of ANOVA: *Std. Dev.*:0.045, *Mean*: 0.57, *CV* %:7.91, *Press*: 0.12, *Adj.* R^2:0.8261, *R sq.*: 0.9131, *Pred. R –sq.*: 0.5874 and *Adq. Precision*: 11.046.

The generated regression equation offers the scope to study the behaviour of system by quantifying the association and there by exploring their individual effects and interaction effects of parameters on the responses. The corresponding regression equation is shown in equation 2.

$$\textbf{GRG}=0.45+0.014*A+0.017*B+0.035*D-0.013*E-0.089*A*B+0.058*A*D-0.037*A*E+0.12*B*C-0.091*B*D-0.016*B*E-8.839E-003*D*E+0.025*A^2+0.030*E^2 \quad (2)$$

3.3. Response Surface Plots

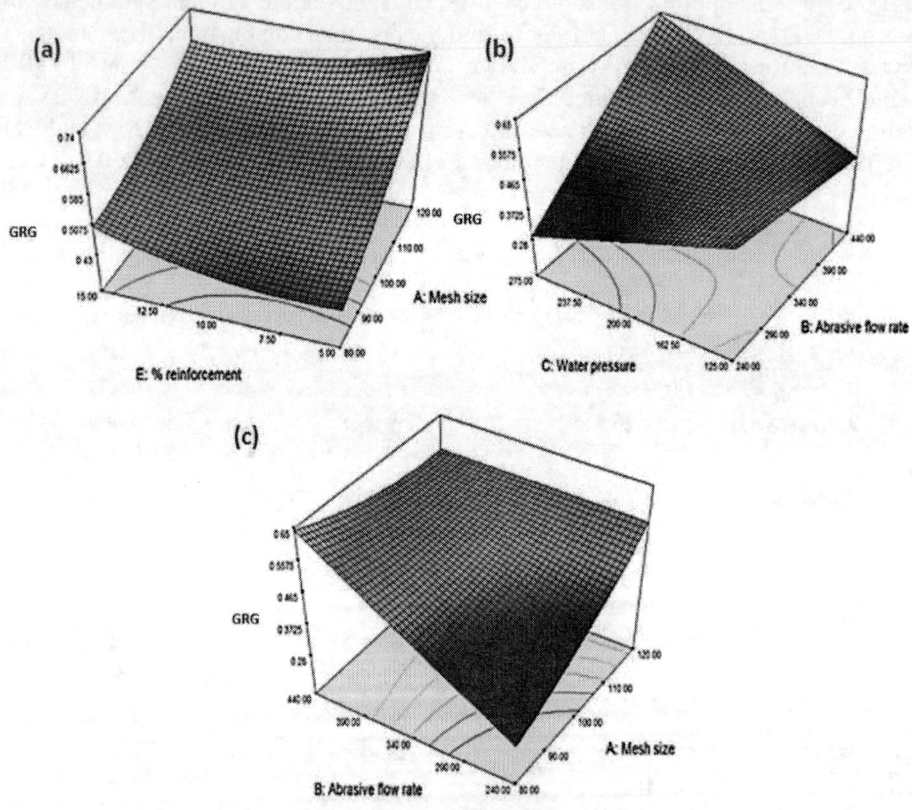

Figure 3(a-c). 3D surface plots.

Response surface plots based on GRG with parameters are shown in the figures 3(a-c). Figure 3a shows the effect of % of reinforcement (E) and Mesh size (A) on the GRG values. The maximum GRG value was obtained at 120 mesh size and 15% reinforcement. Also, it illustrates that GRG and Mesh size are directly proportional. Figure 3 (b) represents the effects of Abrasive flow rate (B) and Water pressure (C) on GRG values. Maximum GRG obtained at highest abrasive flow rate and water pressure. At highest abrasive flow rate, and increase in water pressure increases the GRG. Figure 3 (c) represents the effects of mesh size (A) and abrasive flow rate (B) on GRG. The highest GRG values obtained at low mesh size and highest Abrasive flow rate. A significant reduction in GRG can be observed as the abrasive flow rate decreases.

3.4. Desirability analysis

The results of desirability analysis are shown in figure 4. In the figure 4, the dots indicate the level of the parameter and heights indicate the desirability. The obtained desirability value form the analysis is

0.950. This indicates a higher degree of closeness, as 1 is the desirable value. The optimal level of AWJM parameters obtained is Mesh size - 80, abrasive flow rate - 340, water pressure- 200, Transverse speed-60.6 and reinforcement %- 5.01.

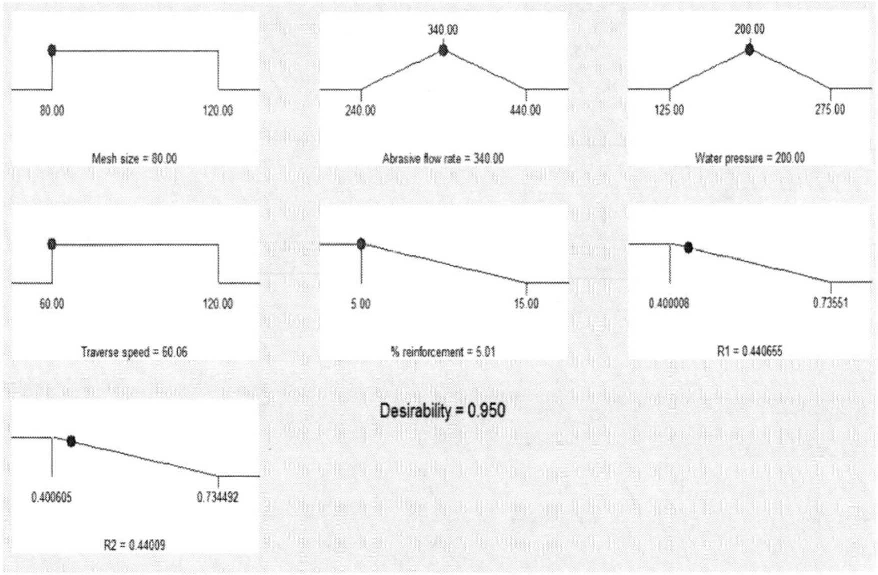

Figure 4. Desirability analysis.

4. Conclusion

Fabrication and parametric optimization of AWJM process on three different AA6061-B_4C-hBN hybrid composites were successfully carried out using the analytical tools G-RSM. The predicted and the experimentally obtained values of grey relational grade were matched well demonstrating a better model fitness. The optimal parameter setting for AWJM machining process using Grey RSM methodology were: Mesh size=80, Abrasive flow rate=340 g/min, Water pressure=200 MPa, Traverse speed =60.06 mm/min, % reinforcement=5.01 Vol.%.

Acknowledgement

This research work was supported by Science and Engineering Research Board (SERB), Govt. of India (Grant No.: EEQ/2017/000382).

References

[1] Nicholls C J, Boswell B, Davies I J, Islam M N, "Review of machining metal matrix composites", International Journal of Advanced Manufacturing Technology, vol. 90(9-12), pp. 2429-2441, 2017.

[2] Michael Oluwatosin Bodunrin, Kenneth Kanayo Alaneme, Lesley Heath Chown, "Aluminium matrix hybrid composites: A review of reinforcement philosophies; mechanical, corrosion and tribological characteristics", Journal of Materials Research and Technology, vol. 4(4), pp. 434-445, 2015.

[3] Kaikai Wang, Xiaopei Li, Qiulin Li, Guogang Shu, Guoyi Tang, "Hot deformation behavior and microstructural evolution of particulate-reinforced AA6061/B4C composite during compression at elevated temperature", Materials Science and Engineering: A, vol. 696, pp. 248-256, 2017.

[4] Pagidi Madhukar, N Selvaraj, C S P Rao, G B Veeresh Kumar, "Tribological behavior of ultrasonic assisted double stir casted novel nano-composite material (AA7150-hBN) using Taguchi technique", Composites Part B: Engineering, Vol. 175, Article 107136, 2019.

[5] Ramanathan Arunachalam, Pradeep Kumar, Krishnan Rajaraman, Muraliraja, "A review on the production of metal matrix composites through stir casting – Furnace design, properties, challenges, and research opportunities", Journal of Manufacturing Processes, vol. 42, pp. 213-245, 2019.

[6] S B Supriya, S Srinivas, "Machinability studies on stainless steel by abrasive water jet - Review", Materials Today: Proceedings, Vol. 5(1), pp. 2871-2876, 2018.

[7] R Senthil Kumar, S Gajendran, R Kesavan, "Estimation of Optimal Process Parameters for Abrasive Water Jet Machining Of Marble Using Multi Response Techniques", Materials Today: Proceedings, Vol.5(5), pp. 11208-11218, 2018.

[8] Padmakar J Pawar, Umesh S Vidhate, Mangesh Y Khalkar, "Improving the quality characteristics of abrasive water jet machining of marble material using multi-objective artificial bee colony algorithm", Journal of Computational Design and Engineering, Vol. 5(3), pp. 319-328, 2018.

[9] A Gnanavelbabu, K Rajkumar, Saravanan P, "Investigation on the Cutting Quality Characteristics of Abrasive Waterjet Machining on AA6061-B4C-hBN Hybrid Metal Matrix Composites", Materials and Manufacturing Processes, vol. 33(12), pp. 1313-1323, 2018.

[10] Jagadish, Sumit Bhowmik, Amitava Ray, "Prediction and optimization of process parameters of green composites in AWJM process using response surface methodology", International Journal of Advanced Manufacturing Technology, vol. 87(5-8), pp. 1359-1370, 2016.

[11] M Manoj, G R Jinu, T Muthuramalingam, "Multi Response Optimization of AWJM Process Parameters on Machining TiB2 Particles Reinforced Al7075 Composite Using Taguchi-DEAR Methodology", Silicon, vol. 10(5), pp. 2287-2293, 2018.

[12] K Ravi Kumar, V S Sreebalaji, T Pridhar, "Characterization and optimization of Abrasive Water Jet Machining of aluminium/tungsten carbide composites", Measurement, vol. 117, pp. 57-66, 2018.

[13] M Santhana kumar, R Adalarasan, M Rajmohan, "Parameter design for cut surface characteristics in abrasive waterjet cutting of Al/SiC/Al2O3 composite using grey theory based RSM", Journal of Mechanical Science and Technology, vol. 30(1), pp. 371-379, 2016.

IOP Conf. Series: Materials Science and Engineering **764** (2020) 012012 doi:10.1088/1757-899X/764/1/012012

Thermal and nonlinear optical properties of semiorganic single crystal: L-Threonine p-nitrophenolate (LTPNP)

V M Mahadevan[1*], T Thangeeswari[2] and G Parthipan[2]
[1]Department of Physics, Adhiparasakthi Engineering College, Melmaruvathur - 603 319, India.

[2]Department of Physics, Vel Tech Multi Tech DR.Rangarajan DR.Sakunthala Engineering College, Chennai- 600 062, India.

*Corresponding author email: devanphysics@gmail.com

Abstract. The single crystal of L-Threonine 4-nitrophenolate developed by uncomplicated solution growth process. The grown crystal is analysed by using various characterization techniques. The structural analyses have been carried out by using Powder XRD method. Fourier Transform Infrared and Nuclear Magnetic Resonance spectra analysis confirms the functional groups exist in grown crystal. The thermal steadiness of L-Threonine 4-nitrophenolate crystals analysed by Thermo-gravimetric (TG) and Differential Thermal analyses (DTA). The sharp diffraction peaks in XRD pattern confirms the well crystalline nature of the synthesized sample. TG and DTA analysis prove that the synthesized sample possess good thermal constancy up to150°C. The UV-visible absorption studies show that the crystal is clearly visible in the whole visible region of the spectrum. Kurtz -Perry powder technique and Z-scan technique reveals the excellence of the synthesized samples for non linear optical (NLO) uses.

1. Introduction

Recent research trend move towards the growth of nonlinear optical crystals with emission of blue and green light for various uses like to store the data via optics, optical amplifiers, image processing and frequency conversion. In general the semi-organic NLO materials show large potential nonlinearities and give immediate response in electro-optic effect. Also semi-organic NLO possess the unique properties like very high nonlinearity, high electronic susceptibility and fast response [1]. The aromatic organic molecules having good optical nonlinearity property due to its unique structure such as acceptor and donor group bonded of the molecule. They are used in many industrial applications, such as optical switches and optical data storage [2]. Due to the demands in various technologies, many non linear optical (NLO) organic crystals have been introduced, like DAST (1-methyl-4-{2-[4-(dimethylamino) phenyl]ethenyl}pyridinium p-toluenesulfonate) and other grown crystals are reported in the previous research results [3-6]. The synthesized crystal L-arginine 4-nitrophenolate 4-nitrophenol dehydrates (LAPP); show the high second order nonlinear coefficient [7, 8]. In this research work, we look into the optical and thermal possessions of L-threonine p-nitro phenolate crystal which was grown from aqueous solution of L-therionine and 4-nitrophenol. The growth process, its crystal structure, thermal possessions and linear and nonlinear optical studies are reported.

2. Experimental Method

2.1. Fabrication of Crystal development Method

IOP Conf. Series: Materials Science and Engineering **764** (2020) 012012 doi:10.1088/1757-899X/764/1/012012

The compound of L-threonine p-nitrophenolate (LTPNP) was fabricated by taking equimolar ratio of L-Threonine and 4-nitrophenol in water as solvent. In a beaker, a calculated amount of L-Threonine is dissolved in double distilled de-ionized water using magnetic stirrer. Then, the amount of 4-nitrophenol is added slowly and stirred vigorously for three hours. To steer clear of impurities, the homogeneous solution is filtered and then transferred to beaker. Then the solution is kept at room temperature and allowed to evaporate slowly. After the evaporate process, three times re-crystallized material is used further to grow as single crystal. After 35 days, semitransparent yellow colored grown single crystal of dimension around 10×4×3 mm^3 in the solution were cautiously cropped. The grown crystals were tested by various techniques and found that they are L-Threonine 4-nitrophenolate single crystals. Figure 1a & figure 1b represents the pictures of grown single crystal LTPNP.

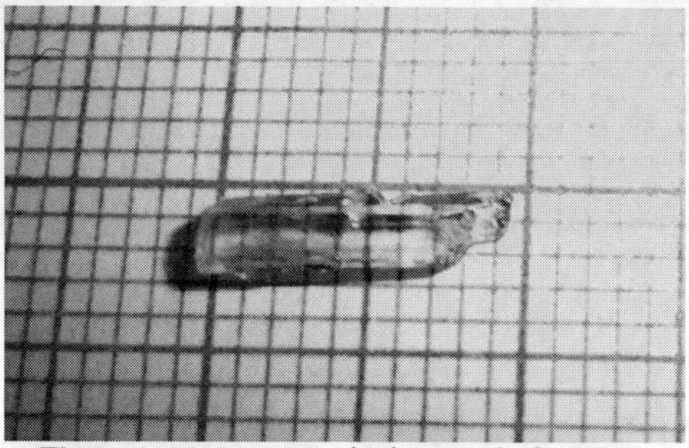

Figure 1. (a) As grown single crystal of LTPNP.

Figure 1. (b) Single crystal of LTPNP.

2.2 Characterization Techniques

The phase analysis of grown crystal L-Threonine 4-nitrophenolate was done by using X'PERT PRO diffraction. The Nuclear Magnetic Resonance spectrum of the grown crystal is evidenced from AMX NMR spectrometer. Fourier transform infrared spectroscopy spectrum of the L-Threonine 4-nitrophenolate was traced to confirm the existence of many functional groups in synthesized crystal. NMR spectrum of LTPNP material was recorded. The thermal characteristics of the crystal were approved by TGA/DTA. The optical property of the sample was characterized by UV-Vis-NIR Spectrophotometer and nonlinear optical studies performed by Kurtz and Perry powder method [11].

IOP Conf. Series: Materials Science and Engineering **764** (2020) 012012 doi:10.1088/1757-899X/764/1/012012

The Z-scan experiment was done by using a 532 nm neodymium yetrium aluminium garnet Laser, which was focused by a lens with focal length of 3.5-cm, Rayleigh length is 1.48 mm and laser beam waist at 15.84 μm. A 1 mm wide optical cell containing LTPNP in distilled water is translated across the focal region along the axial direction that is the direction of the propagation laser beam.

3. Results and Discussions

3.1 Phase Analysis Study

The phase analysis was carried out between 10° and 50° of diffraction angles with the wave length of 1.5460 Å. Synthesized powder sample of LTPNP was subjected to analysis with a monochromatic CuKα radiation (λ=1.54060 Å). The angle of 2θ and the scan step time in the range of 0.0289° and 30.6° seconds. The well defined diffraction peaks ensures well crystalline nature of the synthesized sample. The (hkl) parameters were intended and XR diffraction peaks were indexed from obtained lattice parameters and d spacing [9]. The figure 2 represents the phase analysis of LTPNP.

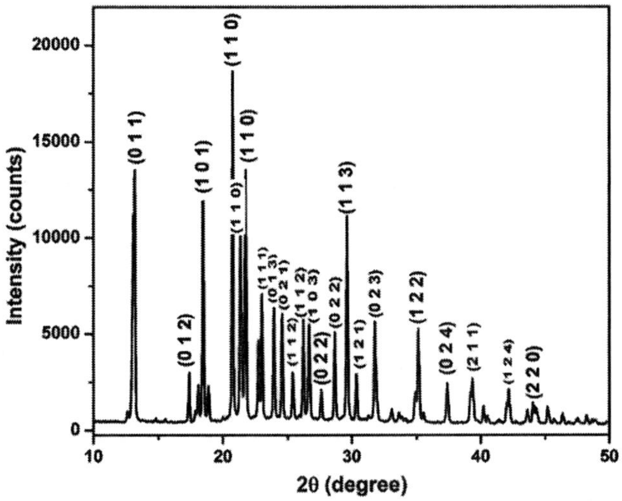

Figure 2. Powder X-ray diffraction pattern of LTPNP.

3.2 FTIR Spectral analysis

The Fourier infrared spectrum is performed to investigate the various elemental groups exist in the crystal. Figure 3 shows the FTIR of synthesized sample in the frequency range between 4000 and 480 cm^{-1} of. The existence of modes of vibrations makes the spectrum to be more complex [10]. Broad peak near 3400 cm^{-1} corresponds to the symmetric and asymmetric stretching vibrations of hydroxyl bond in water. The peak at 1687 cm-1 shows the H-OH of water molecule. The peak at 1112 cm^{-1} is due to C-O vibration of stretching mode. Also another vibration peak also observed around 856 cm^{-1}. The peak at 1310 cm^{-1} represents the NO_2 vibration and the asymmetric stretching vibration is observed at 1590 cm^{-1}. The peak at 500 cm^{-1} confirms the existence of hydrates. The vibration of the C-H is obtained at 1490 cm^{-1}. Based on obtained bands the functional groups in synthesized sample were deep-rooted.

IOP Conf. Series: Materials Science and Engineering **764** (2020) 012012 doi:10.1088/1757-899X/764/1/012012

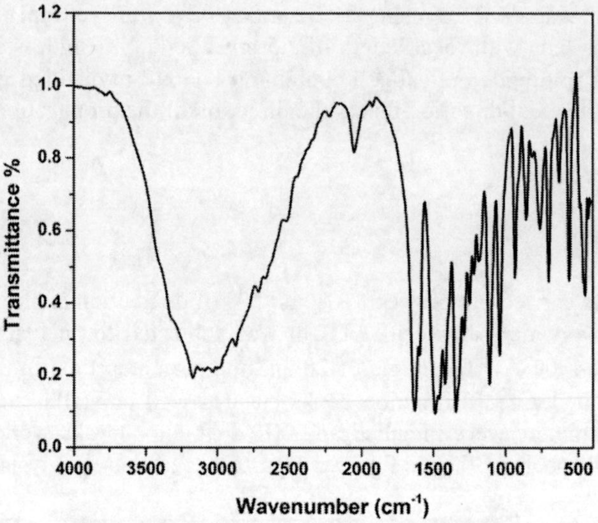

Figure 3. FTIR spectrum of LTPNP.

3.3 Proton NMR Spectral Study

Nuclear magnetic resonance (NMR) spectroscopy is an additional instrument to recognize the fabricated crystal. The LTPNP crystal is undergone to proton NMR spectral analyser. The various chemical elements gives their signal at different chemical shift (δ) position due to presence of hydrogen atoms.. The ortho hydrogen bond exits in the nitro phenol associated to parallel or in opposite and end to resonance. Hence the ortho hydrogens appear as distorted doublet and give their signals at 8.05 and 6.82 ppm [11- 13]. The presence of amine groups in the L-threonine molecules attributes multiple splitting in these signals. The CH and CH_2 groups in the samples produce the signals at different δ value according to their shielding nature. The Proton NMR spectral study has also been performed to corroborate the configuration of the crystal. The obtained spectrum is given in figure 4. Along with the peak at 4.18 ppm, two signals also noticed. The ortho hydrogen in the aromatic structure is associated to parallel or antiparallel and usually ends to resonance twofold. In this study we observed the outputs with multiple splitting at 6.8. This attributes to the ortho hydrogen of benzene ring and the signal at 6.9 represent the meta hydrogen in benzene ring [11].

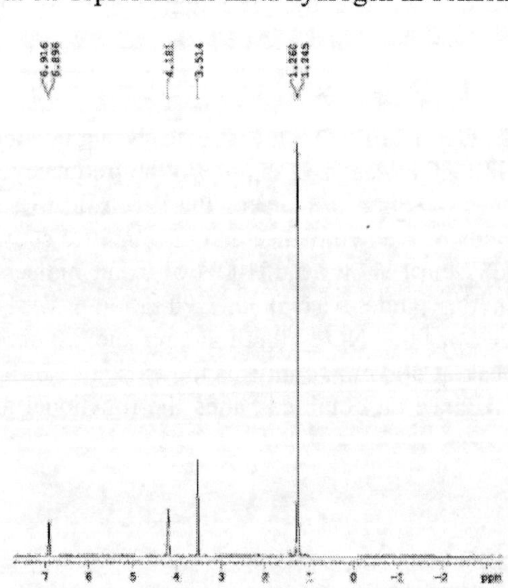

Figure 4. Proton NMR spectrum of LTPNP

IOP Conf. Series: Materials Science and Engineering **764** (2020) 012012 doi:10.1088/1757-899X/764/1/012012

3.4 Linear and Nonlinear Optical properties

The NLO crystal must possess good optical transparency in green visible region. Figure 5 represents the UV absorption spectra from the wavelength between 400 – 800 nm. It is observed that characteristic absorption band of LTPNP occurs at 435.2 nm. Also the crystal is clearly visible in the entire visible region of the spectrum. From the ultra violet absorption spectra, the energy band gap is obtained as 4.5eV.

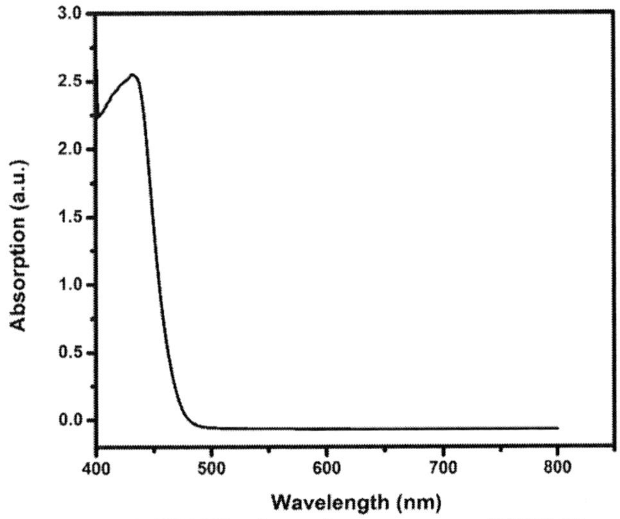

Figure 5. UV-Vis absorption spectrum of LTPNP.

The second order harmonic generation studies were carried out for the LTPNP crystal by Kurtz and Perry technique [12-14]. The second harmonic generation (SHG) effectiveness of material is two times more than that of standard reference KDP and four times greater than Urea materials. Hence this crystal is promisimg materials for various applications in electronics.

3.5 Third order nonlinear characterization

The Z-scan performance is used to establish nonlinear absorption coefficient β and nonlinear index of refraction n_2 [15]. Real part of the third-order susceptibility, [Re $\chi^{(3)}$] is proportional to the nonlinear index of refraction and the [Im $\chi^{(3)}$] is proportional to nonlinear absorption coefficient.

The Z-scan experiments were conducted by using neodymium Laser at 532 nm, which was focused using a 3.5 cm^{-1} focal length lens. A 1 mm wide optical cell containing LTPNP in distilled water is translated across the focal region along the axial direction which is the direction of the propagation of laser beam. Rayleigh length was found out to be 1.48 mm and since our path length is 1 mm, thin sample approximation holds good. The transmission of the beam was measured by an aperture placed in the far field using photo detector connected to the digital power meter (Field master GS-coherent). For an open aperture Z-scan, a lens is used to collect the entire transmitted beam through the sample. The Closed, open and ratio of the closed-to-open normalized Z-scan of LTPNP sample in distilled water at 63% transmittance is as given in figure 6. Both the refractive and absorptive nonlinearities of the sample have been studied by doing the Z-scan with and without aperture. The sign of negative refractive nonlinearity observed from the pattern of peak followed by a valley, i.e., self-defocusing. Due to the continuous exposure to laser, local dissimilarity in the refractive index with the temperature happens resulting in self-defocusing effect. The figure 6b, open aperture curve shows the material to exhibit saturable absorption [15]. The evaluated nonlinear parameters are shown in Table 1.

IOP Conf. Series: Materials Science and Engineering **764** (2020) 012012 doi:10.1088/1757-899X/764/1/012012

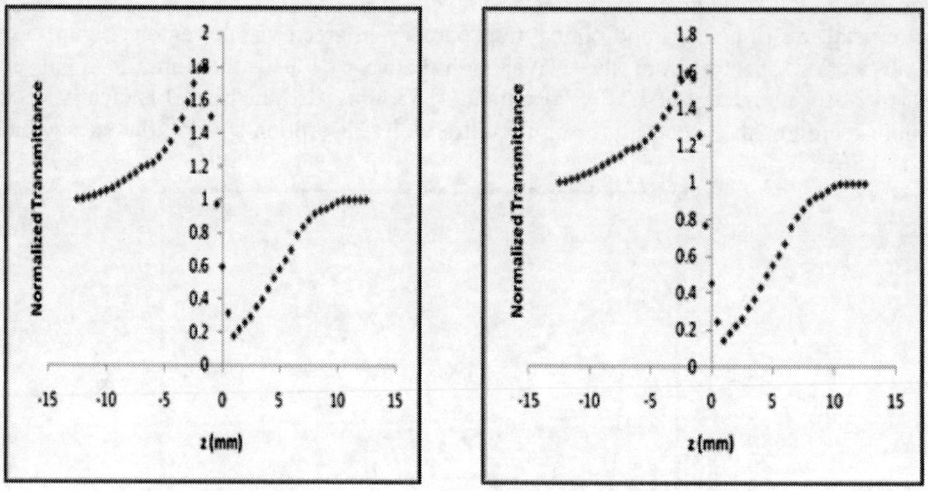

Figure 6. (a) Closed aperture; **(b)** Open aperture z-scan.

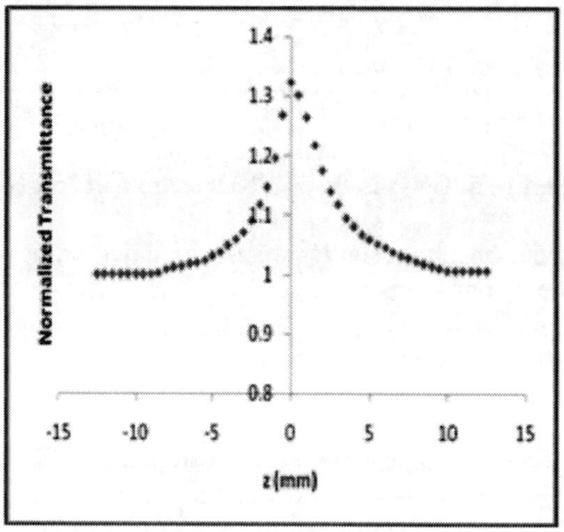

Figure 6. (c) Ratio of closed to open aperture z-scan.

Table 1. Non-linear parameters of LTPNP in distilled water.

n_2 x 10^{-8} cm^2/W	β x 10^{-4} cm/W	Re $\chi^{(3)}$x 10^{-6}esu	Im $\chi^{(3)}$ x 10^{-6}esu	$\chi^{(3)}$ x 10^{-6}esu
3.88	0.28	1.15	1.76	1.17

3.6 Thermal Analysis

To aware of the thermal performance such as thermal constancy and melting point of the material, thermal analysis was carried out. The thermal properties of LTPNP have been analyzed by obtaining TG/DTA curves as represent in figure 7. From the TG curve it is observed that less than 3 % of weight reduction bellows 150°C due to loss of physically absorbed moisture by the crystal surface. A careful observation in the DTA curve infers that there are no phase transitions below 150°C. The first transition occurs at 168.8°C due to the melting of the material and it is associated with the major weight loss of around 72 %. This study confirmed that the sample is steady till 150°C which agreed with the result of literature [9]. Hence, this material has good thermal constancy up to150°C and makes the requirement for the many electronic apparatus production.

IOP Conf. Series: Materials Science and Engineering **764** (2020) 012012 doi:10.1088/1757-899X/764/1/012012

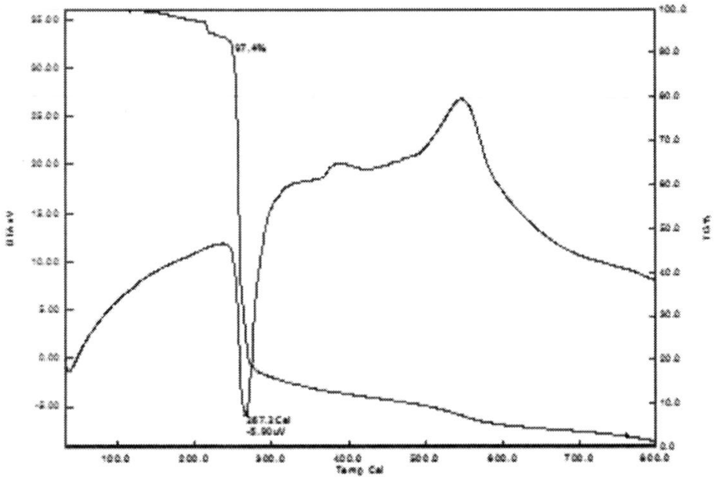

Figure 7. Thermo gravimetric (TG) and Differential Thermal Analysis (DTA) curves of LTPNP.

4. Conclusion

Semi-organic NLO crystal with high quality of L-Threonine para nitrophenolate (LTPNP) was successfully developed by slow evaporation method. The phase analysis of the LTPNP was performed. The functional groups present in the synthesized sample were confirmed by FTIR and NMR spectral analysis. Thermal stability of the L-Threonine para nitrophenolate crystal was analyzed by TG/DTA analysis and melting point of the synthesized sample was recognized to be 168 °C which shows that the materials can be effectively used in various device applications. The UV-Vis-NIR spectrum of the grown sample shows blue shift in the absorbance near visible region 435.2 nm and entire near NIR region. The second harmonic competence value of the LTPNP is two times elevated than that of referenced KDP and four times higher than value of Urea. The z-scan characterization results confirmed that the synthesized LTPNP crystal is suitable for various non linear optical applications.

References

[1] Vijayalakshmi A, Vidyavathy Balraj, Vinitha G, "Structure and characterization of a new organic crystal for optical limiting applications, isonicotinamide bis-p-aminobenzoic acid", Ukrainian Journal of. Physical Optics, Vol.17(3), pp.98-104, 2016.

[2] Chemla D S, Zyss J(Eds.) "Nonlinear Optical Properties of Organic Molecules and Crystals", Academic Press, New York, 1987.

[3] Jothi L, Vasuki G, Ramesh Babu R, Ramamurthi, K, "Synthesis, crystal growth and characterization of organic NLO material: 4-Bromo-4′-hydroxybenzylidene aniline", Optik, Vol.125(9), pp.2017-2021, 2014.

[4] Karuppasamy P, Muthu Senthil Pandian, Ramasamy P, Sunil Verma, "Crystal growth, structural, optical, thermal, mechanical, laser damage threshold and electrical properties of triphenylphosphine oxide 4-nitrophenol (TP4N) single crystals for nonlinear optical applications", Optical Materials, Vol.79, pp.152-171, 2018.

[5] Baumert J C, Twieg R J, Bjorklund G C, Logan J A, Dirk C W, "Crystal growth and characterization of 4-(N,N-dimethyl amino)3-acetamido nitro benzene, a new organic material for nonlinear optics", Applied Physics Letters, Vol.51, pp.1484-1487, 1987.

[6] Indira J, Prakash Karat P, Sarojini B K, "Growth, characterization and nonlinear optical property of chalcone derivative", Journal of Crystal Growth, Vol.242(1-2), pp.209-212, 2002.

[7] Vincent crasta, Ravindrachary V, Bhajantri R F, Richard Gonsalves, "Growth and characterization of an organic NLO crystal: 1-(4-methylphenyl)-3-(4-methoxyphenyl)-2-propen-1 one" Journal of Crystal Growth, Vol.267(1-2), pp.129-133, 2004.

[8] Wang L N, Wang X Q, Zhang G H, Liu X T, Sun Z H, Sun G H, Wang L, Yu W T, Xu D, "Single crystal growth, crystal structure and characterization of a novel crystal: l-arginine 4-nitrophenolate 4-nitrophenol dehydrate (LAPP)", Journal of Crystal Growth, Vol.327(1), pp.133-139, 2011.

[9] Mahadevan M, Ramachandran, P. Anandan, Arivanandhan M, Bhagavannarayana G, Hayakawa Y, "Crystal growth, structural, thermal and mechanical behavior of L-arginine 4-nitrophenolate 4-nitrophenol dihydrate (LAPP) single crystals", Spectrochimica Acta Part A: Molecular and Biomolecular Spectroscopy, Vol.133, pp.396-400, 2014.

[10] Dhanalakshmi B, Ponnusamy S, Muthamizhchelvan C, "Growth and characterization of a solution grown, new organic crystal: L-histidine-4-nitrophenolate 4-nitrophenol (LHPP)", Journal of Crystal Growth, Vol.313(1), pp.30-36, 2010.

[11] Subhashini V, Ponnusamy S, Muthamizhchelvan C, Dhanalakshmi B, "Growth and characterization of piperazinium 4-nitrophenolate monohydrate (PNP): A third order nonlinear optical material", Optical Materials, Vol.35(7), pp.1327-1334, 2013.

[12] Kurtz S K, Perry T T, "A Powder Technique for the Evaluation of Nonlinear Optical Materials", Journal of Applied Physics, Vol.39, 1968.

[13] Anandan P, Vetrivel S, Jayavel R, Vedhi C, Ravi G, Bhagavannarayana G, "Crystal growth, structural and photoluminescence studies of L-tyrosine hydrobromide semi organic single crystal", Journal of Physics and Chemistry of Solids, Vol.73(11), pp.1296-1302, 2012.

[14] Aravindan P, Srinivasan N. Vijayan R, Gopalakrishnan P, Ramasamy P "Investigations on the growth, optical behaviour and factor group of an NLO crystal: L-alanine alaninium nitrate", Crystal Research Technology, Vol.42(11), pp.1097-1103, 2007.

[15] Sheik-Bahae M, Said A A, Wei T, Hagan D J, Van Stryland E W, "Sensitive measurement of optical nonlinearities using a single beam", IEEE Journal of Quantum Electronics, Vol.26(4), pp.760-769, 1990.

IOP Conf. Series: Materials Science and Engineering **764** (2020) 012013 doi:10.1088/1757-899X/764/1/012013

Investigation on various construction material LECA, thermal insulated and conventional blocks

Nandhakumar R[1*], Parthipan G[2], Thangeeswari T[2], Karthikraja S[1] and Gangadurai B[1]
[1]Department of Civil Engineering, Vel Tech Multi Tech Dr.Rangarajan Dr.Sakunthala Engineering College, Chennai - 600 062 Tamil Nadu, India.
[2]Department of Physics, Vel Tech Multi Tech Dr.Rangarajan Dr.Sakunthala Engineering College, Chennai - 600 062 Tamil Nadu, India.

[*]Corresponding author email: nandhagopi.kumar@gmail.com

Abstract. Among the various construction materials, LECA (light weight expanded clay aggregate) is a versatile material due to its unique properties and is utilized in many applications. LECA blocks are highly impermeable and have high performance properties. Thermal insulated blocks are well suited for high rise buildings to withstand high temperature. It requires less steel and concrete for structural members due to its lower density. The LECA is mixed with grade M30 with 50% and 100% of replacement. The molded concrete blocks were tested to determine the strength and workability of the blocks under various experimental conditions and different time periods such as 7 days and 28 days. The tests such as compression test and water absorption test were conducted. From the experimental investigation, we will be able to conclude the best among these blocks based on their strength.

1. Introduction

The historical development of house facilities reveals that man has been modeling his environment throughout the ages for more comfortable living. Through producing and using inexpensive but powerful locally available building materials, these targets can be accomplished in part. It is important to pursue ways in developing countries to reduce construction costs, especially for low-budget housing, as well as to implement easy and efficient solutions for their repair and maintenance. Thermal block is a material's ability to absorb heat energy and store it. Changing the temperature of high-density materials such as cement, bricks and tiles requires a lot of heat energy. Thermal building blocks are a commercial solution for partition, floor and subversive constructions. Its unique properties are high thermal and noise insulation, good compressive strength, easy to handle and moisture resistance. Thermal building blocks recommended are first-rate thermal insulation which is inbuilt to get U-value targets easier in construction and diminish the carbon dioxide emissions from building. LECA is a unique construction material with light aggregates which is made by fired clay in rotary kiln at very high temperature. After the calcinations process, the organic compounds are destroyed by fire and forcing the bubbles to enlarge and form the product in honeycombed structure. The resulting ceramic pellets are lightweight, porous and highly resistant to crushing. Due to its weightless structure, it possesses an excellent thermal and sound insulation and fire-resistant stability [1]. Hence LECA material can be applied in the construction of under-ground areas, detachment walls inside the room and ceilings. In this research work, the aim is to produce an alternate for the

IOP Conf. Series: Materials Science and Engineering **764** (2020) 012013 doi:10.1088/1757-899X/764/1/012013

conventional block, in terms of strength, low cost, weight of the blocks and hence to promote sustainable development in the field of construction. Conventional blocks can be used only in ordinary environment and it will definitely fail in the point of excessive loading and also handling energy is more. Also we studied about the behaviour of solid blocks and compressive strength of solid blocks.

2. Experimental Work

2.1. Insulation Material

The insulation material used in thermal blocks is polystyrene and its physical properties are mentioned in the table 1 below. Polystyrene material used in this study is shown in figure 1 (a-b) below.

Table 1. Physical properties of polystyrene.

S. No	Physical Property	Value
1	Specific Gravity	1.04
2	Flexural Strength	83 MPa (12000 PSI)
3	Tensile Strength	53 MPa (7700 PSI)
4	Melting Temperature	210-249 °C (410-480 °F)
5	Solubility in water	Insoluble

Figure 1 (a-b). Insulation material.

2.2. LECA Material

The physical properties of LECA [1, 5-7] are mentioned in the below table 2. The figure 2 shows the sample LECA material used in this study.

Table 2. Physical properties of LECA.

S.No	Physical Properties	Value
1	Specific gravity	0.56
2	Water absorption	18%
3	Impact value	49.68

Figure 2. LECA material.

2.3. General Information

- Coarse aggregate: Light weight expanded clay aggregate
- Fine aggregate: M-sand
- Insulation material: Polystyrene
- Size of mould: 200x200x400mm
- Size of the insulation material: 150x80x300mm
- Grade of concrete used: M30
- Mix ratio: 1:1.9:1.7
- Water cement ratio=0.45 (IS 456 TABLE-5)

2.4. Preparation of Materials

The lightweight construction blocks are prepared by proper relative amount of the concrete mix with LECA. The amount of aggregate proportion is changes from various nature and place of work. For construction within room in which the place not affected by weather, the 100% lightweight proportions is adequate. Similarly, for construction in outdoor projects, medium mix is needed. But important point to be noticed is, the lighter the concrete, the weaker the block [8-11]. Figure 3 and figure 4 depicts the mould and m-sand used in the study respectively.

Figure 3. Mould used in the study (200x200x400mm).

Figure 4. M-Sand used in the study.

2.5. Testing Process

Experimental test is carried out on the blocks to determine its various parameters which are governing the strength of solid blocks. Test procedure is based on the instruction provided by the standard codes. The compressive strength and water absorption test are performed. The number of sampling required for each test is taken. Average of sampling is taken the results of each test. The compressive strength is determined using compressive testing machine, 3.5N/s to 4N/s rate of loading is applied while applying load to the blocks [12-16].

3. Results and Discussion

The table 3, represents the compressive strength of various building blocks tested as per IS 516-1959 standard on day 7. It indicates that due to the partial addition of LECA material in conventional material block gives better results than other building blocks.

Table 3. Compressive strength on 7th day of solid block.

S. No	Name of the Block	Weight of Blocks (kg)	Area of Blocks (mm²)	Load on Testing (kN)	Stress (N/mm²)
1	LECA (Partially Replaced)	30.13	80000	1107	13.83
2	LECA (Fully Replaced)	26.49	80000	1069	13.36
3	Thermal Insulated Block	28.31	80000	490	6.125
4	Conventional Block	32.4	80000	739	9.26

In the below table 4, compressive strength of various kinds of building blocks tested as per IS 516-1959 standard on day 28. It is clear that the addition of LECA material in conventional material block gives better results.

IOP Conf. Series: Materials Science and Engineering **764** (2020) 012013 doi:10.1088/1757-899X/764/1/012013

Table 4. Compressive strength on 28th day of solid block.

S. No	Name of the Block	Weight of Blocks (kg)	Area of Blocks (mm^2)	Load on Testing (kN)	Stress (N/mm^2)
1	LECA (Partially Replaced)	30.3	80000	2067	25.81
2	LECA (Fully Replaced)	26.72	80000	1855	23.18
3	Thermal Insulated Block	28.53	80000	1072	13.4
4	Conventional Block	33.54	80000	1020	12.75

Comparison of the compressive strength of various building blocks which were tested as per IS 516- 1959 standard on day 7 and 28 were shown on the table 5 below. It indicates that the partial addition of LECA material in block gives better results than fully replaced LECA material blocks and thermal insulated blocks. Figure 5 and figure 6 shows the images of the blocks loaded for compression testing and images of the blocks after testing respectively.

Table 5. Comparison of the compressive strength at different days.

S. No	Name of the block	Load on testing - 7 days (KN)	Load on testing - 28 days (KN)
1	LECA (Partially Replaced)	1107	2067
2	LECA (Fully Replaced)	1069	1855
3	Thermal Insulated Block	490	1072
4	Conventional Block	739	1020

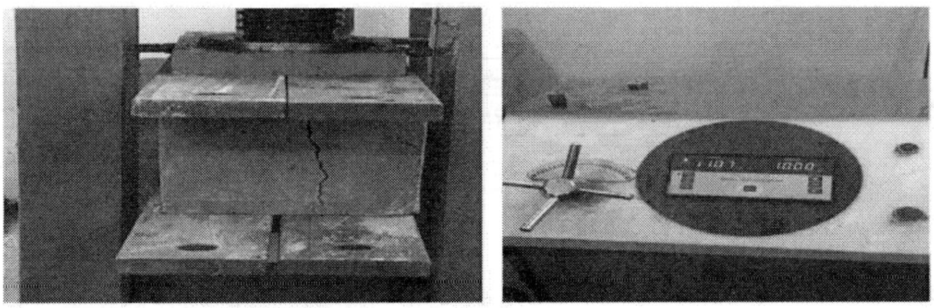

Figure 5. Compression test on blocks.

Figure 6. Blocks after testing.

The table 6, represents water absorption test results of different building blocks in terms of percentage. The water absorption tests were conducted as per IS 1124-1974 standard. The images of the water absorption tests were shown in the below figure 7.

Table 6. Water absorption test on 28th day of solid block.

S. No	Name of the Block	Dry weight of the blocks (kg)	Wet weight of the blocks (kg)	Water Absorption (%)
1	LECA (Partially Replaced)	30.3	39	28.71
2	LECA (Fully Replaced)	26.72	35.95	35.03
3	Thermal Insulated Block	28.18	34.37	22.15
4	Conventional Block	33.54	45.75	36.4

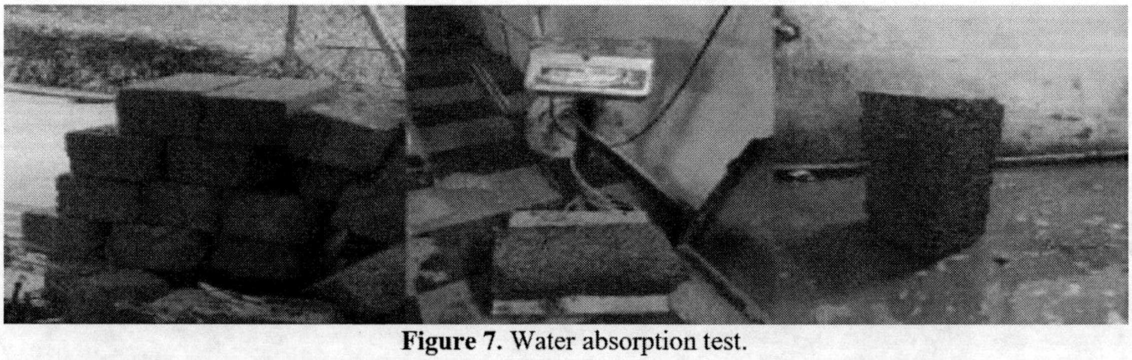

Figure 7. Water absorption test.

4. Results and Discussion

The presented research results show that the compressive strength of partially replaced LECA is 13.83 N/mm^2 and for fully replaced LECA is 13.36 N/mm^2 at seventh day. Similarly, the compressive strength of partially replaced LECA is 25.81 N/mm^2 and for fully replaced LECA is 23.18 N/mm^2 at twenty eighth day. On other hand, thFe obtained results shows the enhancement in compressive strength for partially replaced LECA than thermal insulated blocks (125% and 92.82% higher than thermal insulated blocks at day 7 and day 28. The water absorption percentage of partially replaced and fully replaced LECA is less than conventional blocks (26% and 3.9% lesser at day 28) and less than 64% compared with thermal insulated blocks. From this we can conclude that the reduction in the weight of concrete will reduce the dead weight of concrete. Hence, the construction cost gets decreased by reducing the amount pay out for construction of buildings.

References

[1] R.N. Raj Prakash, A.Krishnamoorthi, "Experimental Study On Light Weight Concrete Using Leca", International Journal of chem. Tech research, Vol.10(8), 98-109, 2017.

[2] Axel Berge, Pär Johansson, "High Performance Thermal Insulation", Chalmers University of Technology, 2012.

[3] Marta Cianfrini, Roberto de Lieto Vollaro and Emanuele Habib, "Dynamic Thermal Features of Insulated Blocks: Actual Behavior and Myths", Energies, Vol.10(11), 1807, 2017.

[4] Marta Cianfrini, Roberto de LietoVollaro and EmanueleHabib, "Thermal Inertia of hollow blocks: Actual Behavior and Myths", Proceedings of CISBAT 2015 International Conference on Future Buildings and Districts - Sustainability from Nano to Urban Scale, pp.149-154, 2015.

[5] R. Vijayalakshmi and S. Ramanagopal., "Structural Concrete using Expanded Clay Aggregate: A Review", Indian Journal of Science and Technology, Vol.11(16), pp.1-12, 2018.

[6] Osama A. B. Hassan, Petra Jonsson, "Lecablock, an Alternative Construction Material for the Exterior Walls of Passive House", Journal of Building Construction and Planning Research, Vol.2(1), pp.96-108, 2014.

[7] V Geza, A Jakovics, S Gendelis, I Usiļonoks and J Timofejevs, "Thermal conductivity of disperse insulation materials and their mixtures", IOP Conference Series: Materials Science and Engineering, 251, 012012, 2017.

[8] Chan-Ki Jeon, Jae-Seong Lee, Hoon Chung, Ju-Ho Kim, and Jong-Pil Park, "A Study on Insulation Characteristics of Glass Wool and Mineral Wool Coated with a Polysiloxane Agent", Advances in Materials Science and Engineering, vol. 2017, Article ID 3938965, 2017.

[9] Hendry, A. W., Fairbairn, D. R., & Khalaf, F. M., "Mechanical properties of materials used in concrete blockwork construction", Magazine of Concrete Research, Vol.44(158), pp.1-14, 1992.

[10] Khalaf, F. M., "Factors influencing compressive strength of concrete masonry prisms", Magazine of Concrete Research, Vol.48(175), pp.95-101, 2015.

[11] Köksal, H. O., Karakoç, C., & Yildirim, H., "Compression Behavior and Failure Mechanisms of Concrete Masonry Prisms", Journal of Materials in Civil Engineering, Vol.17(1), pp.107-115, 2005.

[12] Mohamad, G., Lourenço, P. B., & Roman, H. R., "Mechanics of hollow concrete block masonry prisms under compression: Review and prospects", Cement and Concrete Composites, Vol.29(3), pp.181-192, 2007.

[13] Steadman, M., Drysdale, R. G., & Khattab, M. M., "Influence of Block Geometry and Grout Type on Compressive Strength of Block Masonry", Proceedings of the 7th Canadinan Masonry Symposium, pp. 1116–1127, 1995.

[14] W. Sai Deepak, G. Tirupathi Naidu, "Effect On Compressive Strength Of Concrete Using M Sand And Foundry Sand As A Partial Replacement For Fine Aggregate", International Journal of Research in Engineering and Technology, Vol4(6), pp.180-183, 2015.

[15] D. A. R. Dosage, M. G. S. Dias and C. T. Ariyawansa, "Offshore Sand As A Fine Aggregate for Concrete Production", British Journal of Applied Science & Technology, Vol.3(4), pp.813-825, 2013.

[16] Robert G.Varady, Katherine B.Hankins, Andrea Kaus, Emily Young, Robert Merideth, "to the Sea of Cortés: nature, water, culture, and livelihood in the Lower Colorado River basin and delta—an overview of issues, policies, and approaches to environmental restoration", Journal of Arid Environments, Vol.49(1), pp.195-209, 2001.

IOP Conf. Series: Materials Science and Engineering **764** (2020) 012014 doi:10.1088/1757-899X/764/1/012014

Water flow controller for precision agriculture

Haritha J*, Navina B and Prakash K
Department of Electronics and Instrumentation Engineering, Bannari Amman Institute of Technology, Coimbatore-638 401, India.
*Author for correspondence E-mail: haritha.jeganathann@gmail.com

Abstract. In recent years, water crisis in Asia is reaching an alarming proportion that might very soon attain the worldwide global crisis and there is an utmost need to save water. The water crisis occurs due to the increase of population, poor water harvesting, and water wastage in the manufacturing industries, process industries and agriculture sector. Therefore, each sector requires a mechanism to reduce the water wastage and save it for future needs. In this paper, an efficient irrigation system is designed to reduce the water wastage in agriculture fields. Due to the farmer's irregularity and negligence, they leave the motor turned ON for several hours which will lead to wastage of water and more power consumption. Similarly, they forget to switch ON the motor, which also leads to the damage of crops due to insufficient water supply. The proposed system controls the flow of water in the fields in accordance with the intensity of sunlight and moisture content present in the soil. This system uses the microcontroller (Arduino UNO) assisted with LDR (Light Dependent Resistor), capacitive type moisture sensor and DHT 11 (temperature and humidity sensor). Based on the sunlight intensity and moisture content present in the soil, the Arduino UNO is programmed to turn ON/OFF solenoid valve connected to the pipeline that controls the flow of water in the agricultural fields. Thus the proposed system can operate automatically and supply water to the fields whenever needed without any human intrusion. Further, a Wi-Fi module ESP8266 is connected such that the solenoid valve can be operated manually through mobile phones/laptops based on the measured parameters if required.

1. Introduction

The rapid growth of industries, residential areas and agriculture in vast scope leads to increase in water supply facility. In certain areas water supply is required continuously or at regular intervals. Water irrigation to crops is becoming quite difficult nowadays for the agriculturalists due to irregularity of water supply and their negligence as they leave the motor turned ON for hours together which will ultimately lead to water wastage. Similarly, they forget to switch ON the motor when required which again leads to the damage of the crops. To overcome the issue of water wastage in agricultural field, an alternate solution for irrigation has to be developed. In this paper, the water flow in the field is controlled using a solenoid valve that is operated based on the moisture content in the soil and light intensity of the environment measured using soil moisture sensor and LDR circuit respectively through the Arduino UNO microcontroller. The Wi-Fi module ESP8266 is installed with the microcontroller so that the entire system can be monitored as well as controlled using mobile phones/laptops from elsewhere. As a result, water wastage in the fields can be minimized without human intervention and it is done with more accuracy and precision. The proposed irrigation system is cost efficient and does not require any specific knowledge to operate it.

2. Literature survey

IOP Conf. Series: Materials Science and Engineering **764** (2020) 012014 doi:10.1088/1757-899X/764/1/012014

Zhao Liqiang et al. [1] introduced a cloud based irrigation system that controls the flow of water in the field by measuring the soil parameters like moisture and temperature. At present, water and electricity are in high demand. Pavithra D et al. [2] used an embedded controller to control the water flow and used solar energy to operate the motor in the field to conserve power and water.

Robert W et al. [3] worked on remote sensor systems with integrated control that has been led in scholarly research. To advance economically reachable remote detecting and control systems, valve control equipment and programming were created to be perfect with a business remote sensor hub. The effort was led as a team with a remote system seller to such an extent that the examination result and the item itself could eventually be accessible to producers. The valve control organize was assessed to have a restitution time of around 3.5 4.5 years, however some vineyard cultivators professed to see compensation in under 3 years.

Chandankumarsahu et al. [4] introduced wireless sensor network in irrigation. The primary work of wireless sensor network is to execute two sorts of hubs and building sensor. The equipment stage is comprised by information process unit, radio module, sensor control grid, information stockpiling streak, control supply unit, simple interfaces and broadened computerized interfaces. The objective of a scattering convention is to reliably convey a bit of control and synchronization directions to each hub in the system. The trial results demonstrates that wireless sensor network have wider applications in the agriculture.

Singh et al. [5] have proved that IoT can be used in irrigation to control the water flow in the field by measuring the required parameters in the field even though the farmer is away from the field. G. Eragamreddy et al. [6] developed an automated irrigation system. In this system, the water flow is controlled using the moisture content in the soil. Further the status of the water flow is also indicated via a website.

R. N. Rao et al. [7] introduced a low cost irrigation system, wherein the water flow is controlled automatically and indicated to the user via a text message to the mobile phone and also to the users Gmail account. E. Avar et al. [8] developed an automated irrigation system interfaced with a GSM module and an idea to implement the android application in the mobile phones to monitor and control the status of water flow in the field was developed.

3. Block diagram

The block diagram of the experiment setup is shown in figure 1. It consists of a capacitive type moisture sensor, a Light Dependent Resistor (LDR), temperature and humidity sensor (DHT11), solenoid valve, relay to operate the solenoid valve, Arduino, ESP8266 Wifi module and mobile/laptop to monitor the values.

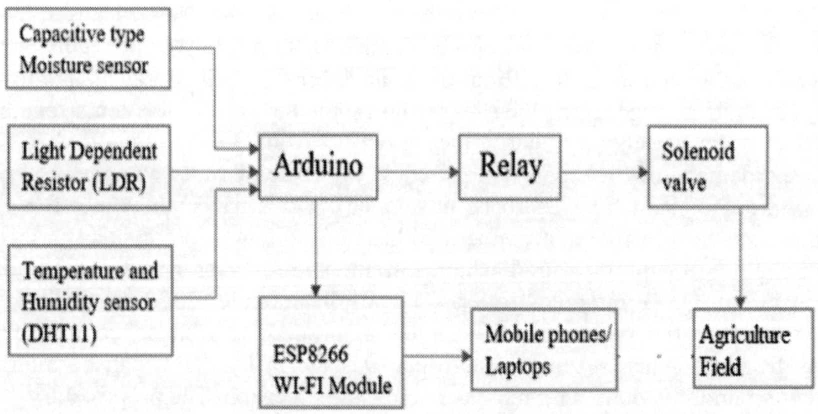

Figure 1. Block diagram of experimental setup.

3.1. Solenoid valve

A solenoid valve is an electro-mechanical device which uses electric current to produce magnetic field. It differs in the mechanism of fluid they control and the characteristic of generated magnetic field strength, the electric current they use. The functions of solenoid valve are to open, close, distribute and to mix fluids. In fluid flows, the mostly used control element is solenoid valve. Solenoid valve is used in the industries, domestic washing machines and dishwashers to control the water from entering into the machines. It is used in automatic irrigation sprinkler systems. It is also used in paintball gun triggers to actuate Carbon dioxide hammer valve.

3.2. Arduino UNO

It is an open source microcontroller based on Atmega328P microcontroller. It consists of analog and digital input/output pins which can be used for interfacing various kinds of sensors and may be interfaced to other boards. It can be used to interface with huge variety of motors and other actuators for different applications. The board has 14 Digital pins, 6 Analog pins.

The Arduino is programmed using the software ARDUINO IDE via type B USB cable. It accepts voltages between 7 and 20 volts. It can be powered by a USB cable or by an external 9-volt battery. The general pins are LED, 5V, VIN, 3.3V, GND and IOREF. The special pins are Serial, PWM, SPI, TWI, AREF and External interrupts. To design an automatic irrigation system using solenoid valve the following components are used viz., Solenoid valve, Arduino UNO, ESP8266 Wi-Fi module, Light Dependent Resistor (LDR), capacitive type moisture sensor, DHT 11 sensor, 5V Relay, Regulated Power Supply (RPS),10 K Ohm Resistor, Breadboard, Jumpers.

The circuit shown in figure 2 describes the connections which are to be done with the solenoid valve and relay. In this circuit, the output voltage from the Arduino is amplified by the amplifier circuit to operate the relay connected to the solenoid valve to perform ON/OFF operations.

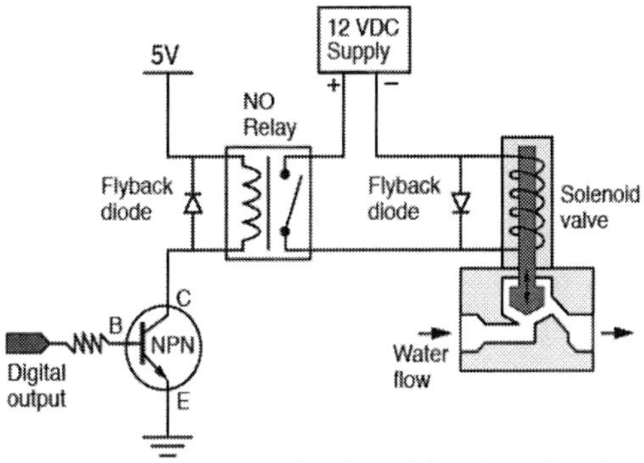

Figure 2. Relay and solenoid valve circuit.

4. Results and Discussion

The proposed system is implemented using the Arduino UNO board. The relay and solenoid valve are controlled automatically and cab be controlled manually using the Blynk Application on an android mobile phone. The android phone is connected to the board using the ESP8266 Wi-Fi module. The four parameters namely moisture, temperature, humidity and intensity is monitored and controlled using the ON/OFF switch button displayed on the Blynk Application. The sensor results viz., humidity, temperature, moisture, sunlight intensity and the status of the motor is linked with the Blynk app shown in figure 3 is the reading taken from the field model created in an area of about 30cm×30cm. When the moisture is above 250% the motor is turned OFF automatically or manually through the Blynk Application. Also the device is programmed such that if the light intensity is above 25% the motor will be still ON until the moisture reaches 300% to compensate the effect of sunlight.

IOP Conf. Series: Materials Science and Engineering **764** (2020) 012014 doi:10.1088/1757-899X/764/1/012014

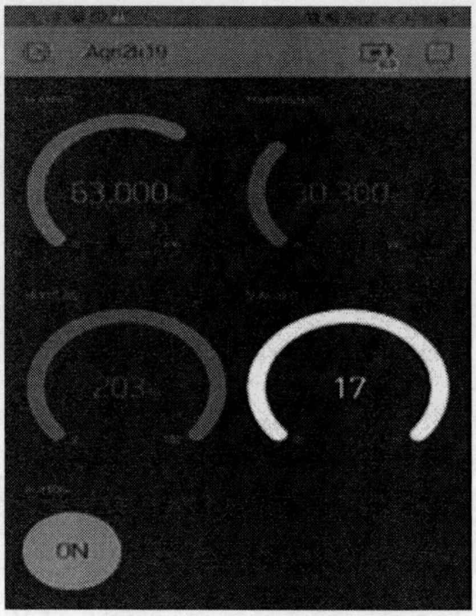

Figure 3. Output using Blynk app.

5. Conclusion and future scopes

In this project, wireless control of irrigation system is done for precision agriculture. The major application of this project is for the farmers, who want to water their crops during their absence in the field. It was found that this system is cost effective and feasible. It can be used in the areas where there is water scarcity thereby improving the water sustainability. In future, this prototype can be extended to various purposes such as feeding the animals without any human intrusion from anywhere.

References

[1] Zhao Liqiang, Yin Shouyi, Liu Leibo, Zhang Zhen, and Wei Shaojun, "A crop monitoring system based on wireless sensor network", Procedia Environmental Sciences, Vol.11, Part B, pp.558-565, 2011.

[2] Robert W. Coates, Michael J. Delwiche, Alan Broad, and Mark Holler, "Wireless sensor network with irrigation valve control", Computers and Electronics in Agriculture, Vol.96, pp.13-22, 2013.

[3] Pavithra D. S, M. S .Srinath, "GSM based Automatic Irrigation Control System for Efficient Use of Resources and Crop Planning by Using an Android Mobile", IOSR Journal of Mechanical and Civil Engineering, Vol.11(4),pp.49-55, 2014.

[4] Chandan kumar sahu, Pramitee Behera, "A Low Cost Smart Irrigation Control System", in 2nd International conference on electronics and communication system (ICECS), Coimbatore, pp.1146-1152, 2015.

[5] Singh, Pushkar, Saikia, Sanghamitra, "Arduino-based smart irrigation using water flow sensor, soil moisture sensor, temperature sensor and ESP8266 WiFi module", in 2016 IEEE Region 10 Humanitarian Technology Conference (R10-HTC), Agra, pp.1-4, 2016.

[6] G. Eragamreddy and K. R. Sree, "Solar powered auto watering system for irrigation using embedded controller", in 017 International Conference on Energy, Communication, Data Analytics and Soft Computing (ICECDS), Chennai, pp.2424-2428, 2017.

[7] R. N. Rao and B. Sridhar, "IoT based smart crop-field monitoring and automation irrigation system", in 2nd International Conference on Inventive Systems and Control (ICISC), Coimbatore, pp.478-483, 2018.

IOP Conf. Series: Materials Science and Engineering **764** (2020) 012014 doi:10.1088/1757-899X/764/1/012014

[8] E. Avar, K. Bulu, M. A. Sarida, and B. Kapur, "Development of a cloud-based automatic irrigation system: A case study on strawberry cultivation", in 7th International Conference on Modern Circuits and Systems Technologies (MOCAST), Thessaloniki, pp.1-4, 2018.

IOP Conf. Series: Materials Science and Engineering **764** (2020) 012015 doi:10.1088/1757-899X/764/1/012015

An F-PROMETHEE technique for analysing the risk factors in green manufacturing

R K A Bhalaji, S Bathrinath* and S Saravanasankar
Department of Mechanical Engineering, Kalasalingam Academy of Research and Education, Krishnankoil-626126, Tamilnadu, India.

*Corresponding author email: bathri@gmail.com

Abstract. In the contemporary globalization, green manufacturing have achieved the more significance. Recently, many firms know about the importance of green because of the environmental and social concerns involved. Green manufacturing implementation is vital one for environment as well as improving the business. The goal of the green manufacturing is to protect the environment, prevent the pollution and also for waste reduction. But the adoption of green manufacturing is challenging for industrial managers because of the several risk factors involved in the firm. For the green manufacturing implementation, this paper aims to recognize and analyse the risk factors of green manufacturing and also to identify the critical alternative in Indian medical equipment manufacturing firm by using MCDM (Multi-Criteria decision making method) like F- PROMETHEE (Fuzzy Preference Ranking Organization method for Enrichment Evaluations). The outcomes of the paper will be surely useful for industrial managers for the implementation of green manufacturing in their firms.

1. Introduction

Pollution from environment, resource allocation and growth of population are the three key issues in global world environment. Environmental pollution plays a major role for affecting both industrial and society development and also in manufacturing firms. Therefore, all the firms are looking for reducing the environmental impact. ISO has suggested some international standards for the environmental management system as well as management system for quality products. The highlight of the paper is to lessen the environment impact. They need to implement the process of Green manufacturing with sustainable development. For future point of view, they consider the characteristics of sustainable consumption and production [1]. For the past decade, manufacturers should manufacture the product with consumption of resource and also low environmental impact during the process such as quality, transport, packaging, production and design. Manufacturers who manufacturing the product with mass production, they should focus only on the green manufacturing. Lot of waste and energy are used on the routine basis, it affects the health and well-being of employees. Development of firm is affected by several risk factors in the firm. This paper mainly focuses on increasing sustainable energy, waste reduction and environmental pollution. To solve this issue, we analysed the risk factors of green manufacturing in medical equipment manufacturing firms and to identify the critical alternative by using MCDM methods like Fuzzy PROMETHEE. This paper outcome will definitely assist the industrial managers for the implementation of green manufacturing. Green manufacturing implementation is not only for the environment and also for increasing the business profit [2]. In the process of manufacturing, material intensiveness and reducing energy is mainly achieved by green manufacturing implementation.

The remainder section of the paper is discussed below. Section 2 represents relevant literature about risk factors in the process of green manufacturing and gap of literature. Flow chart of the paper is demonstrated in Section 3. Section 4 defines the procedure of Fuzzy PROMETHEE method. Case study of the paper is described in Section 5. Conclusion and scope for future research is discussed in Section 6.

2. Relevant Literature

This paper has classified the relevant literature into two sub-segments namely a) Risk factors involved in green manufacturing b) Literature gap

2.1. Risk factors involved in green manufacturing

Paul et al. [3] conducted a review in manufacturing industry related to the significance of green manufacturing, techniques and its application. The findings of the paper showed that implementation of green manufacturing are essential for reducing the waste from products and prevent the pollution from environment. Digalwar et al. [4] evaluated the performance measures for execution of green manufacturing and also case study is conducted in nine different manufacturing firms. This study gives useful tips to improve the business as well as achieving the practices of green manufacturing. Deif [5] created a framework for green manufacturing for the efficiency of production in the department of paint in wood products manufacturing firm. The outcomes of the paper is mainly for accomplishing the green manufacturing implementation as well as level of greenness in the paint dept. Govindan et al. [6] assessed the green manufacturing risk factors and also identify the most influential risks in the Indian industry by using Fuzzy Analytic Hierarchy Process. The consequence depicts that regulation compliance is the most influential factor in the industry and they need to implement some set of standards related to environment. Liu et al. [7] created a replica for green manufacturing risks in various projects and also performed a case study in refrigerator manufacturing firm by using MADM (multi-attribute decision making) with fuzzy approach. This paper shows project 3 has the critical green manufacturing risks and this model is useful for partial data and different elements in decision making of green manufacturing. Mittal et al. [8] evaluated the green manufacturing barriers and drivers in both emerging and developing nations like Germany and India and also to compare the both drivers and barriers. For the successful green manufacturing implementation, this paper surely helps the managers and decision makers from industries in both nations. Sangwan [9] analysed the green manufacturing benefits of both qualitative and quantitative in Indian enterprises by using SPSS software. The outcomes of the paper identified that most important benefits are reducing treatment, handling and disposal cost as well as packaging and transport cost. Seth et al. [10] recognized the most influencing performance measures and risk factors of green manufacturing in cement industry. Commitment from top management, culture in organizations, supply chain management and human resources are the key factors in implementation of green manufacturing.

2.2. Literature gap

As of now, there are only limited papers related to risk factors of green manufacturing by using MCDM methods. But there are no papers related to risk factors of green manufacturing in Indian medical equipment manufacturing firms by using Fuzzy PROMETHEE method. In order to fulfil this gap, we analysed the risk factors in this paper. Table 1 shows the risk factors in green manufacturing process.

Table 1. Risk factors in green manufacturing.

S. No.	Risk Factors	Notation
1	Requirements in supply chain	F1
2	Demand by employees & customers	F2
3	Benefits in Finance	F3
4	Status of the firm	F4
5	Shareholders	F5
6	Environmental conservations	F6
7	Organizational resources	F7
8	Consumers	F8
9	Challengers	F9
10	Commitment from top management	F10

3. Flow chart of the paper

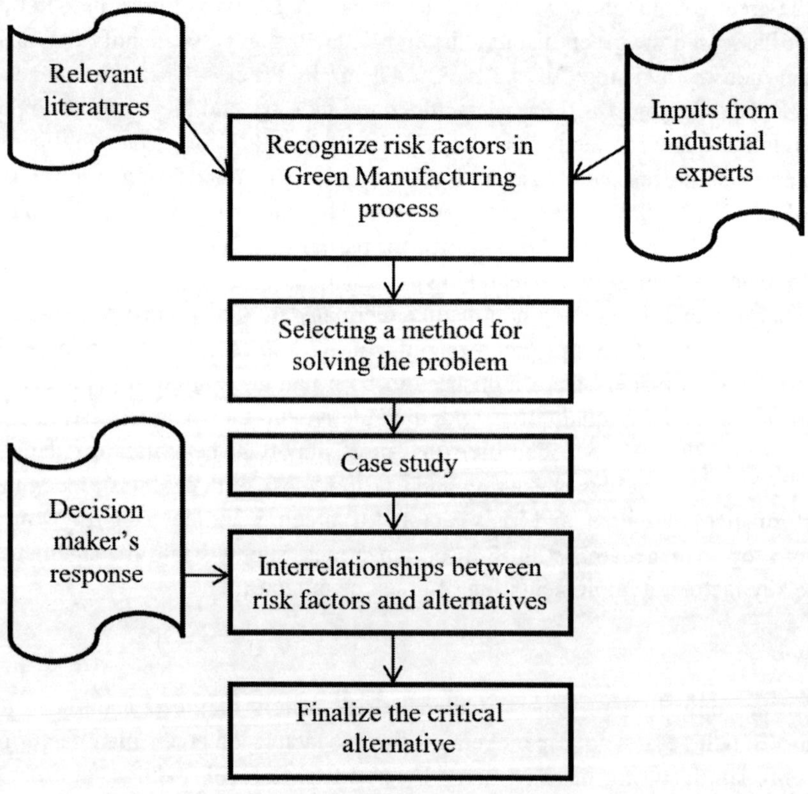

Figure 1. Flowchart of the paper.

The above figure 1 shows the flowchart of this paper work for better clarity and understanding.

4. Fuzzy PROMETHEE method

Brans et al. [11] was first initiated the PROMETHEE method for ranking the number of alternatives. PROMETHEE method is used for understanding the concept very easily and it has more applications in routine life and also it is technique of outranking. Ulengin et al. [12] said that both incomplete and final rankings are provided by PROMETHEE I and II. For processing the fuzzy data in problems of decision making, this method has lack of ability. Fuzzy set theory was initiated to manage the ambiguity of individual decision. To develop the problem with both solution and formulation by fuzzy set theory, which are complicated for the analysis by traditional methods. For this purpose, we integrate both PROMETHEE method and fuzzy set theory and it is termed as Fuzzy PROMETHEE method and this method was first initiated by [13] and they implemented in the steel industry. Many researchers recently used Fuzzy PROMETHEE method for solving the problems [14-16]. The procedure for fuzzy PROMETHEE method is discussed below:

Phase 1: In the first step, to find the decision makers, risk factors and alternatives and they are denoted as m, f and a.

Phase 2: Describe the related fuzzy no's and linguistic values
This paper used the five point linguistic scales suggested by Chen et al., (2011) and they are stated in TFN (Triangular fuzzy no's) and to evaluate the significance weight of risk factors.
Phase 3: Collect the opinion of decision makers
For every risk factor, the average weight is

$$We_i = \frac{1}{m}\left[\sum_{f=1}^{m} We_i^f\right] = \frac{1}{n}\left[We_j^1 + We_j^2 + \dots We_j^n\right] \tag{1}$$

Under every risk factor i (f_i), the assessed value of the alternative j (A_j) is

$$y_{ij} = \frac{1}{m}\left[\sum_{f=1}^{m} y_{ij}^m\right] = \frac{1}{m}\left[y_{ij}^1 + y_{ij}^2 + \dots y_{ij}^n\right] \tag{2}$$

Phase 4: Constructing the fuzzy decision matrix and calculate the average fuzzy weight

$$S = \left[y_{ij}\right]_{a\times f} = \begin{bmatrix} y_{11} & y_{12} & \cdots & y_{1k} \\ y_{21} & y_{22} & \cdots & y_{2k} \\ \cdots & \cdots & \cdots & \cdots \\ y_{a1} & y_{a2} & \cdots & y_{ak} \end{bmatrix} \tag{3}$$

Under every risk factor i (f_i), y_{ij} denotes the assessed value of the alternative j (a_j)

$$We = \left[we_1, we_2, \dots, we_k\right] \quad j = 1,2,\dots,f \tag{4}$$

We_i indicates the weight of the risk factor i.

For the triangular fuzzy number, We_i and y_{ij} are the linguistic variable.

Phase 5: Build the function of fuzzy preference
There are two set of alternatives (A) are p and q, the function of preference Pr (p, q) is described below:

$$\Pr(p, q) = 0 \quad , y_{bj} \leq y_{aj} \qquad j = 1,2,\dots,f \tag{5}$$
$$y_{bj} - y_{aj},\ y_{bj} > y_{aj}$$

$\Pr(p, q)$ preference function signifies the intensity of outranking that p is larger to q. From the pairwise comparison alternatives, the outranking relation is constructed. $y_{bj} > y_{aj} \Leftrightarrow p\Pr q$(p outranks q)

$$y_{bj} = y_{aj} \Leftrightarrow p \Pr q \text{ (p is dissimilar to q)} \tag{6}$$

Phase 6: For describing the relation of valued outranking, firstly to define the function of multi-criteria preference $\pi(p,q)$ and it can define below.

$$\pi(p,q) = \frac{1}{f}\sum_{i=1}^{f} \Pr_i(p,q) \tag{7}$$

$$\pi(p,q) = \sum_{i=1}^{f}[We_i \Pr_i(p,q)] / \sum_{i=1}^{f} Wr_i \tag{8}$$

Phase 7: For the alternatives preordering to calculate the flows

Fuzzy PROMETHEE I: Partial disorder is used to show alternatives and it is difficult to contrast each other.

Outgoing flow

$$\phi^{+}(p) = \sum_{x \neq p} \pi(q,x), \forall p, x \in A \tag{9}$$

Here $\phi^{+}(p)$ is the summation of preference that p is larger to various alternatives. For the better alternatives, $\phi^{+}(p)$ will be higher.

Incoming flow

$$\phi^{-}(p) = \sum_{x \neq p} \pi(q,x), \forall p, x \in A \tag{10}$$

Here $\phi^{-}(p)$ is the summation of preference that various alternatives are larger to p. For the better alternatives, $\phi^{-}(p)$ will be smaller.

Next one is defuzzify the maximizing and minimizing sets.

C (Maximizing sets) = $\{(y, e_c(y)) | y \in C\}$ and

$$e_c(y) = (y - y_1) / (y_2 - y_1),\ y1 \leq y \leq y2 \tag{11}$$

0, otherwise

N (Minimizing sets) = $\{(y, e_n(y)) | y \in N\}$ and

$$e_n(y) = (y - y_2) / (y_1 - y_2),\ y1 \leq y \leq y2 \tag{12}$$

0, otherwise

Fuzzy PROMETHEE I is the incomplete preorder of (A_1, J_1, C) acquired by deliberating the connection of two preorders, and is affirmed by the incomplete preorder which is denoted in eq (13). Outgoing and incoming flows will be recognized in PROMETHEE.

pA_1q (p outranks q) if $pA^{+}q$ and $pA^{-}q$,

Or $pA^{+}q$ and $pA^{-}q$,

Or $pA^{+}q$ and $pA^{-}q$; (13)

pJ_1q (p is dissimilar to q) if $pJ^{+}q$ and $pJ^{-}q$;

pCq (p and q are unequalled) if $pJ^{+}q$ and $pJ^{-}q$;

Else.

Fuzzy PROMETHEE II: It gives the impartial preorder and is persuaded by the total flow and in the equation 14. The impartial preorder appears in eq (20):

Hence, Total stream is

$$\phi(p) = \phi^{+}(p) - \phi^{-}(p) \tag{14}$$

$pA_{11}b$ (p outranks q) if $\phi(p) > \phi(q)$ (15)

$pJ_{11}q$ (p is similar to q) if $\phi(p) = \phi(q)$

From the outgoing and incoming flows, the incomplete preorder is obtained by using Fuzzy PROMETHEE I and the final ranking is based on total flows by using Fuzzy PROMETHEE II.

Phase 8: Building the outranking graph for evaluates the ranking of every alternative.

5. Case study

In the modern day business scenario, green manufacturing is the key one for improving the production as well as turnover. Regarding this scenario, this paper conducted the case study in leading medical equipment manufacturing firm with more than 500 workers employed. It is located in southern part of India. Current case firm has a problem for not protecting the environment from the pollution as well as producing the product with more waste. To overcome this issue, they need implementation of green manufacturing. But there are more no of risk factors in the process of green manufacturing. Therefore risk assessment is needed for controlling the risk factors. The main aim of the paper is to examine the risk factors and to select the critical alternative in the firm by using F-PROMETHEE method. A team of decision makers like $(DM1, DM2, DM3)$ is framed to select the most preferable alternative for the firm related to green manufacturing such as Environmental standards $(A1)$, Innovation in green technologies $(A2)$ and Trend in external markets $(A3)$. The findings of the paper will definitely assist the industrial managers for the green manufacturing implementation. The scope diagram of the paper is presented in Figure 2.

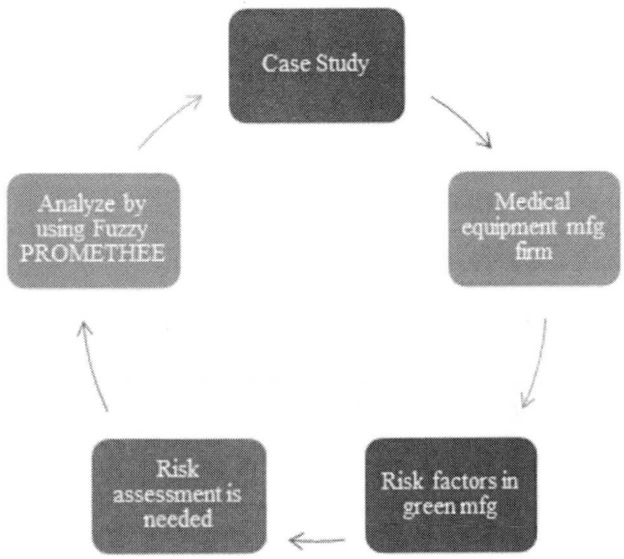

Figure 2. Scope diagram of the paper.

5.1. Analyzing the risk factors in green manufacturing

Step 1: Decision makers provide the ratings for risk factors and alternatives by using Table 2 in the form of questionnaire.

IOP Conf. Series: Materials Science and Engineering **764** (2020) 012015 doi:10.1088/1757-899X/764/1/012015

Table 2. Linguistic scales and their related fuzzy no's.

Fuzzy No	Alternatives rating	Risk Factor weights
(0.75,1,1)	Best (B)	Very High (VH)
(0.5,0.75,1)	Good (G)	High (H)
(0.25,0.5,0.75)	Fair (F)	Medium (M)
(0,0.25,0.5)	Poor (P)	Low (L)
(0,0,0.25)	Worst (W)	Very Low (VL)

Step 2: Decision makers given the rating of both risk factors and alternatives are given in Table 3 and table 4.

Table 3. Weight of every risk factor given by decision makers.

	DM1	DM2	DM3	DM4	Weights of Fuzzy (We_j)
F1	M	H	H	M	(0.375,0.625,0.875)
F2	H	H	M	VH	(0.5,0.75,0.9375)
F3	M	M	L	M	(0.75,1.75,2.75)
F4	H	H	H	M	(1.75,2.75,3.75)
F5	M	H	H	H	(1.75,2.75,3.75)
F6	L	M	M	L	(0.5,1.5,2.5)
F7	H	M	VH	H	(2,3,3,75)
F8	M	M	M	L	(0.75,1.75,2.75)
F9	H	M	H	M	(1.5,2.5,3.5)
F10	M	M	L	M	(0.75,1.75,2.75)

Table 4. For every alternative rating given by decision makers.

	A1	A2	A3
F1	F	P	F
F2	P	P	G
F3	F	F	F
F4	F	P	P
F5	G	W	P
F6	P	W	P
F7	P	P	F
F8	F	P	F
F9	W	F	P
F10	F	P	P

Step 3: Next step is to convert the linguistic evaluation into TRFN (Trapezoidal fuzzy numbers) for generating the fuzzy decision matrix and to find fuzzy weight of every risk factor. Table 5 and table 6 shows the normalized and weighted normalized Fuzzy decision matrix.

Table 5. Normalized fuzzy decision matrix.

	A1	A2	A3
F1	(0.25,5,0.75)	(0,0.25,0.5)	(0.25,0.5,0.75)
F2	(0,0.25,0.5)	(0,0.25,0.5)	(0.5,0.75,1)
F3	(0.25,0.5,0.75)	(0.25,0.5,0.75)	(0.25,0.5,0.75)
F4	(0.25,0.5,0.75)	(0,0.25,0.5)	(0,0.25,0.5)
F5	(0.5,0.75,1)	(0,0,0.25)	(0,0.25,0.5)
F6	(0,0.25,0.5)	(0,0,0.25)	(0,0.25,0.5)
F7	(0,0.25,0.5)	(0,0.25,0.5)	(0.25,0.5,0.75)
F8	(0.25,0.5,0.75)	(0,0.25,0.5)	(0.25,0.5,0.75)
F9	(0,0,0.25)	(0.25,0.5,0.75)	(0,0.25,0.5)
F10	(0.25,0.5,0.75)	(0,0.25,0.5)	(0,0.25,0.5)

Table 6. Weighted normalized fuzzy decision matrix.

	A1	A2	A3
F1	(0.094,0.313,0.656)	(0,0.156,0.438)	(0.094,0.313,0.656)
F2	(0,0.188,0.469)	(0,0.188,0.469)	(0.25,0.563,0.938)
F3	(0.047,0.219,0.516)	(0.047,0.219,0.516)	(0.047,0.219,0.516)
F4	(0.109,0.344,0.703)	(0,0.172,0.469)	(0,0.172,0.469)
F5	(0.219,0.516,0.938)	(0,0,0.234)	(0,0.172,0.469)
F6	(0,0.094,0.313)	(0,0,0.156)	(0,0.094,0.313)
F7	(0,0.188,0.469)	(0,0.188,0.469)	(0.125,0.375,0.703)
F8	(0.047,0.219,0.516)	(0,0.109,0.344)	(0.047,0.219,0.516)
F9	(0,0,0.219)	(0.094,0.313,0.656)	(0,0.156,0.438)
F10	(0.047,0.219,0.516)	(0,0.109,0.344)	(0,0.109,0.344)

Step 4: Hamming distance technique is used for calculating the distance among two alternatives 1 and 2 regarding every risk factors. For instance, to compare the first and second alternative A12 is (0.094, 0.156, and 0.219). Therefore the alternative 1 is ideal than the alternative 2.

Step 5: By using the weight of risk factors and preference function value to calculate the function of fuzzy preference and it is depicted in table 7. By using equation (9) and (10), the fuzzy flows for every alternative are found and listed in the below table 8.

Table 7. Function of Fuzzy preference.

	A1	A2	A3
A1	-	(0.272,0.824,1.511)	(0.19,0.463,0.831)
A2	(0.093,0.312,0.437)	-	(0.125,0.187,0.234)
A3	(0.25,0.468,0.673)	(0.332,0.829,1.354)	-

Table 8. Fuzzy flows.

	Fuzzy outgoing flows	Fuzzy incoming flows
A1	(0.462,1.289,2.342)	(0.343,0.604,0.315)
A2	(0.218,0.5,0.671)	(0.781,1.653,0.651)
A3	(0.582,1.297,2.028)	(1.111,2.866,1.065)

Step 6: By using equation (1), defuzzifying the value of fuzzy outgoing and incoming flows. Table 9 shows the crisp values of both outgoing and incoming flows and also final rankings are based on the total flows.

IOP Conf. Series: Materials Science and Engineering **764** (2020) 012015 doi:10.1088/1757-899X/764/1/012015

Table 9. Crisp flows.

	Outgoing flow	Incoming flow	Total flow	Final Ranking
A1	1.364	0.421	0.9434	1
A2	0.463	1.028	-0.5651	3
A3	1.302	1.681	-0.3783	2

Incomplete and final rankings are provided by Fuzzy PROMETHEE I and II and it is shown in Figure 3 and figure 4.

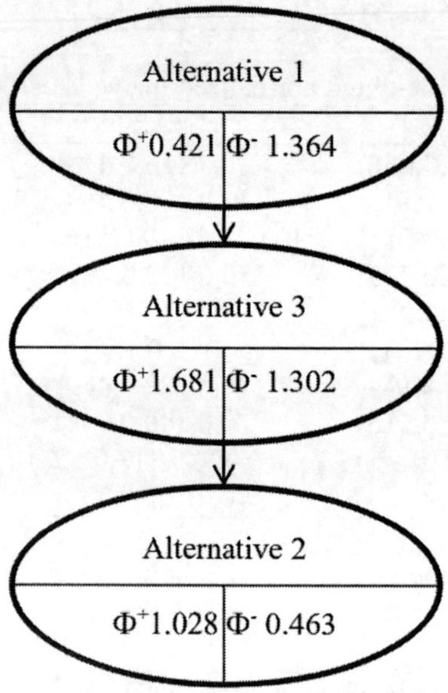

Figure 3. PROMETHEE I - incomplete ranking.

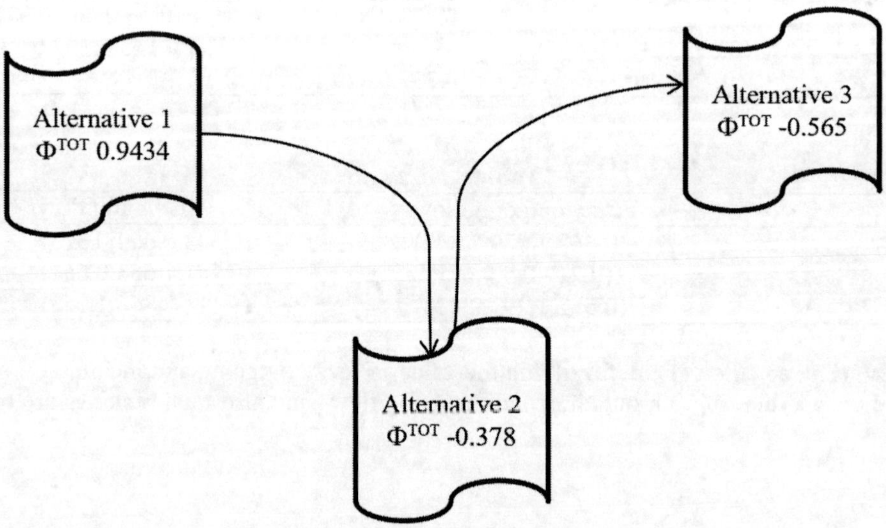

Figure 4. PROM II – final ranking.

IOP Conf. Series: Materials Science and Engineering **764** (2020) 012015 doi:10.1088/1757-899X/764/1/012015

6. Conclusions and scope for future work

This paper used the multi criteria decision making method like F-PROMETHEE for analyzing the risk factors of Green manufacturing in medical equipment manufacturing firms and for choosing the critical alternative in the firm. For the production managers as well as officials in the firm, implementation of environmental management system (EMS) and green manufacturing is a key issue. This paper recognizes ten risk factors and three alternatives from industrial experts as well as gathered from relevant literature. From these alternatives, environmental standards (Alternative 1) with a total flow of 0.9434 are the critical alternative in the firm. The outcomes showed that lack of rules and regulations related to the environment for the green manufacturing implementation. In the strategic manufacturing decisions, environmental standards act as a major role. Therefore the firm should implement the environmental management system with ISO 14001 standards for lessening the environmental impact. For fulfilling the regulations of government, implementation of green manufacturing is the vital one. Currently based on the report from Indian government in 2019, nearly 30,000 organizations are certified with ISO 14001. Green manufacturing implementation is mainly for protecting the environment from pollution, controlling risk, producing the product with less waste and low cost of the product regarding the needs of customer. Green manufacturing is very useful for increasing the performance of the firm. For the importance of green practices, every month to conduct the audit from both internal and external officers related to ISO certification and adoption.

Managers from industries will use our outcomes for the implementation of green manufacturing and also to know and understand about what are the risks in green manufacturing. Final outcomes will also help third parties such as stakeholders, customers, suppliers, advocacy groups related to an environment, politicians, governments and regarding the concerns of employees. Based on the results, production managers conduct the program for employees regarding the importance of green.

This paper has several limitations and it can be explored later on. First one, there are only ten risk factors and three alternatives are identified in this paper, for the future point of view, more no of risk factors and alternatives will be identified. Second one, this paper examining the risk factors by using F-PROMETHEE method. Therefore to analyse the risk factors by using different MCDM ranking methods like TOPSIS, VIKOR, and ANP. Finally, this paper identifies the factors from medical equipment manufacturing firm only. For the future perspective, you will do the research in other firms also.

7. Appendices

Appendix A: Questionnaire

Note:

For a review, a questionnaire was framed and it is provided to the experts in the field to evaluate risk factors in lean implementation of three medical equipment manufacturing firms in the southern region of India. Gathered data will be used only for the purposes of research. Comments from you will not share to third party or any social media and it will be kept secret. Your input will help in bringing about positive outcomes in this research and it is very important.

For providing a rating, our heartfelt thanks for spending your time and effort.

IOP Conf. Series: Materials Science and Engineering **764** (2020) 012015 doi:10.1088/1757-899X/764/1/012015

Table A1. Linguistic scales and their related fuzzy no's.

Fuzzy No	Alternatives rating	Risk Factor weights
(0.75,1,1)	Best (B)	Very High (VH)
(0.5,0.75,1)	Good (G)	High (H)
(0.25,0.5,0.75)	Fair (F)	Medium (M)
(0,0.25,0.5)	Poor (P)	Low (L)
(0,0,0.25)	Worst (W)	Very Low (VL)

Please tick [✓] in any one of rating that you feel suitable for each item. (Refer to Table A1)

	DM1	DM2	DM3	DM4
F1				
F2				
F3				
F4				
F5				
F6				
F7				
F8				
F9				
F10				

	A1	A2	A3
F1			
F2			
F3			
F4			
F5			
F6			
F7			
F8			
F9			
F10			

Please tick [✓] in any one of rating that you feel suitable for each item. (Refer to Table A1)

Profile of the expert:

1. Name: ……………………………………..
2. Experience in Medical equipment manufacturing firms (in years): ……….
3. Name of organization: ………………………..
4. Current position in the organization:
5. Mobile No. & Email: …………………………..

IOP Conf. Series: Materials Science and Engineering **764** (2020) 012015 doi:10.1088/1757-899X/764/1/012015

References

[1] A P S Sethi, I P Ahuja, A Singla, "Shifts Between Technology Push and Market Pull Strategies for Sustainable Development in Manufacturing Industries", Global Value Chains, Flexibility and Sustainability, pp. 319-331, 2018.

[2] K S Sangwan, K Choudhary, "Benchmarking manufacturing industries based on green practices", Benchmarking: An International Journal, Vol. 25(6), pp.1746-1761, 2018.

[3] I D Paul, G Bhole, J R Chaudhari, "A review on green manufacturing: it's important, methodology and its application", Procedia Materials Science, Vol.6, pp.1644-1649, 2014.

[4] A K Digalwar, A R Tagalpallewar, V K Sunnapwar, "Green manufacturing performance measures: an empirical investigation from Indian manufacturing industries", Measuring Business Excellence, Vol.17(4), pp.59-75, 2013.

[5] A M Deif, "A system model for green manufacturing", Journal of Cleaner Production, Vol.19(14), pp.1553-1559, 2011.

[6] K Govindan, A Diabat, K M Shankar, "Analyzing the drivers of green manufacturing with fuzzy approach", Journal of Cleaner Production, vol.96, pp.182-193, 2015.

[7] H Liu, W Chen, Z Kang, T Ngai and Y Li, "Fuzzy multiple attribute decision making for evaluating aggregate risk in green manufacturing", Tsinghua Science & Technology, Vol.10(5), pp.627-632, 2005.

[8] V K Mittal, P Egede, C Herrmann, K S Sangwan, "Comparison of drivers and barriers to green manufacturing: a case of India and Germany", Re-engineering Manufacturing for Sustainability, pp.723-728, 2013.

[9] K S Sangwan, "Quantitative and qualitative benefits of green manufacturing: an empirical study of Indian small and medium enterprises", Glocalized Solutions for Sustainability in Manufacturing, pp. 371-376, 2011.

[10] D Seth, R L Shrivastava, S Shrivastava, "An empirical investigation of critical success factors and performance measures for green manufacturing in cement industry", Journal of Manufacturing Technology Management, Vol. 27(8), pp.1076-1101, 2016.

[11] J P Brans, P Vincke, B Mareschal, "How to select and how to rank projects: The PROMETHEE method", European journal of operational research, Vol. 24(2), pp.228-238, 1986.

[12] F Ülengin, Y I Topcu, O ŞahinŞ, "An integrated decision aid system for Bosphorus water-crossing problem", European Journal of Operational Research, vol.134(1), pp.179-192, 2011.

[13] J Geldermann, T Spengler, O Rentz, "Fuzzy outranking for environmental assessment. Case study: iron and steel making industry", Fuzzy sets and systems, vol.115(1), pp.45-65, 2000.

[14] M Gul, E Celik, A T Gumus, A F Guner, "A fuzzy logic based PROMETHEE method for material selection problems", Beni-Suef University Journal of Basic and Applied Sciences, Vol.7(1), pp.68-79, 2011.

[15] E Tabaraee, S Ebrahimnejad, S Bamdad, "Evaluation of power plants to prioritise the investment projects using fuzzy PROMETHEE method", International Journal of Sustainable Energy, vol.37(10), pp.941-955, 2018.

[16] A P Vavatsikos, O E Demesouka, K P Anagnostopoulos, "GIS-based suitability analysis using fuzzy PROMETHEE", Journal of Environmental Planning and Management, pp.1-25, 2019.

Moisture absorption studies on Kenaf composites at various temperatures

R Naveen[1,*], C Rakeshkumar[2], L Sanjana[1], C Shwetha[1] and S Vignesh[1]
[1]Department of Aeronautical Engineering, Bannari Amman Institute of Technology, Coimbatore - 638 401, India.
[2]Department of Aeronautical Engineering, Vel Tech Rangarajan Dr. Sagunthala R&D Institute of Science and Technology, Avadi, Chennai, India.
*Author for correspondence E-mail: aeror.naveen@gmail.com

Abstract. In the world of manufacturing and production industries, composites are emerging to be the best of all the other available sources. Their lightweight, high strength, ease of manufacture and usage makes them superior to other materials. Nowadays, the applications of composites is almost everywhere. From household applications to automotive and aviation industries they are used. The composition of most composites mainly consist of glass, carbon fibers with epoxy resin, all these materials provide good mechanical properties to the composite but at the same time, they are very hard to dispose of as they are non - biodegradable. As these composites are not eco - friendly we can replace these synthetic fibers with natural fibers. Comparing different natural plant fibers, kenaf fibers show immense use in the automotive industry. Kenaf fibers are obtained from the plant kenaf (Hibiscus cannabinus) which is commonly called as Deccan hemp. In this paper, the kenaf fibers are studied for their flexural properties, moisture-absorbing capacity, and diffusion coefficient at different environmental and physical conditions. The weight of the fiber in the composite is varied for different percentages such as 10, 20, 30, 40 and 50. The experiments are conducted at two different temperature conditions, at 30 °C and 50 °C and the results are noted down separately. The flexural properties are tested under three conditions, at 30°C, 50°C and at the dry condition. Experimental results show that the flexural properties are optimum for 20% weight of fiber at the dry condition. The water-absorbing properties were found to be low for lower weight ratios at 30°C. The fiber exhibited good diffusion properties at 50°C for almost all weight ratio.

1. Introduction

The Over the span of 50 years, there is a developing eagerness in the utilization of plant fibers in the composite world. Plant fiber composites gives several points of interest in contrast with the synthetic fibers. The supremacy of these composites is that the tool wear is low, less density, low cost of fiber, easy accessibility, furthermore, eco-friendly nature. They possess splendid insulation of thermal and acoustic properties. The most well-known natural plant utilized in applications is bast fibers, for example, jute, hemp, fiax, sisal, and kenaf. The major reason for this increasing interest is that the plant fibers have a greater specific quality than that of glass fibers and the comparative specific modulus. Among them, the kenaf fiber is receiving more recognition among the production centers and experimenters to make use of it in various polymer composites. Kenaf fibers when reinforced with epoxy resin results in improved firmness for the composites.

A specialized report TR2101-ENV [1] concluded that the absorbing nature of kenaf core can be used for the oil spills cleaning. The examination demonstrated that the other common absorbent materials have been beaten by the core of the fiber by a critical margin. Also ACI Committee 544 [2] stated that the tensile characteristics of these plant fibers are practically identical to the other plant-based fibers, such as bamboo, flax, and jute, which have been used earlier in the yielding of natural fiber reinforced concrete. A researcher studied about the improvement of composites utilizing natural fibers and inferred that the geometric properties and firmness of plant-based fibers predominantly rely on the development condition, area of emergence and different attributes which impact the mechanical properties of these composites. The same researcher [3] assured that mechanical quality and thermal properties of kenaf composite are superior to other plant-based composites, therefore it is appropriate for high-performance cases. In an experiment, it is observed that kenaf can develop to an extent of around 350 cm to 450 cm in a time period of 5 months with a yearly yield of fiber, 5000 to 10000 kg of withered fiber per acre of land. Whereas the southern pine trees produce a yield of only one-fourth of kenaf [4]. Another discovery states that the CO_2 absorbed by the plant is nearly one and a half times of its weight. Of many plants studied by the researchers, kenaf exhibits the highest level of absorption [5]. Plant-based composites are hydrophilic in nature so they absorb moisture. The moisture enters composite material between polymer chains through the micro pores [6]. The height, fiber quantity and fiber alignment of this fiber which influence intensive and extensive properties of fiber strengthened soy related organic composites [7]. It was discovered that kenaf composites can be used as an alternative for gasoline related composite materials in huge numbers of the industrial implementation and suggesting creative natural, farming, production and purchaser benefits detailing strategies [8-10]. The kenaf plant fiber is obtained from the shoot of plants variety Hibiscus, Malvaceae family and the Hibiscus Cannabinus species. It needs only a small amount of water to develop on the grounds that kenaf fiber has a life span of 5 to 6 months with a normal product of 1700 kilogram/hectare [11]. With the likelihood of growing the plant two times annually, kenaf has an incredible capacity to be utilized as a supplement derivation of fiber for medium-density fiber board and particle board production [12]. The tensile modulus of Kenaf fibers reduced drastically when immersed in seawater, acidic rainwater and distilled water [13]. The utilization of kenaf fibers for the advancement of eco-friendly, the mechanical-accommodating resources for the automotive, sports enterprises, food packaging and carpentry trade line is emerging fast [14]. Researchers state that the bast fibers of kenaf have good tensile and flexural properties that make them as a substitution to glass fibers, a reinforcing component in composites as they are obtained from various parts of that plant and is appropriate for different uses [15-17]. Kenaf fibers are broadly contemplated as appropriate organic assets and alternative for non-renewable energy sources and wood-pulps [18]. The mechanical properties of flax-kenaf mixed breed composites and inferred that hybrid composites have preferable properties over single fiber composites. The most fascinating part of natural fibers is their positive ecological effect [19-20]. Natural fibers are sustainable assets, eco-friendly and their generation needs only less effort [21]. A noteworthy downside of plant fibers in comparison to polymer fibers is their dissimilarity, assortment of measurements, and their mechanical properties [22]. In this manner, the significant assignment to be fathomed, so as to help the acknowledgment of plant fibers as a standard option in comparison to traditional fibers, is to grow elite natural fibers reinforced composites [23-24].Natural fibers contain a lot of the OH^- group, this makes the fiber opposed and deliquescent. Plastics, in general, are water-hating in nature. When adding hydrophilic natural fibers to water-hating plastic it will bring about an end product with low tensile strength owing to the uneven scattering of fiber in the matrix and in the interphase [25]. This polar nature likewise brings about increased moisture absorption in natural fibers based composites, prompting fibers bulging and minute pores in the interphase of the fiber matrix. The water content, if not expelled from natural fibers before intensifying by withering, will bring about a permeable product. Increased moisture absorbing capacity could likewise result in low tensile and flexural strength and cause instability in the dimension [26-27]. About one-third of kenaf is bast fibers, which is appropriate for commercial products and about two of it is core [28]. A mono fiber of kenaf can have a tensile strength of 11900 MPa and tensile modulus of 60000 MPa [29]. The capability of kenaf as a reinforcing fibers in a polypropylene matrix and contrasted the mechanical properties and other ordinarily utilized composites [30].

The prime aim is to study the moisture-absorbing property and flexural properties of the kenaf fiber reinforced composite by varying the weight ratio of kenaf fiber in the composite at various temperatures.

2. Materials and methods

The materials used in the making of kenaf fiber composite are Kenaf fiber, epoxy resin (LY-556) and hardener (HY-556). The composite is made by using the hand layup method under room temperature. The hand layup method is an open molding method where fiber reinforcements are kept by human hand and then it is made wet using resin. This method is ideal for lower volumes as it exhibits low production volume per mold. First of all, the matrix is made by mixing epoxy resin with hardener in the ratio 10:1. The matrix is mixed well using a mechanical stirrer. Then the kenaf fiber with various weight percentages (10%, 20%, 30%, 40% and 50%) is added to the matrix and the composites are made. They are allowed to cure under room temperature. No heating process is carried out throughout the preparation of composites. The cured composite is then subjected to various methods of characterization. The characterization carried out for the kenaf fiber composite includes the following methods.

2.1. Moisture absorption

Moisture absorbing capacity of natural fiber reinforced composite is an important affair predominantly when it comes to outdoor applications. Moisture absorption depends on various factors such as the quantity of the fiber, fiber alignment, temperature, area of open surface, the porosity of fiber, vacant space and deliquescence of every single component [31-34]. Moisture absorbing capacity in kenaf fibers takes place mainly by two methods namely, through microspores and capillary action. Here the kenaf fiber reinforced composites were tested for moisture absorption under two conditions - by varying the weight percentage of kenaf fiber, by changing the temperature (30°C and 50°C). These tests were done over a period of immersion time. It is calculated from the following formula,

$$\%\,absorption = \frac{Wo - W}{W} \qquad \text{Eqn. (1)}$$

Where,
W_0 is the weight after absorption,
W is the initial weight before absorption.

2.2. Flexural strength

Flexural strength is a material property characterized as the failure stress in bending of the structure. This is equivalent or slightly greater than the stress at failure during tension. Flexural strength of a construction and manufacturing materials is usually measured so as to understand the bending and load-carrying capacity of the material. As this Kenaf fiber reinforced composite is to be used for such purposes, flexural strength is measured for various weight percentages and at different temperatures. The flexural strength is determined using the following formula,

$$\sigma = \frac{3FL}{2bd^2} \qquad \text{Eqn. (2)}$$

Where, F is the force applied, L is the length, b is the breadth and d is the depth of the composite material.

2.3. Flexural modulus

The flexural modulus is a property which is figured as the proportion of stress to strain during the flexural distortion. It is the propensity of bending resistance of a material. The value from the slope of a stress-strain curve of the testing of flexural gives the value of the flexural modulus. It is on a very basic level equivalent to the elastic modulus, which is actually the elasticity of the material. Specifically, in polymeric materials, these two properties generally have different values. Flexural modulus of kenaf fiber reinforced composite material is studied and measured under various weight percentage of kenaf fibers in the composite at two different temperatures.

2.4. Diffusion coefficient

The diffusion coefficient can be given as the volume of a substance that spreads out from one area to another and passes through every area of cross-section per unit of time at unit volume-concentration gradient. It is an unchanging physical factor which depends on the sub-atomic size and different properties of the diffusing substance just as on temperature and pressure. Diffusion coefficient of Kenaf fiber reinforced composite material is measured under various weight percentage of kenaf fibers in the composite at two different temperatures and is represented graphically.

3. Results and discussion

3.1. Moisture absorption at 30°C

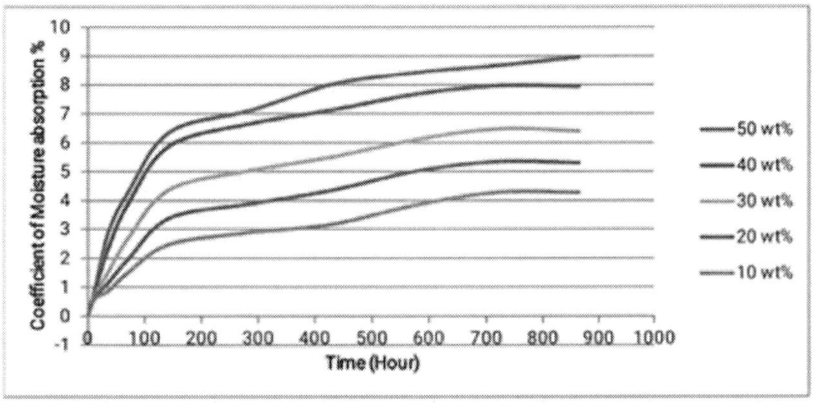

Figure 1. Co-eff. of moisture absorption % of the composite for varied wt. over the time of immersion of the composite materials in water at 30°C.

3.2. Moisture absorption at 50°C

The block diagram of the experiment setup is shown in figure 1. It consists of a capacitive type moisture sensor, a Light Dependent Resistor (LDR), temperature and humidity sensor (DHT11), solenoid valve, relay to operate the solenoid valve, Arduino, ESP8266 Wifi module and mobile/laptop to monitor the values.

IOP Conf. Series: Materials Science and Engineering **764** (2020) 012016 doi:10.1088/1757-899X/764/1/012016

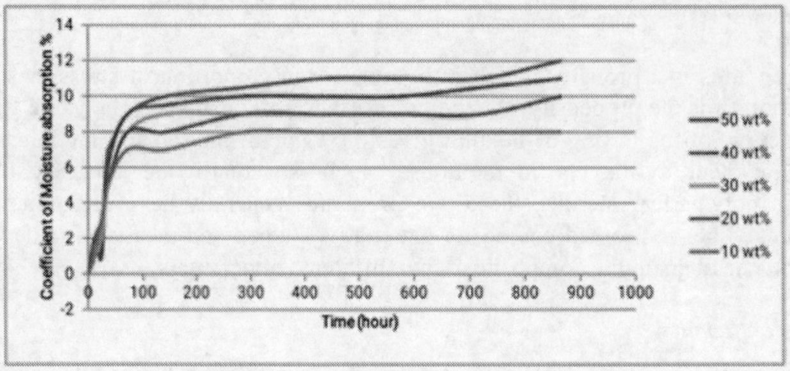

Figure 2. Co-eff. of moisture absorption % of the composites over the time of immersion for varied wt. % of the composite material in water at 50°C.

From the above graphs shown in figure 1 and figure 2, it is observed that the moisture absorption coefficient at the initial hours increased drastically and after that, it showed a gradual increase. But in the case of less weight percentage (10 wt. %), there is a gradual increase from the starting to the ending. And finally in the later hours when the fibers obtained saturation moisture absorption stopped. The moisture absorption is comparatively greater for composite with high weight percentage and decreases accordingly.

In the second case there is not much of a difference in concern to the weight of fiber in composite, the composites exhibited drastic increase for the first 75 hours (approx.) then the value maintained over the same value for the rest of the hours until saturation. Interestingly, it is detected that the kenaf fiber reinforced composites illustrates an alternate extent of assimilation for the above-stated conditions.

3.3. Flexural strength at various temperatures

From the graphs shown in the figure 3 and figure 4, it is noticed that three sets of graphs can be observed there the first set shows the value of flexural strength of fiber composite for various weight percentages at 30°C, second set at 50°C and the third at dry conditions. Clearly, the flexural strength is maximum at dry condition. Moreover, among various weight percentages 20% weight ratio seems optimum. Flexural modulus value at different temperature conditions namely 30°C, 50°C and at dry condition are observed for various weight percentages. Flexural modulus is found to be maximum at dry condition.

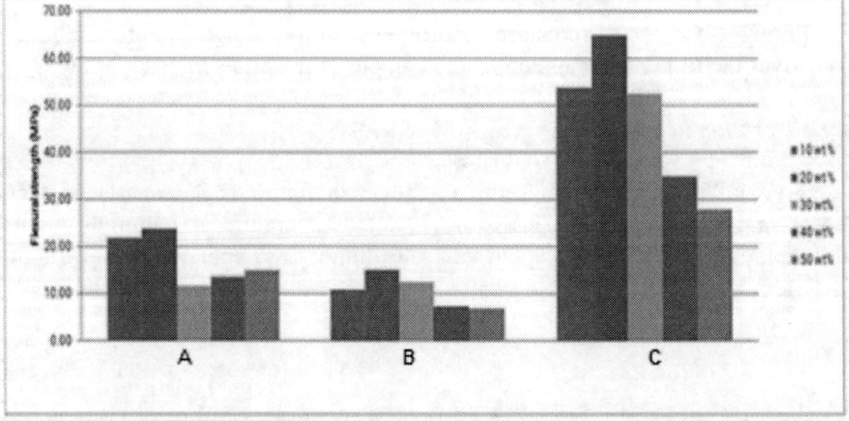

Figure 3. Differentiation of flexural strength of kenaf fiber reinforced composites of various wt. % A- at 30 °C, B- at 50 °C and C- at dry conditions.

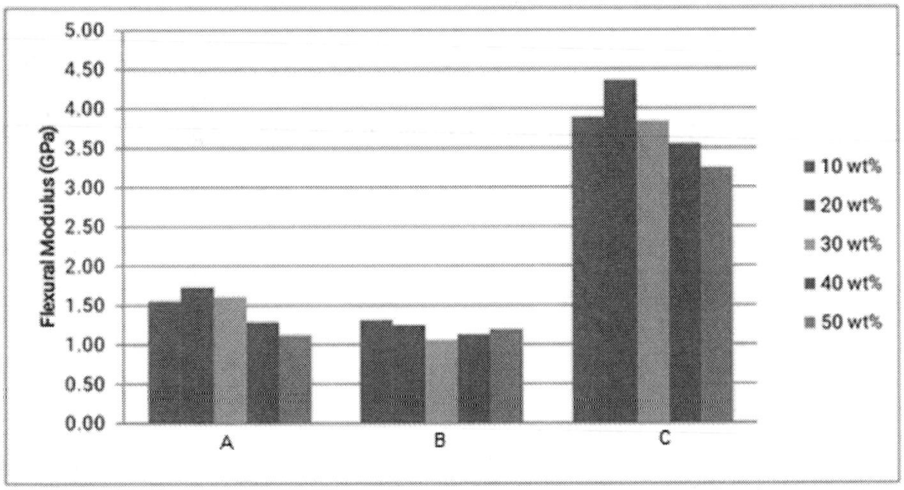

Figure 4. Comparison of flexural modulus of the composites of various wt. % at different temperature A- at 30 °C, B- at 50 °C and C- at dry conditions.

3.4. Diffusion coefficient curve

From the figure 5, the diffusion coefficient is observed for composites with varying fiber weight percentages at two different temperatures 30°C and 50°C. At 30°C the diffusion coefficient stays almost the same for all weight percentages. And at 50°C it is minimum for 10% weight and a drastic increase is observed for 20% and after that a gradual increase is observed.

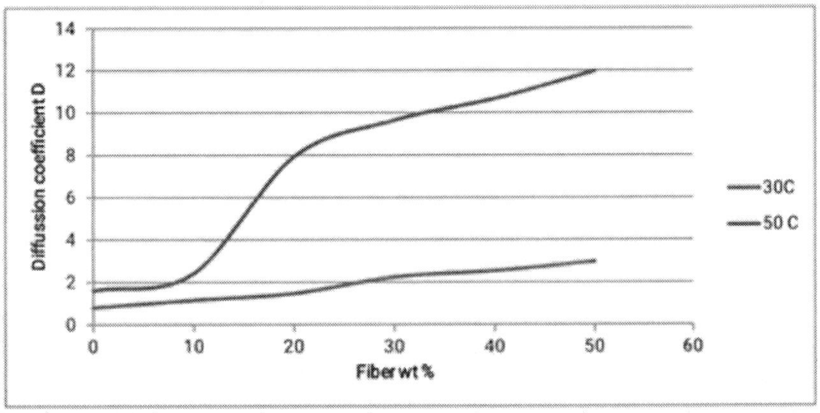

Figure 5. Diffusion coeff. for various wt. % of the composite at 30°C and 50°C.

4. Conclusion

Researches on kenaf fiber reinforced composites are getting high scrutiny owing to its extremely good properties and environmental deliberations. As these plant fiber composite materials take an important position in automobile and construction industries, the study of their properties is necessary. This paper includes the experimental studies such as flexural strength and flexural modulus, moisture absorption property and diffusion coefficient of kenaf fiber reinforced composite materials. It is noticed that the moisture absorbing capacity of kenaf fiber reinforced composite is found to be different at different temperature conditions. Moreover, the moisture absorption increased drastically over the initial period of immersion, later it increased gradually and then attained a stable condition when it is saturated. While comparing the flexural strength and modulus at different conditions, the composite exhibited high results during dry conditions. The diffusion coefficient is also compared at two different temperatures and the diffusion coefficient curves are plotted.

References

[1] Bharath, VR Raman, B. VijayaRamnath, and N. Manoharan, "Kenaf fibre reinforced composites: A review", Materials and Design, Vol.31, pp.4960-4964, 2006.

[2] ACI Committee 544, "State of the art report on fiber reinforced concrete", American Concrete Institute, Farmington Hills, pp.66, 2002.

[3] Nishino, Takashi, Koichi Hirao, Masaru Kotera, Katsuhiko Nakamae, and Hiroshi Inagaki, "Kenaf reinforced biodegradable composite", Composites science and technology, Vol.63(9), pp.1281-1286, 2003.

[4] Zaveri, Mitul Dilip, "Absorbency characteristics of kenaf core particles", North Carolina state University, 2004.

[5] Mohanty, Amar K., Manjusri Misra, and Lawrence T Drzal, "Natural fibers, biopolymers, and biocomposites", CRC press, 2005.

[6] Kim, Hyo Jin, and Do Won Seo, "Effect of water absorption fatigue on mechanical properties of sisal textile-reinforced composites", International Journal of Fatigue, Vol.28(10), pp.1307-1314, 2006.

[7] Liu, W., Drzal, L.T., Mohanty, A.K. and Misra, M., "Influence of processing methods and fiber length on physical properties of kenaf fiber reinforced soy based biocomposites", Composites Part B: Engineering, Vol.38(3), pp.352-359, 2007.

[8] Krishnaprasad R., Veena NR, Maria HJ, Rajan R., Skrifvars M. and Joseph K, "Mechanical and thermal properties of bamboo micro fibril reinforced polyhydroxybutyrate biocomposites", Journal of Polymers and Environment, Vol.17, pp.109-114, 2009.

[9] Han G, Lei Y, Wu Q, Kojima Y and Suzuki S, "Bamboo fiber filled high density polyethylene composites: effect of coupling treatment and nanoclay", Journal of Polymers and Environment, Vol.16, pp.123-130, 2008.

[10] Elsaid A, Dawood M, Seracino R and Bobko C, "Mechanical properties of kenaf fiber reinforced concrete", Construction and Building Materials, Vol.25, pp.1991-2001, 2011.

[11] Norlin Nosbi, Hazizan Md Akil, Z.A. MohdIshak and A. Abu Bakar, "Degradation of compressive properties of pultured kenaf fiber reinforced composites after immersion in various solutions", Materials and Design, Vol.31, pp.4960-4964, 2010.

[12] Mossello A.A, Harun J, Tahir P.M, Resalati H, Ibrahim R, Fallah Shamsi S.R. and Mohammed A. Z, "A review of literature related to using kenaf for pulp production (beating fractionation and recycled fiber)", Modern Applied Science, Vol.4(9), pp.21-29, 2010.

[13] H. M. Akil, M. F. Omar, A. A. M. Mazuki, S. Safiee, Z. A. M. Ishak and A. Abu Bakar, "Kenaf fiber Reinforced Composites. A Review", Material and Design, Vol.32(8-9), pp.4107-4121, 2011.

[14] Anuar H and Zuraida A, "Improvement in mechanical properties of reinforced thermoplastic elastomer composite with kenaf bast fiber", Composites Part B: Engineering, Vol. 42, pp.462-65, 2011.

[15] Karimi S, Tahir PM, Karimi A and Dufresne A Abdulkhani, "A Kenaf bast cellulosic fibers hierarchy; a comprehensive approach from micro to nano", Carbohydrate Polymers, Vol.101, pp.878-885, 2014.

[16] Paridah MT, Basher AB, Saiful Azri S. and Ahmed Z, "Retting process of some bast plant fibers and its effect on fiber quality: a review", Bio Resourses, Vol.6, pp.5260 81, 2011.

[17] Faruk O., Bledzki AK, Fink HP and Sain M, "Biocomposites reinforced with natural fibers", Progress in Polymer Science, Vol.37, pp.1552-96, 2012.

[18] Juliana A, Paridah M, Rahim S, Nor Azowa I and Anwar U, "Properties of particleboard made from kenaf (Hibiscus cannabinus L.) as a function of particle geometry", Materials and Design, Vol.34, pp.406-411, 2012.

[19] V.S. Srinivasan, S. Rajendra Boopathy and B. Vijaya Ramnath, "Fabrication and Evaluation of Tensile Properties of Kenaf- Flax Hybrid Composite", International Journal of Applied Mechanics and Materials, Vol.680, pp.50-53, 2014.

[20] V.S. Srinivasan, S. Rajendra Boopathy and B. Vijaya Ramnath, "Investigation on Shear Behavior of flax-kenaf Hybrid Composite", Advanced Materials Research, Vol. 1051, pp.139-142, 2014.

[21] Mohanty A, Drzal L, Misra M, "Engineered natural fiber reinforced polypropylene composites: influence of surface modifications and novel powder impregnation processing", Journal of Adhesion Science and Technology, Vol.16(8), pp.999–1015, 2012.

[22] Bismarck A,Mishra S,Lampke T, "Plant fibers as reinforcement for green composites", Natural fibers, Biopolymers and Biocomposites, 2005.

[23] Mohanty AK, Mishra M, Drzal LT, editors, "Natural fibers, biopolymers and bio composites", Boca Raton (FL): CRC Press, 2005.

[24] John MJ, Thomas S, "Biofibers and biocomposites", Carbohydrate Polymers, Vol.71(8), pp. 343–364, 2008.

[25] Baillie, Caroline, "Green composites: polymer composites and the environment", Green Composites, Polymer Composites and the Environment, CRC Press, 2005.

[26] G. Mehta, A. K. Mohanty, M. Misra, L. T. Drzal, "Effect of novel sizing on the mechanical and morphological characteristics of natural fiber reinforced unsaturated polyester resin based bio-composites", Journal of Material Sciences, Vol.39(8), pp.2961–2964, 2004.

[27] Alvarez V, Fraga A, Vazquez A, "Effects of the moisture and fiber content on the mechanical properties of biodegradable polymer–sisal fiber bio composites", Journal of Applied Polymer Science, Vol.91, pp.4007–16, 2004.

[28] Massimo Baiardo, Elisa Zini, Mariastella Scandola, "Flax fiber–polyester composites", Composites Part A: Applied Science and Manufacturing, Vol.35(6), pp.703–10, 2004.

[29] Lee, S.A and Eiteman, M.A, "Ground kenaf core as a filtration aid", Industrial crops and products, Vol 13(2), pp. 155-161, 2001.

[30] Karnani R, Krishnan M, Narayan R., "Biofiber reinforced polypropylene composites". Polymer Engineering and Science, vol 37, pp. 476–483, 1997.

[31] Ramaswamy GN, Sellers T, Tao W, Crook LG, "Kenaf nonwovens as substrates for laminations", Industrial Crops and Products, Vol.17(1), pp.1–8, 2003.

[32] Dhakal H, Zhang Z, Richardson M, "Effect of water absorption on the mechanical properties of hemp Fiber reinforced unsaturated polyester composites", Composites Science and Technology, Vol.67(7-8), pp.1674–1683, 2007.

[33] Chow C, Xing X, Li R, "Moisture absorption studies of sisal fiber reinforced polypropylene composites", Composites Science and Technology, Vol.67(2), pp.306–313, 2007.

[34] M.D.H. Beg, K.L. Pickering, "Mechanical performance of kraft fiber reinforced polypropylene composites: influence of fiber length, fiber beating and hygrothermal ageing", Composites Part A: Applied Science and Manufacturing, Vol 39(11), pp.1748–1755, 2008.

[35] Nosbi N, Akil HM, MohdIshak ZA, Abu Bakar A, "Degradation of compressive properties of pultruded kenaf fiber reinforced composites after immersion in various solutions", Journal of Materials and Science , Vol.31(10), pp.4960–4964, 2010.

IOP Conf. Series: Materials Science and Engineering **764** (2020) 012017 doi:10.1088/1757-899X/764/1/012017

Parametric studies on thin plates with holes

Preetha V[*], Vignesh P S, Indhumathi L and Senthilkumar V

Department of Civil Engineering, Bannari Amman Institute of Technology, Sathyamangalam-638 401, India.

[*]Author for correspondence E-mail: preethav@bitsathy.ac.in

Abstract. Plates are most important structural members finding applications in composite construction. Plates usually provided with holes meant of structural connections, service ducts, and ventilation services. The impact of holes in thin plates induces higher stress concentration and stress distribution parameters around the holes. This paper aims to analyze the impact of holes on thin plate subjected to uniform tensile load at its ends and its stress concentration factor. The parameters involved for the studies are number and the shape of the holes. The comparative study is made with analytical and numerical investigation on the plates with different d/D ratios. The variations in stress concentration factor for different d/D ratios are determined based on finite element analysis using ANSYS.

1. Introduction

The rectangular plate with holes finds its application in various structural engineering initiatives of steel design. The steel plates are often provided with holes for service ducts, connections causes pitting corrosion which reduces the mechanical strength leading to the failure of structure under service loads. Therefore, it is essential to analyze the state of stress around the holes to ensure safety and to evaluate the load bearing capacity of these structures. The solution for stress state around the hole is as shown in figure 1 and figure 2 below.

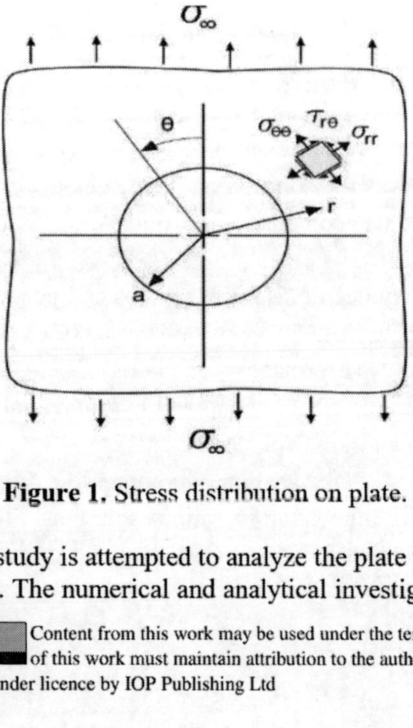

Figure 1. Stress distribution on plate.

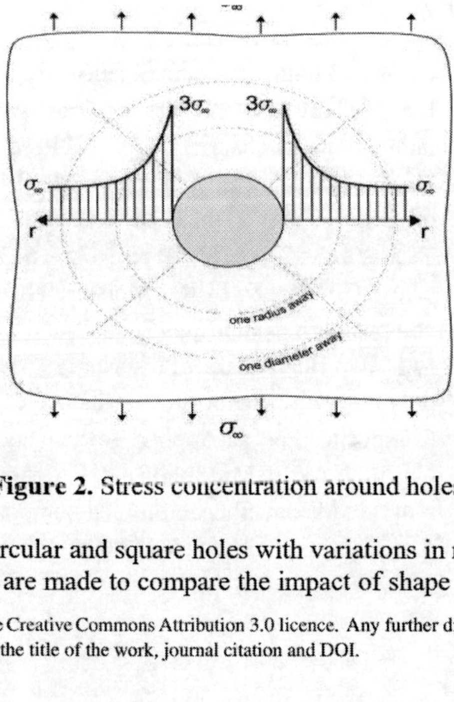

Figure 2. Stress concentration around holes.

The study is attempted to analyze the plate with circular and square holes with variations in number of holes. The numerical and analytical investigations are made to compare the impact of shape of hole

IOP Conf. Series: Materials Science and Engineering **764** (2020) 012017 doi:10.1088/1757-899X/764/1/012017

and the stress concentration factor on a rectangular plate subjected to tensile load. The rectangular plate of size 1000mm x 500mm x 15mm provided with diameter of hole varies from 100mm to 400mm respectively.

2. Literature survey

Gokul [1] investigated the analysis of thin plates with holes. The analysis includes the shape and size of the hole on the stress distribution of plate. The diameter of hole increases with increase in stress distribution, whereas the circular holes are inducing less concentration when compare to square opening. Ai-Azzawi and sadeq Aziz Abed [2] attempted on numerical analysis of reinforced concrete hollow core slabs, the hollow holes are used to reduce the cracking load and ultimate capacity of about (13.5%) and increased deflection of (39.5%). The finite element technique to assess stress distribution within a rectangular plate with two holes subjected to in-plane loading is also made. Durgadevagi and NA jabez [3] studied the comparative analysis of slab with different shape of cut out for various end conditions using ANSYS. They aimed to know the variation of displacement, stress and strain with different end conditions of the slab comparing the slabs with different boundary conditions with opening. Based on the result, square slab with circular opening is found to be efficient by providing fixed support on all edges. Lotfi Toubal et al. [4] investigated the stress concentration in a circular hole in composite plate. In this study, stress concentration has been calculated, in which Stress near holes in experiment is lesser when compare to analytical and numerical model. Babulan K S et al., [5] investigated the isotropic rectangular plate with centrally located holes which is subjected to axial uniform tension of 50Mpa along the thickness direction. From this analysis the stress around the holes is larger than the stress at other region of the plate, which was defined that in terms of stress concentration factor. The holes in the plate includes d/D ratio of 0.2, 0.4, 0.6. By finite element analysis shows that deformation of plate depends upon d/D ratio of hole. From comparison of manual and software analysis the percentage variation of stress in manual analysis have been 4.8 %, 2.9%, and 4.7% higher than software analysis for 0.2, 0.4 &0.6 respectively. The maximum stress occurred around the holes of the plate. Saksham Dhanjal [6] described that plate without hole subjected to loads causes the uniform stress employed on the all region of the hole but presence of holes, notches and keyways often leads to stress concentration around the hole. The three dimensional finite analysis was conducted on rectangular plate with hole at centre was subjected to uniform tension at two adjacent edges. The comparisons were made for different sizes on same plate. From the obtained results size of holes increases with increasing in stress concentration. Dheeraj Gunwant [7] investigated the comparison of stress and displacement on different ratio of major to minor axis of elliptical holes on rectangular plate subjected to uniform tension. The ratio from minor axis to major axis is assumed as 0.2, 0.4, 0.6, 0.8 & 1.0. From this analysis the maximum deflection of plate when the ratio of elliptical hole is 1.0. The maximum stress is situated at all four corners of elliptical hole.

3. Analytical approach

The analytical investigation is performed to study the stress concentration factor (K_t), a dimensionless factor used to locate the concentrated stress in a plate material and defined as the ratio of the maximum principal stress in the element to the normal stress.

$$\mathrm{Kt} = \frac{\sigma max}{\sigma nor}$$

3.1. Rectangular plate with single and two holes

The rectangular plates of uniform size and thickness as shown in figure 3 and figure 4 are taken for the study in which the numbers of holes are varied to predict the stress concentration parameters. The variation of d/D ratios for different diameter of holes are shown in table 1 and table 2 respectively for single and double holes.

IOP Conf. Series: Materials Science and Engineering **764** (2020) 012017 doi:10.1088/1757-899X/764/1/012017

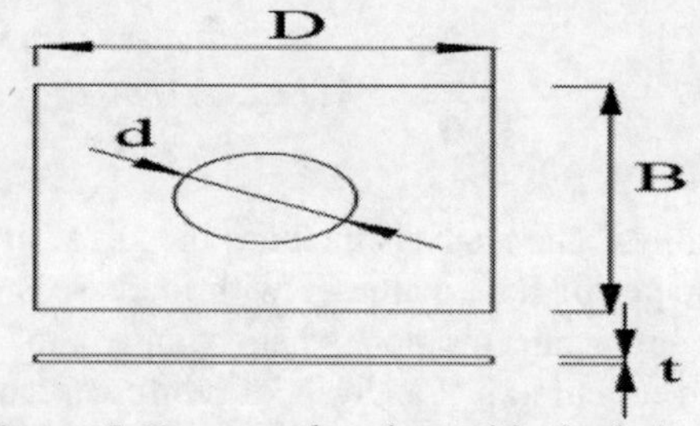

Figure 3. Rectangular plate with single hole.

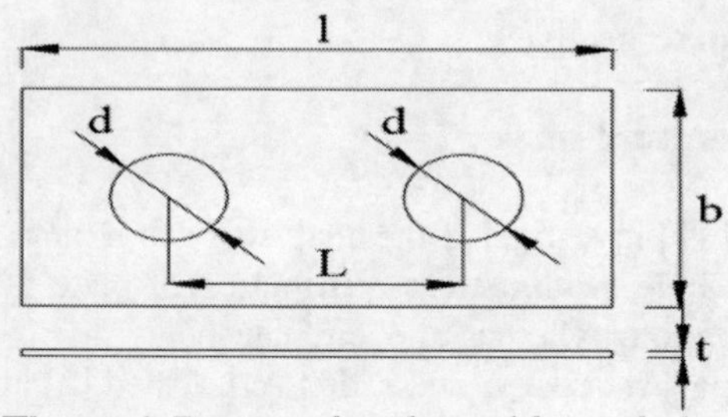

Figure 4. Rectangular plate with two holes.

Table 1. Rectangular Plate with single hole.

S. No.	Size of plate	Diameter of hole	d/D ratio
1		0	0
2		100mm	0.2
3	500×500×15mm	200mm	0.4
4		300mm	0.6
5		400mm	0.8

The stress concentration factor (Kt) for rectangular plate with single hole is given by the below equation (1),

$$K_t = 3.0 - 3.14(^d/_D) + 3.667\ (^d/_D)\ 2 - 1.527(^d/_D)\ 3$$

Eqn. (1)

Where, $0 < d / D < 1$

Table 2. Rectangular Plate with two hole.

S. No.	Size of plate	Diameter of holes	d/D ratio
1		0	0
2		100mm	0.2
3	1000×500×15mm	200mm	0.4
4		300mm	0.6
5		400mm	0.8

The stress concentration factor (Kt) for rectangular plate with two holes is given by the equation (2) below,

$$K_t = 3.0 - 0.712\ (^d/_L) + 0.271\ (^d/_L)\ 2$$

Eqn. (2)

Where, $0 < d / L < 1$

The variation of maximum stresses based on the analytical approach and the numerical approach using ANSYS for plate with single and two holes are compared in the table 3 and table 4 respectively.

Table 3. Comparison of stresses for plate with single hole.

S. No.	d/D ratio	Kt	Stress by analytical approach σmax (Mpa)	Stress by numerical approach σmax (Mpa)
1	0	3	20.01	20.34
2	0.2	2.506	20.87	21.53
3	0.4	2.233	25.00	24.35
4	0.6	2.106	35.14	30.63
5	0.8	2.026	68.00	47.59

Table 4. Comparison of stresses for the plates with two holes.

S. No.	d/D ratio	Kt	Stress by analytical approach σmax (Mpa)	Stress by numerical approach σmax (Mpa)
1	0	3	20.01	20.27
2	0.2	2.87	23.92	21.53
3	0.4	2.76	30.65	23.33
4	0.6	2.67	44.51	27.28
5	0.8	2.6	86.67	48.26

3.2. Rectangular plate with square opening

In case of rectangular plate with square opening, the stress concentration is too high near the sharp edges. The rectangular plate of size 500×500×15mm with square opening of size 100mm, 200mm and 300mm respectively are analysed. The variation in Stress concentration factor for plate with square hole are calculated by the equation (3) below,

$$Kt = C1 + C2\,(a \backslash b) + C3\,(a \backslash b)^2 + C4(a \backslash b)^3$$

The variation of kt for the plate with square opening is too high when compared to circular opening. The maximum stresses for square hole plate based on the analytical approach and the numerical approach using ANSYS are compared as shown in Table 5.

Table 5. Comparison of stresses for the plates with square hole.

S. No.	d/D ratio	Kt	Stress by analytical approach σmax (Mpa)	Stress by numerical approach σmax (Mpa)
1	0.2	7.075	58.68	21.56
2	0.4	7.075	78.26	26.219
3	0.6	7.075	67.41	47.526

4. Results and Discussion

The variation of stress concentration factor with d/D ratio for rectangular plates with single and two holes were simulated using ANSYS workbench 19.2. The analytical results were compared with the numerical finite element simulation with the variation of 15%. The variation of stresses increases with increases in number of holes in the plate as shown in figure 5, wherein the stress concentration factor increases with reduction in cross section of plates.

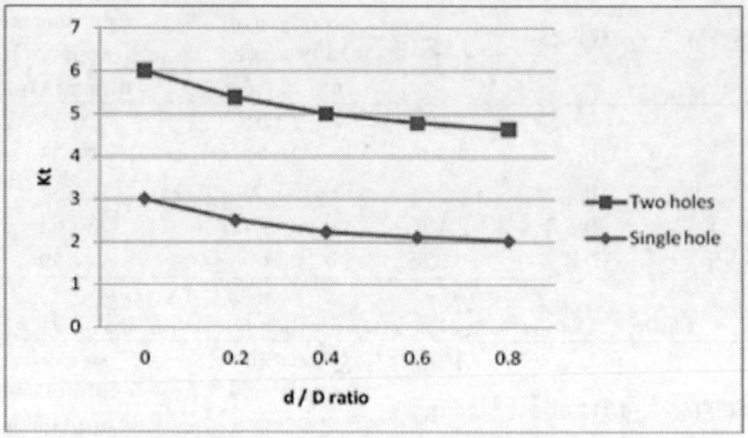

Figure 5. Variation of stress concentration factor.

It indicates that the diameter of holes increases with increase in stress concentration factor due to reduction on cross section. Thus large variation of stresses in the plate around the holes. The maximum stress concentration factor under comparative analysis had a slight variation with numerical approach. The variation gets reduced by finer mesh using finite element analysis. The variation in d/D ratio for rectangular plate with two holes increases linearly with increase in Kt values gradually as shown in figure 6.

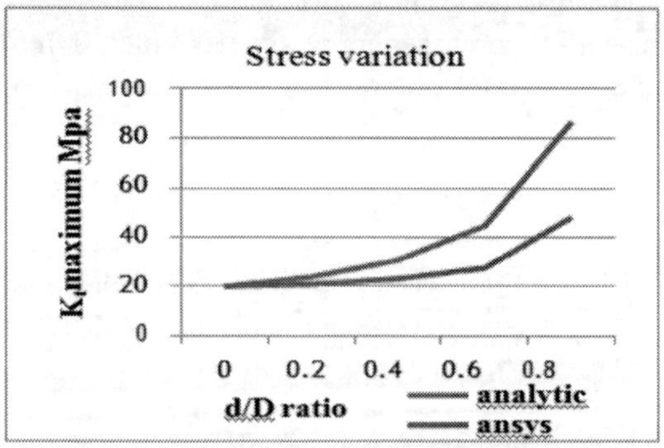

Figure 6. Stress variation with two hole.

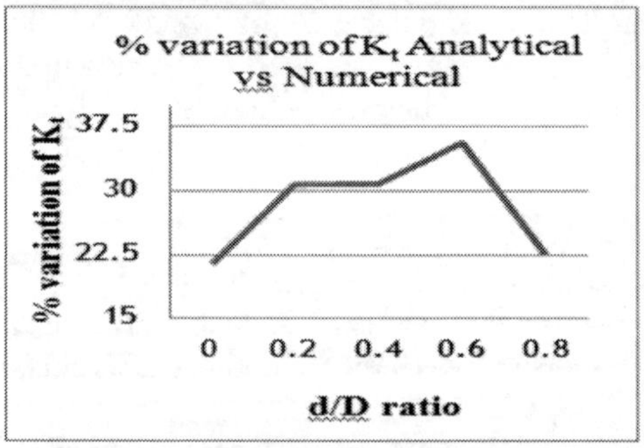

Figure 7. Percentage variation of Kt.

The above figure 7 shows that the percentage variations of stress concentration factor (Kt) on analytical and numerical analysis. When the d/D ratio increases gradually with increase in percentage variation of Kt upto certain limit wherein the optimum percentage variation adopted at d/D ratio from 0.4 to 0.6 respectively for the plates with two holes.

The variations of stress distribution are compared for the plates with single holes as shown in the below figure 8 and figure 9. Higher the d/D ratio, stress variation on the plate is also higher. This variation may be minimized by increases the mesh size in the ANSYS software. The variation increases from 1.28 % to 34.02% with increase in d/D ratio of 0.8.

IOP Conf. Series: Materials Science and Engineering **764** (2020) 012017 doi:10.1088/1757-899X/764/1/012017

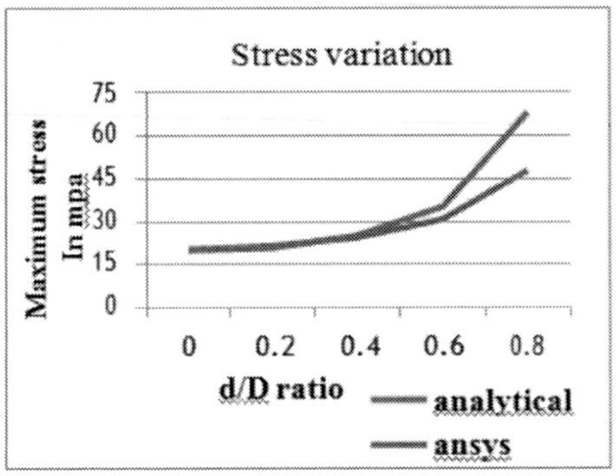

Figure 8. Stress variation with single hole.

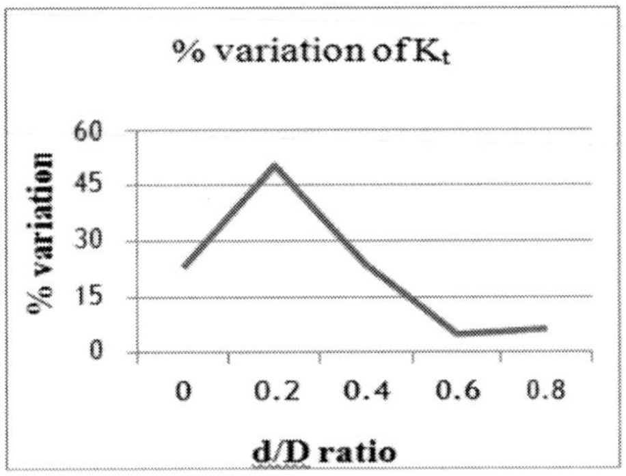

Figure 9. Percentage variation of Kt.

5. Numerical Approach

From the figure 10 to 22 shows the simulation studies using finite element analysis for variation of rectangular plate without hole, with single hole and two holes. Comparative analysis on stress concentration factor (Kt) for d/D ratio increases with increase in % variation. The stress concentration near the holes are so negligible for the plate with single hole, where in it increases with increase in diameter of the hole to the boundary, and certainly negligible at plate with two hole distance.

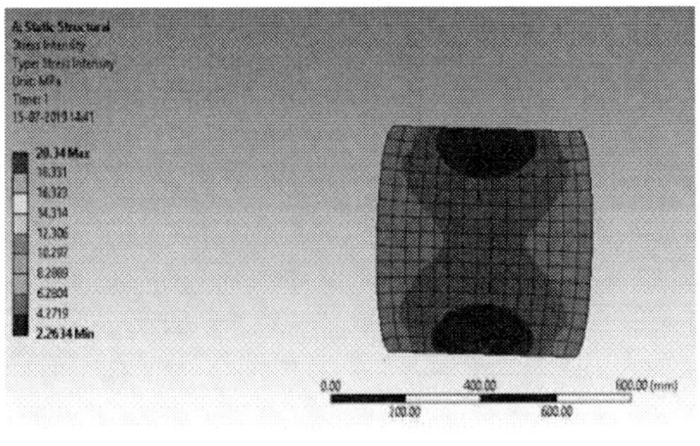

Figure 10. Single hole, d/D=0.

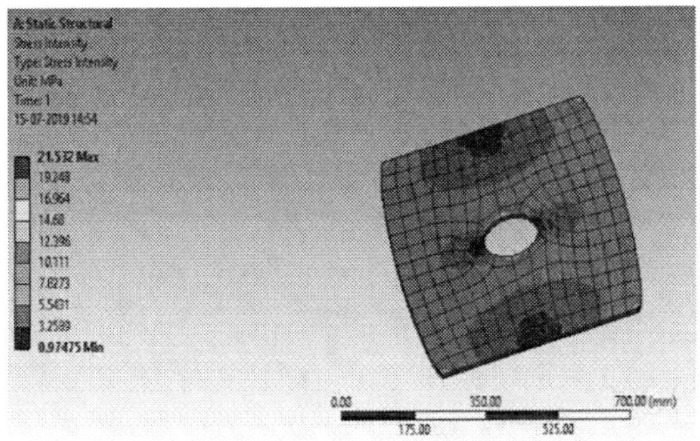

Figure 11. Single hole, d/D=0.2.

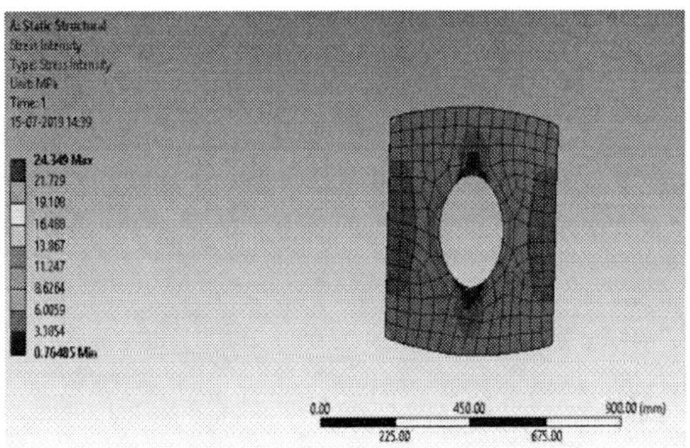

Figure 12. Single hole, d/D=0.4.

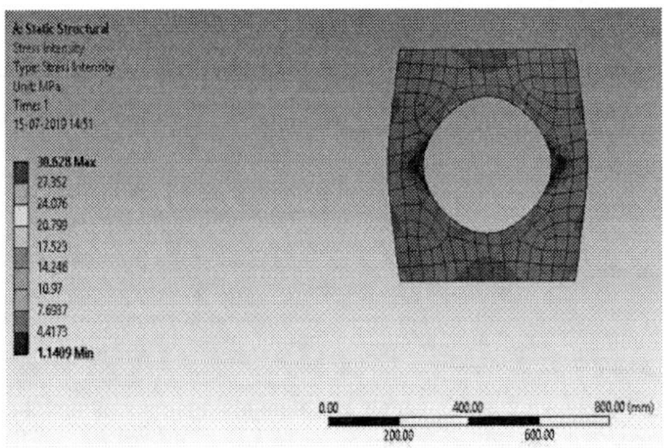

Figure 13. Single hole, d/D=0.6.

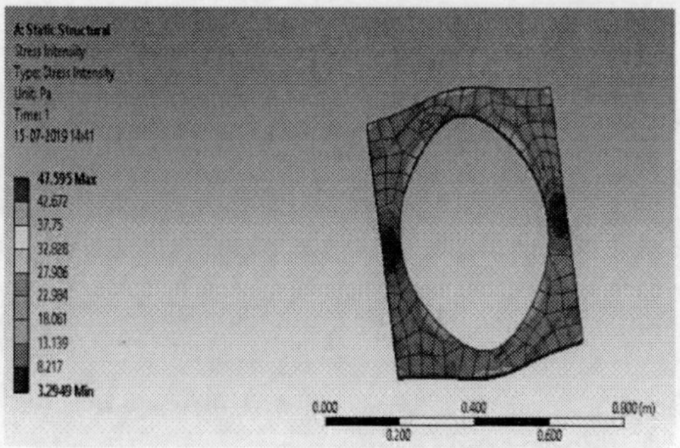

Figure 14. Single hole, d/D=0.8.

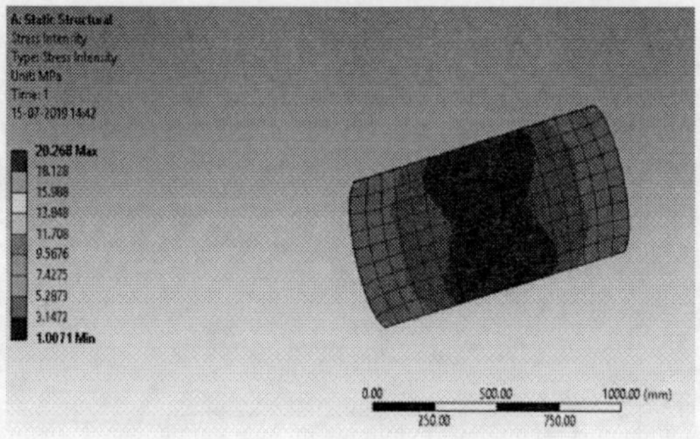

Figure 15. Two hole, d/D=0.

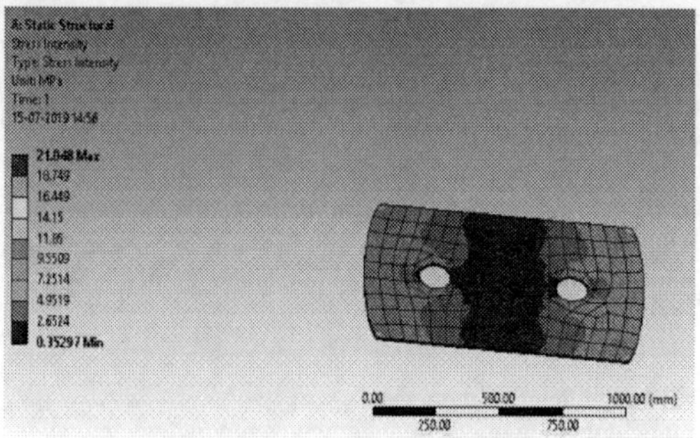

Figure 16. Two hole, d/D=0.2.

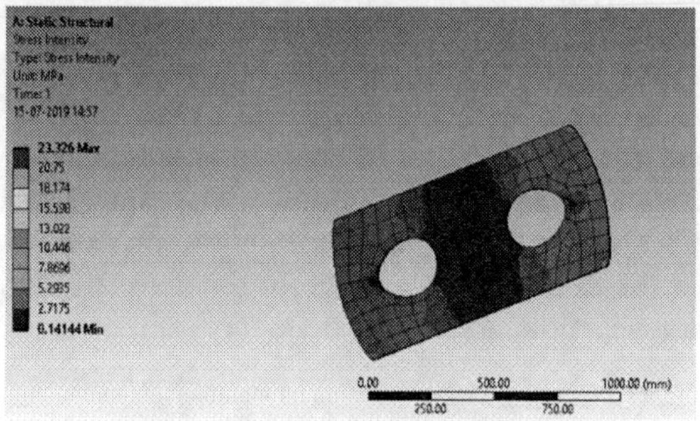

Figure 17. Two hole, d/D=0.4.

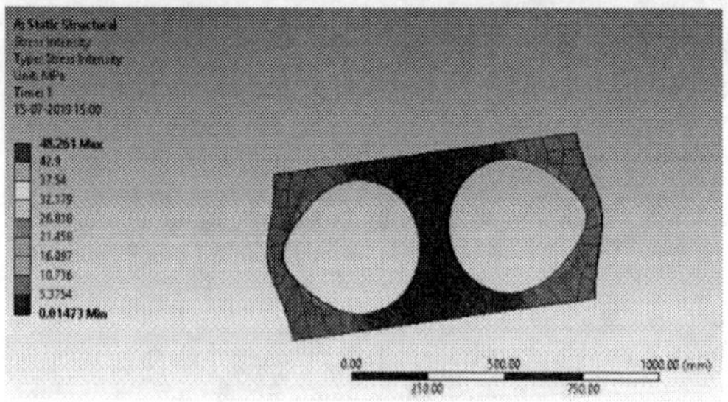

Figure 18. Two hole, d/D=0.6.

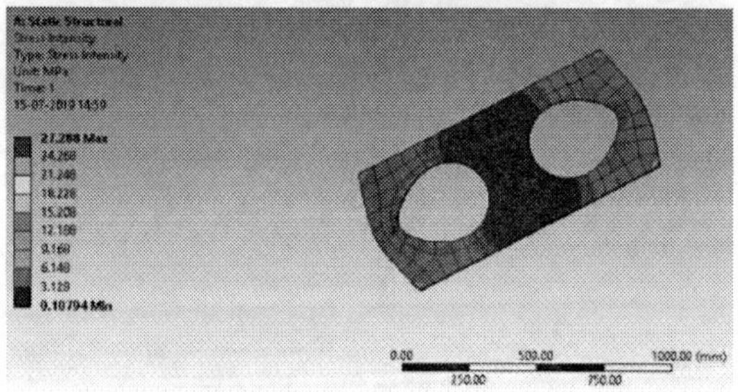

Figure 19. Two hole, d/D=0.8.

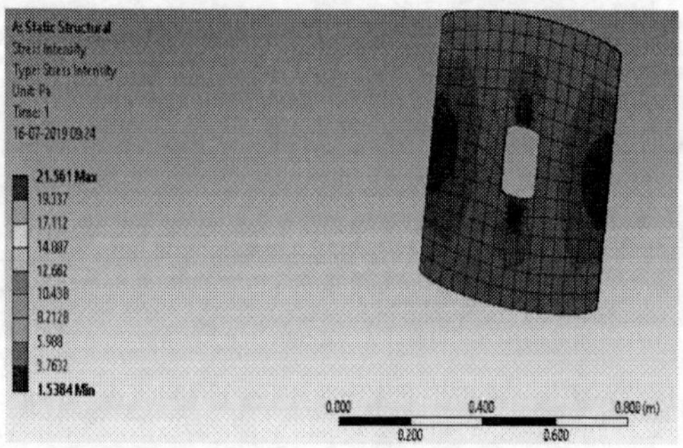

Figure 20. Single square hole, d/D=0.2.

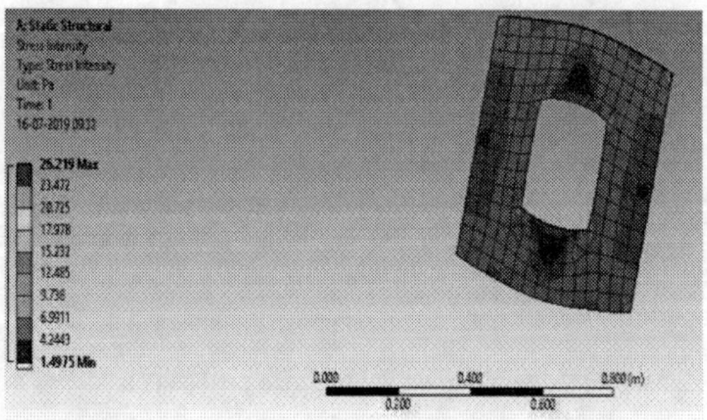

Figure 21. Single square hole, d/D=0.4.

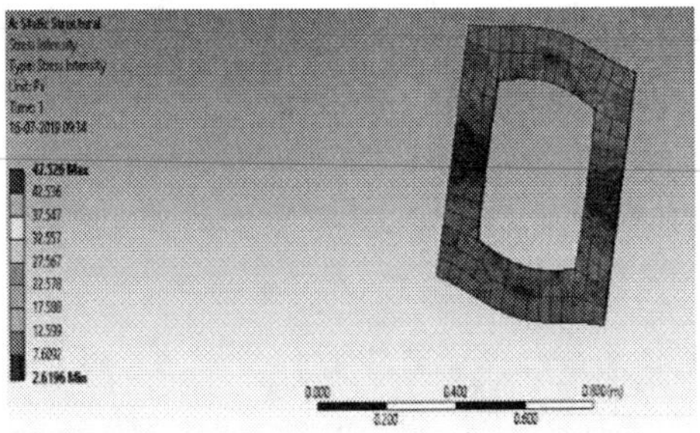

Figure 22. Single square hole, d/D=0.6.

6. Conclusions

Following conclusions are reported based on analytical and numerical investigation using finite element analysis,

- The maximum stress concentration occurs around the hole for all cases of plates irrespective of diameter and the shape of the holes.
- The stress concentration is much smaller near one diameter distance from the edge of the hole, and negligible at two diameters distance from the edge of the hole.
- The rectangular plates with two holes have higher stress concentration between the two holes wherein in the number of holes increases with increase in stress variation due to reduction in cross sectional area.
- ANSYS provides a very suitable way of determining stresses and the obtained results are not very accurate as compared to the analytical results but they can be used for the simulation of complex geometries. The accuracy can be achieved by means of reducing the mesh size to suitable limits.
- The percentage variation on analysis of rectangular plate with single hole results in reduction in variation of Kt for 0.6 to 0.8 d/b ratio.
- The percentage variation on analysis of rectangular plate with two holes results in reduction in variation of Kt for 0.8 d/b ratio.
- When the shape of hole varies form square to circle the stress concentration factor also varies invariably as for square hole, sharp corners leads to abrupt changes as per d/b ratio.

References

[1] Gokul, "Numerical Analysis of thin plates with holes", Applied Mechanics and materials, Vol.857, pp.154-158, 2016.

[2] Adel A. Al-Azzawi and Basma M. Abdul Al-Aziz, "Numerical Analysis of Reinforced light weight aggregate Concrete Hollow - core slabs", ARPN Journal of Engineering and Applied Sciences, Vol.12(6), pp.1743-1757, 2017.

[3] Durgadevagi, N A Jabedz, "Comparative Analysis of slab with different shape of cut out for various end condition by using ANSYS", Journal of Chemical and Pharmaceutical Research, Vol.8(12), pp.180-184, 2016.

[4] Lotfi Toubal,Moussakarama,Bernard Lorrain, "Stress concentration in a circular hole in composite plate", Composite structures, Vol.68(1), pp.31-36, 2005.

[5] Babulala K S, Ashenafiaduguna, Vimalkannan, A.Pradeep, "Stress concentration on rectangular plate with multiple Opposite Semi circular Notches Using Finite Element Analysis", International Journal of Mechanical and Industrial Technology, Vol.4(2), pp.7-12, 2017.

[6] H. Saksham Dhanjal, "Stress analysis of rectangular plate with circular hole using three dimensional finite element model", International journal of Engineering, business, enterprise application, Vol.12(1),pp.77-80, 2015.

[7] Dheeraj Gunwant, "Stress and displacement analysis of rectangular plate with central elliptical hole", International journal of Engineering and Innovative technology, Vol.3(3), 2013.

IOP Conf. Series: Materials Science and Engineering **764** (2020) 012018 doi:10.1088/1757-899X/764/1/012018

IoT based detection of adulteration in Gold using ANN

[*]S.Lakshmipriya, [#]S. Bhuvaneswari, [1]V. Parvathy, [2]C.R. Suganya Devi, [3]M.Abinaya.

[*,#,1,2,3] Assistant professor, [*,1,2,3] Department of Electronics and Communication Engineering, [#]Department of Mechanical Engineering, Bannari Amman Institute of technology.

*Author for correspondence Mail Id: lakshmipriyas@bitsathy.ac.in

ABSTRACT. *According to the old sayings, "All the glitters are not gold". We should be very aware of this quote. Nowadays we don't get a pure crystalline of gold (Au). The Gold (Au) cannot itself be made into proper jewellery, so it needs some impurities as such to make and sell them into a product. So, in olden days to till now the gold (Au) is mixed with Copper (Cu) to make it into a nice wearing's and products. The Copper (Cu) is not only mixed with Gold (Au) it is also mixed with silver, Palladium, Platinum and more. Nowadays the iridium is mixed with gold to enhance the weight of the gold. Thus the gold sellers sell the iridium (Ir) mixed gold (Au) to the traders. Unknowingly the traders get the gold (Au) as of such in a high cost. The iridium (Ir) cannot be identified. It can be identified only when the gold (Au) is melted. The nature and property of gold remains the same when mixed with iridium, but the mass of the gold (Au) gets inflated. Thus the proposed system uses the x-ray to identify the Proportion of iridium (Ir) in gold (Au). The Controller senses the image captured the by the x-ray machines and matches them with the other samples taken from the pure Gold (Au) through Back Propagation algorithm and finds the error. The identification of such these will not cost more. It is highly efficient and traces the impurity in gold and provides the traders with at most satisfaction.*

Keywords: X-ray, Processing units, Internet of things, artificial neural networks, Iridium (Ir), Gold (Au).

1. INTRODUCTION

The Gold (Au) is the most valuable ornament and most alluring possession of the people. Many of us like to wear Gold ornaments and love to have them. It shows the prestige, courage, love and more. Meanwhile, people like to have them because of its attraction.

The proposed system is a frame work for many other applications of finding adulteration in metals, not only in gold (Au) but also in the other metals. The figure: 1 represents the Colours of Au-Ag-Cu alloys as a function of composition.

Basically the gold (Au) is so soft to handle with so, it is been allowed to combine with the other metals to make it hard to process them into bars or coins or sheets.

IOP Conf. Series: Materials Science and Engineering **764** (2020) 012018 doi:10.1088/1757-899X/764/1/012018

$$3Cu + 2AuCl_3 \quad 3CuCl_2 + 2Au \quad \textbf{(1)}$$

Copper+ Gold Chloride → Copper chloride+ Gold precipitate

Maximum of the reactions comes with copper (Cu) as like the copper is cheap than the silver (Ag) and gives more brittleness. So, the malleability property is also been enhanced by the same. Thus the most of the alloying of gold is used with the copper (Cu). We will get the range of 22 karat gold (Au) as pure; because the 2 karats is added with impurities to make them malleable and ductile, i.e., we have 91.66% of gold (Au) and the rest is the impurities.

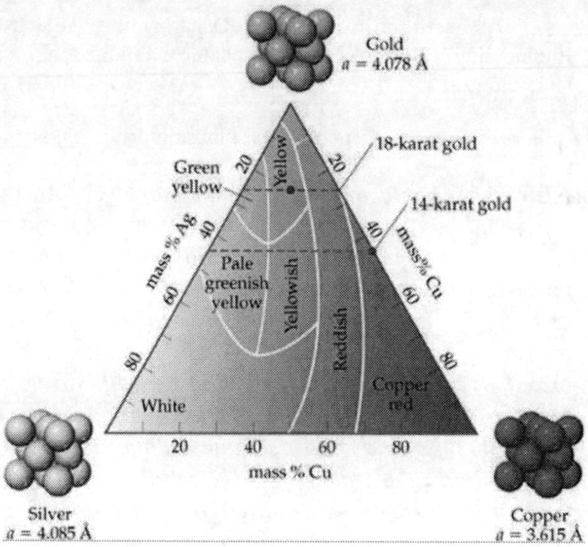

Figure: 1 Colours of Au-Ag-Cu alloys as a function of composition.

For an instance an 1Kg gold is purchased for casting we could find 70 – 80 gms of Iridium (Ir) is mixed within them. The iridium cost less than the copper, and once it is mixed with the liquid gold we cannot identify them, after once they get melted they gets separated by its way. The iridium has high melting point and cannot be easily identified during the aqua regia treatment also. Many cases the osmium, ruthenium is also used for the adulteration but not as like the Iridium (Ir).

$$\mathbf{Ircl_3 + 2Auo_3 \rightarrow AuIr + 3Clo_2 \rightarrow (2)}$$

Iridium chloride + Gold Oxide → Gold Iridium + Chloride

Mikkelsen, S. L., & et.al., (1988), in his proposed work found the adulteration by the aqua regia process. The aqua regia is a mixture of HCL and HNO_3. In which the Hydro chloric acid erodes some of the gold Particles. This causes the loss of the pure metal [2,11-17]. Cody, (1989) in his propsed work identified the adulteration in drugs by the transmission electron microscope. But the TEM is so costly to afford and it needs an expert to deal with [8]. Novakowski, W., & et.al., (2011) in his work proposed the manufacturing of the gold with copper electrodes for the sensitive element identification. Thus in their work the property of the gold is mainly discusses such that how they gets reacted to other metals and form its own property. [5]

Ali, M. E & et.al., (2012) in his work found the visual detection of adulteration of the gold is identified by the scanning electron microscope. He could identify the internal structure of the metal.

But cannot identify how many percentage of the impurities were and what kind of the impurities id been added to the Gold (Au) [1]. Mecker, L. C & et.al., (2012) in their work found the adulteration in gold by the Raman spectroscopy techniques. They used the light scattering technique, but it is not much efficient and the method is too costly to implement [3, 18-21]. Zheng, D & et.al., (2012) in their work the internal atomic structure and the molecular arrangements were been discussed briefly, such that the metallurgical ways are been identified by the molecular bindings and thus they could expose the adulterers in the metals [6].

Kumar, N., & et.al., (2014) in his paper, the attainment of the milk adulteration is identified by the gold nano particles. In his proposed work the adulteration is been identified by the calorimeter. By the same we use the X-ray spectroscopy for identifying the Adulteration in an efficient way [4, 25]. Kumar, P & et.al., (2015) in their paper, the urea adulteration in gold Nano particles is been identified by the bios sensors called the apt sensors. Thus the sensors are used to find the most remarkable adulterations in the metallurgical surfaces [7, 22]. Guo, (2015) in his proposed work, the Raman's high speed spectrograph methodology is used. Thus the efficiency is negligible [9]. Bransford & et.al.,(2000). The appreciation of the high-tech resources is discussed and the maintenance of their approval is been activated through the artificial neural networking techniques [26]. Raichle, M. E., & et.al., (2001) thus the back propagation methodology is used to consolidate the work being done and produce the output to a greater extent of efficiency [27-30].

Gubbi, J., Buyya & et. al., (2013), the Internet of things is mainly used for the data storage in the random view of the pulsating clouds. The clouds are the highly secured modes of data collecting arenas that make use of the storage of the data's. Thy deserving part of the Internet of things is that makes use of the data acquisition in the clouds [32]. Kelly, S. D. T., & et.al., (2015) in his paper iteratively manages to store the bulk data for the higher classification of ranges . The data's are taken into account and make sure to get damaged but the security system is reliable. [33-40]

2. METHODOLOGY:

Gold is the most precious metal of all times, to which the sellers adds the impurities of all kind. The impurities cannot be detected in the normal eyes. It can only be detected when melted or treated with aqua regia. Thus the proposed system uses the x-ray modelling to enhance the output of the crystallographic view. The input image is taken in the form of predefined x-ray images through the Back propagation algorithm. The algorithm used here may be so simple but it enhances the feature of the use.

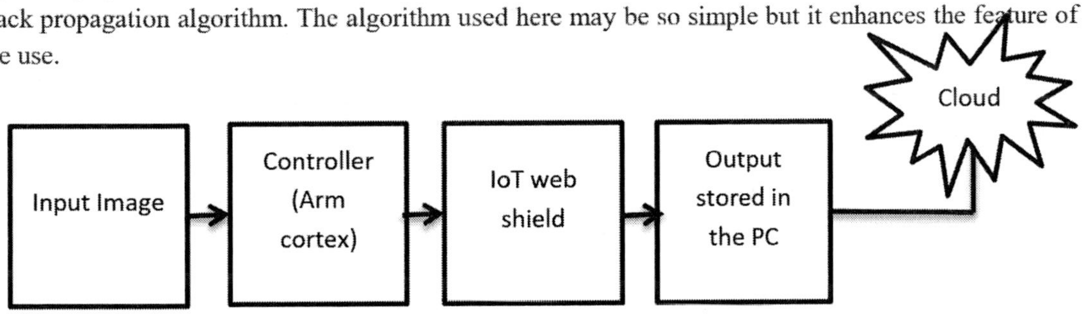

Figure: 2 block diagram of the proposed system.

The featured system blocks are the simple layout of the proposed system. The input image is fed from the X-ray scanned images which were already been taken into account by the Artificial neural networks. Thus when the ornament to be tested is placed under the system, it starts scanning by the x-rays provided with them and analyses them through the Predefined stored images in the cloud nad makes appositive result of the output containing the impurities. The sustainable impurities were been

IOP Conf. Series: Materials Science and Engineering **764** (2020) 012018 doi:10.1088/1757-899X/764/1/012018

identified by the RPROP algorithm. The controller unit used here is an Arm processor which processes the data by the usual C language.

The RPROP algorithm uses the adaptive learning techniques. Most probably the step size tries to overcome the choice of the connections during the learning processes. This technique provides the accuracy, reliability and the robustness to the output. The weight of the network is been calculated and fast forwarded to the next layers and amended as the output. The weight of the network is been denoted as W_{ij} where the weight denoted from the neuron i to the neuron j.

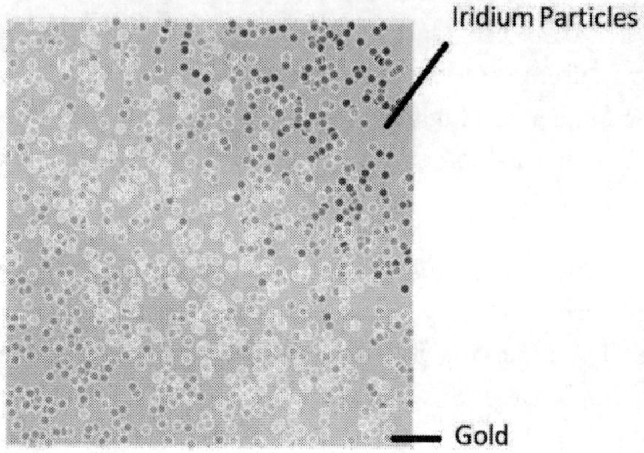

Input 1

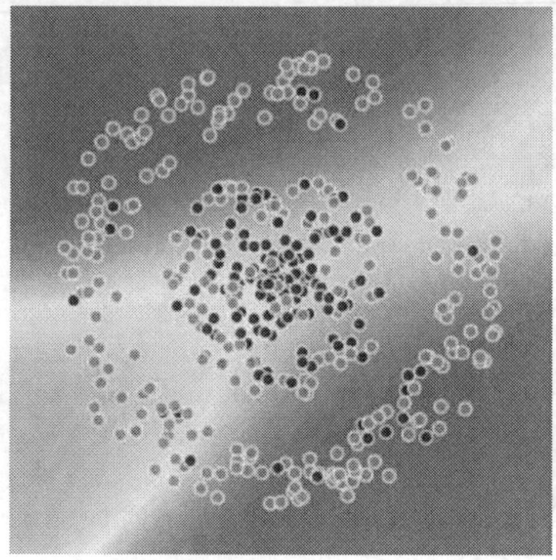

Input 2

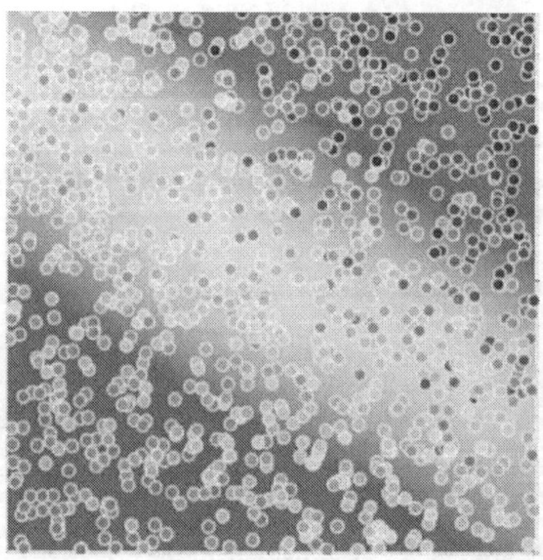

Input 3

Figure: 3 the input of the proposed system for training the Neural Network.

The input images are taken into account for training the Neural Network through the RPROP algorithm. The algorithm is based on the following rule:

$$\Delta_{ij}^{(t)} := \begin{cases} \eta^{+} \cdot \Delta_{ij}^{(t-1)} & , \text{ if } \frac{\partial E}{\partial w_{ij}}^{(t-1)} \cdot \frac{\partial E}{\partial w_{ij}}^{(t)} > 0 \\ \eta^{-} \cdot \Delta_{ij}^{(t-1)} & , \text{ if } \frac{\partial E}{\partial w_{ij}}^{(t-1)} \cdot \frac{\partial E}{\partial w_{ij}}^{(t)} < 0 \\ \Delta_{ij}^{(t-1)} & , \text{ else }, \end{cases} \quad (1)$$

where ij are the two networks of the systems

$$\text{if } \frac{\partial E}{\partial w_{ij}}^{(t-1)} \cdot \frac{\partial E}{\partial w_{ij}}^{(t)} \geq 0 \text{ then}$$

$$\Delta w_{ij}^{(t)} := -\operatorname{sign}\left(\frac{\partial E}{\partial w_{ij}}^{(t)}\right) \cdot \Delta_{ij}^{(t)} , \quad (2)$$

where w is the weight of the network and E is the Activation factor of the input and the hidden layers.

$$\text{if } \frac{\partial E}{\partial w_{ij}}^{(t-1)} \cdot \frac{\partial E}{\partial w_{ij}}^{(t)} < 0 \text{ then}$$

$$\Delta w_{ij}^{(t)} := -\Delta w_{ij}^{(t-1)} \text{ and } \frac{\partial E}{\partial w_{ij}}^{(t)} := 0 . \quad (3)$$

when the conditions gets more than the 0 value, it tends to operate in a passive mode and again we need

$$w_{ij}^{(t+1)} := w_{ij}^{(t)} + \Delta w_{ij}^{(t)} . \quad (4)$$

$$\text{if } \frac{\partial E}{\partial w_{ij}}^{(t-1)} \cdot \frac{\partial E}{\partial w_{ij}}^{(t)} < 0 \text{ and } E^{(t)} > E^{(t-1)}$$

$$\text{then } \Delta w_{ij}^{(t)} := -\Delta w_{ij}^{(t-1)} . \quad (5)$$

thus the defined equation profoundly illuminates the Equivalency through the Network layers.

The output equation of the system is defined to be

$$E_a(\boldsymbol{w}) = \sum_{i=1}^{n} \left(a^{\frac{i-1}{n-1}} (\boldsymbol{w}, \boldsymbol{o}_i)\right)^2 .$$

IOP Conf. Series: Materials Science and Engineering **764** (2020) 012018 doi:10.1088/1757-899X/764/1/012018

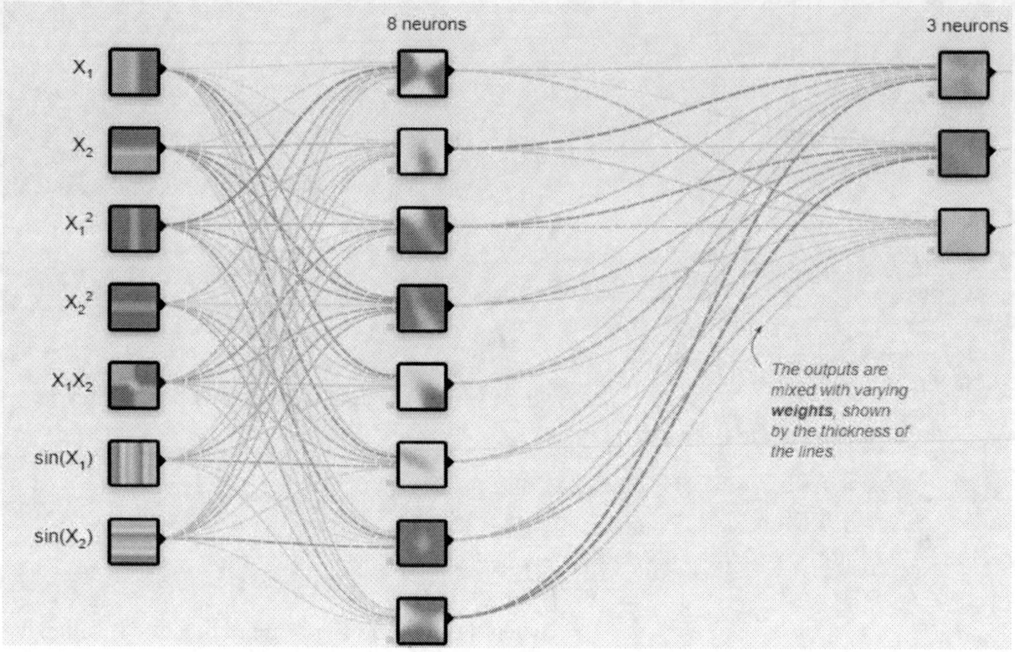

Figure: 4 neural network layers of the proposed systems.

The input layers of the network provide a meshy looping area and that trains the system by the RPROP algorithm. The algorithm trains the system to get a higher efficient and gradient system of all. The proposed system model is given below.

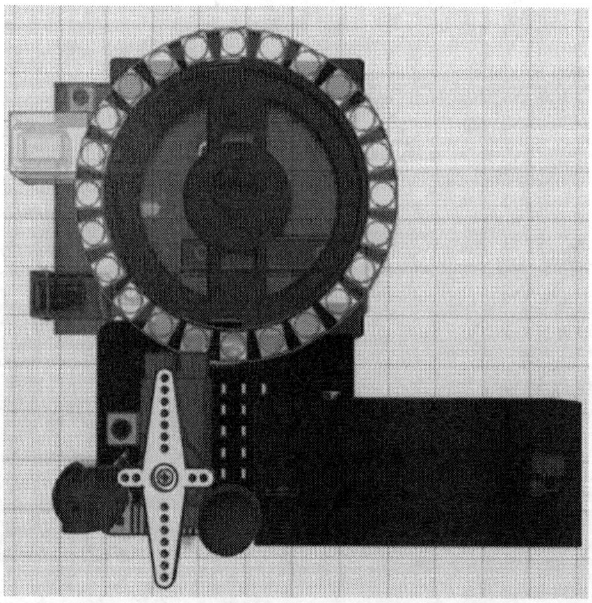

Figure: 5 proposed system model.

The output is been updated to the cloud for further updating of the inputs through the Artificial neural networks. Thus Internet of things is been playing a vital role for collecting the data base for the inputs. Thus the Iot is been Accesses by the Open foam, an Open source platform for accessing the data's.

IOP Conf. Series: Materials Science and Engineering **764** (2020) 012018 doi:10.1088/1757-899X/764/1/012018

3. RESULT AND DISCUSSION:

The Proposed system model identifies the Iridium mixing in the Gold. The iridium cannot be separated easily until it gets melted or then it gets into the aqua regia. The iridium, the rarest metal in the earth but costs very less than the other impurities is been used prominently used in the adulteration of gold. Thus the output of the system provides a good result for proving the adulteration of iridium. Thus the iridium causes cancer and prolong usage of the adulterer material causes the person with many diseases. Thus the output of the system is defined below:

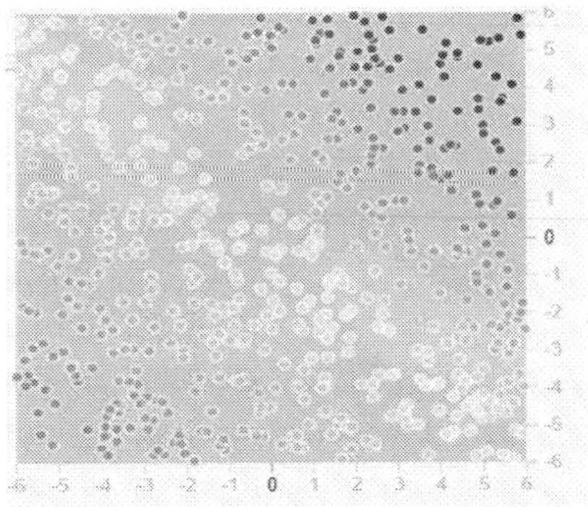

Figure: 6 Output of the proposed system.

4. CONCLUSION:

In an attempt to enhance the output stability and its performance the RPROP algorithm is used. Thus the Artificial Neural network is the main factor of unrevealing the issues in the metal. The system adheres to the greater extent of the arena that everything is been adultered. Most probably the Internet of things is used for the Data storage and back upping the outputs. The total backing up makes the system to work more consistently with higher range of efficiency with efficacy.

REFERENCES:

1. Ali, M. E., Hashim, U., Mustafa, S., Man, Y. B., & Islam, K. N. (2012). nanoparticle sensor for the visual detection of adulteration in gold formulation. *Journal of Nanomaterials*, *2012*, 1.
2. Mikkelsen, S. L., & Ash, K. O. (1988). Adulterants causing false negatives in illicit gold testing. *Clinical Chemistry*, *34*(11), 2333-2336.
3. Mecker, L. C., Tyner, K. M., Kauffman, J. F., Arzhantsev, S., Mans, D. J., & Gryniewicz-Ruzicka, C. M. (2012). Selective melamine detection in multiple sample matrices with a portable Raman instrument using surface enhanced Raman spectroscopy-active gold nanoparticles. *Analytica chimica acta*, *733*, 48-55.
4. Kumar, N., Seth, R., & Kumar, H. (2014). Colorimetric detection of melamine in milk by citrate-stabilized gold nanoparticles. *Analytical biochemistry*, *456*, 43-49.
5. Novakowski, W., Bertotti, M., & Paixão, T. R. (2011). Use of copper and gold electrodes as sensitive elements: *Microchemical Journal*, *99*(1), 145-151.
6. Zheng, D., Zou, R., & Lou, X. (2012). Label-free fluorescent detection of ions, proteins, and small molecules using structure-switching aptamers, SYBR Gold, and exonuclease I. *Analytical chemistry*, *84*(8), 3554-3560.
7. Kumar, P., Lambadi, P. R., & Navani, N. K. (2015). Non-enzymatic detection of urea using unmodified gold nanoparticles based aptasensor. *Biosensors and Bioelectronics*, *72*, 340-347.

8. Cody, J. T., & Schwarzhoff, R. H. (1989). Impact of adulterants on RIA analysis of gold for drugs. *Journal of analytical toxicology*, *13*(5), 277-284.
9. Guo, Z., Cheng, Z., Li, R., Chen, L., Lv, H., Zhao, B., & Choo, J. (2014). One-step detection of melamine in milk by hollow gold chip based on surface-enhanced Raman scattering. *Talanta*, *122*, 80-84.
10. Liu, J., & Lu, Y. (2004). Accelerated color change of gold nanoparticles assembled by DNAzymes for simple and fast colorimetric Pb2+ detection. *Journal of the American Chemical Society*, *126*(39), 12298-12305.
11. Lou, T., Chen, L., Zhang, C., Kang, Q., You, H., Shen, D., & Chen, L. (2012). A simple and sensitive colorimetric method for detection of mercury ions based on anti-aggregation of gold nanoparticles. *Analytical Methods*, *4*(2), 488-491.
12. Kalluri, J. R., Arbneshi, T., Afrin Khan, S., Neely, A., Candice, P., Varisli, B., ... & Singh, A. K. (2009). Use of gold nanoparticles in a simple colorimetric and ultrasensitive dynamic light scattering assay: selective detection of arsenic in groundwater. *Angewandte Chemie International Edition*, *48*(51), 9668-9671.
13. Jiang, Y., Zhao, H., Zhu, N., Lin, Y., Yu, P., & Mao, L. (2008). A simple assay for direct colorimetric visualization of trinitrotoluene at picomolar levels using gold nanoparticles. *Angewandte Chemie International Edition*, *47*(45), 8601-8604.
14. Jiang, Y., Zhao, H., Zhu, N., Lin, Y., Yu, P., & Mao, L. (2008). A simple assay for direct colorimetric visualization of trinitrotoluene at picomolar levels using gold nanoparticles. *Angewandte Chemie International Edition*, *47*(45), 8601-8604.
15. Baptista, P., Doria, G., Henriques, D., Pereira, E., & Franco, R. (2005). Colorimetric detection of eukaryotic gene expression with DNA-derivatized gold nanoparticles. *Journal of biotechnology*, *119*(2), 111-117.
16. Cao, R., & Li, B. (2011). A simple and sensitive method for visual detection of heparin using positively-charged gold nanoparticles as colorimetric probes. *Chemical Communications*, *47*(10), 2865-2867.
17. Qin, L., Zeng, G., Lai, C., Huang, D., Zhang, C., Xu, P., ... & Hu, L. (2017). A visual application of gold nanoparticles: simple, reliable and sensitive detection of kanamycin based on hydrogen-bonding recognition. *Sensors and Actuators B: Chemical*, *243*, 946-954.
18. Lindfors, K., Kalkbrenner, T., Stoller, P., & Sandoghdar, V. (2004). Detection and spectroscopy of gold nanoparticles using supercontinuum white light confocal microscopy. *Physical review letters*, *93*(3), 037401.
19. Li, J. F., Huang, Y. F., Ding, Y., Yang, Z. L., Li, S. B., Zhou, X. S., ... & Wang, Z. L. (2010). Shell-isolated nanoparticle-enhanced Raman spectroscopy. *nature*, *464*(7287), 392.
20. Jans, H., & Huo, Q. (2012). Gold nanoparticle-enabled biological and chemical detection and analysis. *Chemical Society Reviews*, *41*(7), 2849-2866.
21. Cai, W., Gao, T., Hong, H., & Sun, J. (2008). Applications of gold nanoparticles in cancer nanotechnology. *Nanotechnology, science and applications*, *1*, 17.
22. Ho, J. A. A., Chang, H. C., Shih, N. Y., Wu, L. C., Chang, Y. F., Chen, C. C., & Chou, C. (2010). Diagnostic detection of human lung cancer-associated antigen using a gold nanoparticle-based electrochemical immunosensor. *Analytical chemistry*, *82*(14), 5944-5950.
23. Liu, X., Xiang, J. J., Tang, Y., Zhang, X. L., Fu, Q. Q., Zou, J. H., & Lin, Y. (2012). Colloidal gold nanoparticle probe-based immunochromatographic assay for the rapid detection of chromium ions in water and serum samples. *Analytica chimica acta*, *745*, 99-105.
24. Razo, S. C., Panferov, V. G., Safenkova, I. V., Varitsev, Y. A., Zherdev, A. V., & Dzantiev, B. B. (2018). Double-enhanced lateral flow immunoassay for potato virus X based on a combination of magnetic and gold nanoparticles. *Analytica chimica acta*, *1007*, 50-60.

25. Taranova, N. A., Urusov, A. E., Sadykhov, E. G., Zherdev, A. V., & Dzantiev, B. B. (2017). Bifunctional gold nanoparticles as an agglomeration-enhancing tool for highly sensitive lateral flow tests: a case study with procalcitonin. *Microchimica Acta, 184*(10), 4189-4195.
26. Bransford, J. D., Brown, A. L., & Cocking, R. R. (2000). *How people learn* (Vol. 11). Washington, DC: National academy press.
27. Raichle, M. E., MacLeod, A. M., Snyder, A. Z., Powers, W. J., Gusnard, D. A., & Shulman, G. L. (2001). A default mode of brain function. *Proceedings of the National Academy of Sciences, 98*(2), 676-682.
28. Hecht-Nielsen, R. (1992). Theory of the backpropagation neural network. In *Neural networks for perception* (pp. 65-93). Academic Press.
29. Riedmiller, M., & Braun, H. (1993, March). A direct adaptive method for faster backpropagation learning: The RPROP algorithm. In *Proceedings of the IEEE international conference on neural networks* (Vol. 1993, pp. 586-591).
30. Vogl, T. P., Mangis, J. K., Rigler, A. K., Zink, W. T., & Alkon, D. L. (1988). Accelerating the convergence of the back-propagation method. *Biological cybernetics, 59*(4-5), 257-263.
31. Pineda, F. J. (1987). Generalization of back-propagation to recurrent neural networks. *Physical review letters, 59*(19), 2229.
32. Gubbi, J., Buyya, R., Marusic, S., & Palaniswami, M. (2013). Internet of Things (IoT): A vision, architectural elements, and future directions. *Future generation computer systems, 29*(7), 1645-1660.
33. Lee, I., & Lee, K. (2015). The Internet of Things (IoT): Applications, investments, and challenges for enterprises. *Business Horizons, 58*(4), 431-440.
34. Kelly, S. D. T., Suryadevara, N. K., & Mukhopadhyay, S. C. (2013). Towards the implementation of IoT for environmental condition monitoring in homes. *IEEE sensors journal, 13*(10), 3846-3853.
35. Huh, S., Cho, S., & Kim, S. (2017, February). Managing IoT devices using blockchain platform. In *2017 19th international conference on advanced communication technology (ICACT)* (pp. 464-467). IEEE.
36. Furber, S. (1995). Computing without clocks: Micropipelining the ARM processor. In *Asynchronous Digital Circuit Design* (pp. 211-262). Springer, London.
37. Thom, S., Cox, J., Linsley, D., Nystrom, M., Raj, H., Robinson, D., ... & Wolman, A. (2013). *U.S. Patent No. 8,375,221*. Washington, DC: U.S. Patent and Trademark Office.
38. Boggs, D., Brown, G., Tuck, N., & Venkatraman, K. S. (2015). Denver: Nvidia's first 64-bit ARM processor. *IEEE Micro, 35*(2), 46-55.
39. Iturbe, X., Venu, B., Ozer, E., & Das, S. (2016, June). A triple core lock-step (TCLS) ARM® Cortex®-R5 processor for safety-critical and ultra-reliable applications. In *2016 46th Annual IEEE/IFIP International Conference on Dependable Systems and Networks Workshop (DSN-W)* (pp. 246-249). IEEE.
40. de Oliveira, Á. B., Tambara, L. A., & Kastensmidt, F. L. (2017, April). Exploring performance overhead versus soft error detection in lockstep dual-core ARM cortex-A9 processor embedded into Xilinx Zynq APSoC. In *International Symposium on Applied Reconfigurable Computing* (pp. 189-201). Springer, Cham.

IOP Conf. Series: Materials Science and Engineering **764** (2020) 012019 doi:10.1088/1757-899X/764/1/012019

Fabrication and mechanical behaviour evaluation of A7075 basalt fly ash MMCs prepared by stir casting method

T Ramkumar Venkatasamy*, D Venkatesa Prabu, S Baskaran
Department of Mechatronics Engineering, Dr.Mahalingam College of Engineering and Technology, Pollachi, Tamilnadu, India - 642 003.

*Corresponding author email: ramkumarv@drmcet.ac.in

Abstract. In this work an attempt is made to evaluate the behaviour of A7075 MMC strengthened using fly ash and basalt prepared by means of stir-casting method. Combination of strength with better resistance to corrosion, makes metal matrix composites as an attractive choice for engineering applications like connecting rod, pistons, cylinder liner and etc. The relatively economical stir casting technique is used for the production of fly ash/Basalt ash/aluminium alloy MMCs. Three Combinations of compound MMCs are processed by changing the weight fraction of the additives (3% basalt + 7% fly ash, 5% basalt + 5% fly ash, 7% basalt + 3% fly ash). The effect of strengtheners (reinforcements) upon the mechanical properties of the compounded composites such as tensile strength, Wear and hardness is analysed. The outcomes affirm that tensile strength and wear is increased proportional to weight fraction of fly ash and basalt, results also suggests that hardness is proportional to weight fraction of the basalt. Microscopic Examination confirms the even dispersion of the basalt and flyash reinforcements in the Al7075.

1. Introduction

MMCs are known to significantly improve the properties like strength & stiffness. It also results in weight savings compared to other conventional compatriots. Availability of relatively cheaper reinforcements and developments in material processing, boosts its use as MMCs for automotive vehicles, aerospace and other structural applications [1]. Amidst the plenty of available reinforcements, fly ash is the cheapest reinforcement as it enhances the tensile strength and low density [2]. Fly ash and basalt can be effectively implemented to fabricate hybrid composite using stir casting method. Improvements are observed in tensile, compressive and hardness with increase in wt. % of fly ash and basalt [3]. The aluminium alloys 7075 are commonly used in aircraft structures and automobiles due to their notable strength-to-weight ratio, machine ability, improved wear resistance and low cost. In the current scenario reinforced aluminium composites were capturing attention in light of their low cost with advantages like tensile strength properties and wear properties [4]. Cost was the major thing for the producing components of even complex shapes. Casting is wide technology for key thing to rectify this problem. For achieving high mechanical properties, it is important to achieve the even dispersion of reinforcement within the matrix [5]. Reinforced aluminium matrix fragment have higher specific strength, modulus and good wearing resistance in comparison with unreinforced alloys. Due to the compositions these alloys were susceptible to corrosion in structural integrity of aircraft structures and automobile parts [6].The investigation was to study the mechanical and physical properties of A7075/basalt fiber with various percentages of reinforcements [7]. The hybrid MMCs were build using the stir casting method. Mechanical

IOP Conf. Series: Materials Science and Engineering **764** (2020) 012019 doi:10.1088/1757-899X/764/1/012019

properties like tensile, hardness, impact strength were investigated and microstructure of hybrid MMCs is also analysed [9].

2. Material selection

2.1. Al 7075 alloy

Al 7075 alloy with zinc has excellent fatigue strength and reasonable machinability, but very less wield ability and it is prone to corrosion compared to other alloys. The aluminium 7075 matrix base metal is shown in Figure 1. The stir-casting approach was used to fabricate the aluminium 7075 metal matrix composites as it is being one of the cost effective methods in industrial field. The results showcases nearly an even distribution and good distribution of reinforced particles within aluminium metal matrix composite. Both tensile strength and hardness of composite is improvised by incorporation of additive particles into the matrix. The heat treatment of reinforcement particles enhanced its wet ability with molten Al 7075 alloy.

Figure 1. Al7075 alloy.

2.2. Fly ash particle

Fly ash or flue-ash is added as strengtheners in molten metal for its unique property to reduce the overall weight and density. The combination of fly ash being used is given in the Figure 2 given below. The waste residue from thermal power plant is fly ash has been used as reinforcement in geo polymers. Fly-ash can be applied as filler materials, the following key properties such as Tensile Strength, Compressive Strength, Hardness, Wear rate and density were improved. Combination of fly ash [wt. %] is shown in Table 1.

Figure 2. Fly ash.

Table 1. Composition of fly-ash (wt. %).

Elements	Wt. %
Al_2O_3	30.1
SiO_2	57.93
Fe_2O_3	8.54
TiO_2	2.78
Carbon / Loss on Ignition	1.98

2.3. Basalt particle

Basalt is a common volcanic rock formed by decompression melting of the earth's mantle. It contains a fine matrix of quartz and used to create alternative building material to metal reinforcements like steel and aluminium. Basalt mesh is used for frameworks, structural reinforcement and material integrity. The Various Composition of Basalt as shown in the Table 2.

Table 2. Chemical composition of basalt (wt. %).

Elements	Wt. %
Oxides(O)	31.2
Sodium oxide(Na)	2.60
Magnesium (Mg)	3.77
Silicon (Si)	27.1
Potassium (K)	0.49
Calcium(Ca)	6.89

2.4. Fabrication of MMCs

Basalt and fly ash with aluminium alloy (Al7075) composites, prepared using stir casting method (Liquid metallurgy route) was used in this work. With meticulous care 3 different compositions of composites were prepared from the casting process as shown in the Table 3.

Table 3. Ratios of basalt and fly ash with Al7075.

S. No	Composites with Various ratio of reinforcements
1	Sample 1- Al7075 +3% Basalt + 7% Fly ash
2	Sample 2- Al7075 +5% Basalt + 5% Fly ash
3	Sample 3- Al7075+ 7 % Basalt + 3%. Fly ash

Another name for stir casting technique is vortex method, which is the most popular as well as relatively low cost liquid processing method to fabricate Metal Matrix Composites. This technique is straightforward, effortless as well as very attractive to fabricate large number quantity of production. This method is generally used for producing composite castings, with pure matrix metal interface. This type of processing assures the securing of intact reinforcement materials along with high strength. The stir casting operation of equipment setup is shown in Figure 3 (a-b). Aluminium alloy is processed inside the container at temperature of 750°C for 1 hour.

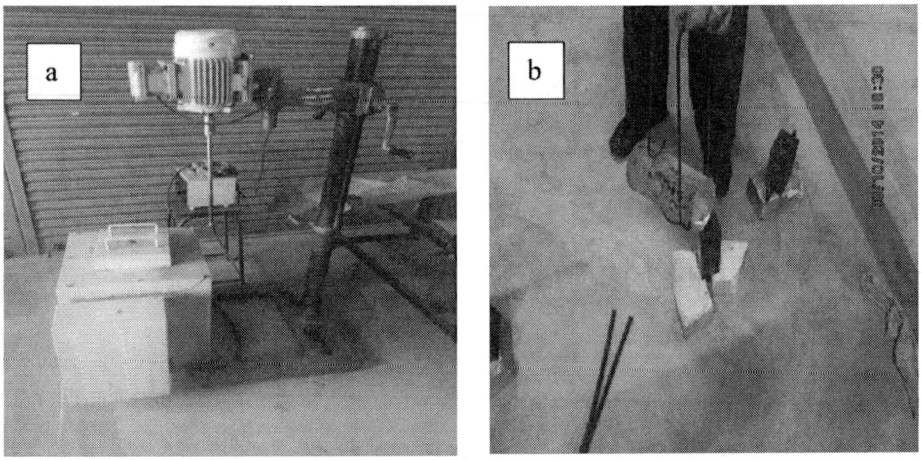

Figure 3(a-b). Stir casting process.

The reinforced particles (fly ash and basalt) are preheated initially at 500C for one hour inside the furnace to demoisturise the particles surface for better adhesiveness. Researchers established that preheated fly ash and basalt removes the efficient surface impurities, desorption of gases and modifies surface compositions due to the accumulation of an oxide layer on the surface. The molten MMC is trembled for the 2 minutes to create a vortex after that it is added to preheated particulates. Vortex method is recommended to disperse the particles evenly among the metal matrix. Post vortex molten matrix is mixed for 3 minutes at the speed of 200rpm in order to alleviate void formation and impurities on the surface to make it porous. Stirring rpm is the significant process parameter as stirring is necessary to help in promoting wet ability, i.e. bonding between matrix and reinforcement. As solidifying rate is faster it will increase the percentage of wet ability. The stirring temperature is a major important process parameter to fabricate the new product. It is related to the melting temperature of metal matrix composites, i.e. Aluminium. Aluminium generally melts at 600°C - 650°C. The processing temperature is determined by viscosity of melting point of Al matrix composite. The viscosity influences the particle distribution in the matrix of the composite.

2.5. Mechanical testing

Mechanical behaviour of composite material is investigated using required standard dimension specimens which were prepared for tensile test, hardness test, wear test and microstructure analysis. The tensile test was taken using UTM as per ASTM E8. The specimens were prepared for measuring Vickers hardness indenter by polishing them with suitable grades of emery and etching them finally. Wear testing was done using pin on disc tribometer for the Al 7075 MMC composite specimens. ASTM G99 standards was followed for conducting wear test. The composite alloy pins were made up of circular cross section of 9 mm diameter and 25 mm length. The composites prepared were examined under optical microscope after they were belt grinded and polished by different grades of emery papers.

3. Results and discussion

3.1 Tensile test

Computerized UTM (Universal Testing Machine) is used to find out the tensile strength. The samples were machined as per the ASTM E8M-13A specification. The tensile testing of Aluminium with the different % of basalt and fly ash reinforcement is depicted in Figure 4.

Addition of weight % of fly ash and basalt increases tensile strength of composites. The specimen prepared to be a two end jaws of the universal testing machine. Point load is progressively applied on the specimen with movable cross head, until the specimen breaks. Figure 6 Composites Vs. Ultimate Tensile Strength. Volume fraction affects the tensile strength. The tensile strength of sample1 is 88.437N/mm2, sample2 is 104.221N/mm2, sample3 is 175.619 N/mm2, and this effect is due to the hardening of the aluminium alloy in composition of fly ash and basalt particles. The composites Vs % Elongation graph as shown in the Figure 6.

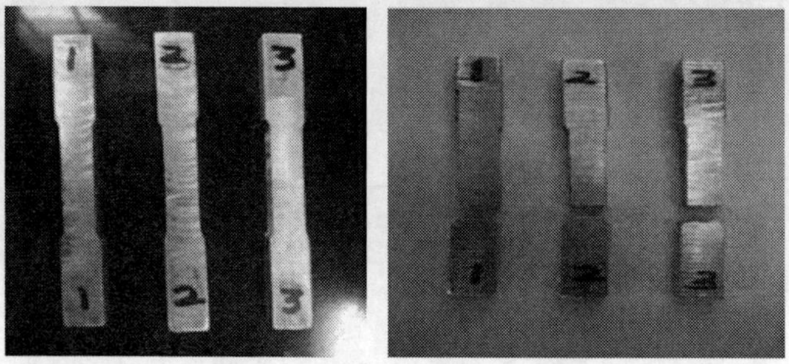

Figure 4. Tensile test specimens before and after testing.

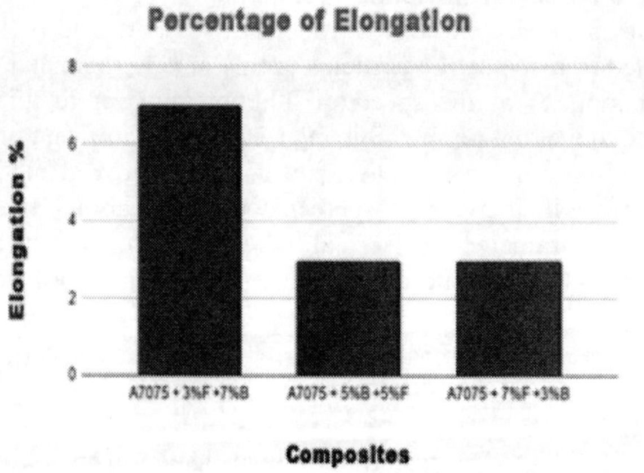

Figure 5. Percentage of elongation for different composites.

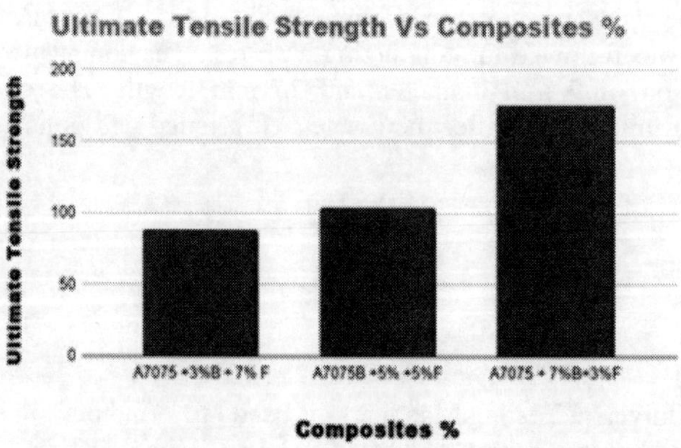

Figure 6. Different composites % vs. ultimate tensile strength (Mpa).

IOP Conf. Series: Materials Science and Engineering **764** (2020) 012019 doi:10.1088/1757-899X/764/1/012019

3.2 Chemical composition

Chemical constitution is verified before and after reinforcement (sample 2 composite) of aluminium composites as shown in the Table 4. Basalt and fly ash reinforcement maximizes the percentage of silicon, copper, ferrous and Zinc. Silicon increases the hardness value up to 70Mpa and the ultimate tensile strength is increased due to addition of copper. The presence of ferrous increases the mechanical property such as ductility and presences of zinc having high density precipitates.

Table 4. Chemical composition of A7075 before and after reinforcement (sample 2).

Element %	Before Reinforcement	After Reinforcement
Si	0.138	0.181
Fe	0.334	0.336
Cu	1.49	1.53
Mn	0.0830	0.079
Mg	2.68	2.58
Zn	5.86	5.66
Al	Balance	Balance

3.3 Hardness test

Hardness test pieces were made as per the dimensions shown in the Figure 7. Hardness testing using Vickers machine were used to find the hardness value with applied load of 500gm and 10sec as dwell period. The maximum value of hardness occur at the composition of Al7075+7%basalt+3% fly ash as shown in the Figure 8.

Figure 7. Hardness specimens.

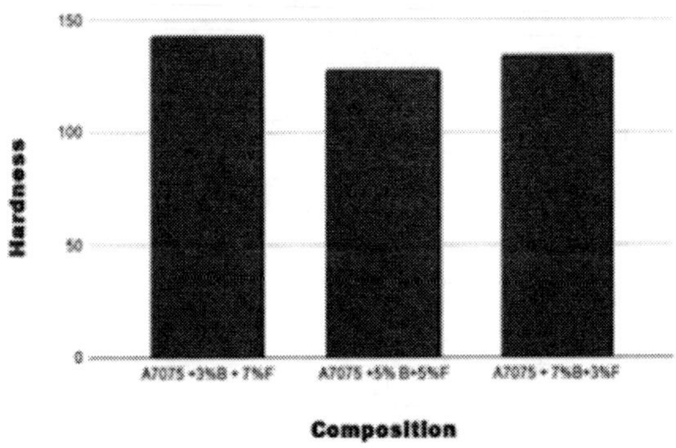

Figure 8. Different composites vs. hardness.

IOP Conf. Series: Materials Science and Engineering **764** (2020) 012019 doi:10.1088/1757-899X/764/1/012019

3.4 Wear test

Figure 9 depicts the pin on disc wear testing apparatus on which the wear testing is carried out, the sample for wear test was prepared with standard dimension of 9mm diameter and 50mm height was shown in the Figure 10. Wear test were conducted using pin on disc machine. The wear test was conducted at three different loads, with increments 1Kg, 2Kg then the speed, time and track radius keeping constant. The minimum value wear occur at the Al7075+3%basalt+7%fly ash as shown in the Table 5.

Figure 9. Pin on disc apparatus.

Figure 10. Wear test specimens.

Table 5. Wear rate and coefficient of friction.

Compositions	Wear Rate (µm)		Coefficient of Friction (µ)	
	Load (1 Kg)	Load (2 Kg)	Load (1 Kg)	Load (2 Kg)
Sample 1	45	157	0.44	0.49
Sample 2	57	165	0.49	0.57
Sample 3	44	149	0.45	0.49

4. Microstructure

The voids are obtained by the amount of gas rejected by the melting. If the gas is released in large amount, it results in sphere shaped pores, it continues to grow through diffusion before coherency of dendrite is finally reached as shown in the Figure 11 (a) & 11 (b). Their size will be the order of tenths of millimeter to 1 millimeter. This kind of defect is termed as gas porosity. A lower amount of gas rejection softens the metal. After dendrite coherency pores formation occurs during subsequent stages of solidification. The concurrent build-up of shrinkage cavities nucleated by pores results in micro shrinkage.

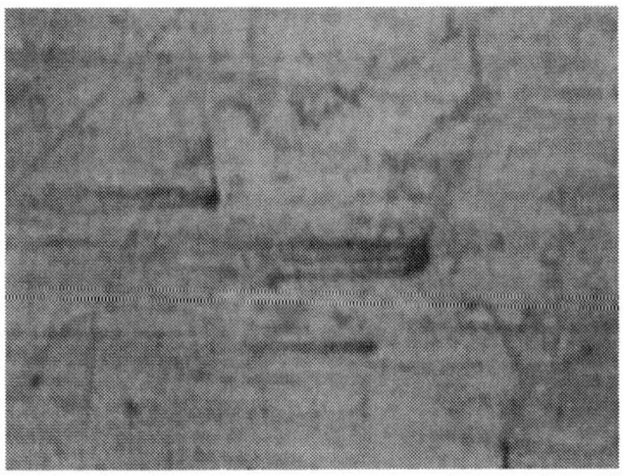

Figure 11 (a). Microstructure of Al7075 alloy.

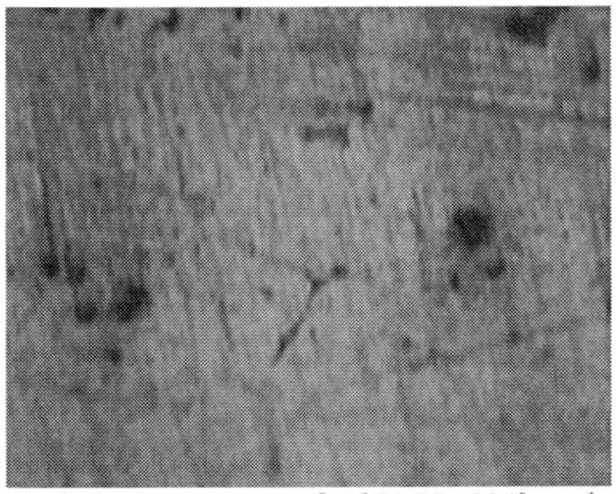

Figure 11 (b). Microstructure of Al7075+3%basalt+7%fly ash.

5. Conclusion

Combination out of Aluminium 7075-fly ash and basalt blends were fabricated by means of Stir-casting method along with perfect diffusion of fly ash and basalt particles spread evenly through the specimen. MMCs (7075) reinforced with fly ash and basalt particles considerably improved the mechanical properties.

- In comparison to the Al7075 + 3% basalt + 7% fly ash and Al7075+5%basalt+5%fly ash, the reinforced Al7075 + 7% basalt + 3% fly ash exhibits more tensile strength. It is due to the hardening of aluminium alloy by basalt and fly ash particles.
- With Al7075 + 3% basalt+7% Fly Ash composite, hardness of composite is maximized. It happens because of the presence of basalt along with addition of 7% of flyash helps to enhance the hardness value of the material.
- The third composition (Al 7075+3% Basalt Ash +7% Fly Ash) reduced the wear rate value of the composite material. Existence of fly Ash and basalt helps to increasing wear resistance value of hybrid material in both 1kg and 2kg loading condition.
- Microscopic evaluation affirms that interfacial bonding between Al 7075+basalt 3%+flyash7%.

IOP Conf. Series: Materials Science and Engineering **764** (2020) 012019 doi:10.1088/1757-899X/764/1/012019

References

[1] D.Singla and S. R.Mediratta, "Evaluation of mechanical properties of Al 7075-fly ash composite material", International Journal of Innovative Research in Science, Engineering and Technology, Vol.4(2), pp.951-959, 2013.

[2] G.Venkatachalam, A.Kumaravel, N.Arun Kumar and Dhanasekaran Rajagopal, "Effect of Li and Zr addition on microstructure and mechanical properties of modified LM25 aluminium alloy", International Journal of Applied Engineering Research, Vol.10, pp.647-651, 2015.

[3] R.Dhanasekaran and K.Sathishkumar, "Microstructure, Mechanical properties of A356/Li aluminum alloy fabrication by stir casting method", International Journal of Applied Engineering Research, Vol.10, pp.416- 419, 2015.

[4] S.R.Umashankar, "Effect of water chill in metallic mould during solidification of Al alloy (7075) reinforced with bottom ash, magnesium", International Journal of Science Technology Engineering, Vol. 1(5), pp.2-9, 2014.

[5] Srisaran Venkatachalam, S.Baskaran, R.S. Karrthik, Deepak B. Thimiri Govinda Raj, T. Ramesh Kumar, "Titanium diboride reinforced aluminum composite as a robust material for automobile applications", AIP Conference Proceedings, Vol.2128, pp. 020010-1–020010-10, 2019.

[6] S.Mohanty and Y.P.Chugh, "Development of fly ash-based automotive brake lining", Tribology International, Vol.40, pp.1217-1224, 2007.

[7] T.Senthilvelan, S.Gopalakannan, S.Vishnuvarthan and K.Keerthivaran, "Fabrication and characterization of SiC, Al2O3 and B4C reinforced Al- Zn-Mg-Cu alloy (AA 7075) metal matrix composites: a study", Advanced Materials Research, Vol.622, pp. 1295-1299, 2012.

[8] M.S.Reddy, S. V.Chetty, and S.Premkumar, "Effect of reinforcements and heat treatment on tensile strength of Al-Si-Mg based hybrid composites", International Journal of Applied Science and Engineering Research, Vol.1, pp.176-183, 2012.

[9] P.Sharma, G.P.Chauhan and G.Sharma, "Production of AMC by stir casting an overview", International Journal of Contemporary Practices, Vol.2, pp.23-46, 2011.

IOP Conf. Series: Materials Science and Engineering **764** (2020) 012020 doi:10.1088/1757-899X/764/1/012020

Analysis of different fixed and flexible dielectric materials on microstrip patch antenna for ISM band applications

M Prasad, P Ramya*, C Poongodi, R Prasanna, S Raj Kamal and M Praveen Kumar

Department of Electronics and Communication Engineering, Bannari Amman Institute of Technology, Sathyamangalamm-638401, India.

*Corresponding author email: ramyap@bitsathy.ac.in

Abstract. Recently, advancement in wearable technologies, one challenging improvement is textile antenna. The most requirements of textile antennas are dielectric materials which include fabric substrate and the structure of the antenna. The characteristics of the antenna depends on the substrate material .The dielectric value of the fabric material can be calculated from the operating frequency of patch antenna. In this paper the micro strip patch antenna is designed for ISM band applications. The main objective of this paper is to design and compare the performance of the micro strip patch antenna on different fixed and flexible dielectric material like FR-4, RT DUROID 6010, Polystyrene, Teflon, Quartz, Cotton, Alumina, Polyester with different constant in terms of gain, return loss, radiation patterns The proposed antenna operate at the resonant frequency 2.45GHz(ISM band) with low return loss and high gain and directivity.The size of the designed antenna has been also miniaturized.

1. Introduction

The development of antenna technology for man-machine boundary has taken quantum hop in employing textile dielectric materials as antenna substrates. In that, textile materials form attractive substrates hence textile antennas can be simply integrated into clothes. Fabric substrates have a very low dielectric constant, which reduces the skin effect and improves the performance of the antenna. Wearable antennas for medical applications are design by textile materials, and these materials are easily available. The antenna design depends on characterization of their electric and electromagnetic properties. A wearable system is always on, does not restrict the user's activities and is aware of the user's situation. The wearable antenna integrates fabric into the wireless communication system. The fast advancement of wireless communication technology, many researchers are paying greater attention to the design of antenna system for body area networks (WBANs). Various electronic devices are interfaced with human body by wireless body are network. Several frequency bands have been allotted for WBAN systems, such as; the Medical Implant Communication System (MICS: 400 MHz) band, the Industrial Scientific Medical (ISM: 2.4 and 5.8 GHz) band, and the Ultra-wideband (UWB: 3–10 GHz).There exist various antennas for medical applications, printed textile antennas are always find way due to its low dielectric value, ease of fabrication, compactness and low profile. Though, the major deal with the designing antennas for healthcare applications is its shape and size of the device. Various textile materials are used for the design of wearable antenna in Ref [1-4]. Diverse techniques are used to reduce the size of the antenna for medical application in Refs

IOP Conf. Series: Materials Science and Engineering **764** (2020) 012020 doi:10.1088/1757-899X/764/1/012020

[5-6].Different feeding methods and bandwidth enhancing techniques are used to enhance the performance of the antenna [7-9]. Textile antenna is a better candidate for wearable applications, as it can be made embedded into clothing [10-13].The primary objective of this paper is to analyze and compare the performance of the antenna on different dielectric materials

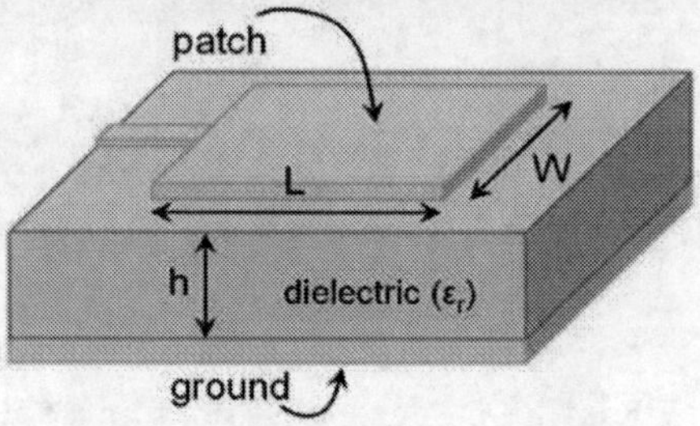

Figure 1. Micro strip patch antenna.

Patch antennas are employed for wireless applications. Microstrip Patch antennas are low profile easy to construct and analyze. Compact size, radiation pattern and resonance frequency of antenna show the major attractions. Prior to develop the proposed antenna basic structure of Microstrip patch antenna is considered as shown in figure 1. A dielectric material has conducting material on one side and other side it has a ground plane. A rectangular patch antenna is a fashionable one due to its compactness and it has low profile and can be constructed on printed circuit board using printed circuit (photolithographic) techniques. Line feed is connected to the patch it gives the best impedance match to a 50Ω system it has the one-half wavelength of transmission line. There is no hard rule for selection of antenna for different applications but the cost and size of the antenna plays a crucial role so before designing an antenna the characteristics of the antenna should be analyzed using RF momentum.

2. Design of Patch Antenna

The dielectric constant and thickness of the substrate, length and width of the patch are the key parameters for designing an antenna. The design computation steps are listed below

2.1. Calculation of width (W)

$$Width = \frac{c}{2f_0}\sqrt{\frac{2}{\varepsilon_r + 1}} \qquad \text{---------------- (1)}$$

Where
c - free space velocity of light, 3 x 108 m/s
f_0 - frequency of operation
ε_r - dielectric constant

2.2. Determination of effective dielectric constant (εeff)

$$\varepsilon_{reff} = \frac{\varepsilon_r + 1}{2} + \frac{\varepsilon_r - 1}{2}\left(1 + 12\left(\frac{h}{W}\right)\right)^{-\frac{1}{2}} \qquad \text{------------- (2)}$$

Where;
ε_r - dielectric constant
h - Height of dielectric substrate
W - Width of the patch

IOP Conf. Series: Materials Science and Engineering **764** (2020) 012020 doi:10.1088/1757-899X/764/1/012020

2.3. *Determination of physical length of patch (L)*

$$L = L_{eff} - 2\Delta L \qquad \text{------------------ (3)}$$

$$L_{eff} = \frac{1}{2 f_0 \sqrt{\mu_0 \varepsilon_0 \varepsilon_{reff}}} \qquad \text{------------------ (4)}$$

Where;

c - Free space velocity of light, 3 x 108 m/s

f_0 - frequency of operation

ε_{reff} - effective dielectric constant

2.4. The extension of length calculation (ΔL)

$$\frac{\Delta L}{h} = \frac{0.412\left(\varepsilon_{reff} + 0.3\right)\left(\frac{Width}{h} + 0.264\right)}{\left(\varepsilon_{reff} - 0.258\right)\left(\frac{Width}{h} + 0.8\right)} \qquad \text{------------ (5)}$$

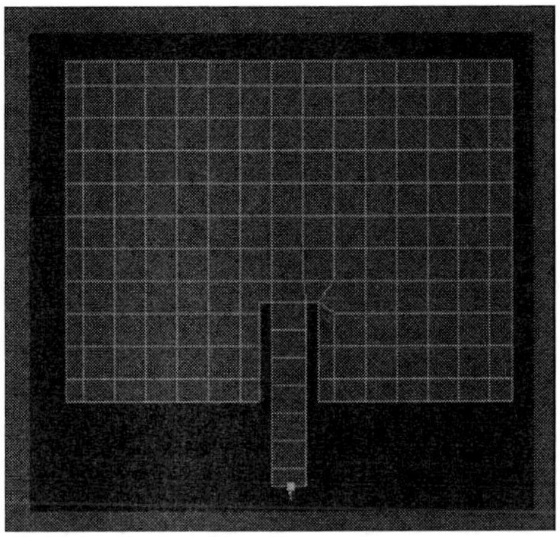

Figure 2. Structure of the proposed antenna.

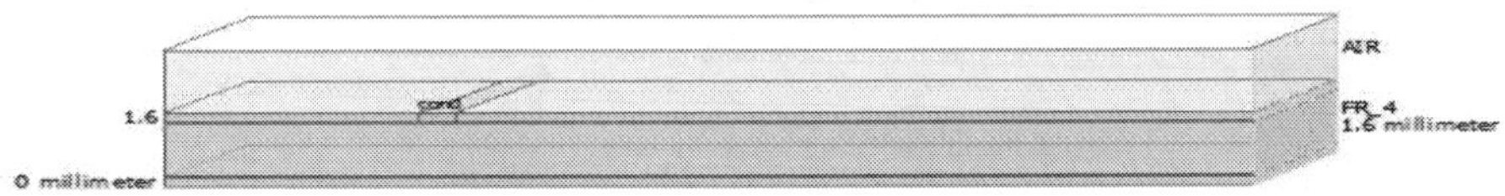

Figure 3. Substrate of the antenna.

The size of the proposed antenna are obtained using the equations [1-5] and its optimized parameters and the antenna structure are shown in figure 2 and figure 3 and table 1.

IOP Conf. Series: Materials Science and Engineering **764** (2020) 012020 doi:10.1088/1757-899X/764/1/012020

Table.1 Optimized Parameters of the proposed antenna.

Parameters	Value (mm)
Length	20
Width	30
Feed width	2.98
Feed length	14
Substrate (constant)	4.6
Substrate Thickness	1.6
Conductor Thickness	0.35

2.5. Analysis

The microstrip patch antenna is analyzed using transmission line method. In this currents are equally distributed on the edges of the patch antenna. Antenna acts as a resonator where it has electric and magnetic field that varies along the length of the antenna .The fringing field around the edges of the radiator induce the antenna to radiate. The size of the antenna and substrate thickness has control the fringing field. The main cause of fringing effect is due to width and position of feed in antenna. Thus it is essential to consider effective substrate material for analysis. FR4, RT DUROID, Alumina, Quartz, Cotton, Teflon, Polyester as substrate materials for simulation. The properties of the above materials are referred to in [1]. Where cotton and FR34 has low dielectric values compared to other materials.

3. Results and discussions

3.1. *Return Loss Characteristics*

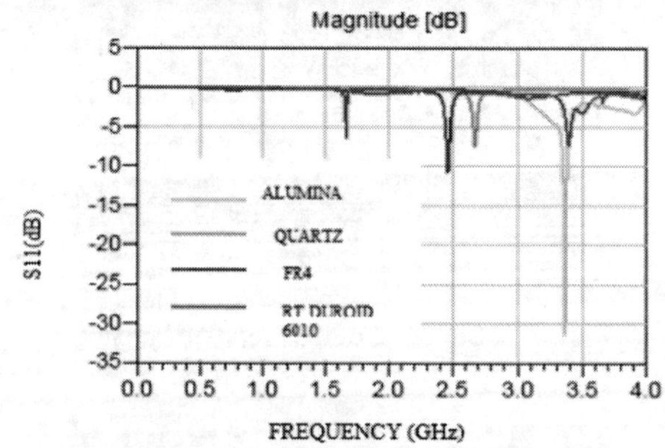

Figure 4. Return loss of fixed substrate.

IOP Conf. Series: Materials Science and Engineering **764** (2020) 012020 doi:10.1088/1757-899X/764/1/012020

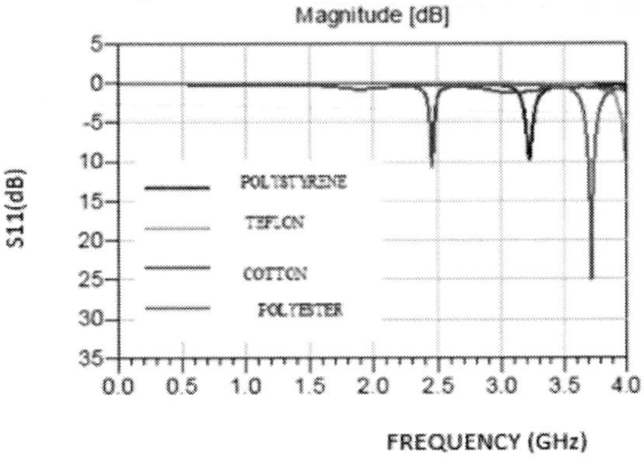

Figure 5. Return loss of flexible substrate.

The return loss is the proportion of a signal that is reflected as a result of an impedance mismatch. Figure 4 shows the return loss characteristics of different fixed substrates .In this four types of fixed substrate is considered with different dielectric constants and same thickness .the substrates like Quartz, Alumina, RT Duroid 6010 resonate at different frequencies only the substrate flame retardant (FR4) is resonate at required frequency2.45GHZ with the return loss of -10dB.

The above figure 5 shows the return loss characteristics of different flexible substrates. In this four types of flexible substrate is considered with different dielectric constants and same thickness. The substrates like Polyester; Polystyrene; Teflon resonate at different frequencies only the cotton substrate is resonate at the appropriate frequency of 2.45GHZ with the return loss of -10dB. Therefore both FR4 and Cotton materials are suitable for ISM band applications

3.2. Radiation Pattern

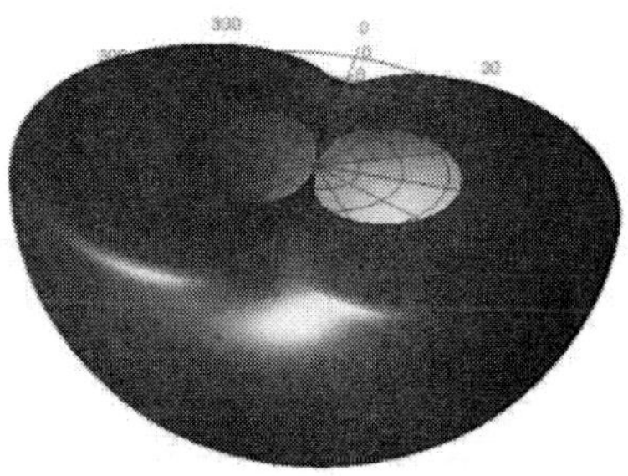

Figure 6 (a). 3D Radiation pattern of antenna at 2.45GHZ for FR4 Substrate.

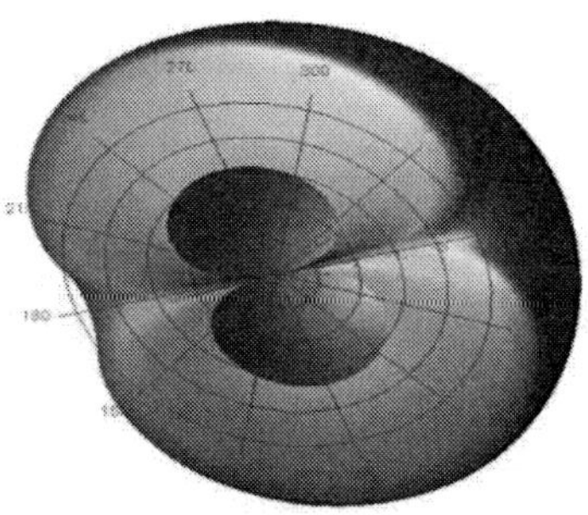

Figure 6 (b). 3D Radiation pattern of antenna at 2.45GHZ for cotton Substrate.

IOP Conf. Series: Materials Science and Engineering **764** (2020) 012020 doi:10.1088/1757-899X/764/1/012020

A radiation pattern defines the variation of the power radiated by an antenna as a function of the direction away from the antenna.it is important to determine the radiation characteristics of the antenna such as directivity,gain and beamwidth. Figure 6 (a) and figure (b) shows the three dimensional radiation pattern of proposed antenna for fixed and flexible substrate .In this, the red colour indicates the radiation field it has high current distribution and green colour indicates the reactance field which store the energy it depends the distance of the antenna. Both FR4 and cotton provides the high gain and directivity of 4.56 dB and 5.63 dB at 2.45GHz which is shown in table 2.

3.3. Field Pattern

Figure 7 below, shows the two dimensional field pattern of an antenna for different theta and phi values for theta it provides eight shape pattern and for phi it gives omnidirectional pattern.

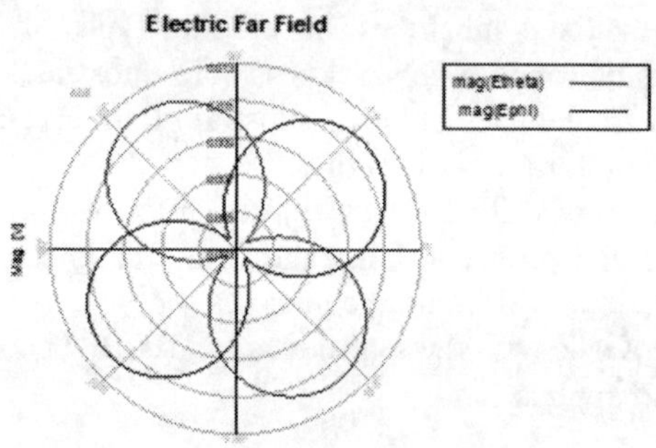

(a) ϴ=0.

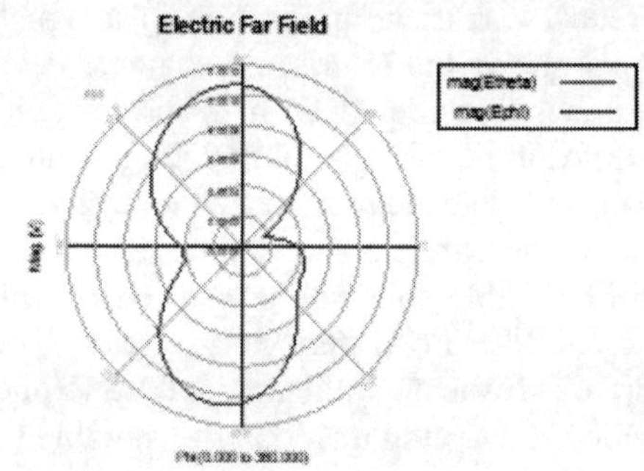

(b) ϴ=90.

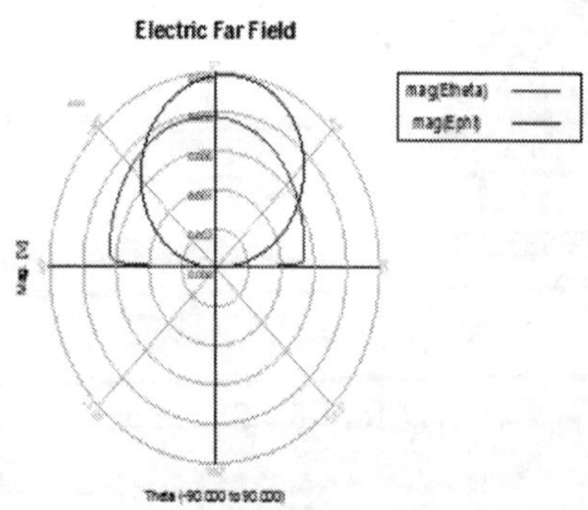

(c) Φ=0.

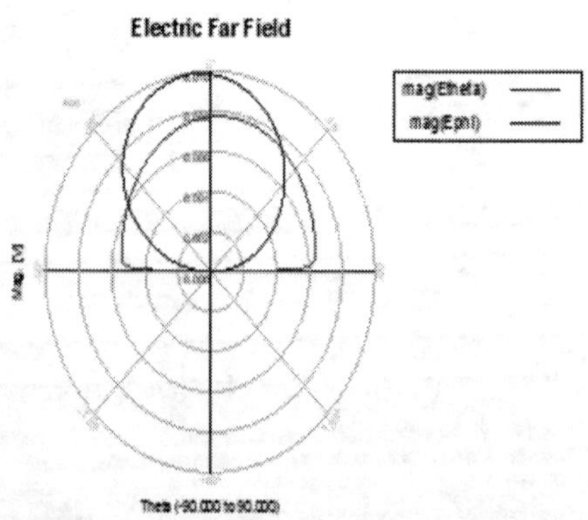

(d) Φ=180.

Figure 7. Field pattern of an antenna.

3.4. Comparision of Design Parameters of Proposed Antenna

Table 2. Performance parameters of the antenna.

Parameters	Values
Frequency (GHz)	2.45 GHZ
Directivity (dB)	5.63619 dB
Gain (dB)	4.56 dB
Radiation Efficiency (%)	81

Table 3. Comparision of performance parameters for fixed substrate.

Dielectric Substrate	Dielectric Constant	Return Loss(dB)	Loss Tangent	Frequency (GHz)
FR4	4.36	-10.73	0.01	2.45
RT Duroid 6010	10.2	-7.85	0.0023	3.4
Alumina	9.9	-32.34	0.0001	3.35
Quartz	3.78	-8.63	0.001	2.65

Table 4. Comparision of performance parameters for textile substrate.

Dielectric Substrate	Dielectric Constant	Return Loss(dB)	Loss Tangent	Frequency (GHz)
Teflon	2	-10	0.0002	4
Cotton	1.60	-10.73	0.002	2.45
Polyester	1.90	-2.5	0.0045	3.75
Polystyrene	2.5	-9.97	0.0004	3.25

Table 3 and Table 4 provides The summary of the simulated results of different substrate materials in terms of Return loss, Loss Tangent and Center frequency (CF).In that FR4 is operate at the 2.45GHZ frequency with high return loss of 10.73dB and in flexible material cotton is reasonate at same band of frequency with the return loss of -10.73 dB.So these materials can be suitable for ISM band applications.

4. Conclusions

The eight different dielectric substrate materials – FR4, RT Duroid 6010, Alumina, Quartz, Teflon, Cotton, Polyester, and Polystyrene are used as a substrate in micro strip patch antenna and its performance is analysed. Cotton and FR-4 substrate gives better efficiency with improved bandwidth compared to the other substrate also the dielectric value of these substrate is very low among the eight substrates which increases the bandwidth because the bandwidth is inversely proportional to the dielectric permittivity. It decreases the reflection coefficient as -10.73.Also; FR-4 has highest tensile strength and breakdown voltage due to which it does not succumb to the electrical pressure easily. So it does not lose its insulating property. Water absorption changes dissipation factor or loss tangent of the dielectric substrate. Loss tangent value of FR-4 is very low due to its water absorption property. Bandwidth and radiation pattern depends on loss tangent also, so FR-4 is good dielectric substrate for microstrip patch antenna. Thus the simulation results in terms of Frequency, gain, directivity, electric far field. Shows that the FR4 and Cotton provides high return loss and gain with reduced thickness and dielectric constant. Therefore these two substrates can be used for ISM band applications.

Acknowledgment

The authors acknowledge the support of Bannari Amman Institute of Technology, Sathyamangalam, India for providing the laboratory facilities to carry out this work.

References

[1] Rita Salvado, Caroline Loss, Ricardo Gonçalves and Pedro Pinho, "Textile Materials for the Design of Wearable Antennas: A Survey", Sensors, Vol.12(11), pp.15841–15857, 2012

[2] Nikhil Singh, Ashutosh Kumar Singh, and Vinod Kumar Singh, "Design and Performance of Wearable Ultra Wide Band Textile Antenna for Medical Application", Microwave Optical Technology Letter, Vol.57(7): pp.1553–1557, 2015.

[3] Mai A. R. Osman, M. K. A. Rahim, M. Azfar N. A. Samsuri, F. Zubir, and K. Kamardin., "Design, Implementation and Performance of Ultra-wideband Textile Antenna", Progress in Electromagnetic Research. B, Vol.27, pp.307-325, 2011.

[4] S.S.Yavalkar, R.T.Dahatonde, Dr.S.S.Rathod, Dr.S.B.Deosrkar, "Parametric Study For Rectangular Microstrip Patch Antennas", IOSR Journal of Electronics and Communication Engineering, Vol.5, pp.49-53, 2013.

[5] Rajeshkumar V, Priyadarshini K, Glory Devakirubai D, Ananthi C, Snekha P, "Design and Comparative Study of Pin feed and Line feed Microstrip Patch Antenna for X-band Applications", International Journal of Applied Information Systems, Vol.1(5), pp.21-25, 2012.

[6] A.B. Mutiyara, R. Refianti, Rachmansyah, "Design of Microstrip Antenna for Wireless Communication at 2.4 GHz", Journal of Theoretical and Applied Information Technology, Vol.33(2), pp.184-192, 2011.

[7] D.Rakesh, P.Rakesh Kumar, Prof. Habibulla Khan, K Ch Sri Kavya, B.T.P.Madhav, K.Prabhukumar, S Baladurga Prasad, "Performance Evaluation Of Microstrip Square Patch Antenna On Different Substrate Materials", Journal of Theoretical and Applied Information Technology, Vol. 26(2), pp.97-106, 2011.

[8] S.S.Yavalkar, R.T.Dahatonde, Dr.S.S.Rathod, Dr.S.B.Deosrkar, "Comparative Analysis of Bandwidth Enhancement of Microstrip Patch Antenna using Various Geometries", IOSR Journal of Electronics and Communication Engineering, Vol.3, pp.15-18, 2012.

[9] Hatem H. Abbas, Jabir S. Aziz, "Bandwidth enhancement of Microstrip patch antenna", Journal of mobile communication, Vol.4, pp.54-59, 2010.

[10] Locher I. Klemm, M.Kirstein, T.Tröster, G, "Design and Characterization of Purely Textile Patch Antennas", IEEE Transactions on Advanced Packaging, Vol.29, pp.777–788, 2006.

IOP Conf. Series: Materials Science and Engineering **764** (2020) 012020 doi:10.1088/1757-899X/764/1/012020

[11] Morton W.E, Hearle J.W.S., Physical Properties of Textile Fibres, 4th Ed., Wood head Publishing, Cambridge, UK, 2008.

[12] Carla Hertleer, Anneleen Tronquo, Hendrik Rogier and Lieva Van Langenhove, "The Use of Textile Materials to Design Wearable Microstrip Patch Antennas", Textile Research Journal, Vol.78, pp.651-658, 2008.

[13] Pekka Salonen, Yahya Rahimat-Samii, "Textile antennas: Effects of antenna bending on input matching and impedance bandwidth", IEEE Aerospace and Electronic Systems Magazine, Vol.22(3), pp.10–14, 2007.

IOP Conf. Series: Materials Science and Engineering **764** (2020) 012021 doi:10.1088/1757-899X/764/1/012021

Enhancing the material properties of EN24 pinion material by heat treatment process

P Tamilarasu[1,*], J Jones Praveen[2], B Siddharthan[3], U Vivek[1] and M Giridharan[1]
[1]Department of Mechanical Engineering, K.S.Rangasamy College of Technology, Tiruchengode, India.

[2]Department of Mechanical Engineering, SRM TRP Engineering College, Tiruchirappalli, India.

[3]Department of Mechatronics, Bannari Amman Institute of Technology, Sathyamangalam, India.

[*]Corresponding author email: tamilarasu147@gmail.com

Abstract. The failure analysis was carried out on a failed pinion gear which is used for power transmission system in the automobile sector. The pinion gear is made of case carburized steel with low nickel and chromium content and it also has undergone extensive corrosion due to an environmental problem. The teeth are badly damaged and corroded at the surface. However, the crack has occurred due to stress and it is found only in certain places. To improve the corrosion resistance, nickel and chromium content are added and also heat treatment is carried out. The mechanical properties and wear resistance are improved in EN24 steel when compared to failure gear and EN24 steel before heat treatment.

1. Introduction

A Wear and fracture of gears is the major problem in all industries. The failure occurs due to chemical composition and material selection etc. Wear has been defined as the material removed from solid surfaces, which causes the failure of industrial components. Many investigators investigate the wear for many years but wear could not be avoided but it was minimized. The wear and wear rate extensively depend on chemical composition, microstructure, surface properties of materials and the load level. At the time of contact between the disc and the pin heavy wear occurs in low load level. As the result, high wear rate occurs and after some period of time, wear mode changes to steady.

The process of heat treatment is carried out by heating the base material above critical temperature and then cooling with salt water, air and oil. The purpose of the heat treatment is to increase the strength, to change the grain size, to modify the structure of the base material and to relieve the internal stress in the material after hot and cold working processes.

In this experiment, gas nitriding process was carried out on EN24 steel at 515°C for 180 minutes followed by oil quenching at 535°C for 60 minutes and tempering at 250°C for 90 minutes.

2. Literature Review

The pinion is the critical component in power transmission systems in the automobile sector. The study concludes that the failure is due to the compromise made by the manufacturer in raw material composition because nickel and chromium costs are high, which is evident by the presence of low nickel and chromium content.

Bhat et al. [1] studied the strengthening and apparently intrinsic brittleness of secondary hardening steel. Finally, the enhancement of toughness from the partial substitution of silicon by aluminum is considered. Bhaumik [2] analyzed the failure of an intermediate gearbox of a helicopter. The gear had fractured into pieces. The broken pieces of the gear were examined to identify the mode of fracture. The hardness survey of the gear tooth was carried out using a microhardness tester. The results indicate that the fatigue cracks initiated at the gear.

Das et al. [3] analyzed the failure for the fracture of the pinion. The fracture resistance of the material was reduced by the presence of elongated sulphide particles. Misalignment also led to the generation of high stress. Finally, the results indicated that the fracture was caused by fatigue and the fatigue crack initiated from one of the pinion teeth. Enver et al. [4] conducted abrasive wear tests on a wear cup model device. Corrosion tests were undertaken in a 10% H_2 〖SO〗_4 solution at a temperature of 56°C. Finally, the result determines that the boride layer had toothed structure and the structure was homogeneously distributed over the surface.

Jiwang Zhang et al. [5] enhanced the microstructure, mechanical properties and fracture behavior by using a heat treatment process. Finally, the austempering temperature is decreased and yield strength and tensile strength of the Austempered Ductile Iron (ADI) is increased. A decrease in both elongation and impact energy are found. Khorsand et al. [6] investigated the fatigue test and wear test for wear, tensile and compression mode. Finally, the result shows that fatigue strength increased and wear decreased by heat treatment processes.

Kiyoshi Funatani [7] enhanced the effective use of materials and to achieve the desired properties of the component, bake hardening technology is used to improve body strength. Various corrosion resistant coatings are used for panels and case hardening is used to improve hardness. This paper presented an overview of the materials as well as the heat treatment and surface technologies. Liu sheng-Fa et al. [8] investigated the new developed Mn-Cu alloyed by optical microscope, scanning electron microscope and performance measurements. Finally, the results indicated an improvement in the wear resistance, performance and fatigue behavior of the alloy.

Myounggu Park [9] analyzed the failure of a bevel gear installed on a turbojet engine. There is no surface defect on the failed bevel gear. The reason for failure is improper surface heat treatment. As a result, tooth bending fatigue was caused by improper case hardening. Siti Khadijah Alias et al. [10] enhanced the strength and hardness properties. Samples are prepared and polished according to ASTM standards. The result depicted that paste carburized samples provided a significant improvement in both tensile strength and hardness values compare to uncarburized samples. Ugur Cavda et al. [11] analyzed the powder metal compacts prepared by mixing iron powder. The compacts used a conventional sintering furnace. The results showed that, in carbonitrided samples, there is highest bending strength and in boronized samples, there is highest surface hardness and the densities increased by heat treatment processes.

3. Experimental Setup

The experimental investigation is conducted on failure pinion case-carburized steel. The following methodology is used to analyze and improve the tribological properties.

3.1. Visual examination

Visual examination is done on the fracture surface of the failed pinion, after cleaning with acetone, figure 1 shows the presence of cracks on the edge of a spur gear tooth. In a failed pinion gear, the front side of the spur gear tooth had a smooth surface, while the other side had a rough surface due to improper contact during operation. Small pieces were cut from the failure pinion and prepared for metallographic examination.

Figure 1. Visual examination of pinion.

3.2. Material identification

Material identification is carried out using an optical emission spectrometer. The failure pinion indicated that the pinion material was case carburized steel and its chemical composition is shown in Table 1. The chemical elements show that there is a lack of nickel and chromium content. Then the alternative material is EN24 steel. The chemical composition of EN24 steel is shown in Table 2.

Table 1. Chemical composition of steel.

Elements	Composition %
Carbon	0.16
Silicon	0.25
Manganese	0.78
Chromium	1.03
Nickel	1.212
Molybdenum	0.124
Vanadium	0.005
Sulphur	0.023
Phosphorous	0.014
Aluminium	0.030
Cobalt	0.002
Copper	0.237
Niobium	0.003
Titanium	0.003
Tungsten	0.01
Antimony	0.004
Tin	0.015
Iron	Remainder

Table 2. Chemical composition of EN24 steel.

Elements	Composition %
Carbon	0.447
Silicon	0.255
Manganese	0.483
Chromium	1.023
Nickel	1.360
Molybdenum	0.211
Vanadium	0.025
Sulphur	0.013
Phosphorous	0.017
Aluminium	0.005
Copper	0.122
Titanium	0.004
Tungsten	0.040
Vanadium	0.025
Iron	95.99
Lead	0.005

3.3. Hardness

The hardness of the core and case of the failed pinion (base material), before and after heat treatment of EN24 steel are measured using a Rockwell hardness with 150 kg load, was found to be 30-40 RC, 42-50 RC before heat treatment (BHT) and 60-50 RC after heat treatment (AHT). The surface hardness locations are shown in figure 2. The hardness values at the different locations of the specimen are plotted as a graph as shown in figure 3. The graph shows that the hardness value is different between BHT and AHT and hardness value of AHT is higher than that of BHT.

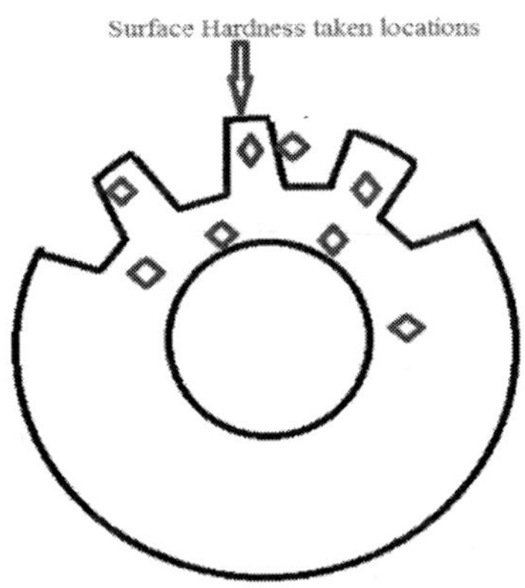

Figure 2. Rockwell hardness sample location.

IOP Conf. Series: Materials Science and Engineering **764** (2020) 012021 doi:10.1088/1757-899X/764/1/012021

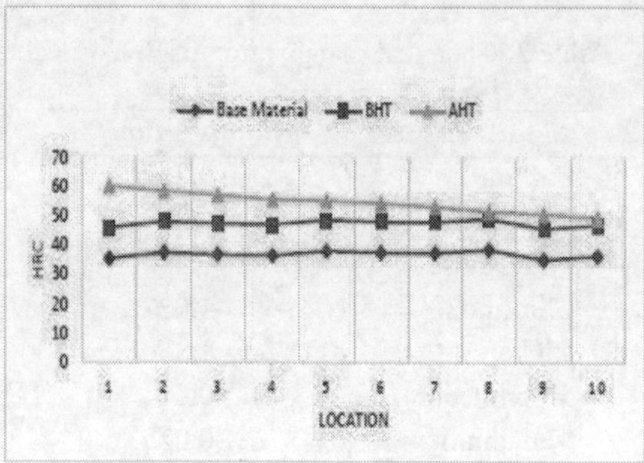

Figure 3. Rockwell hardness result.

3.4. *Vickers Hardness Test*

Micro hardness measurements with 200 g load, is performed on the carburized steel and the readings are taken at least five times. The hardness sample location is shown in figure 4 and the results are shown in figure 5. In the failed pinion, hardness of the surface was different for inner tooth and outer tooth. From figure 5, it has been found that the hardness value goes on decreasing with increase in depth. The comparative results of BHT and AHT shows that the hardness values are higher in case of AHT.

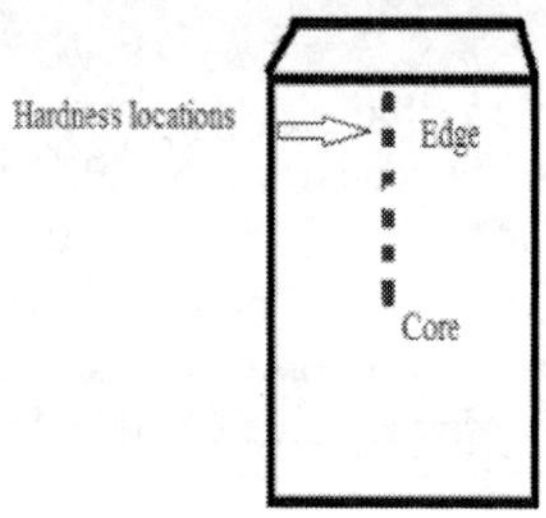

Figure 4. Vickers hardness sample location.

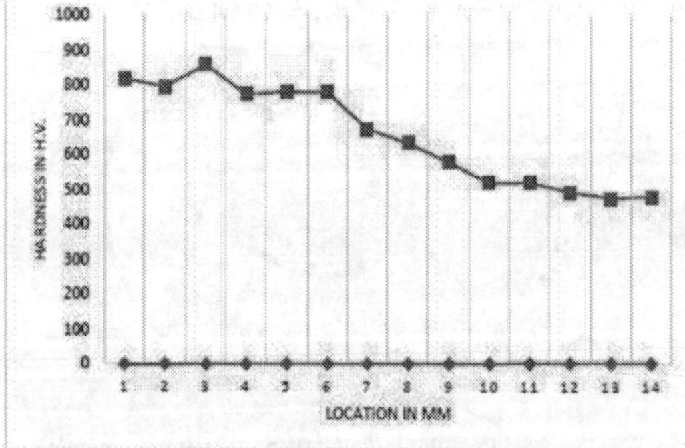

Figure 5. Vickers hardness result.

3.5. *Microstructural examination*

The microstructure of the failed gear tooth showed a dark etched area close to the pinion surface clearly indicating that, the pinion material was surface hardened and the microscope image is shown in the below figure 6. The high carbon content present in the case caused faster etching, and therefore it was confirmed that the pinion material was case carburized which is clear from the figure 7.

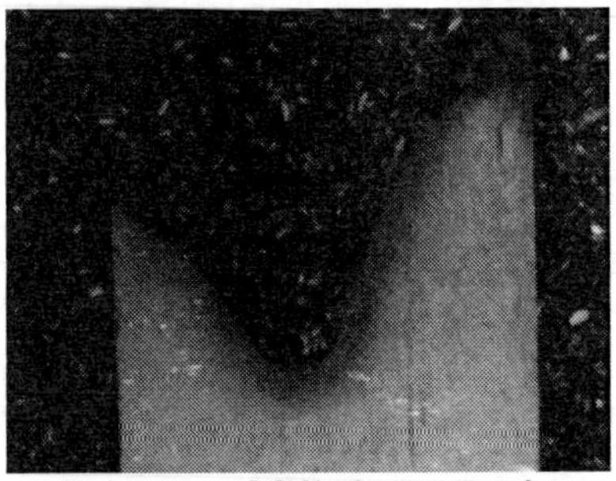

Figure 6. Microscopic image of failed gear tooth - magnification: 20x.

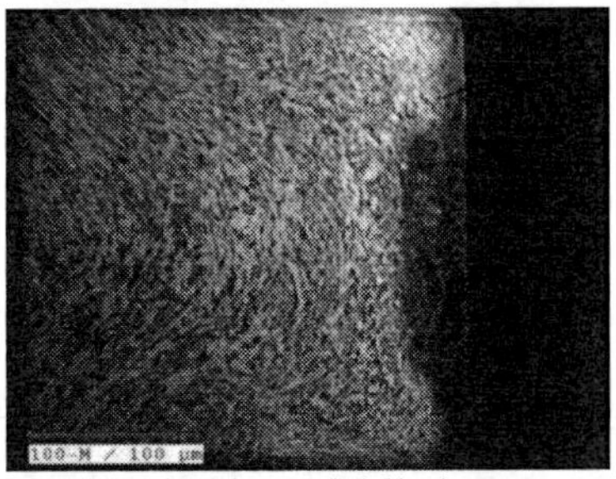

Figure 7. Microstructure examination of failed gear tooth - magnification: 60x

The fractured sample was prepared for the micro-examinations. The microstructure at different locations of the fracture was determined by sectioning samples from the cracked surface both in the longitudinal and transverse directions as shown in figure 8 and figure 9.

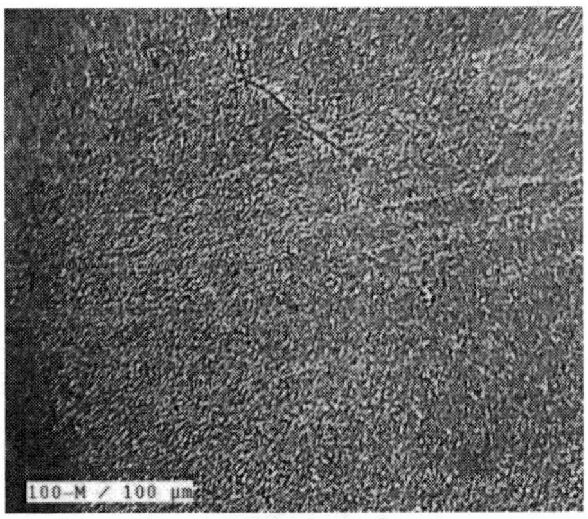

Figure 8. Edge of gear.

IOP Conf. Series: Materials Science and Engineering **764** (2020) 012021 doi:10.1088/1757-899X/764/1/012021

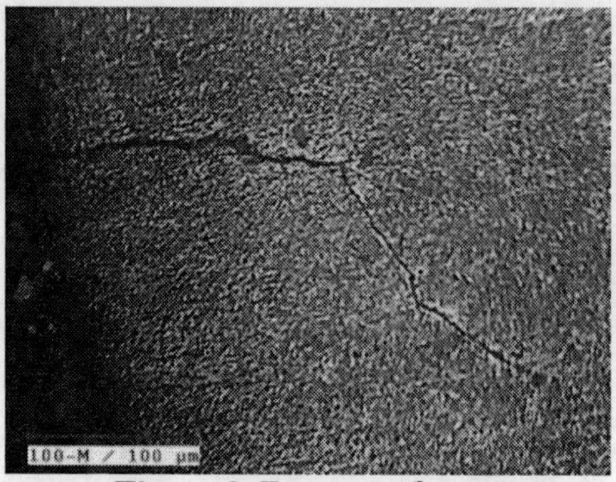

Figure 9. Fracture of gear.

Similar micro cracks were also observed at the corners of the tooth of the failed pinion, with one of the micro cracks initiating from the fillet of the failed pinion. Figure 10 shows the core with coarse tempered martensite and figure 11 shows the case with erosion due to corrosion and the microstructure is fine tempered martensite of the alternative material EN24 steel after heat treatment.

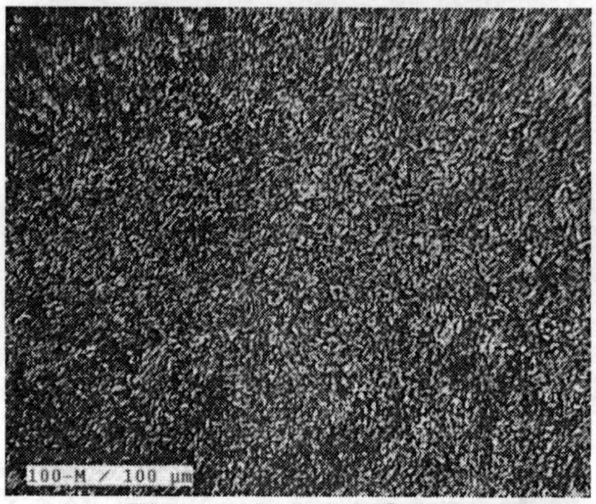

Figure 10. Core of gear.

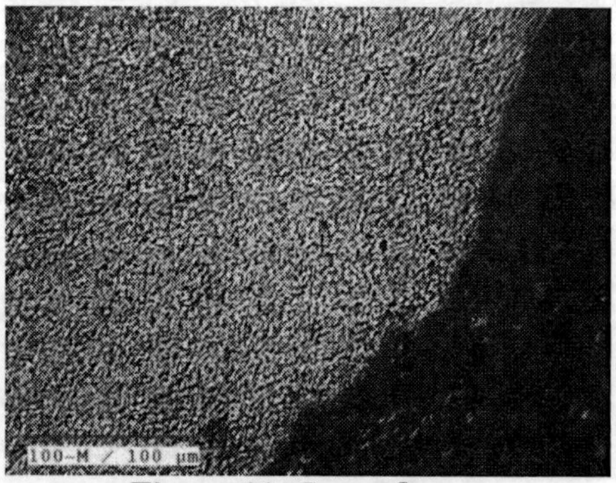

Figure 11. Case of gear.

3.6. Wear-Test (Pin on Disc Method)

Wear test has been conducted on the pin on disc with a load capacity of 1 kg and 2 kg respectively in figure 12 and figure 13.

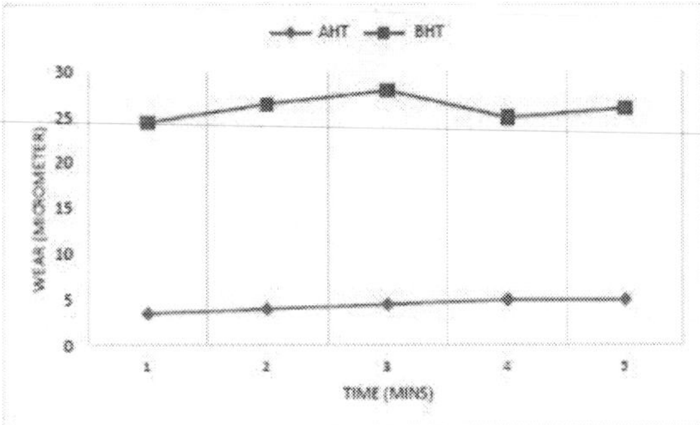

Figure 12. Wear rate at 1 kg.

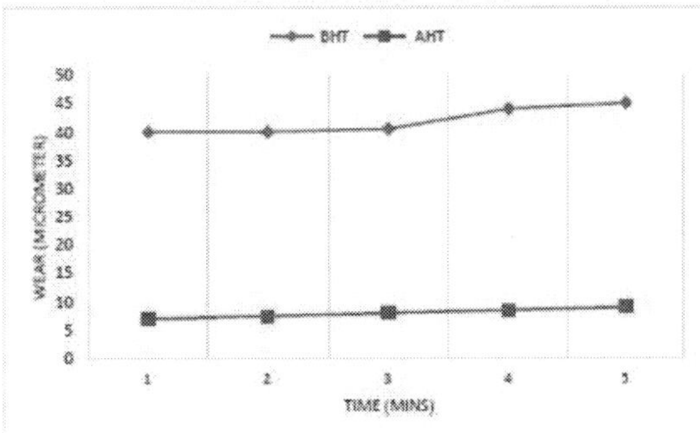

Figure 13. Wear rate at 2 kg.

From figure 12 and figure 13, it is clear that the wear rate in case of BHT is higher than that of AHT. The same condition is seen in both the cases of 1 kg and 2 kg loading. From the results, it is clear that the heat treatment of the material increases the wear properties.

4. Conclusion

The failure analysis of pinion is made up of case carburized steel with low nickel and chromium content and also it has undergone extensive corrosion due to an environmental problem. The failure analysis of pinion shows the teeth are badly corroded at the surface. However, the crack is due to stress corrosion cracking and found only at certain places. High nickel and chromium content steel (EN24) are selected to overcome the failure. The gas nitriding process improves the mechanical properties of steel. The mechanical properties such as hardness, wear and tensile are improved when compared to case carburized steel and EN24 steel before heat treatment.

References

[1] M.S.Bhat, W.M.Garrison Jr, V.F.Zackay, "Relations between Microstructure and Mechanical Properties in Secondary Hardening steels", Materials Science and Engineering, vol.41(1), pp.1–15, 1979.

[2] S.K.Bhaumik, M.Sujata, M. Suresh Kumar, M.A.Venkataswamy, M.A.Parameswara, "Failure of an intermediate gearbox of a helicopter", Engineering Failure Analysis, vol.14(1), pp. 85–100, 2007.

[3] C.R.Das, S.K.Albert, A.K.Bhaduri, S.K.Ray, "Failure analysis of a pinion", Engineering Failure Analysis, vol. 12(2), pp.287–289, 2005.

[4] Enver Atik, Umut Yunker, Cevdet Meric, "The effects of conventional heat treatment and boronizing on abrasive wear and corrosion of SAE 1010, SAE 1040, D2 and 304 steels", Tribology International, vol.36(3), pp. 155–161, 2003.

[5] Jiwang Zhang, Ning Zang, Mintang Zhang, Liantao Lu, Dongfang Zeng, "Microstructure and mechanical properties of austempered ductile iron with different strength grades", Materials Letters, vol. 119(15), pp.47–50, 2014.

[6] H Khorsand, S.M.Habibi, H Yoozbashizadea, K Janghorban,S.M.S. Reihani, H Ranhmani Seraji, M. Ashtari, "The role of heat treatment on wear behavior of powder metallurgy low alloy steels", Materials and Design, vol. 23(7), pp.667–670, 2002.

[7] Kiyoshi funatani, "Heat treatment of automotive components: current states and future trends", Transactions- Indian Institute of Metals, vol.57(4), pp. 381–396, 2004.

[8] Sheng-fa LIU, Yang CHEN, Xin CHEN, Hua-ming MIAO, "Microstructure and Mechanical Properties of Helical Bevel Gears Made by Mn-Cu alloyed Austempered Ductile Iron", Journal of iron and steel research, International, vol.19(2), pp.36-42, 2012.

[9] Myounggu Park, "Failure analysis of an accessory bevel gear installed on a J69 turbojet engine", Engineering failure analysis, vol.10(3), pp. 371-382, 2003.

[10] Siti Khadijah Alias, Bulan Abdullah, Ahmed Jaffar, Syazuan Abdul Latip, SalmiahKasolang, MohdFaizulIzham, MuhdAziminAbdZhani, "Mechanical properties of paste carburized ASTM A516 steel", Procedia Engineering, vol.68, pp. 525–530, 2013.

[11] Ugur Cavda, BekirSadikUnlu, Ahmet Murat Pinar, EnverAtik, "Mechanical properties of heat treated iron-based compacts", Material and Design, vol.65, pp.312–317, 2015.

IOP Conf. Series: Materials Science and Engineering **764** (2020) 012022 doi:10.1088/1757-899X/764/1/012022

Toughness study of Borided, Borided and induction modified AISI 4340 steel

T Sivakumar[1,*], M Prince[2], P Gopalakrishnan[3]

[1]Department of Mechanical Engineering, Malla Reddy College of Engineering and Technology, Hyderabad.
[2]Department of Mechanical Engineering, Sri Krishna College of Technology, Kovaipudur, Coimbatore.
[3]Department of Metallurgical Engineering, PSG College of Technology, Coimbatore.

*Author for correspondence E-mail: tsivat@rediffmail.com

Abstract. In this study, pack boriding is done on AISI 4340 steel in steel containers for 3 hours at 950°C. Boriding results in the formation of FeB or Fe_2B columnar microstructure at the case. Such microstructure brings about a high brittleness of the boride layer. Induction surface modification is done on Borided samples to bring down the surface hardness. We put an effort to enhance the toughness of boride layer using 30 kW power high frequency induction heat source. The toughness of borided, borided and induction surface modified specimens are evaluated by shear punch test. Cylindrical punch of 4 mm diameter is used in this test. The toughness is calculated from the load-displacement curve. Induction modified specimens exhibit 38% improvement in toughness as compared with borided specimens. This improvement in toughness may be due to blunting of acicular boride structure in to a globular structure at the interface, smooth hardness gradient and formation of single phase Fe_2B microstructure as a result of induction surface modification.

1. Introduction

The AISI 4340 is a low alloy steel, widely used as a material for manufacturing of different types of gears, shafts owing to the exceptional combinations of good mechanical and fatigue properties [1]. However, this steel has inadequate surface hardness when it is used in tribological applications [2]. Thus, its surface modification is necessary to improve the hardness. Boriding is one of the capable surface modification techniques to produce very hard surfaces on steels [3-4]. Boriding on steels can offer higher surface hardness (1400 – 2100 HV), as compared to carburizing (700 – 850 HV), nitriding (600 – 1100 HV) and carbo-nitriding (940 – 1000 HV) [5]. Borided steels offer wear resistance analogous to that of sintered carbides [6].

As Boriding on steels can be done through different methods like pack, molten salt and gas boriding [7-8]. This process is carried out in the temperature range of 700°C to 1000°C for 1-12 hours in the presence of suitable boriding mixture [9-10]. During this process distribution of boron into the surface of the steel takes place, two phase microstructure (Fe_2B and FeB) and superficially, directional saw tooth microstructure is framed [11]. Eyre [12] reported that Fe_2B and FeB phases formation based on the boron potential. If the boron potential is 8.83%, Fe_2B phase only forms in the case; at 16.3% of boron, FeB phase also forms over and above Fe_2B. The Fe_2B phase forms nearby to the core and the FeB phase forms in close proximity to the surface. Two phase microstructure (FeB+Fe_2B), pores in the case, anisotropic distribution of boron in the case, high process time, high process temperature result in poor toughness and ductility [13]. The effective use of borided components is restricted due to high brittleness.

For improving the toughness of the boride layer or/and borided steel, several kinds of surface engineering techniques have been tried by the researchers. The full disappearance of FeB layer is observed as a result of post diffusion annealing on borided steels, by heating the borided steels at 950^{0}C for 6-8 hours [14-16]. Multicomponent boriding is suggested as one of the effective method to enhance the toughness of borided steels by carefully choosing the alloying elements such as copper, aluminium, chromium, silicon and nickel along with diffusion of boron [17-19]. Superplastic boronising is reported as an effective method to result in equiaxed boride grains with high toughness [20-24]. However, this is comparatively harder technique to be used for standard engineering components. Interrupted boriding is suggested as another alternative method to enhance the toughness and ductility of borided steels with fewer disturbances to the core [25]. Laser surface modification of as-borided steels results in significant improvement in the toughness due to the formation globular microstructure and better hardness gradient [26]. By doing laser treatment, surface hardness decreases and simultaneously toughness and ductility increases appreciably by reducing the degree of non-uniformity of the borided layer and generating residual compressive stresses [27].

In this investigation, an attempt is made to study the improvement in toughness of borided 4340 steel using induction surface modification.

2. Experimental details

2.1. Base Material

A medium carbon low alloy steel (0.39 C, 1.4 Ni, 0.8 Cr, 0.6 Mn, 0.3 Si, 0.4 Mo, balance Fe) is used as a base material (AISI 4340) for boriding. The steel specimens are normalized before boriding. The steel has a hardness of 449±5 HV (at 100 g load). Specimens of 12 mm diameter and 0.6 mm thickness are used for shear punch test.

2.2. Boriding and Induction surface modification

Boriding is done on AISI 4340 steel sheets of 18 mm diameter, 0.600 mm thickness using pack process (paste method) in a furnace at 950^{0}C for 3 hours.

The borided surfaces are modified using 30 kW power high frequency induction heating machine. The gap between the coil and specimen is kept constant as 3.5mm. The input voltage is kept constant as 210V. The induction coil is fixed stationary. The specimens are moved at four different feed rates. A protective paste is applied on the exterior (surface) of the specimen to avoid oxidation.

After this treatment, the specimens are studied metallographically. The microstructures are observed using Zeiss axiovert optical microscope at suitable magnifications. The microhardness along the case was measured using Mitutoyo microhardness tester at a load of 100 g. The phase analyses are carried out using Shimadzu X-ray diffractometer, using Cu Kα radiation. The toughness of borided, borided and induction surface modified specimens are evaluated by shear punch test using small specimen of about 18 mm diameter and 0.6 mm thickness as per ASTM D732-10. The shear punch test is performed on a computerized materials testing system (FINE Testing Industries, Model TFNP 25KN) with a load cell of 25 kN capacity and an optical encoder with 1000 pulse per minute for uninterrupted monitoring of the applied load and a data acquisition system measuring punch head displacement. The shear punch experimental setup utilizes a flat cylindrical punch of about 4 mm in diameter. The load on the punch is calculated as a function of the punch journey and the graph of load vs. displacement is made. The material selected for both the die and punch is A2 air hardened tool steel (RC 58).

3. Results and discussions

The optical microstructure of borided specimen is shown in Figure.1. The microstructure shows typical two phase (Fe_2B – FeB) acicular boride needles for an average length of 100 μm.

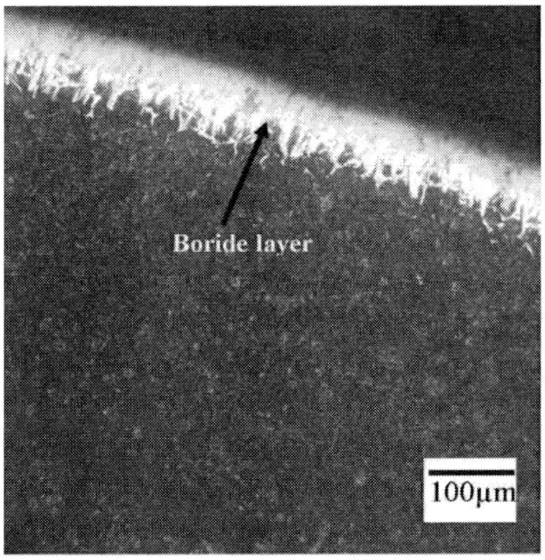

Figure 1. Optical microstructure of borided AISI 4340 steel.

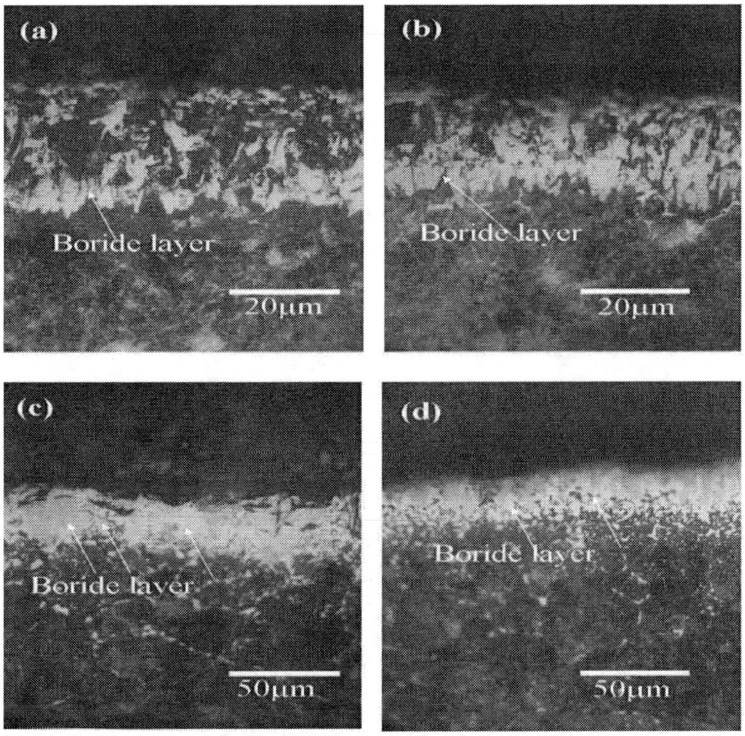

Figure 2. Optical microstructures of borided and induction modified steel at different feed rates, **(a)** 8.7 mm/s; **(b)** 9.7 mm/s; **(c)** 10.5 mm/s; **(d)** 11.5 mm/s.

The above figure 2 (a-d) give details an optical microstructures of borided and induction modified specimens treated at a feed rates of 8.7, 9.7, 10.5, 11.5 mm/s. It is found that induction surface modification of borided specimens treated at a feed rates less than or equal to 9.7 mm/s results in the deterioration of the case microstructure. Figure 2 (c & d) show the improved modification of the case with increasing feed rates, with apparent sign of rounding of acicular boride needles at the interface

IOP Conf. Series: Materials Science and Engineering **764** (2020) 012022 doi:10.1088/1757-899X/764/1/012022

resulting in possible improvement of toughness of boride layer. It is observed that by increasing the induction power, acicular nature of boride layer disappears significantly. The compactness of the boride layer increases with increase in induction power. Figure 2 (d) also demonstrate the development of spherical boride particles proximity to the surface. The average case depth of induction modified specimen treated at 11.5 mm/s was 72 µm. The case depth decreases may be due to dilution of boron from case to core. Further, it is apparent from the present investigation that a minimum feed rate of 11.5 mm/s is necessary to cause effective microstructural modification. Characteristic directional growth of the thermo-chemical diffusion based treatment vanished due to induction treatment as similar like laser treatment [27].

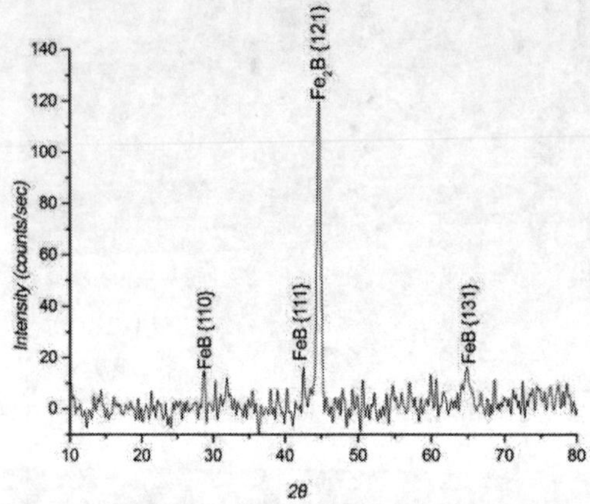

Figure 3. XRD pattern of borided specimen.

Figure 3, shows XRD pattern of borided specimen. XRD pattern illustrates the presence of Fe_2B and FeB phase in the borided case. This two phase boride microstructure exhibits inferior mechanical properties as reported by many researchers [3, 6, 11 and 13].

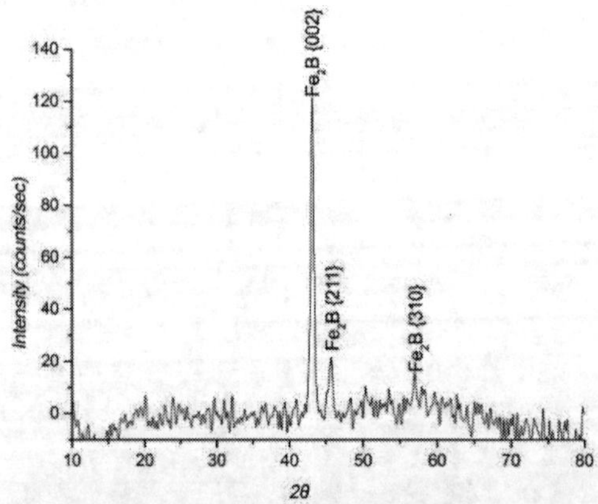

Figure 4. XRD Pattern of borided and induction modified specimen at a feed rate 11.5mm/s.

Figure 4 shows XRD pattern of induction modified specimen treated at a feed rate of 11.5 mm/s. The absence of hard brittle phase (FeB) has been observed from the XRD pattern. XRD pattern explains the presence of single phase Fe_2B in the induction modified surface. The single phase Fe_2B microstructure exhibits better ductile behavior as compared with dual phase microstructure as reported by many researchers [3, 6, 11 and 13].

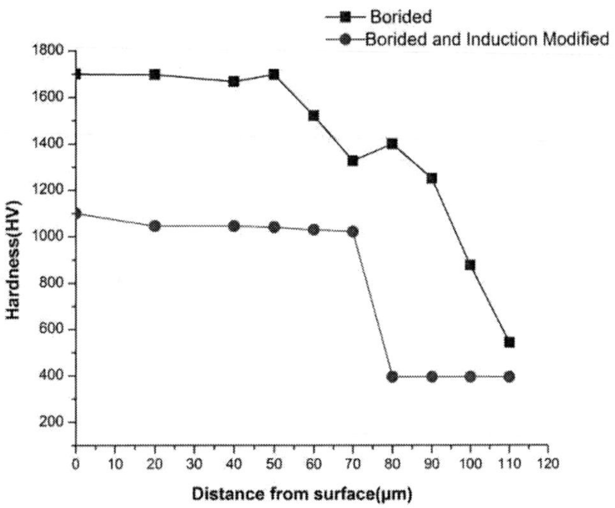

Figure 5. Microhardness profiles of borided, induction modified at a feed rate of 11.5 mm/s.

The above figure 5, shows the microhardness profiles (ASTM E384) of the borided specimen and borided and induction modified specimen with feed rate of 11.5 mm/s. The average peak surface hardness measured from borided specimen is 1700 HV. This peak surface hardness is caused by the presence hard FeB phase present on the case. The microhardness is observed in the range between 800 HV – 1700 HV approximately. There is a great variation in the microhardness of the borided case indicates greater inhomogeneity of boron distribution in the case. The average peak surface hardness of induction modified specimen is 1100 HV. The micro hardness is found to vary between 1000 HV – 1100 HV approximately. Microstructural refinement due to induction modification results in better distribution of boron and other alloying elements in the case. The decrease in surface hardness and less scatter in hardness across the section are due to effective homogenization happened during induction modification.

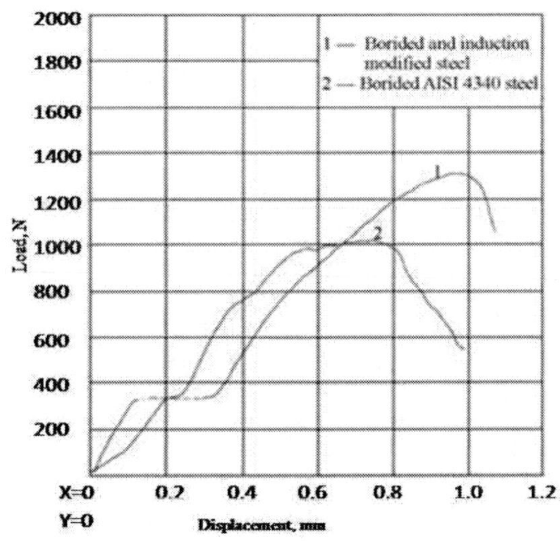

Figure 6. A comparison of load vs punch displacement plot from a shear punch test of borided specimen and borided, induction modified specimen treated at 11.5 mm/s.

The mechanical properties such as ductility, strength and toughness of borided and borided and induction modified were estimated from the load-displacement curve shown in the figure 6. The shear punch test is preferred method for the evaluation of mechanical properties of thin coatings. Shear punch test is based on blanking process, consisting of clamping a small thin sheet between the die and punch [26].

It is found that induction modification is the cost effective method to break down the acicular needle of conventionally borided steel as compared to laser modification. Borided and induction modified specimens exhibit lesser tensile strength as compared to the only borided specimens. The strength of the induction modified borided specimen is lowered be due to drop in the surface hardness. The notch angle between boride layer and iron substrate in the case of conventional boriding is around 900°C. It is observed from the optical microstructures that induction modification results in reducing the notch angle significantly. Because of reduction in notch angle, toughness is improved significantly. 38% improvement is estimated from the shear punch curve.

4. Conclusions

Acicular microstructure is replaced by globular microstructure in the induction surface modified borided AISI 4340 steels. Due to effective homogenization, lesser hardness scatter is observed. Such lesser hardness gradient leads to single phase of Fe_2B microstructure on the surface. Thus, toughness in induction modified boride steel is enhanced by 38%.

Acknowledgments

The authors thankfully recognize All India Council for Technical Education (20/AICTE/RIFD/RPS (Policy II) 6/2012-13, dated 20.01.2013), New Delhi for providing the financial support for procuring shear punch test machine. The authors wish to show gratitude to Mr.Madhu, managing partner of M/s. Sri Vishnu Induction Hardening, Coimbatore for providing facilities to carry out induction modification trials. Also, the authors are grateful to the management of PSG college of Technology, Coimbatore for allowing us to carry out all the research facilities available in the Department of Metallurgical Engineering.

References

[1] K.J. B.Ribeiro, R.R.M.de Sousa, F.O.De Araújo, R.A.de Brito, J.C.P. Barbosa, and C.Alves, "Industrial application of AISI 4340 steels treated in cathodic cage plasma nitriding technique", Material Science and Engineering: A, Vol.479(1), pp.142-147, 2008.

[2] Sirin, Sule Yildiz, Kahraman Sirin, and Erdinc Kaluc, "Effect of the ion nitriding surface hardening process on fatigue behavior of AISI 4340 steel", Material Characterization, Vol.59 (4), pp.351-358, 2008.

[3] Hwang, Jiun-Ren, and Chin-Ping Fung, "Effect of electron beam surface hardening on fatigue crack growth rate in AISI 4340 steel", Surface and Coatings Technology, Vol.80(3), pp.271-278, 1996.

[4] A Graf von Matuschka, "Boronizing", 1st edition, Carl Hanser Verlag Munchen Wien, Philadelphia, pp.11, 1980.

[5] O.Allaoui, N.Bouaouadja and G.Saindernan, "Characterization of boronized layers on a XC38 steel", Surface and Coatings Technology, Vol.201(6), pp.3475-3482, 2006.

[6] J.P.Deebasree, R.Thirumurugesan, P.Shankar, R.V.Subba Rao, P.Gopalakrishnan, and M.Vijayalakshmi, "Laser boriding of austenitic type 304L stainless steel", International Symposium for Research Scholars on Materials Science and Engineering, Department of Metallurgical and Materials Engineering, Indian Institute of Technology Madras, Chennai, 2004.

[7] A.K.Sinha, "Boriding (Boronizing of Steels)", ASM Handbook, OH, USA, Heat treatment, Vol. 4, pp. 978-1000, 1991.

[8] E.Atik, U.Yunker and C.Meric, "The effects of conventional heat treatment and boronizing on abrasive wear and corrosion of SAE 1010, SAE 1040, D2 and 304 steels", Tribology International, Vol.36(3), pp.155-161, 2003.

[9] A.Balandin, "Surface hardening of die steels by diffusion boronizing, borocopperizing and

borochromizing in fluidized bed", Metal Science and Heat Treatment, Vol.47(3-4), pp.27-30, 2005.
[10] W.Fichtl, "Boronizing and its practical applications", Material and Design, Vol.2(6), pp.276-286, 1981.
[11] R.H. Biddulph, "Boronizing for erosion resistance",Thin Solid Films, Vol.45(2), pp.341- 347, 1977.
[12] I.Campos, O.Bautista, G. Ramirez, M.Islas, J. De La Parra and L.Zuniga, "Effect of boron paste thickness on the growth kinetics of Fe_2B boride layers during the boriding process," Applied Surface Science, Vol.243(1-4), pp.429-436, 2005.
[13] T.S Eyre, "Effect of boronising on friction and wear of ferrous metals", Wear, Vol.34(3), pp.383-397, 1975.
[14] Kulka, M. and Pertek, A., "Microstructure and properties of borided 41Cr4 steel after laser surface modification with re-melting", Applied Surface Science, Vol.214(1-4), pp.278-288, 2003.
[15] V.I.Dybkov, W.Lengauer and K.Barmak, "Formation of boride layers at the Fe-10% Cr alloy-boron interface", Journal of Alloys and Compounds, Vol.398(1-2), pp.113-122, 2005.
[16] V.I.Dybkov, W.Lengauer and P.Gas,"Formation of boride layers at the Fe-25% Cr alloy- boron interface", Journal of Material Science, Vol.41(15), pp.4948-4960, 2006.
[17] K. C.Bindal and A.Erdemir, "Ultralow friction behavior of borided steel surfaces after flash annealing," Applied Physics Letters, Vol.68(7), pp.923-925, 1996.
[18] F.S.Chen and K.L.Wang, "The kinetics and mechanism of multi-component diffusion on AISI 1045 steel", Surface and Coatings Technology, Vol.115(2-3), pp.239-248, 1999.
[19] G.V.Zemskov and N.G.Kaidash, "Borosiliconizing of iron and steel", Metal Science and Heat Treatment, Vol.6(3), pp.179-181, 1964.
[20] T.Wierzchoh, P.Biefliski and K.Sikorski, "Formation and properties of multicomponent and composite borided layers on steel", Surface and Coatings Technology, Vol.73(1-2), pp.121-124, 1995.
[21] C.H.Xu, W.Gao and Y.L.Yang, "Superplastic boronizing of a low alloy steel - microstructural aspects, Journal of Material Processing Technology, Vol.108(3), pp.349-355, 2001.
[22] C.H.Xu, J.K.Xi and W.Gao, "Improving the mechanical properties of boronized layers by superplastic boronizing", Journal of Material Processing Technology, Vol.65 (1-3), pp.94-98, 1997.
[23] C.H.Xu, J.K.Xi and W.Gao, "Isothermal superplastic boronizing of high carbon and low alloy steels", Scripta Materialia, Vol.34(3), pp.455-46, 1996.
[24] I.Jauhari, H.A.M Yusof, and R.Saidan, "Superplastic boronizing of duplex stainless steel under dual compression method", Material Science and Engineering: A, Vol.528(28), pp.8106-8110, 2011.
[25] P.Gopalakrishnan, P.Shankar, M.Palaniappa and S.S. Ramakrishnan, "Interrupted boriding of medium-carbon steels", Metallurgical and Materials Transactions A, Vol.33A (5), pp.1475-1485, 2002.
[26] Kulka, M. and Pertek, A., "Microstructure and properties of borocarburized 15CrNi6 steel after laser surface modification", Applied Surface Science, Vol.236(1-4), pp.98-105, 2004.
[27] M.Kulka, and A.Pertek, "Characterization of complex (B+C+N) diffusion layers formed on chromium and nickel-based low carbon steel", Applied Surface Science, Vol.218 (1-4), pp.114-123, 2003.

IOP Conf. Series: Materials Science and Engineering **764** (2020) 012023 doi:10.1088/1757-899X/764/1/012023

Processing and wear properties of aluminium surface composites using friction stir processing

A Thangarasu[1] and N Murugan[2]

[1]Sri Ramakrishna Institute of technology, Coimbatore, Tamilnadu, India 641010
[2]PSG College of Technology,Coimbatore, Tamilnadu, India 641004

[1]Corresponding author's e-mail ID: athangarasu@gmail.com

Abstract: Friction stir processing (FSP) was employed to fabricate Aluminium Surface Composites (SC) by reinforcing boron carbide (B_4C) particles with AA6082 aluminium matrix alloy. The role of reinforced particles on the microstructure and mechanical behaviour of the prepared surface composites was studied. The microstructure of the produced composite was examined utilising optical microscopy, scanning electron microscopy (SEM) and electron backscattered diffraction (EBSD). The micrographs of the processed zone revealed fine grains and uniform dissemination of B_4C particles with good interfacial bonding. Microhardness and wear resistance tests were conducted on the matrix alloy, FSPed matrix and produced surfaced composites with the different volume fraction of B_4C particles. The microhardness and wear resistance showed the higher values compared to aluminium matrix and friction stir processed aluminium without B_4C particulates. It was seen that the wear resistance of the FSPed surface composite was improved significantly as compared to that of the AA6082 matrix alloy.

Key words: Surface composites - Rotational speed - Processing speed – Microstructure - Microhardness - Wear rate

1. Introduction

Aluminium Matrix composites (AMCs) are most wanted materials in aircraft, marine and automobile industrials. Generally, AMCs are having some important properties such as their higher specific Strength, low density, and wear resistance. However, reinforced hard ceramic particles reduced the ductility and toughness [1-3]. Research is being carried out to retain the interior properties and enhance the mechanical properties of the surface layer of materials in the desired thickness level (Surface Composites).

Several methods such as powder metallurgy, cast sintering, electron beam irradiation, plasma spraying, and high energy laser melt injection, have been used to produce surface composites. These processes are performed under the elevated temperatures, which causes the interfacial reaction between particles and alloy, and produced intermetallic products. A simple novel process, Friction stir processing (FSP) is introduced to fabricate surface composites using the principles of Friction stir welding (FSW). The reinforcement particles are packed in grooves or holes on the top surface of the material. A non-consumable rotating tool with a pin is plunged in to the top of the material and travelled longitudinally along the groove. Due to the rotation of the tool, heat is generated and plasticize material and surface composites is produced with reinforced particles [4, 5]. Maxwell et al [6] fabricated TiC and B_4C particles reinforced hybrid surface composites with various volume ratios. The increased fraction of volume ceramic particles increased the mechanical properties. Qiang et al [7] stated the influence of different vol. % of multi-walled carbon nanotubes incorporated into the aluminium matrix using FSP. You et al [8] fabricated aluminium surface composites with Mg and CuO are reinforcements using FSP and enhanced the tensile strength of the matrix alloy. Hashemi & Hussain et al [9] fabricated TiN Nanoparticles reinforced with AA7075 surface composites with different tool profiles using FSP. TiN particulates and threaded profile tool enhanced the wear properties. Shahraki et al. [10] fabricated surface composites using AA5083 as matrix and ZrO_2 as reinforcing particles and observed the improved microstructure and mechanical properties.

Bahrami et al. [11] fabricated AA7075/SiC composite and evaluated the effect of tool pin shape on microstructure and mechanical properties. Maza-heri et al. [12] fabricated A356/Al_2O_3 surface composites and examined the microstructure and tribological behaviours. Khodabakhshi et al. [13] developed AA5052-TiO_2 composites and observed the role of annealing on the solid state chemical reactions. Zhao et al. [14] processed Aluminium Surface composites reinforced with boron carbide particles and observed that how the number of passes affect the distribution of particles. Therefore, in this work, an endeavour is made to produce aluminium surface composites with addition of B_4C particles as reinforcement and investigate the microstructural and mechanical features.

2. Materials and Methods

The chemical composition of AA6082 aluminium alloy plate size 100 mm x 50 mm x 10 mm was used for this study is shown in Table 1. A rectangle 5.5 mm deep groove and different size width was created along the middle portion line of the plate by wire cut Electrical Discharge Machining. The B_4C powder was packed in a groove. A tool made of H13without pin was used to close the top layer of the rectangle groove after packing with B_4C particles to inhibit the particles from sprinkling during FSP. An aforesaid material tool which has shoulder diameter of 18 mm, pin diameter of 6 mm and 5.7 mm pin length was plunged into the closed groove along the longitudinal direction. The process was performed with the tool rotating speed of 1200 rpm, 60 mm/min processing speed and 10 kN axial force. The aforesaid process parameters were chosen according to the results inferred from the past works and trial experiments conducted by the corresponding author. The procedure for the production of surface composites was explained elsewhere [15].

Table 1. Chemical composition of matrix alloy (AA6082)

Element	Si	Mg	Mn	Fe	Cu	Cr	Zn	Ti	Al
wt.%	1.06	0.78	0.55	0.21	0.09	0.03	0.06	0.01	97.20

The samples under study were attained from the middle line of the processed zone and polished as per metallographic procedure. The Keller's reagent was used for the purpose of etching. An Olympus-BX51M optical microscope was used to obtain the micrographic images. The reinforced particle dissemination was observed by FESEM, Carl Zeiss-SIGMA HV and EBSD. The microhardness of the stir zone was measured using a Mitutoyo MVK-H1Vickers microhardness testing machine at 500 grams load applied for 15 seconds. A pin-on-disc wear apparatus (DUCOM TR20 - LE) was used to evaluate the wear behaviour of the processed surfaced composites.

3. Results and Discussion

The optical micrographs of the parent material and processed zone are shown in figure. 1(a) - (d). From the micrograph of the processed matrix material, it is noticed that the reinforced particles influence the grain size of the aluminium matrix. Aluminium matrix grain refinement takes place due to the effect of pinning of hard B_4C particles. The irregular grain development of matrix alloy is hindered by sliding of the grain boundary. Moreover, aluminium matrix grains are further reduced, because of nucleation locations that are developed by the dynamic recrystallization process. The dynamic recrystallization phenomenon contributes to the grain refinement and modification of grain structure. The size and shape of the incorporated B_4C is shown in figure 1(b). The FESEM images reveal that B_4C particles are uniformly disseminated in the stir zone. The load - posture capacity of the material depends on the interface between matrix and reinforcements. Figure 1(c) depicts a very good interface bonding between matrix and ceramics particle is obtained. There is no pores and cavities found in the microstructure as in the case of liquid casting processes. Figure 2 (a & b) depicts the EDS spectrum analysis of surface composites. The results of the EDS analysis suggest that the layer formation at the stir zone interface are Inter - Metallic Compounds (IMCs) layer with the average chemical composition (wt. %) of 70.99% B/14.43% C/10.57% Al/3.42% Si, and 0.6% O. It can be noticed that the microvoids are not located around the reinforced particles. In the processed zone, there is no evidence for the presence of third phase particles.

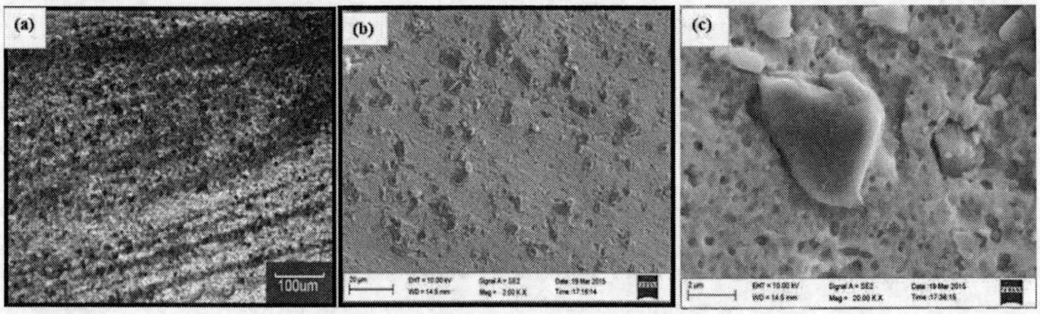

Figure 1 : (a) Micrograph AA 6082/B_4C (b) FESEM image of AA6082/ B_4C surface composite (c) Higher magnification of AA6082/B_4C

Moreover, the microstructure was further studied by EBSD is shown in figure 2 (c &d). The figure clearly shows the EBSD grain boundary map images of the stir zone. It is noticed that the grain structure is recrystallized and equiaxed fine grains which are finer than matrix alloy. The heat generated during the process and plastic deformation leads to dynamic recrystallization and ensures the fine structure of grain.

The average microhardness and wear rate of surface composites are shown in Table 2. The average microhardness value of matrix alloy was found to be 62 HV. FSPed matrix alloy average hardness value had enhanced and a maximum of 75 HV was achieved. The increased hardness value is due to the grain refinement and fragmentation and dissemination of intermetallic particles. These dispersed particles may enhance the mechanical properties of the alloy through Orowan strengthening mechanism (16). The reinforced B_4C particles improved the hardness of the composite. The maximum hardness value of 125.6 HV is reached in the composites. Furthermore, the B_4C particles have better Cohesion between the aluminium grains. The applied load is transmitted to B_4C particles and has got higher load posture capacity compared to aluminium alloy.

IOP Conf. Series: Materials Science and Engineering **764** (2020) 012023 doi:10.1088/1757-899X/764/1/012023

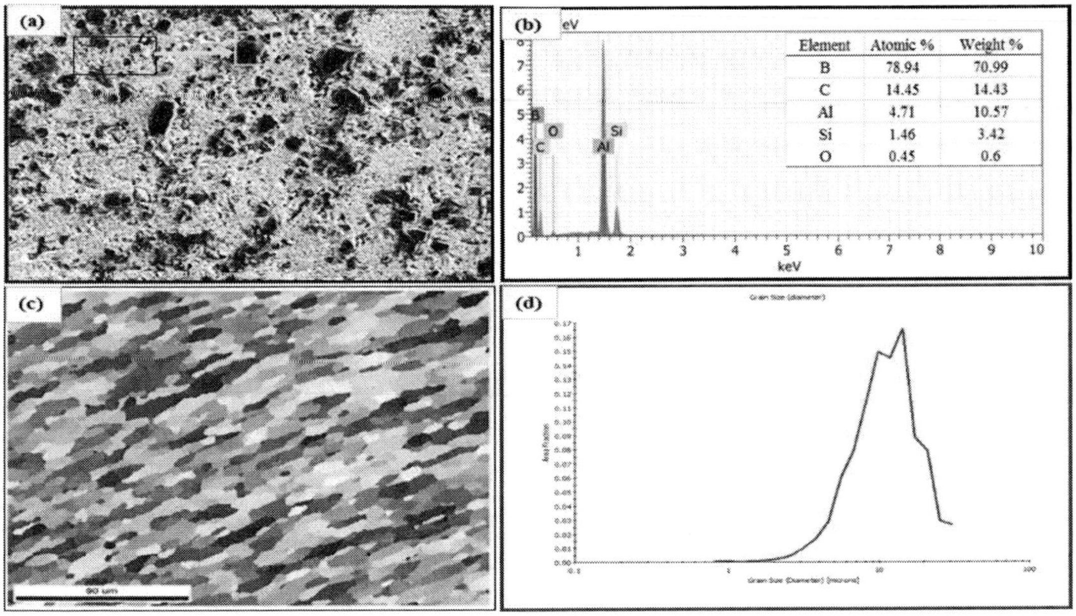

Figure 2: (a & b) FESEM image and EDS analysis of AA6082/B_4C composites,(c & d) EBSD grain boundary map of AA6082/B_4C surface composite

From the observations made on Table 2, the wear resistance of the surface composite has significantly improved. The wear rate of the composites depended on the microstructural characteristics and wearing condition such as applied load, sliding distance, temperature and ceramic particle size. In this case, the incorporation of B_4C particles and microstructure modification done by the processing play a vital role in wear resistance improvement. The wear rate was evaluated to be 536 x 10^{-5} mm^3/m at 0 vol. % and 352.3 x 10^{-5} mm^3/m at 22.9 vol. %. The wear rate was evaluated to be 536 x 10^{-5} mm^3/m at 0 vol. % and 352.3 x 10^{-5} mm^3/m at 22.9 vol. %.

Table 2. Influence of volume fraction on mechanical properties

Sample No.	Volume fraction %	Microhardness (VHN)	Wear rate (x 10^{-5} mm^3/m
1	0	75.30	536.17
2	3.8	88.40	463.87
3	8.5	103.88	424.50
4	14.6	117.62	384.40
5	22.9	125.60	352.30

4. Conclusions

Aluminium alloy AA6082 reinforced with boron carbide composites were produced with help of FSP and the influence of B_4C particles and its fraction of volume on microstructure, hardness and sliding wear properties were analysed. The incorporated particles were uniformly dispersed and a clean

IOP Conf. Series: Materials Science and Engineering **764** (2020) 012023 doi:10.1088/1757-899X/764/1/012023

interface was observed between B_4C particles and matrix alloy. B_4C particles strengthened the surface composites. Both the hardness and the wear resistance were enhanced when the B_4C particles volume fraction was increased.

5. References

[1] Rahimian M, Ehsania N, Parvin N, Baharvandi HR, *J Mater Proc Technol 2* (2009) 5387.
[2] Sharifi EM, Karimzadeh F, Enayati MH, *Mater Des* **32** (2011) 3263.
[3] Hemanth J, *Mater Des* (2009) 30323.
[4] Mishra RS, ZY. Charit MI, *Mater Sci Eng A* **341**(2003) 307.
[5] Morisada Y, Fujii H, Nagaoka T, Fukusumi M, *Scrip Mater* **55** (2006)1067.
[6] Maxwell Rejil A, Dinaharan I, Vijay SJ, Murugan N, *Mater Sci Eng A* **552** (2012) 336.
[7] Qiang Liu, Liming Ke, Fencheng Liu, Chunping Huang, *Mater Des* **45** (2013) 343.
[8] You GL, Ho NJ, Kao PW, *Mater Let* **100** (2013) 219.
[9] Hashemi, R, Hussain, G, *Wear,* **324**(2005) 45.
[10] Shahraki S, Khorasani S, Behnagh RA, Fotouhi Y,Bisadi H, *Metall Mater Trans B* **44** (2013) 1546.
[11] Bahrami M, Givi MKB, Dehghani K, Parvin N, *Mater Des* **53** (2014) 519.
[12] Mazaheri Y, Karimzadeh F, Enayati MH, Metall *Mater Trans A* **45** (2014) 2250.
[13] Khodabakhshi F, Simchi A, Kokabi AH, Gerlich AP, Nosko M, *Mater Des* **63** (2014) 30
[14] Zhao Y, Huang X, Li Q, Huang J, Yan K, *Int J Adv Manuf Technol* **78** (2015)1437.
[15] Lee CJ, Huang JC, Hsieh PJ, *Scr Mater*, **54** (2006) 1415.
[16] Ruff AW, *ASM handbook on friction, lubrication, and wear technology*, (1992) 629.

IOP Conf. Series: Materials Science and Engineering **764** (2020) 012024 doi:10.1088/1757-899X/764/1/012024

An Experimental Investigation on Tribological Behaviour of Jatropha Biolubricant with Pyrolysed Cardanol Biolubricant at Varying Loads

S Suresh[1,a], R S Mohan Kumar[2], G P Arun[3] and M Siva[4]
[1,2]Assistant Professor, Department of Mechanical Engineering, Kumaraguru College of Technology, Coimbatore- 641049, India
[3,4]Student, Department of Mechanical Engineering, Kumaraguru College of Technology, Coimbatore- 641049, India

E-mail: [a]suresh.s.mec@kct.ac.in

Abstract: Nowadays Demand for bio lubricants is increasing due to pollution and depletion of crude oil based fuels in this globe. This paper portraits and compares the tribological behavior of jatropha bio oil with cardanol oil bywear testing machine. SAE20W40 base lubricant is blended with 5%, 10% and 15% ratios of Jatropha and cardanol based bio lubricants. While testing with 2.6m/s sliding velocity friction and wear rate of jatropha and cardanol bio lubricant is accomplished while testing under 50N, 100N and 150N loads. Low wear rate was discerned while mixing 5%and 10% jatropha bio oil with SAE20W40 lubricant, when this limit increases then the wear rate gradually increases. While conducting theexperiment with cardanol oil, addition of 5% and 15% blend with SAE20W40 lubricant has given least rate of wear. 10% blend of Cardanol oil shows increase in wear rate with SAE20W40 lubricant.The mechanical efficiency was raisedduring the usage of CB5 and CB15 at the sliding velocity of 2.5 m/s. So, CB5 and CB15 can act as best alternative lubricant.

1. Introduction

Friction between surfaces which are in mutual contact during motion can be reduced by introducing lubricant. It will reduce friction in components of an IC engine such as liners and rings in cylinder. Due to depletion of non renewable sources and increase of population, need for bio-degradable Lubricants and unhazardous is increasing day by day. So,natural and reproduciblelubricating oils are introduced for meeting these requirements. Oils which are having high flash point and viscosity index have the capabilility to be a better substitute for the biolubricant. Also oils which are selected as a lubricant must have high thermal stability, Hydraulic stability and resistance to corrosion. Less pollution will be emitted while using vegetable oils instead of mineral oils. While using vegetable oil as a lubricant, it shows some limitations such as hydraulic stability and cold flow characteristics. Oil structure has to be changed to meet the need, so there is a necessity to blend vegetable oils with base lubricant. Here, we have used cardanol bio lubricant and jatropha bio lubricant with SAE20W40 base lubricant for testing at varying loads and velocities using tribometer. From the test results, wear rate and frictional characteristics of cardanol bio lubricant and jatropha bio lubricant has been compared to analyze the tribological characteristics. Seeds of Jatropha curcas are the source of Jatropha oil extraction; it is an important oil crop for biodiesel production. Among the components ofCashewnut

IOP Conf. Series: Materials Science and Engineering **764** (2020) 012024 doi:10.1088/1757-899X/764/1/012024

shell liquid oil,cardanol plays a vital role. During the pyrolysis process acid has been converted to alcohol.

2. Experimental Setup

Frictional force and wear rate of the materials which are in contact while in motion will be predicted using pin on disc tribometer. The setup consists of Rotating disc, Weight loaded pin assembly, motor and a recorder.The figure which is given below represents the set up of Pin on disc tribometer.

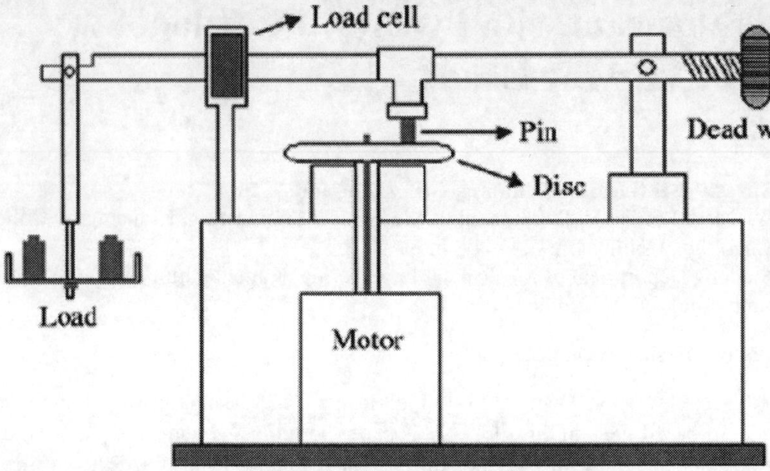

Figure1. Pin on disc Tribometer

The rotating part (disk) which is used in the setup is driven a motor. LVDT sensor is the main component to collect and send rate of wear and friction force to recorder. The pin assembly is kept in contact with the rotating disc using dead weight and the load during the experiment will be given by counter weight. While rotating, due to wear on the pin the lever makes contact with the LVDT sensor. The collected values of wear and frictional force from the LVDT sensor are transferred to the recorder.

Table 1. Experimental Properties

Device	:	Pin on Disk
rpm	:	200 to 2000
Load range	:	1N to 2000N
Frictional force	:	0 to 250N
Wear	:	± 1.8mm
Sliding Speed	:	Min 0.49m/s to Max 9.5m/s

Wear test is carried out in the pin on disc setup at 50N, 100N and 150N load using Jatropha oil mixed with SAE20W40 lubricant in the proportions of 05% JB, 10% JB and 15% JB. The same experiment was conducted using cardanol bio lubricant in the proportions of 5% CB, 10% CB and 15% CB with the same loads which is mentioned above. Base lubricant and the mentioned percentage of bio lubricants stirred by magnetic stirrer. Initially, using sand paper the disc is rubbed to clean the disc surface. The experiment is conducted by varying the loads and sliding velocity and the readings were noted. The charts conspired to observe readings of rate of wear, coefficient of friction at variable loads.

IOP Conf. Series: Materials Science and Engineering **764** (2020) 012024 doi:10.1088/1757-899X/764/1/012024

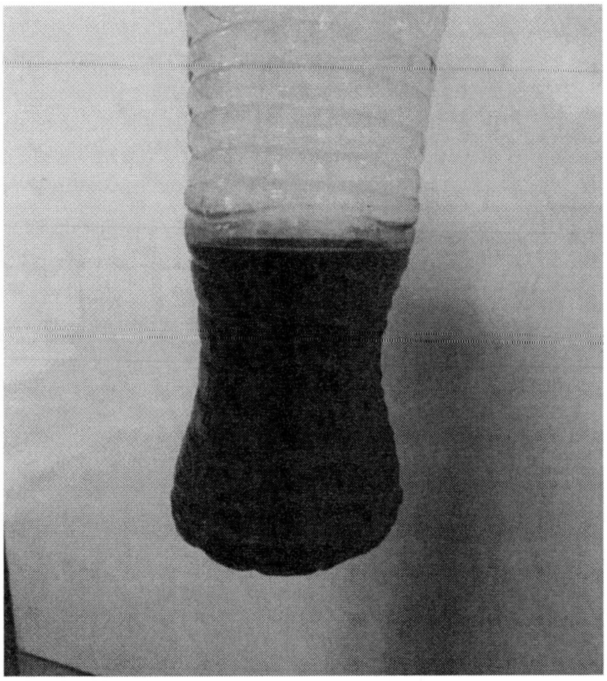

Figure 2. Oil sample of Cardanol biolubricant blended with SAE20W40

The Figure 2shows sample of cardanol biolubricant mixed with SAE20W40 lubricant in specified ratios.

3. Results and Discussion

3.1 Discussion on rate of wear for different proportions of biolubricant

Figure 3 shows loading effect on rate ofwear for different proportions of bio lubricants at different loads. From the observation of readings of wear rate for variant loads, wear rate is increasing with different ratios of bio lubricants. However, while compared to JB 15, Lubricants JB5 and JB10 shows least wear rate. Figure 4 shows effect of wear for Cardanol bio lubricants. Least rate of wear is noted with CB5lubricant while compared to CB10 and CB15.

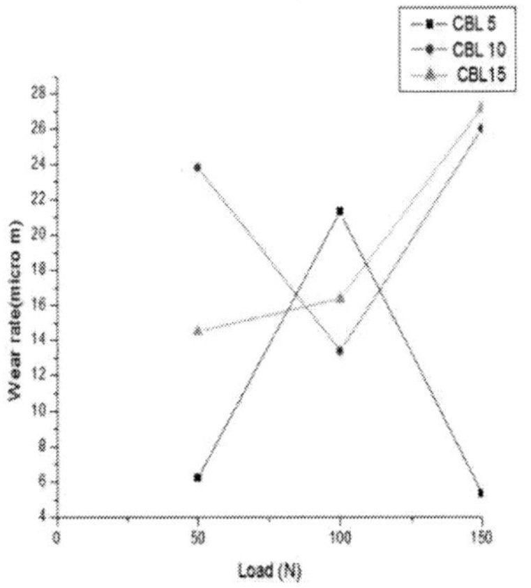

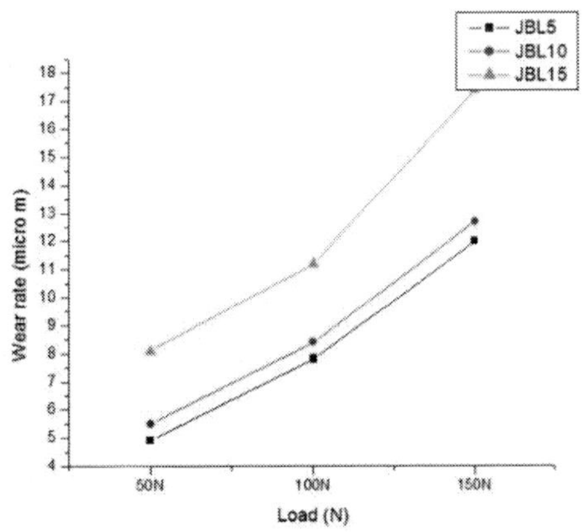

Figure3. Load vsrate of wear CB**Figure 4.**Load vsrate of wear for JB

3.2 Discussionon frictioncoefficientfor different proportions of biolubricant

Figure 5 and figure 6 shows loading effect on friction coefficient for different proportions of bio lubricants at different loads. From the observations of coefficient of friction, JB 5 and JB10 show least value as compared to JB15. While testing with cardanol bio lubricant, test result shows least value of coefficient of friction is observed with CB5 whereas CB10 and CB15 have shown maximum value of coefficient of friction.

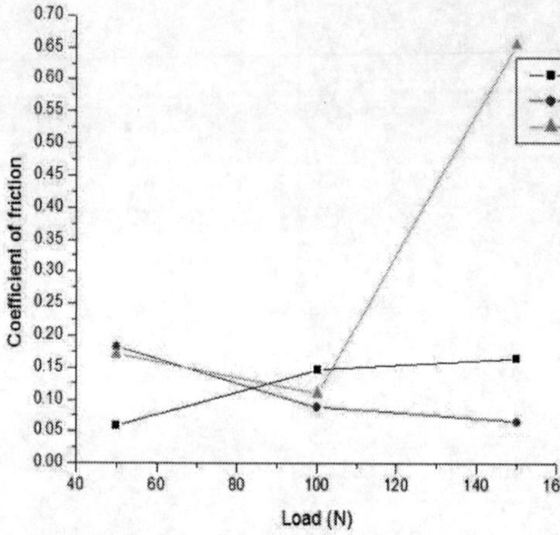

Figure 5. Load vsFrictionCoefficient for CB

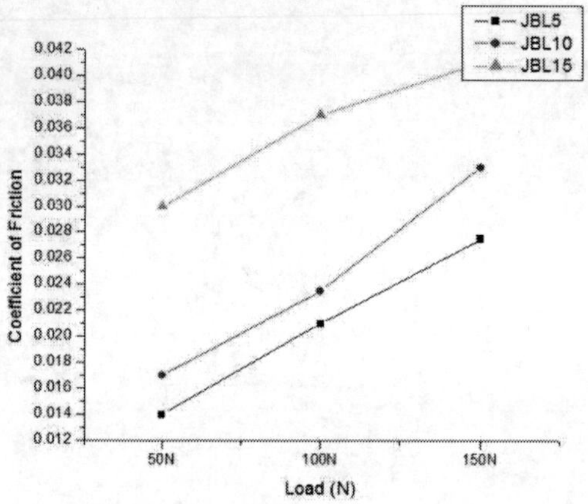

Figure 6. Load vsFriction Coefficient for JB

4. Conclusion

From the observations and discussions, it is summarized that

1. Least rate of wear is noted for the blend CB5 at 50N and 150Nloads because of high viscosity of the blend.
2. JB 5 shows minimum rate of wear at 50N, 100N and 150N loads.
3. Minimum metal contact between the pin and disc was observed due to high viscosity of blend CB5.

ACKNOWLEDGMENT

We (Authors) thank Kumaraguru College of Technology, Coimbatore – India (Management) for offering the opportunities and good research facilities to execute the research.

5. References

[1] Yashvirsingh, AmneeshSingla and SwapnilBhurat, Tribological behavior of Pongamia oil based biodiesel blended lubricant at different loads, Energy Sources, *Part A: Recovery, Utilization, and Environmental Effects* ISSN: 1556-7036 (Print) 1556-7230 (Online) Journal

[2] Alves S M, Barros B S, Trajano M F, Ribeiro K SB,andMoura,E 2013Tribological behavior of vegetable oilbased lubricants with nanoparticles of oxides in boundary lubrication conditions*Tribol. Int.***65**28–36.

[3] Gui M M, Lee K T, and Bhatia S 2008 Feasibility of edible oil vs. non-edible oil vs. waste edible oil as biodiesel feedstock *Energy***33**1646–53.

[4] Maleque M A, Masjuki H H, Haseeb A S M A.2000. Effect of mechanical factors on tribological properties of palm oil methyl ester blended lubricant *Wear***239**117–125.
[5] Shahabuddin M, Masjuki H H, Kalam M A, Bhuiya M M K, Mehat H 2013Comparative tribological investigation of bio-lubricant formulated from a non-edible oil source (Jatropha oil). *Ind. Crops Prod.***47**323–30
[6] Tullao, A. H. 2007. A new life for old plants. *Chem. Eng. News Arch.***85**53–55
[7] He Z, Lu J, Zeng X, Shao H, Ren T, and Liu W 2004 Study of the tribological behaviors of S, P-containing triazine derivatives as additives in rapeseed oil*Wear***257**389–94.
[8] Sivakumar S, VenkatachalamRamasamy, Nedunchezhian Natarajan, SivakumarPandian 2015 Optimization of non transesterifiedcardanol as biofuel based on its physical and chemical properties *Journal of Chemical and Pharmaceutical Sciences* **7**15-18
[9] Zengshe Liu, Jie Chen, Gerhard Knothe, XiaoanNie, and Jianchun Jiang Synthesis of EpoxidizedCardanol and Its Antioxidative Properties for Vegetable Oils and Biodiesel ACS Sustainable *Chem. Eng* 2016 **4** (3) pp 901–906

IOP Conf. Series: Materials Science and Engineering **764** (2020) 012025 doi:10.1088/1757-899X/764/1/012025

Internet of Things (IoT) in smart manufacturing

N Santhosh [1,4,a], M Srinivsan[2,4], K Ragupathy[3,4]
[1]Department of Mechanical Engineering,
[2]Department of Electrical and Electronics Engineering,
[3]Department of Automobile Engineering,
[4]Bannari Amman Institute of Technology, Tamil Nadu, India
E-mail: [a]santhoshmech10@gmail.com

Abstract: Rapid growth in the manufacturing sector has triggered to explore the potentials of Wireless Sensor Networks (WSN) and Internet of Things (IoT) which paves the way to escalate the efficacy of manufacturing. Industries implemented IoT in various domains starting from purchase of raw materials till customer service and support. IoT is considered as a key technology of the industrial revolution 4.0 and this provides a promising prospect to build influential services and applications for manufacturing. Further, it also provides an interactive relation between smart machines to share the data and information, which is essential for the complex systems to take a decision on the real-time working environment. In this paper, we reviewed the impact of IoT for a sustainable development, especially with regard to the manufacturing dimensions. This will further put forth the current scenario of IoT in manufacturing and this leads the researchers for pioneering their research towards the cyber integrated manufacturing.

1. Introduction

From the time of the Industrial Revolution, the manufacturing sector has played a dominant role in the economy of the countries and companies. We are in the new era of modern industrial revolution 4.0 where, the cyber integrated manufacturing is a pressing matter. An IoT enabled smart manufacturing provides an interactive relation between smart machines to share the data and information, which is essential for the complex systems to take a decision on the real-time working environment [1-8]. In the day to day point of view, attaining the growth in the resources and energy effeciency is the key strategy to reach the sustainability in manufacturing [9]. Business organizations face many challenges in the recent days due to technological advancement and extreme global competition. In order to overcome these issues, innovation in their product and processes is necessary for entering into the sustainable development in their future. IoT is the technology through which the companies are focusing to enhance their product and process development. Even though there is a tremendous growth of IoT in various sectors such as health care, energy managing, smart retail, agriculture, etc., the implementation of IoT in the manufacturing sector is in infancy. The majority of the manufacturing process still depends upon the third industrial revolution which is called the digital revolution where there is an emergence of electronics such as transistor, microprocessor, telecommunication and computers.

This revolution led to the development of automation in the production such as programmable logic controllers (PLCs) and robots. These provisions paved the way for more production in industries with less time and economy. But due to high competition and globalization, the industries are looking for the next level of technological advancement. New business models incorporated with IoT enable the organizations to develop innovative products with high productivity. The manufacturing industries are enforced to produce more products by using less raw materials and less energy. The aforesaid survey estimated that smart production line and industries would result in an annual efficiency surge of 3.3 over the next few years resulting in 2.6% annual cost savings. This puts forth the companies to invest in IoT and a recent survey estimated that about 140 billion euro will be endowed by the European companies in renovating the ways they manufacture the products [10]. The well sophisticated IoT network leads to the informed manufacturing in an organization which connects the four basic elements such as products, people, processes and infrastructure [11]. This enhances the entire process of an organization from the supply of raw materials to the customer service. The IoT has many more advantages in the manufacturing sector, such as the empowered connection between manufacturers and machines, smarter business decisions, improved management of the global supply chain [12].

2. Challenges for Smart Manufacturing

IoT leads to the enhancement of the product quality and manufacturing process efficiency, but still there are huge uncertainty issues which the industries face during the implementation [13]. There are three big challenges identified [14], one is general hesitancy of machine builders and end users to actually contrivance this technology. The second one is the security issues, where the collected information about the industry has to be made as big data and to be shared with the co-partners for making it really a IoT cultured organization. Finally, partnering the other industries, where only a few industries make it possible to bring all the supporting firms into one coherent package [15]. System integration is the biggest challenge to implement IoT and development of a connected manufacturing strategy. Various practical challenges [16] in implementing IoT is listed below

2.1 Vertical sector specific needs

Each vertical sector in manufacturing has different process and need of diverse gateways and platforms to collect the data from various sensors of different sensitivity levels. Connecting all the devices is the major challenge encountered in IoT and this will vary the entire communication protocol and the underlying technologies. Presently the centralized paradigm is used to collect the data from different nodes in the network [17]. But it will be complex to maintain the same strategy in the future, if the devices which are to be connected are in the billions. Also, the maintenance cost of cloud computing will be very high to maintain this vast data. In future a decentralized network will be required to connect the different nodes in the network. Each node will communicate with other node with a unique authentication directly without a broker in-between. This communication technique also has a disadvantage of security issues which can be overcome by the recent technologies such as blockchain [18].

2.2 Security

Data retrieved from smart manufacturing should be stored in the cloud which can be utilized by the stakeholders to observe the enactment of every process in manufacturing and ensure a robust control and adapt industrious measures using real-time information. Security plays a vital concern when developing the smart factories. Different frameworks are analyzed by the researchers to ensure the security and privacy in the network [19-20]. Today the vast information available on the internet has facilitated hackers to identify the weak spots and liabilities of "Smart Businesses". The data transfer from industrial unit to the cloud server should be fully secured. The list of security issues and its precautions are given in the table 1[21].

Table 1. Security issues in IoT enabled manufacturing

Security issues	Description	Precautions
Secure constrained devices	Less capability devices (memory, storage, processing capability) are not suitable for carrying out complex encryption and decryption quickly which is very essential for security approaches	Multiple defensive layer should be implemented to separate devices into separate networks and using firewalls, to compensate for these device limitations
Device authentication	Unique identification should be provided to the devices before they can access platforms and upstream services and apps	Two factor authentication (2FA) and imposing the usage of robust passwords or certificates is recommended
Managing device updates	Devices might be temporarily towed from production for updation and also updation will be not available for older devices where these are no more supported by the manufacturer	The device administrator should automatically update the devices and to keep a track on the current level of updating required for devices and also on the retirement of devices
Safe communications	Soon after the device is secured the next is to safeguard their communication	Sending the encrypted communication between the devices. A separate network is to be used to isolated the devices such that the communication will be confidential

Data privacy and integrity	After the data communication, it has to be stored securely and the unwanted communication should be eliminated	Employ the digital signature to confirm that the data is not altered and using the safety technique such as blockchain to preserve the data from hackers
Secure cloud applications	Cloud apps are used to control and process IoT devices and theses are also to be protected as a multi-layer slant to IoT security	Apps should be provided with protected authentication, for both the apps and the users, by providing options such as 2FA and secure password recovery options
Availability	Unavailability of IoT data and app leads to the revenue loss, destruction to the tools and life threatening issues	Ensure availability by including redundancy to avoid one point failure and should be resilient and fault tolerant
Predict and Manage vulnerabilities	Precautions must be implemented to identify the area of fault occurrence and security measures should be applied for both real time issues and future	Use predictive techniques and Artificial Intelligence to predict the future issues from the past performance

2.3 Standards
Connectivity and utility standards are very big challenges to implement the IoT. There are many different proprietary methods and custom integration solutions followed by the manufacturers. Standardization leads to the interaction of machines and assets in a translucent fashion. Standardization involves the creation of standard mechanisms that will allow the devices to interoperate. The mostly used standard in IoT is IEEE 802.15.4 which includes the communication technology such as ZigBee [22]. Continuous research is in progress to develop the sustainable standard for IoT.

2.4 Human Machines Interfaces
Human Machine Interface (HMI) facilitates an interaction between the human operator and machine with a mode of remote control in industries [23]. Mobile applications have a vital role in managing Industrial Control Systems (ICSs) but it is not suited for the task of learning, allocating, and building HMIs. Hence speciality framework such as No Effort Rapid Development (NERD) is developed to achieve a high level HMI [24]. IoT favors the communication between human and machine by providing the real time

IOP Conf. Series: Materials Science and Engineering **764** (2020) 012025 doi:10.1088/1757-899X/764/1/012025

information about the process. 'Informed people' leads to smart design, intelligent operations, maintenance, quality service and safety.

Table 2. IoT in sustainable manufacturing [25]

Sustainability issues	Description
Reducing product miles	Using IoT enabled smart product tracking, detect the optimal route for the products to reach the destination from the starting point and thus making the distribution much more proficient so that customers get the goods at the right time
Lengthening the life of machinery and tools	The life time of the product can be significantly increased by tracking and constantly monitoring the performance. The maintenance requirement of the product is easily determined and further this can enhance the efficiency of the product
Preserving energy and water	IoT enabled manufacturing unit can facilitate the managers about the usage of energy in various units such as heating and cooling systems. This paves the way for making the effective utilization of energy. Also waste management can be effectively tackled by IoT enabled systems
Make supply chains smarter	IoT technology enable to provide remotely monitor and automation of the gadgets which is located in multiple locations including the supplier and customer locations and retrieving the data back to the central database for future purpose.

3. An IoT enabled Smart Manufacturing

To enable an IoT integrated smart manufacturing three major building blocks are required as shown in Figure 1 [26], the first one is the sensing nodes. The sensing devices which are used in IoT environment may vary widely depending upon the application. It may be a camera for image monitoring, Radio Frequency Identification (RFID) reader to sense the existence of an object/person or a simple thermocouple measuring temperature. The second building block is the embedded processing nodes, which may be a hybrid microcontroller or microprocessor which the embedded processing takes place in real time mode. The third one is the wired or wireless communication nodes, which transfer the communication between the two former building blocks to execute a task. It is generally a two way communication node.

IOP Conf. Series: Materials Science and Engineering **764** (2020) 012025 doi:10.1088/1757-899X/764/1/012025

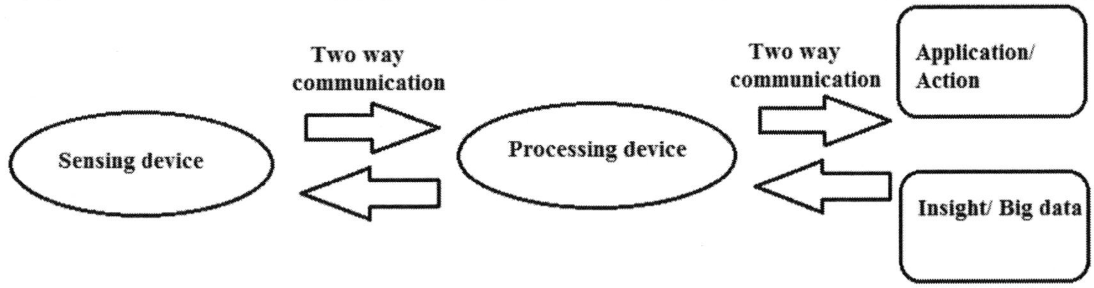

Figure 1. Building blocks of an IoT enabled smart manufacturing

The pervasiveness of IoT enabled manufacturing not only confined to a particular localized industry segment or value chain process, rather it is applicable for multiple segments of manufacturing and logistics as shown in Figure 2. The potential of IoT in various segments is given as follows

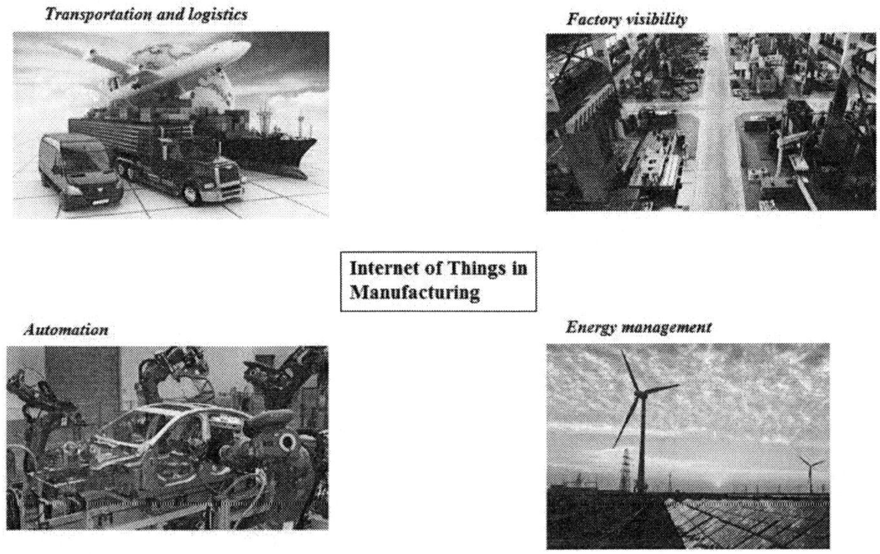

Figure 2. Influence of IoT in manufacturing sector

3.1 Transportation and logistics

In IoT enabled transportation and logistics, the products are incorporated with sensors along embedded tags [27]. When the product moves, the flow can be traced by various participants in the supply chain. Further, the product becomes totally autonomous and various decisions can be taken by the readers from the obtained information. This leads to the end to end visibility in the supply chain and warehouse management, ensuring that the product is in the right place at the right time. Optimal fleet management is also possible where the fleet downtime is considerably reduced.

IOP Conf. Series: Materials Science and Engineering **764** (2020) 012025 doi:10.1088/1757-899X/764/1/012025

3.2 Factory visibility

When the factory is equipped with IoT, then it paves the way to access the real time information and efficient collaboration [28]. A plant manager can be informed the status of machines and production from any locations and this can considerably reduce the downtime for decision and action. The quality control can be effectively achieved by shutting down the production process when the problem is known in advance and the product wastage is totally eliminated. The benefit of factory visibility extends to the suppliers and third party participants.

3.3 Automation

The IoT and Internet Protocol (IP) network is used to connect the entire network and provide connectivity and information sharing across multiple business networks. Flexible and compliant automation needs a complete industrial IoT elucidation that addresses business, technologies, and facility aspects in order to (i) deliver real-time control for optimization and enhanced uptime; (ii) empower the horizontal association between devices; and (iii) offer protected and expected inclusion of service features [29].

3.4 Energy management

Energy plays a vital role in manufacturing and is often the second major operating cost. But many industries are lagging in modeling tools or management tools to optimize energy usage in industrial operations. Certain IoT enabled plants offers integrated energy data and prediction analysis. Hyunjeong Lee et. al. 2016 [30] proposed an energy management framework for smart factory where the smart factory consist of three layers. The first layer is the sensor layer which collects the data and send to the second layer where it analyses the data and infers the context and finaly to the third layer where the energy management is achieved through monitoring and controlling the status of energy consumption.

4. Benefits of IoT in Smart Manufacturing

4.1 Greater Energy Efficiency

Energy plays a significant role in the expenses for manufacturing industries. Industries measure their expenses for energy conception from the bill that is generated. It is very complex or not possible to detect the inefficient power conception from the available data. Nearly 77 % percent of industries retrieve their energy consumption data from their monthly utility bills or energy utilization monitoring facility, where the data available is limited [31]. Implementing IoT can access the data from the device level and it is possible to pinpoint the devices which are under performing. By tracking the power conception of each device separately, managers acquire visibility to benchmark the under performing devices with the best performing devices. Energy utilization is optimized and much of the

4.2 Predictive Maintenance

IoT completely eliminates the conventional method of maintenance by mere guessing from the historical data, but instead realizing the maintenance requirement for the device at the exact moment by obtaining the real data. The sensor gives the condition of each component and if it doesn't need maintenance the replacement is totally eliminated and money and time are saved. These circumstantial perceptions keep the equipment at good running condition and minimize the risk of costly downtime. Recent analysis shows that the market for predictive maintenance applications is poised to grow from $2.2B in 2017 to $10.9B by 2022, at an 39% annual growth rate. According to recent research IoT and advances in analytics leads to an efficiency gain of 25% - 30% in industries implementing the IoT enabled predictive maintenance. Few of the top listed companies in the world, providing IoT solutions in the predictive maintenance [32] is shown in the table 3.

Table 3. World top listed companies, providing IoT solutions in the predictive maintenance

Company	Description
IBM	IBM launched a Predictive Maintenance and Quality (PMQ) tool which has the ability of condition monitoring and reports the same for the purpose of predictive maintenance. Based on this technology Kone has implemented 24/7 PMQ Connected Services
SAP	Top most German based IoT solution provider, It's SAP Leonardo IoT Portfolio which is an intelligent technical solution intended to connect the people and process to attain tangible business outcomes
Siemens	Provided a system which automatically monitors and consecutively preventive maintenance takes place. NASA Armstrong Flight Center (cooling systems) is an example of Siemens work
Microsoft	Microsoft Azure is developed as the public cloud platform of IoT monitored predictive maintenance. The source code for all the solution accelerators is open source and is available in GitHub. This paves the way for developing ones own IoT solutions and the application code in each solution accelerator includes a web app that lets you manage the solution accelerator
GE	GE has two categories towards predictive maintenance one is the GE measurement and the other is GE digital which monitors the hardware and software of the plant. Further, the GE digital also contributes in processing the concept of digital twin, an important basis for predictive maintenance analytics

4.3 Higher Product Quality
IoT can help in quality improvement of product which leads to achieve waste reduction, lowering the costs, increased customer satisfaction and higher sales. One of the major causes for quality is fault equipment.

These kind of problems can be overcome by the IoT technology. The various quality improvement enabled by the IoT is given in the table 4 [33].

Table 4. Quality improvement enabled by the IoT

Quality Improvements	Description
Avoiding performance degradation	An integrated sensor with a remote software to avoid fault in the system and send an instruction note to the customer to handle the particular problem
Refining warranty costs and service contract profitability	Connected device permit the identification of issues in the system and this leads to early detection and prevention of faults. This gains the satisfaction of the customer
Termination of scrap and rework	In a networked manufacturing process, the machines can communicate with each other and if there is any deviation is found out, the system can automatically tune to bring the final product within stipulation.
Reinventing the service contract	Sensors support to determine the appropriate time for service. This optimize the time for servicing the product and minimize the service expenses for the customer.
Out-innovating the competition	With sensors entrenched in all the aspects of factory level, it creates a new opportunity for the business and increase the profit.

4.4 Reduced Downtime

The Analysis says that downtime costs between 5 and 20 percent for a factory of its productive capacity, but nearly more than three fourths of amenities are unable to precisely estimate their total downtime cost [34]. Timely, precise and high-quality manufacture is at the core of profits. When the machine breakdown in the middle it leads to an enormous loss in addition to the traditional downtime expenses. Unexpected downtime is a worst-case situation for manufacturers. Zero downtime (ZDT) helps factories from downtime through predictive maintenance, which is a common IoT approach [35]. With ZDT, maintenance required are founded on real-time usage of the machine. Though ZDT is best known for proactive failure forecast, it has many more applications. Its capability to remotely monitor the machine's health which allows for optimal production, for example, decreasing energy intake, improving production amount, and extending the mechanical life of machines. Two different types of models for IoT downtime analysis is provided [36] in table 5.

Table 5. Models for IoT downtime analysis

IoT in downtime	Conventional analytical models	Determines the occurrence of failure by separating the repairable and non-repairable parts based on their usage. These models influence historical time-to-failure data to detect the future failure of parts. In conditions where manufacturers do not have the privilege of detailed information from sensor data, these models are useful
	Composite models	This is an advanced model where the cause for the machine failure is also predicted. This model is more advance in all levels than the previous one

Among these, two models the selection depends upon the availability of data. When data and information about the parts failure is available, then the first model is applicable. On the other hand when only the claim data is available, then the second model is preferred. Other factors influencing the selection of the model are the nature of parts and the stage of the parts lifecycle (intial, growth and maturity or decline). With the selection of right model, manufacturers can take their predictive maintenance to new levels.

4.5 Informed Decisions

Previously manufacturing industries use statistical process control and data analysis tool for optimizing manufacturing [37]. Now managers can make decisions on proactive approach and this leads to decrease in stress in decision making and increase in production and profit.

5 Conclusion

IoT is broadly accepted as a new technology that plays a significant role in the enhancement of the manufacturing industry. It can integrate the entire components of manufacturing sectors including sensor, processing unit, communication devices and actuation gadgets. This highly integrated smart cyber – physical systems opens the door to create a new business and market opportunities in manufacturing and paves the way for the new industrial revolution 4.0. It brings a huge opportunity in the manufacturing sector in achieving better system performance in globalized and distributed environments. The various advances in manufacturing industry due to the implementation of IoT and their challenges are emphasized as follows:

- IoT enables the sustainable development in various sectors in industries such as reducing product miles, extending the life of machinery and equipment, preserving energy and water and enhancement of the supply chain.
- The decentralized paradigm network is required to connect all the devices with unique authentication for each node without a broker in-between.
- Multiple defensive layer, 2FA, encrypted communication, digital signature is few techniques to keep the data safe and secure.
- IoT completely eliminates the conventional method of maintenance by mere guessing from the historical data, but instead retrieving real-time data to realize the maintenance needs at an exact moment.
- IoT can help in quality improvement of product which leads to achieve reduced waste, lower costs, increased customer satisfaction and higher sales. Also, IoT can reduce the downtime cost significantly.

Moreover, the implementation of IoT in manufacturing is in the primary stage and a high research activity is in demand to enable the IoT technologies for secure, reliable operations.

References

[1] Yee T, Yen N and Onathan S 2017. Internet of Things real-time monitoring of energy efficiency on manufacturing shop floors. *Procedia CIRP* **61** 376-381.
[2] Chen Y, Weiming S and Xianbin W 2016. Application of Internet of Thinga in manufacturing. Proceedings of the 2016 EEE 20th International Conference on Computer supported cooperative work in design.
[3] Zhuming B, Li D, and Chengen W 2014. Internet of Things for enterprise systems of modern manufacturing. *IEEE Transactions on Industrial informatics* **10(2)**.
[4] Atzori L, Lera A and Morabito G 2010. The internet of things: A survey. *Computer networks* **54(15)** 2878-2805.
[5] Miorandi D, Sicari S and Pellegrini F 2012. Internet of things: Vision, application and research challenges. *Ad Hoc Networks* **10(7)** 1497-1516.
[6] Lee J, Bagheri B and Kao H.A 2015. A cyber- physical systems architecture for industry 4.0 – based manufacturing systems. *Manufacturing letters* **3** 18-23.
[7] Kagermann H, Helbig J and Hellinger A 2013. Recommendations for implementing the strategic initiative industrie 4.0: Securing the future of German Manufacturing Industry. Final repor of the Industrie 4.0 working group Forschungsunion.
[8] Wen L, Sami K 2017. Methodology for Monitoring Manufacturing Environment by using Wireless Sensor Networks (WSN) and the Internet of Things (IoT). *Procedia CIRP* **61** 323-328.
[9] IoT in manufacturing. [online] https://www.visualcomponents.com/ (Accessed 20 August 2018).
[10] Design for manufacturing's 'Internet of Things' June 2014, Cognizant.
[11] Finoit. [online] https://www.finoit.com/ (Accessed 20 August 2018).
[12] Anaya L, Ulaimi M and Abdallah S 2015. An investigation into the role of enterprise information systems in enabling business innovation. Business process management **21(4)**.

[13] Lihui W and Albert S 2016. Challenges in smart manufacturing. Journal of Manufacturing Systems **40(1)**.
[14] IoT.ieee.org. [online] https://iot.ieee.org/newsletter/march-2017/three-major-challenges-facing-iot.html/ (Accessed 20 August 2018).
[15] Ihsmarkit. [online] https://www.ihsmarkit.com/ (Accessed 20 August 2018).
[16] Eurostars. [online] https://www.eurostars-eureka.eu/ (Accessed 20 August 2018).
[17] Blog. Apnic.net. [online] https://blog.apnic.net/2015/10/20/5-challenges-of-the-internet-of-things/ (Accessed 20 August 2018).
[18] I-scoop. [online] https://www.i-scoop.eu/blockchain-distributed-ledger-technology/blockchain-iot/(Accessed 20 August 2018).
[19] Mahmoud A, Giovanni R and Bruno C 2018. Internet of Things: A survey on the security of IoT frameworks. Journal of Informattion security and application **38** 8-27.
[20] Minhaj A and Khaled S 2018. IoT security: Review, block chain solutions and open challenges. *Future generation computer systems* **82** 395-411.
[21] Developer.ibm. [online] https://developer.ibm.com/dwblog/2017/iot-security-challenges/ (Accessed 20 August 2018).
[22] Ahmed M and Thorat S 2015. Internet of Things (IoT) Standards, Protocol and Security issues. International Journal of Advanced Research in Computer and Communication Engineering, **4(11)** 491-495.
[23] Harpreet S and Daruwala R 2016. Human Machine Interface in Internet Of Things System. International Conference on Computing Communication Control and automation (ICCUBEA) Pune, India.
[24] Czauski T, White J and Sun Y 2016. NERD—middleware for IoT human machine interfaces. Annals of Telecommunications. **71(3-4)** 109-119.
[25] Ulehssustainability. [online] https://www.ulehssustainability.com/blog/sustainability/4-ways-iot-boosts-sustainability-performance/#sthash.P7qs9YKx.dpbs/ (Accessed 20 August 2018).
[26] Apicsterragrande. [online] http://apicsterragrande.org/images/articles/Machado__Internet_of_Things_impacts_on_Supply_Chain_Shah_Machado_Second_Place_Grad.pdf/ (Accessed 20 August 2018).
[27] Paweł T 2015. Application of Internet of Things in Logistics – Current Challenges. *Economics and Management* **7(4)** 54-64.
[28] iotwf. [online] http://cdn.iotwf.com/resources/6/iot_in_manufacturing_january.pdf/ (Accessed 20 August 2018).
[29] Hongyu P and Kristian S 2015. Internet of Things for Industrial Automation Challenges and Technical Solutions. IEEE International Conference on Data Science and Data Intensive Systems.
[30] Hyunjeong L, Sangkeun Y and Yong-Woon K 2016. An Energy Management Framework for Smart Factory based on Context-awareness. ICACT2016.
[31] Fathym. [online] https://fathym.com/2017/05/5-powerful-benefits-iot-manufacturing-industry/(Accessed 20 August 2018).
[32] IoT-analytics. [online] https://iot-analytics.com/top-20-companies-enabling-predictive-maintenance/(Accessed 20 August 2018).
[33] Industryweek.com. [online] https://www.industryweek.com/IoT-boosts-quality/(Accessed 20 August 2018).
[34] PTC [online] https://www.ptc.com/en/product-lifecycle-report/iot-slashes-downtime-with-predictive-maintenance/(Accessed 20 August 2018).
[35] Newsroom.cisco. [online] https://newsroom.cisco.com/feature-content?type=webcontent&articleId=1764957/(Accessed 20 August 2018).
[36] Blog.iiconsortium. [online] http://blog.iiconsortium.org/2017/07/from-reactive-to-predictive-maintenance-an-iot-enabled-manufacturing-leap.html/(Accessed 20 August 2018).

[37] Intel. [online] https://www.intel.com/content/dam/www/public/us/en/documents/solution-briefs/smart-manufacturing-solutions-top-10.pdf /(Accessed 20 August 2018).

IOP Conf. Series: Materials Science and Engineering **764** (2020) 012026 doi:10.1088/1757-899X/764/1/012026

Numerical study on Longitudinal control of Cessna 172 Skyhawk aircraft by Tail arm length

P Vadivelu[1], D Lakshmanan[1], R Naveen[1], K Mathivannan[1] and G Yoges Kumar[1]

[1] Department of Aeronautical Engineering, Bannari Amman Institute of Technology, Sathyamangalam, India - 638401

vadivelupv@gmail.com

Abstract. This paper covers a new method of longitudinally controlling an aircraft by changing the horizontal tail arm length, without the conventional control surface called 'elevator'. The effects of changing horizontal tail arm length from its steady level trimmed condition are determined numerically for the chosen Cessna aircraft. From the numerical calculations, the possibilities of implementing this type of control in the chosen Cessna aircraft are analyzed. At last, some recommendations are suggested for the effective longitudinal control of the Cessna aircraft by changing horizontal tail arm length. Further, it is also suggested that employing both elevator control and tail arm length control together will make an aircraft to meet any kind of longitudinal controllability and stability requirement at any instant. This optimal aircraft configuration can be achieved by proposing a novel configuration and morphing solutions.

1. Introduction

From the origin of aircraft, stability and control have been a subject of concern that must be considered when designing an aircraft[1], [2]. Similarly fora rocket/missile, or any other type of vehicle. It apparently depicts that without stability and control, a flying vehicle cannot be constructed[3]. The study of stability indicates that whether the vehicle will be easy, difficult or impossible to fly. It has two separate disciplines, namely, static stability and dynamic stability. And, the study of controllability ensures that whether the aircraft will be able to perform the desired flight maneuvers. Stability, as well as control, are identified based on three rotational degrees of freedom namely longitudinal, directional and lateral. Traditionally, an aircraft has been controlled by the primary control surfaces like elevator, rudder, and aileron. Here, the elevator is used for longitudinal control and rudder and aileron are used for directional and lateral control since they are coupled motion.

Currently, there are a lot of research works going on in the field of aircraft's stability and control in subsonic, supersonic and hypersonic speeds. These researches focus on the requirement and required improvement in static stability[4]–[8], dynamic stability [9]–[11]and control[12].

Generally, stability and control characteristics of an aircraft can be analyzed byanalytical (theoretical) methods and experimental methods. Now, Analytical techniques are getting highly advanced with developments in computer modeling. for example, (i)[13] used MATLAB for mathematical modelling for Nonlinear six-degree-of-freedom flight, (ii) [14] used SDSA (Simulation and Dynamic Stability

IOP Conf. Series: Materials Science and Engineering **764** (2020) 012026 doi:10.1088/1757-899X/764/1/012026

Analysis) application for analysing the dynamic characteristics of the aircraft just in the conceptual design stage, (iii) [8] used CFD (ANSYS Fluent) with dynamic mesh for the prediction of the longitudinal dynamic of an UAV. However, these methods have to be correlated with experiments before they can be used in confidence[4], [15]–[17]. [18]discusses the considerations in the determination of stability and control derivatives and dynamic characteristics from flight data. A combination of these two methods is commonly used in today's applications. Whereas, [19]used an integrated (computational and experimental) approach to estimate the X-31 aircraft's stability & controllability.

Literature verifies that continuous studies have been performed and concentrated on the area of stability and controllability. However, all the studies focused only on the traditional control surfaces. None of the researchers did a research on theother way of controlling an aircraft. This paper focuses on the analyzing the possibilities of longitudinally controlling the aircraft by changing the tail arm length of horizontal stabilizer without elevator. Moreover, this paper covers only the numerical study on the effects of horizontal tail arm length on the static longitudinal stability and controllability of the selected aircraft.

2. Static longitudinal stability

In general, to have the ability of returning tolongitudinal trim condition (static longitudinal stability), an aircraft must satisfy the following two conditions[20];

- When AOAincreases due to disturbance, the whole aircraft's center of pressure must move rearward from its center of gravity (cg) and
- When AOA decreases due to disturbance, the whole aircraft's center of pressure must move forward from its 'cg'.

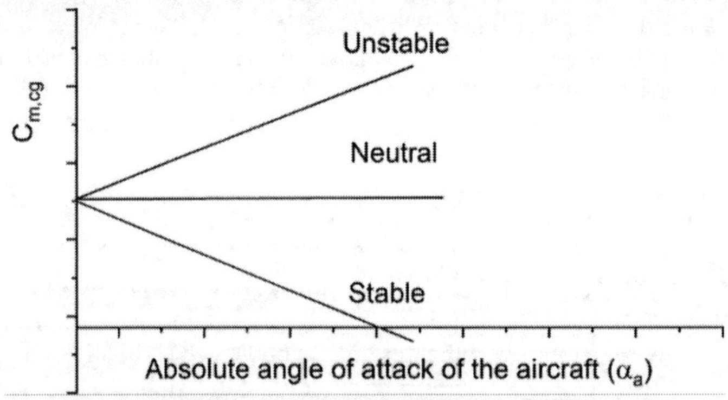

Figure 1. Typical static longitudinal stability graph

Theabovetwo conditions can be represented in terms of pitching moment coefficient about the center of gravity $\left(C_{m,cg}\right)$ with respect to the absoluteAOA (α_a) as shown in figure1[21]. From the figure1, it is observed that the derivative $\left(\partial C_{m,cg}/\partial \alpha_a\right)$ must be negative for the static longitudinal stability. In order to fix the derivative value (static longitudinal stability/ longitudinal stiffness), the contribution of various parts (wing-body and horizontal tail) of the aircraft has been related.

IOP Conf. Series: Materials Science and Engineering **764** (2020) 012026 doi:10.1088/1757-899X/764/1/012026

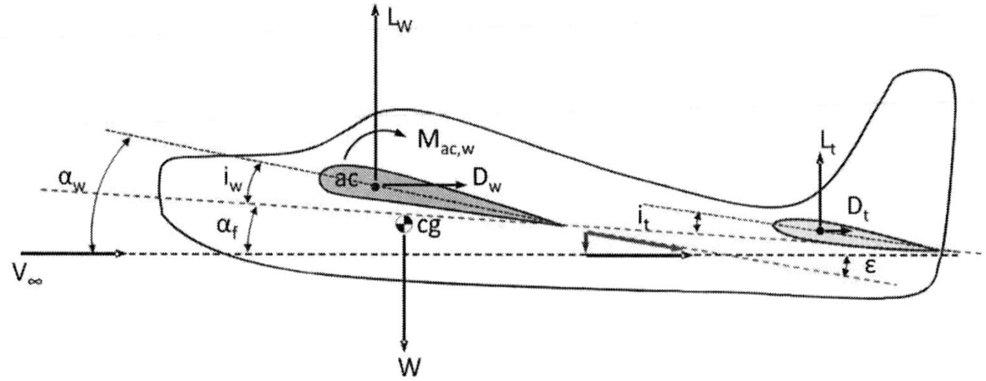

Figure 2. Wing-body and tail contribution to $C_{m,cg}$[22]

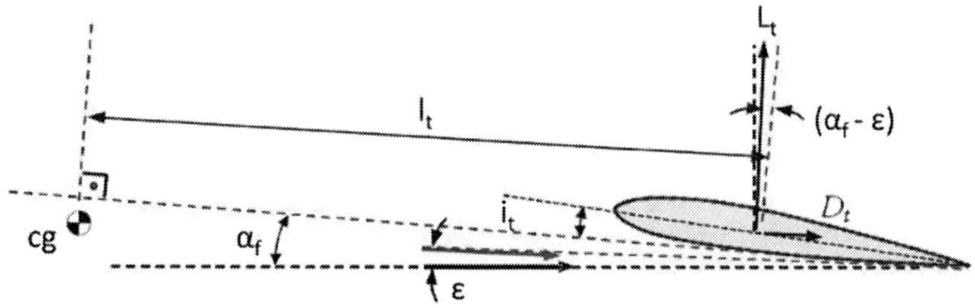

Figure 3. Tail contribution[22]

From the figure 2 and 3, the equation of total pitching moment coefficient about 'cg' in terms of lift coefficient is given by

$$C_{m,cg} = C_{m,ac,w} + C_{L,w}\left[h_{cg} - h_{ac}\right] - V_H C_{L,t} \tag{1}$$

And in terms of AOA

$$C_{m,cg} = C_{m,ac,w} + \left(a \times \alpha_a\right)\left[h_{cg} - h_{ac}\right] - V_H a_t \alpha_t \tag{2}$$

$$C_{m,cg} = C_{m,ac,w} + \left(a \times \alpha_a\right)\left[h_{cg} - h_{ac}\right] - V_H a_t \left(\alpha_a - i_t - \varepsilon\right) \tag{3}$$

$$C_{m,cg} = C_{m,ac,w} + \left(a \times \alpha_a\right)\left[h_{cg} - h_{ac}\right] - V_H a_t \left(\alpha_a - i_t - \left\{\varepsilon_o + \frac{\partial \varepsilon}{\partial \alpha_a}\right\}\right) \tag{4}$$

$$C_{m,cg} = C_{m,ac,w} + \left(a \times \alpha_a\right)\left[h_{cg} - h_{ac} - V_H \frac{a_t}{a}\left\{1 - \frac{\partial \varepsilon}{\partial \alpha_a}\right\}\right] + V_H a_t \left(i_t + \varepsilon_o\right) \tag{5}$$

WhereTail volume coefficient $V_H = \dfrac{l_t \times S_t}{S \times C}$

When α_a is zero, the $C_{m,cg}$can be termed as zero-lift pitching moment coefficient $C_{m,o}$

$$C_{m,o} = C_{m,ac,w} + V_H a_t \left(i_t + \varepsilon_o\right) \tag{6}$$

The slope of the $C_{m,cg}$curve is obtained by differentiating equation (5)with respect to α_a

$$\frac{\partial C_{m,cg}}{\partial \alpha_a} = a\left[h_{cg} - h_{ac} - V_H \times \frac{a_t}{a} \times \left\{1 - \frac{\partial \varepsilon}{\partial \alpha_a}\right\}\right] \tag{7}$$

The equation (7) clearly shows the powerful influences of the location of the 'cg' $\left(h_{cg}\right)$ and the tail volume ratio V_H in determining the pitch stability of an aircraft. It also makes us to establish a certain viewpoint in the design of an airplane.

$$\frac{\partial C_{m,cg}}{\partial \alpha_a} = a\left[h_{cg} - h_{np}\right] = -a\left[h_{np} - h_{cg}\right] \tag{8}$$

$$\frac{\partial C_{m,cg}}{\partial \alpha_a} = -a\,[\text{Static Margin}] \tag{9}$$

The distance $\left(h_{np} - h_{cg}\right)$ is referred as static margin which is a straight measure of static longitudinal stability. For static longitudinal stability, this static margin must be positive. Moreover, the larger static margin leads to have more longitudinal stability for the airplane.

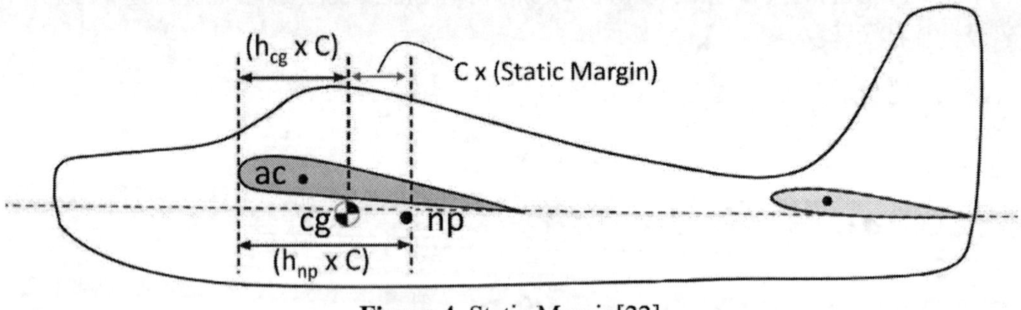

Figure 4. Static Margin[22]

Therefore,

$$C_{m,cg} = C_{m,o} + \left(\frac{\partial C_{m,cg}}{\partial \alpha_a}\right) \times \alpha_a$$

These equations(1) to(9) are derived based on the following assumptions[23];

- $\cos\alpha_{wb} \approx 1$ and $\sin\alpha_{wb} \approx \alpha_{wb}$,
- Center of gravity is assumed very close to the zero lift line,
- $Z_t << l_t$,
- $D_t << l_t$,
- The angle ($\alpha_{wb} - \varepsilon$) is small,
- $\sin\left(\alpha_{wb} - \varepsilon\right) \approx 0$ and $\cos\left(\alpha_{wb} - \varepsilon\right) \approx 1$,

- $M_{ac,t}$ is small in magnitude, and
- $\alpha_{wb} \rightarrow \alpha_a$ (Zero lift line of wing-body and Zero lift line of aircraft are almost same), $a_{wb} \rightarrow a$ and $C_{L,wb} \approx C_L$.

3. Cessna 172 skyhawk

Since Cessna 172 Skyhawk aircraft's design dataare freely available in internet, books and other sources, it has been chosen for this study purpose. It is a most popular single-engine trainer aircraftwhich is favorite for student pilots. It has excellent flight performance, better stall characteristics, and low landing speed.Figure 5 shows the three views diagram of the Cessna 172 Skyhawk aircraft.

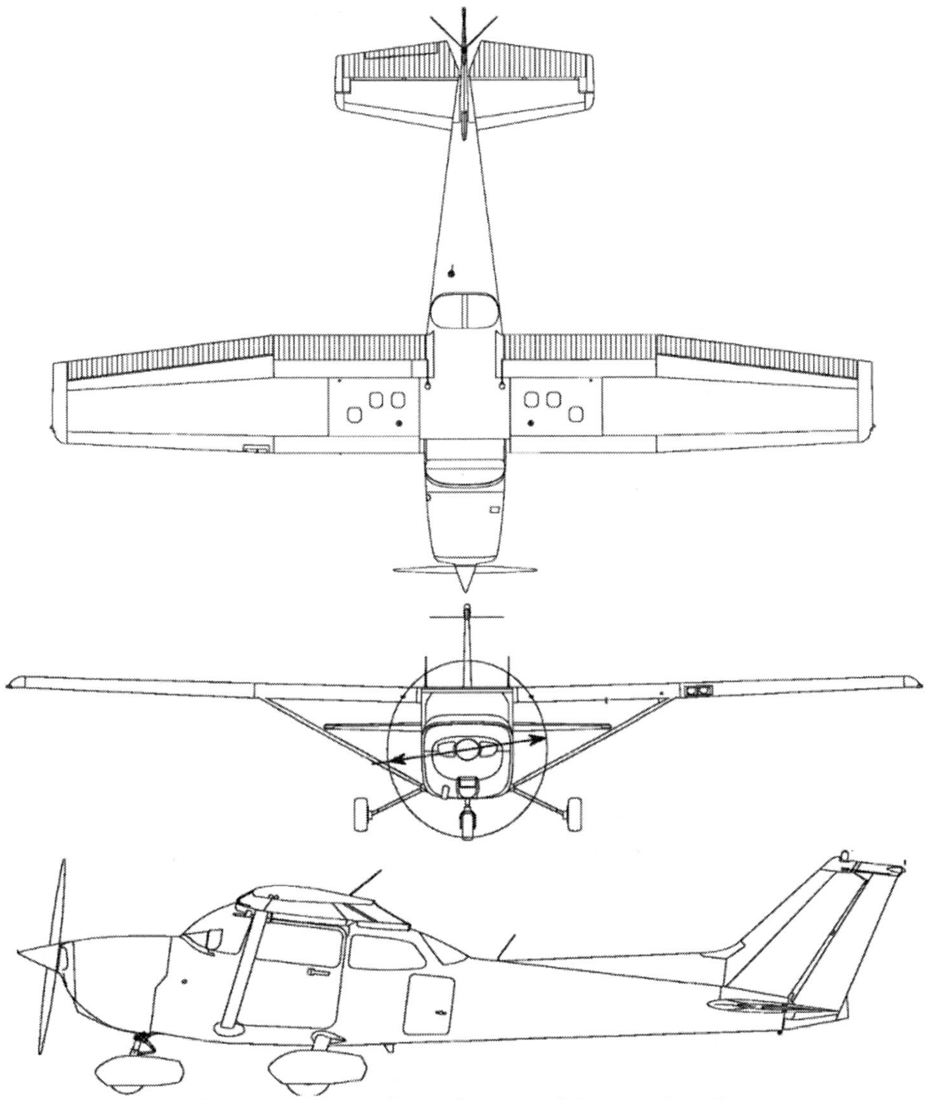

Figure 5. Three views diagram of Cessna aircraft

For numerical calculations of stability, the specifications of Cessna 172 Skyhawk aircraft are referred from Wikipedia and other online sources. The specifications are given below

3.1 General characteristics [24]

- Crew : one
- Capacity : three passengers
- Length : 8.28 m
- Wingspan : 11 m
- Height : 2.72 m
- Wing area : 16.2 m^2
- Aspect ratio : 7.32
- Empty weight : 767 kg
- Gross weight : 1,111 kg
- Fuel capacity : 212 litres

3.2 Performance characteristics [24]

- Cruise speed : 226 km/h
- Stall speed : 87 km/h (power off, flaps down)
- Never exceed speed : 302 km/h (IAS)
- Range : 1,289 km with 45 minute reserve, 55% power, at 12,000 ft
- Service ceiling : 4,100 m
- Rate of climb : 3.66 m/s
- Wing loading : 68.6 kg/m^2

3.3 Design characteristics [24]

- Wing Aerofoil [25] : NACA 2412
- Wing's lift curve slope (a) : 5.143
- Tails's lift curve slope (a_t) : 5
- Wing surface area (S) : 16.2 m^2
- Mean aerodynamic chord (C) of wing: 1.47 m
- Tail surface area (S_t) : 2.83 m^2
- Downwash variation with respect to α_a($d\varepsilon/d\alpha$)[26]:0.25
- Wing leading edge distance from the nose: 1.934 m
- Wing aerodynamic center from leading edge:0.3675 m
- h_{ac}[26] :0.25
- Cl_{max}[27] : 1.3

4. Static longitudinal stability – numerical analysis

Static longitudinal stability of the aircraft can be determined by knowing the 'cg' location and contribution of various aircraft parts. Since mass of the aircraft is not constant during flying time, 'cg' of the aircraft will not occupy a fixed location. Hence, static longitudinal stability also varies based on the 'cg' location during flying. So, it is important to find the extreme locations of the 'cg' in its mission.

4.1 Possible position of the 'cg'

In an aircraft, the masses should be distributed such that it has a defined 'cg' position, which is critical. Also, the mass distribution should be such that on certain situations where some components may be consumed or even removed, its 'cg' movement should be in a controllable manner so that is not compromised. Table 1 shows the possible extreme position of 'cg' during the mission phases of Cessna 172 aircraft.

Table 1. Center of gravity location details of Cessna

Particulars	**Most forward position (i.e, with all the crew and full tank of fuel)**	**Most rearward position (i.e, without all the crew and fuel)**
Centre of gravity position from nose	2.53563 m	2.56053 m
Centre of gravity position from leading edge of the wing	0.60163 m	0.62653 m

4.2 Determination of stability parameters

The stability, control, and handling qualities of the aircraft are analyzed by using tail volume coefficient and position of the 'cg' as some percentage of wing chord length.

From equation (7)

$$\frac{\partial C_{m,cg}}{\partial \alpha_a} = a\left[h_{cg} - h_{ac} - V_H \times \frac{a_t}{a} \times \left\{1 - \frac{\partial \varepsilon}{\partial \alpha_a}\right\}\right]$$

The thrust has three effects, namely the direct moment of the thrust, the propeller or inlet normal force due to turning of the air, and the propeller wash or jet induced flows upon the aircraft parts. These power or thrust effects on static longitudinal stability are excluded in the above equation.

Table 2. Stability parameters of Cessna

Particulars	**Forthe Most forward position of 'cg'**	**Forthe Most rearward position of 'cg'**
h_{cg}	0.40927	0.426
l_t	4.03083 m	4.03069 m
V_H[26]	0.479	0.47899
Longitudinal Stiffness $\left(\partial C_{m,cg} / \partial \alpha_a\right)$	-0.97717	-0.89
Stability curve slope in degree	-44.34°	-41.67°
Cm_o	0.07	0.07
h_{np}	0.6	0.6
Neutral Point from leading edge of the wing	0.882 m	0.882 m
Neutral Point from nose	2.816 m	2.816 m
Static Margin[26]	0.19	0.17305

The calculated static stability curves of Cessna aircraft for the most forward location of 'cg' and for the most rearward location of 'cg' is shown in figure 6.

IOP Conf. Series: Materials Science and Engineering **764** (2020) 012026 doi:10.1088/1757-899X/764/1/012026

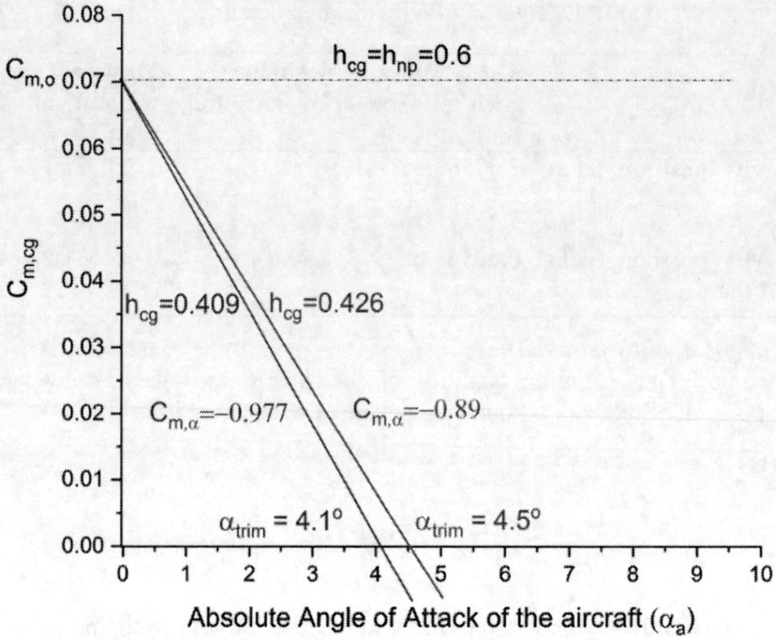

Figure 6. Stability curve of Cessna aircraft

The Cessna aircraft is being controlled longitudinally by a primary control surface called elevators. The elevators are usually hinged to the tailplane or horizontal stabilizer. It controls the pitching of an aircraft, and therefore the α_a and the lift of the wing by altering the pressure distribution of horizontal stabilizer. The effectiveness of elevators on pitching moment coefficient about 'cg' is measured as $Cm_{\delta e}$

From the literature, for Cessna aircraft $Cm_{\delta e}$ = -1.28

Note: *Upward movement of the elevator is assumed as negative and downward movement of the elevator is assumed as positive. The stick force given by a pilot for this type of control is usually very small.*

5. Longitudinal control by tail arm length (without elevator)

Apparently, horizontal tail arm length value (l_t) plays important roles in deciding the longitudinal stability and controllability of an aircraft. It has great influences. So, it is intended to have the longitudinal control by tail arm length. As schematically represented in figure 7, it is proposed to control the pitching of an aircraft by changing the horizontal tail arm length value. This can be achieved by implementing a mechanism to move the horizontal tail. In this paper, the numerical analysis of longitudinal control by 'l_t' has been done for the selected aircraft (Cessna 172 Skyhawk).

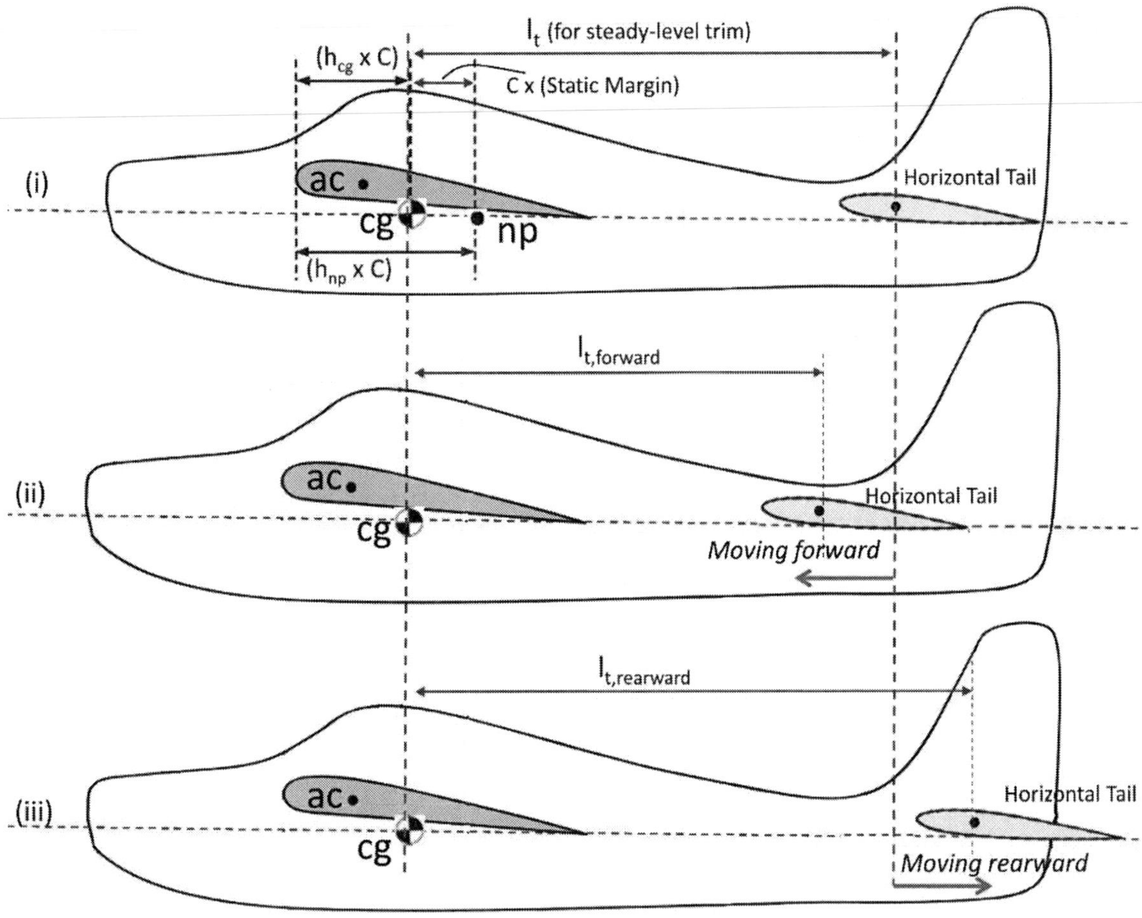

Figure 7. Typical diagram of longitudinal control by tail arm length

5.1 For the most forward location of 'cg'

For the most forward location of the 'cg', the effects of tail arm length variation on Cm_o, longitudinal stiffness and its corresponding change in α_a are tabulated in the table 3.

Table 3. Details of tail arm length (l_t) control for the most forward location of 'cg'

Horizontal tail arm length (l_t)	**Cm_o**	**Longitudinal Stiffness $\left(\partial C_{m,cg}/\partial \alpha_a\right)$**	**Angle of attack (α_a) in degree**	**Stability curve slope in degree**	**Remark**
1.500	-0.0022	0.1507	0.84	8.6	Unstable
1.838	0.0074	0.0000	-11511.66	0.0	Neutral
1.900	0.0092	-0.0276	19.13	-1.6	Stable
2.000	0.0121	-0.0722	9.58	-4.1	Stable
2.500	0.0263	-0.2950	5.12	-16.4	Stable
3.000	0.0406	-0.5178	4.49	-27.4	Stable
3.500	0.0549	-0.7406	4.25	-36.5	Stable
4.031	0.0700	-0.9772	4.11	-44.4	Stable
4.500	0.0834	-1.1863	4.03	-49.9	Stable

IOP Conf. Series: Materials Science and Engineering **764** (2020) 012026 doi:10.1088/1757-899X/764/1/012026

5.000	0.0977	-1.8219	3.07	-61.3	Stable
5.500	0.1119	-2.0447	3.14	-64.0	Stable
6.000	0.1262	-2.2676	3.19	-66.2	Stable
6.500	0.1405	-2.4904	3.23	-68.2	Stable
7.000	0.1547	-2.7132	3.27	-69.8	Stable
7.500	0.1690	-2.9360	3.30	-71.2	Stable
8.000	0.1832	-3.1588	3.33	-72.5	Stable
8.500	0.1975	-3.3817	3.35	-73.6	Stable
9.000	0.2118	-3.6045	3.37	-74.5	Stable
9.500	0.2260	-3.8273	3.39	-75.4	Stable
10.000	0.2403	-4.0501	3.40	-76.2	Stable
15.000	0.3830	-6.2783	3.50	-81.0	Stable
20.000	0.5256	-8.5065	3.54	-83.3	Stable
50.000	1.3815	-21.8758	3.62	-87.4	Stable
100.000	2.8081	-44.1578	3.65	-88.7	Stable
1000.000	28.4860	-445.2349	3.67	-89.9	Stable
10000.000	285.2646	-4456.0059	3.67	-90.0	Stable

The effects of tail arm length on the AOA are shown in figure 8. And the effects of tail arm length on AOA and stability are shown in figure 9. From the figure 8, 9 and table 3, it is observed that the tail arm length change will not create considerable changes in decreasing the AOA from the steady level trim condition since stability increases greatly. Stability curve is almost perpendicular line after the tail arm length value is 20 m. whereas, decrement in tail arm length leads to an increase in AOA as well as a decrement in stability. In this case, even a slight decrement in tail length creates a huge difference in angle of attack. This also creates stability problems. Stability is neutral for the tail arm length value 1.838 m. below this value makes the aircraft inunstable condition. As a maximum control extent, by this way it can be controlled for the AOA form 3.2 degree to stall AOA theoretically. So it is recommended that not to go for the longitudinal control to decrease the α_a from the steady level trim condition by changing tail arm length. Even to increase the α_a from the steady level trim condition, it brings stability problems.

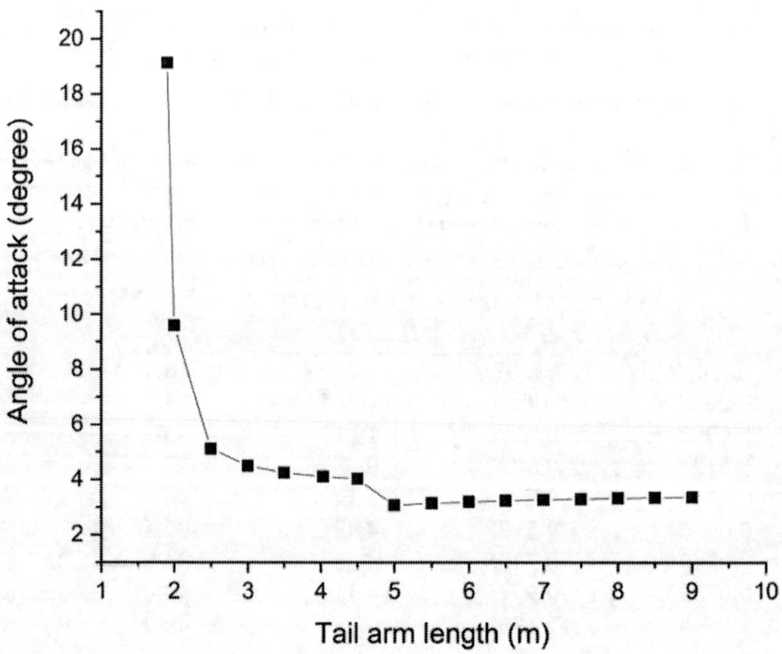

Figure 8.AOA variation with respect to tail arm length (l_t) for the most forward location of 'cg' case

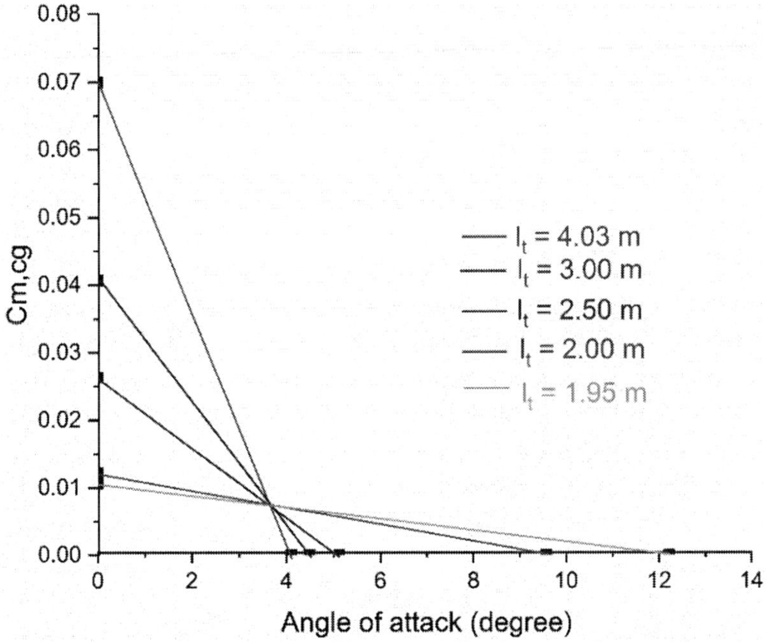

Figure 9. Tail arm length control stability curves for the most forward location of 'cg'

5.2 For the most rearward location of 'cg'

For the most rearward location of the 'cg', the effects of tail arm length variation on Cm_o, longitudinal stiffness and its corresponding change in α_a are tabulated in the table 4.

Table 4. Details of tail arm length (l_t) control for the most rearward location of 'cg'

Horizontal tail arm length (l_t)	**Cm_o**	**Longitudinal Stiffness $\left(\partial C_{m,cg}/\partial \alpha_a\right)$**	**Angle of attack (α_a) in degree**	**Stability curve slope in degree**	**Remark**
2.000	0.0121	0.0139	-49.80	0.8	Unstable
2.031	0.0129	0.0001	-10500.62	0.0	Neutral
2.100	0.0149	-0.0307	27.87	-1.8	Stable
2.200	0.0178	-0.0752	13.54	-4.3	Stable
2.300	0.0206	-0.1198	9.87	-6.8	Stable
2.500	0.0263	-0.2089	7.22	-11.8	Stable
3.000	0.0406	-0.4318	5.39	-23.4	Stable
3.500	0.0549	-0.6546	4.80	-33.2	Stable
4.031	0.0700	-0.8911	4.50	-41.7	Stable
4.500	0.0834	-1.1002	4.34	-47.8	Stable
5.000	0.0977	-1.3230	4.23	-52.9	Stable
5.500	0.1119	-1.5459	4.15	-57.1	Stable
6.000	0.1262	-1.7687	4.09	-60.5	Stable
6.500	0.1405	-1.9915	4.04	-63.4	Stable
7.000	0.1547	-2.2143	4.01	-65.7	Stable
7.500	0.1690	-2.4371	3.97	-67.7	Stable
8.000	0.1833	-2.6600	3.95	-69.4	Stable

IOP Conf. Series: Materials Science and Engineering **764** (2020) 012026 doi:10.1088/1757-899X/764/1/012026

8.500	0.1975	-2.8828	3.93	-70.9	Stable
9.000	0.2118	-3.1056	3.91	-72.2	Stable
9.500	0.2260	-3.3284	3.89	-73.3	Stable
10.000	0.2403	-3.5512	3.88	-74.3	Stable
15.000	0.3830	-5.7795	3.80	-80.2	Stable
20.000	0.5256	-8.0077	3.76	-82.9	Stable
50.000	1.3816	-21.3769	3.70	-87.4	Stable
100.000	2.8082	-43.6590	3.69	-88.7	Stable
1000.000	28.4866	-444.7361	3.67	-89.9	Stable
10000.000	285.2706	-4455.5070	3.67	-90.0	Stable

The effects of tail arm length on the AOA are shown in figure 10. And the effects of tail arm length on AOA and stability are shown in figure 11. From the figure 10, 11 and table 4, it is observed that the tail arm length change will not create considerable changes in decreasing the AOA from the steady level trim condition since stability increases greatly. Stability curve is almost perpendicular line after the tail arm length value is 20 m. whereas, decrement in tail arm length leads to an increase in AOA as well as a decrement in stability. In this case, even a slight decrement in tail length creates a huge difference in angle of attack. This also creates stability problems. Stability is neutral for the tail arm length value 2.031 m. below this value makes the aircraft inunstable condition. As a maximum control extent, by this way it can be controlled for the AOA form 4 degree to stall AOA theoretically. So it is recommended that not to go for the longitudinal control to decrease the α_a from the steady level trim condition by changing tail arm length. Even to increase the α_a from the steady level trim condition, it brings stability problems.

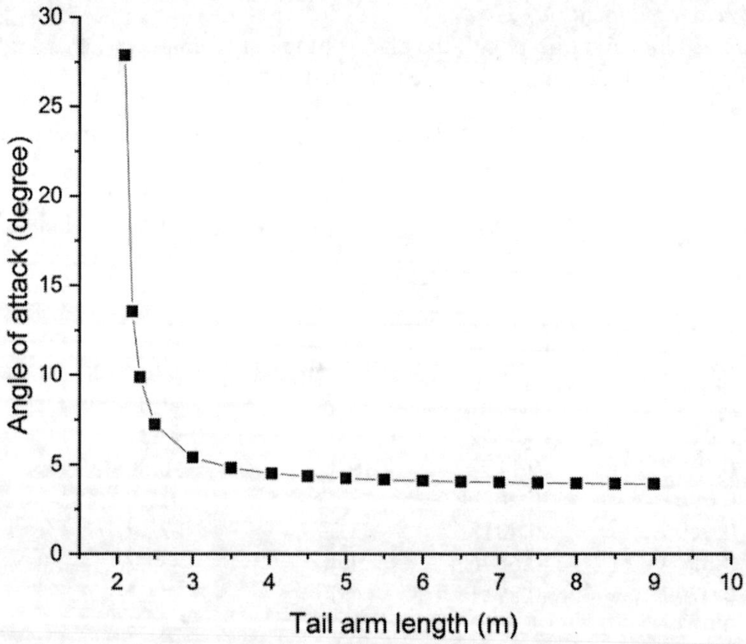

Figure 10.AOA variation with respect to tail arm length (l_t) for the most rearward location of 'cg' case

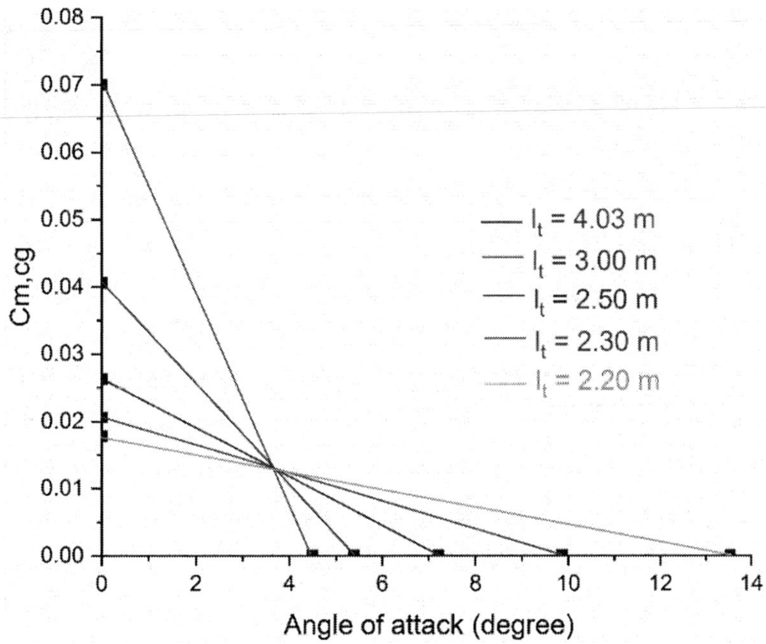

Figure 11. Tail arm length control stability curves for the most rearward location of 'cg'

6. Conclusion

In this numerical study, the effects of changing the horizontal tail (without elevators) arm length on the static longitudinal stability of the chosen Cessna 172 Skyhawk are presented. It shows how the AOA and stability curve varies based on the tail arm length value. The study concludes the following two points (i) Increasing AOA by decreasing tail arm length from the steady level trim condition is possible but it also affects the stability greatly, and (ii) Decreasing AOA by increasing tail arm length from the steady level trim condition is not achievable as expected. Hence, it is recommended that employing both elevator control and tail arm length control together will make an aircraft to meet any kind of longitudinal controllability and stability requirement at any instant especially during dynamic oscillations. This embedded aircraft configuration can be attained by proposing aircraft morphing technologies which have been an emerging field of research.

Symbols and abbreviations

cg	Center of gravity
ac	Aerodynamic center
np	Neutral point
AOA	Angle of Attack
$C_{m,cg}$	Pitching moment coefficient about 'cg' of the aircraft
α_a	Absolute AOA of the aircraft
α_f	AOA of fuselage
α_{wb}	AOA of wing-body
α_w	AOA of wing
α_t	Absolute AOA of tail
$C_{m,ac,w}$	Moment coefficient of the wing about aerodynamic center
$C_{m,o}$	Zero lift pitching moment
h_{cg}	Location of cg from leading edge of the wing in terms of percentage of chord
h_{ac}	Location of wing 'ac'from the leading edge in terms of percentage of chord
h_{np}	Location of 'np' from the leading edge of wing in terms of percentage of chord

V_∞	Free stream velocity
L_w	Lift force produced by the wing
D_w	Drag force produced by the wing
L_t	Lift force produced by the tail
D_t	Drag force produced by the tail
W	Weight of the aircraft
$C_{L,w}$	Coefficient of lift of wing
$C_{L,t}$	Coefficient of lift of wing-body
$C_{L,t}$	Coefficient of lift of horizontal tail
$C_{l,max}$	Maximum coefficient of lift
a	Lift curve slope of aircraft
a_{wb}	Lift curve slope of wing-body
a_t	Lift curve slope of horizontal tail
i_w	Wing incidence angle
S	Wing reference surface area
C	Wing chord length (approximately equal to zero lift chord length of wing)
$M_{ac,w}$	Moment about aerodynamiccenter of wing
$M_{ac,t}$	Moment about aerodynamiccenter of tail
ε	Downwash produced by the wing
V_H	Horizontal tail volume coefficient
i_t	Wing incidence angle
l_t	Tail arm length
z_t	Perpendicular distance between 'cg' of the aircraft and 'ac' of the tail.
S_t	Tail surface area
δ_e	Elevator deflection angle

Acknowledgement
Authors express sincere thanks to Bannariamman Institute of Technology, Sathyamangalam for utilizing its Cessna aircraft to carry out this research work successfully. Further, we acknowledge that no funding or grant received for conducting this research work.

References

[1] Rizzi A 2011 Modeling and simulating aircraft stability and control - The SimSAC project *Prog. Aerosp. Sci.* **47** 8 pp 573–588 2011.
[2] Ammar S, Legros C and Trépanier JY 2017 Conceptual design, performance and stability analysis of a 200 passengers Blended Wing Body aircraft *Aerosp. Sci. Technol.* **71** pp 325–336.
[3] Tuzcu I 2008 On the stability of flexible aircraft *Aerosp. Sci. Technol.* **12** 5 pp 376–384
[4] Stephenson JD and Hardy GH 1989 Longitudinal Stability and Control Characteristics of the Quiet Short-Haul Research Aircraft (QSRA) *Aerosp. Sci. Technol.* **13**
[5] Chun HH and Chang CH 2002 Longitudinal stability and dynamic motions of a small passenger WIG craft *Ocean Eng.* **29** 10 pp 1145–1162
[6] Xu X and Zhou Z 2015 Study on longitudinal stability improvement of flying wing aircraft based on synthetic jet flow control *Aerosp. Sci. Technol.* **46** pp 287–298
[7] Yang W, Yang Z and Collu M 2015 Longitudinal static stability requirements for wing in ground effect vehicle *Int. J. Nav. Archit. Ocean Eng.* **7** 2 pp 259–269
[8] Sandoval P, Cornejo P and Tinapp F 2015 Evaluating the longitudinal stability of an UAV using a CFD-6DOF model *Aerosp. Sci. Technol.* **43** pp 463–470
[9] Liu X, Liu W and Zhao Y 2016 Navier-Stokes predictions of dynamic stability derivatives for air-breathing hypersonic vehicle *Acta Astronaut.* **118** pp 262–285
[10] Liu Y, Huang X, Zhang Y and Zhou Y 2018 Dynamic Stability and Control of a Manipulating Unmanned Aerial Vehicle *Int. J. Aerosp. Eng.* pp 1–13

[11] Mieloszyk J and Goetzendorf-Grabowski T 2017 Introduction of full flight dynamic stability constraints in aircraft multidisciplinary optimization *Aerosp. Sci. Technol.* **68** pp 252–260
[12] Stengel R, Broussard JR, Berry PW and Taylor JH 1977 Modern Methods of Aircraft Stability and Control Analysis
[13] Biber K 2016 Flight Dynamics Modeling of a Propeller-Driven Cargo Airplane in *54th AIAA Aerospace Sciences Meeting* January pp 1–9.
[14] Goetzendorf-Grabowski T, Mieszalski D and Marcinkiewicz E 2011 Stability analysis using SDSA tool *Prog. Aerosp. Sci.* **47** 8 pp 636–646
[15] Tomac M and Stenfelt G 2014 Predictions of stability and control for a flying wing *Aerosp. Sci. Technol.* **39** pp 179–186
[16] Moes TR and Iliff TW 2002 Stability and Control Estimation Flight Test Results for the SR-71 Aircraft With Externally Mounted Experiments
[17] Jing W, Yankui W and Xueying D 2016 An experimental investigation on static directional stability *Chinese J. Aeronaut.* **29** 6 pp 1527–1540
[18] Wolowicz CH 1987 Considerations in the determination of stability and control derivatives and dynamic characteristics from flight data
[19] Schütte A, Cummings RM and Loeser T 2012 An integrated computational/experimental approach to X-31 stability & control estimation *Aerosp. Sci. Technol.* **20** 1 pp 2–11
[20] Tulapurkara E 2012 Longitudinal stick – free static stability and control *Indian National Programme on Technology Enhanced Learning (NPTEL)* [Online]. Available: http://nptel.ac.in/courses/101106043/18. [Accessed: 23-Sep-2017].
[21] 2009 Longitudinal Static Stability *Flight Emergency & Advanced Maneuvers Training Inc. dba Flightlab* [Online]. Available: www.flightlab.net/Flightlab.net/Download.../5_LongitudinalStatic%232BA158.pdf%0A. [Accessed: 03-Aug-2017].
[22] Vural M, 2011 STATIC PITCH STABILITY of an AIRPLANE *Illinois Institute of Technology* [Online]. Available: https://mypages.iit.edu/~vural/Pitch Stability.pdf. [Accessed: 30-Aug-2017].
[23] Anderson JD 2017 *Introduction to Flight* 6th ed. McGraw Hill Education.
[24] 2018 Cessna 172 *Wikipedia* [Online]. Available: https://en.wikipedia.org/wiki/Cessna_172. [Accessed: 08-Aug-2018].
[25] Lednicer D 2014 Airfoils of U S and Canadian Aircraft. [Online]. Available: http://www.aerofiles.com/airfoils.html. [Accessed: 23-Sep-2017].
[26] Raymer DP 2004 *Aircraft Design: A Conceptual Approach* 2nd ed. California: AIAA Education Series
[27] 2017 Airfoil Tools *UIUC Airfoil Coordinates Database* [Online]. Available: http://airfoiltools.com/airfoil/details?airfoil=naca2412-il. [Accessed: 23-Sep-2017].

IOP Conf. Series: Materials Science and Engineering **764** (2020) 012027 doi:10.1088/1757-899X/764/1/012027

Performance analysis of printed antenna on different novel substrates to enhance V2V communication in future

N Susidharan[1], P Hamsagayathri[1], C Poongodi[1], S Shakthisandhanu[1], R Sivabalan[1] and P Kaviyarashu[1]

[1]Bannari Amman Institute of Technology,sathyamangalam.
E-mail: susidharan.ec17@bitsathy.ac.in

Abstract: In recent years, advancement in wireless technologies, communication in automotive vehicles has witnessed a tremendous growth in the market. All the sophisticated wireless services offered depends on the type of antenna used in it. The researchers in antenna community are trying very hard to improve the performance of the systems by making smaller antenna devices; however they face both scientific and technical limits. Other alternative approach is to use of novel materials for patch and substrate to adhere higher performance and better functionality. In this paper rectangular patch antenna with stepped ground structure is proposed and its radiation characteristics are analyzed for different dielectric substrates like Si3N4, Al2O3, BN, Quartz and Silica. This study will facilitate researchers to get a clear idea on the patch and substrate material that can be used for V2V applications. The proposed CNT patch antenna on Si3N4 substrate resonates at 5.9 GHz is designed and simulated results show that the minimum return loss of -37 dB is achieved with maximum gain of 6.87. In Addition, the proposed antenna also achieves an omni-directional radiation pattern with maximum radiation efficiency of 89%.

Keywords: Microstrip patch, graphene, V2V Communication, Nano Materials, defected ground

1. Introduction

Now a days, most of the vehicles are developed with multiple antennas to facilitate wireless communication between moving vehicle with other vehicle or to the outside world. These devices are usually designed to help the drivers to improve their driving experience and increase travel safety. There is no wireless communication without an antenna, where it plays an important role in transmission and reception of signals in free-space. The latest communication technologies like GSM and LTE provides the interpersonal communication in the vehicles. In addition, Wireless fidelity (WiFi) allows people in the vehicle to connect Internet faster. Recently, researchers are fond of incorporating the artificial intelligence to vehicles and that would result in autonomous or unmanned vehicles in future. However, it is highly essential to the undergo communication between vehicles. In addition, using V2V communications, the real time traffic information are provided to drivers to ease traffic congestion. IEEE updated 802.11 standards for V2V communication with 5.9 GHz. In order to have a good reception in a vehicle, the antenna must be

IOP Conf. Series: Materials Science and Engineering **764** (2020) 012027 doi:10.1088/1757-899X/764/1/012027

able to receive the signals from any direction around the car. Hence the radiation pattern of the antenna is required to be omni directional. When compared to single antenna, antenna arrays shall also be considered to fulfill this requirement. The antenna must be easily integrated into the vehicle. The performance of the antenna is affected severely by its surroundings. Both the substrate materials and distances between the antenna and surrounding objects need to be considered during design. A box of 3 times of antenna size around the antenna is considered to obstruct the environment disturbances. The spurious noise generated from the engine disturbs transmission and reception of the signals in the vehicles. Therefore it is essential to maintain the distance between the antenna and engine to use the antenna for Vehicle-2- Vehicle communication at 5.9 GHz.

In recent years, vehicular communication has become a trending topic and it attracts the attention of people. Various antenna structures were proposed and developed for the V2V communication in literature. Authors [3-10] developed monopole antenna arrays and achieved omni directional radiation pattern. These miniaturized monopole antennas are easy to fix in vehicles and attained radiation pattern was most suited for vehicular applications. The four element antenna array with loading with short pins was designed for the frequency of 5.9 GHz [5] .

The printed loop antenna proposed in [11-13] is used for taxis communication. The antenna was uniformly loaded with capacitance. Though, antennas proposed in literature are simple and easy to install, their gain is relatively low. Furthermore, manufacturing cost of some antennas are high and occupies large space and that increases the overall antenna size. The microstrip antenna is light weight, low profile, easy of manufacturing and integration. It also holds good concealing and hence it has wide-ranging applications in the field of automotive antennas [14-15] . A low profile rectangular microstrip antenna loaded with carbon nano tube patch with Si3N4 substrate is proposed to achieve high speed V2V communication. The proposed antenna achieves good impedance and radiation characteristics with etching stepped ground structures. It exhibits omni directional radiation pattern which is most suitable for V2V communication

In general, any Microstrip patch antenna (MPA) has conducting patch on the substrate with perfect ground-plane at the bottom. The substrate is a dielectric material that separates the patch and ground-plane of an antenna[1]. In most of the previous work, researchers had copper patch as radiating element [1-2] . Now a days, in V2V communications, the antenna requires to be operate at high speed, cheaper, smaller in size with improved performance. In addition to copper, Graphene and Carbon Nanotubes can also be used as radiating element because of its significant electronic and material properties. Carbon Nanotube can be either semiconductor or metallic in nature. Both semiconductor and metallic CNTs are very popular in the field of high speed electronics. Graphene is one allotropes of carbon where in their atoms are bonded in the hexagonal form. Nowadays, most of the high speed communication devices are designed and fabricated using CNT and Graphene. The utilization of graphene is still underway where researchers are looking forward to eliminate nano -scale related problems associated with it. The performance of the antenna is analyzed for different substrates Si3N4, Al2O3, BN, Silica, Quartz using basic antenna parameters such as return loss, vswr, gain, directivity, efficiency and radiation pattern.

IOP Conf. Series: Materials Science and Engineering **764** (2020) 012027 doi:10.1088/1757-899X/764/1/012027

2. Antenna modeling

2.1 Design

The substrate dielectric constant, thickness of the substrate and operating frequency are three key parameters required for designing an antenna. The step by step design calculation are listed below

Width of the Patch is calculated as

$$W = \frac{c_0}{2f_r}\sqrt{\frac{2}{\varepsilon_r + 1}} \tag{1}$$

$$\epsilon_{reff} = \frac{\varepsilon_r + 1}{2} + \frac{\varepsilon_r - 1}{2}\left[1 + 12\frac{h}{W}\right]^{\frac{1}{2}} \tag{2}$$

$$\frac{\Delta L}{h} = \frac{0.412\left(\epsilon_{reff} + 0.3\right)\left(\frac{W}{h} + 0.264\right)}{\left(\epsilon_{reff} - 0.258\right)\left(\frac{W}{h} + 0.8\right)} \tag{3}$$

Where ΔL is the function of effective dielectric constant ϵ_reff , height of the substrate h and the width-to-height ratio (W/h).

The length of the patch is calculated by subtracting the length extended on both edges from effective length.

$$L_{eff} = \frac{C}{2f_0\sqrt{\varepsilon_{reff}}} \tag{4}$$

Length of the Patch

$$L = L_{eff} - 2\Delta L \tag{5}$$

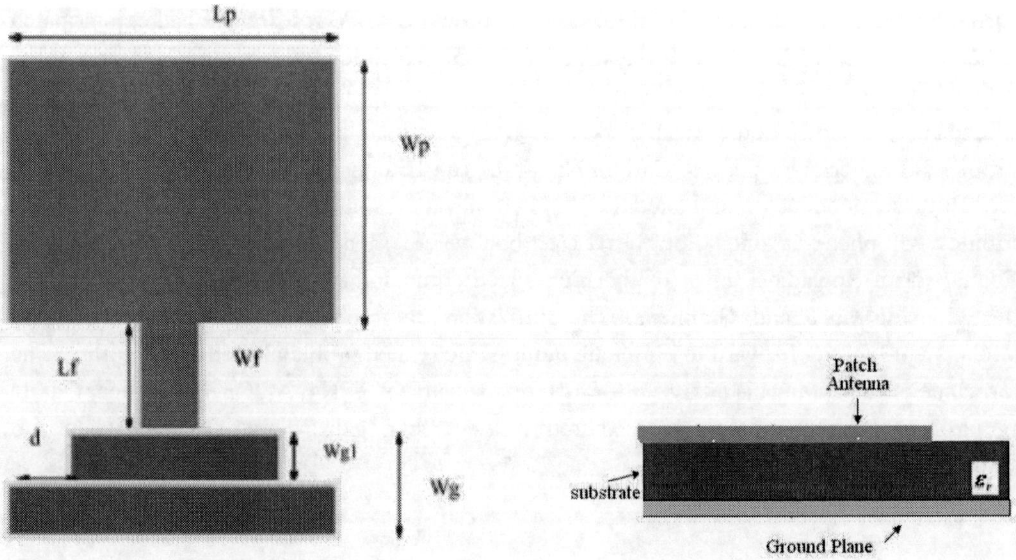

Figure 1. Structure of the proposed antenna

IOP Conf. Series: Materials Science and Engineering **764** (2020) 012027 doi:10.1088/1757-899X/764/1/012027

The dimensions of the proposed antenna are calculated using the above design equations. The layout of the proposed antenna and its parameters are provided in the Table 1 and Figure 1.

Table 1. Layout parameters of the proposed antenna

Layout Parameter	Value (mm)
Lp	12
Wp	15
Wf	3
Lf	10
Wg	2.5
Wg1	1.5
d	3

2.2 Analysis

The transmission line model is used to analysis of microstrip patch antenna. In this method, electric and magnetic currents are uniformly distributed over the two edges of the patch. The radiating patch acts as a resonator where it has transverse field that varies along the length of the radiator. Patch edges allow the electric fields to undergo fringing and these fields induce the antenna to radiate. Both the patch dimensions and substrate height has great influence on it. Due to fringing effect, wave incident on the patch travels in both the substrate and air, thus it is required to consider effective dielectric constant for analysis. The solutions are obtained based on the Maxwell's Integral equations. Copper, Carbon NanoTubes (CNT) and Graphene materials are considered as patch materials for simulation. The properties of the substrate materials are referred to in [3] where graphene has high electrical conductivity of 10^8 compared to other materials. Similarly, dielectric properties of the substrate greatly influence the antenna performance. The simulation is carried out for different substrates like Si_3N_4, Al_2O_3, BN, Silica, Quartz.

3. **Simulation results and discussion**

The proposed antenna is designed and simulated using ADS (2016). Numerical computations are performed using direct dense solver that incorporates Methods of Moments techniques. The performance of the proposed antenna structure is analyzed for different combination of patch and substrate materials and the best suited material is obtained for V2V communications.

3.1 Impedance characteristics

The reflection characteristics of proposed antenna structure with different patch and substrate and patch materials are shown in Figure 2. The antenna shows either inductive or capacitive impedance in non-resonating mode. Therefore, wave travelling in the antenna gets cancel each other and radiation does not occurs. However, inductive and capacitive impedance gets cancel each other at resonant frequency and allows the waves to travel back and forth, which reinforce each other and radiate into free space. Figure 2 explains that the proposed antenna structure with CNT on different substrate, where it resonates at the frequency of 5.9 GHz for V2V communications. The return loss of the proposed antenna for different types of radiating elements on different substrates is shown in Table 2.

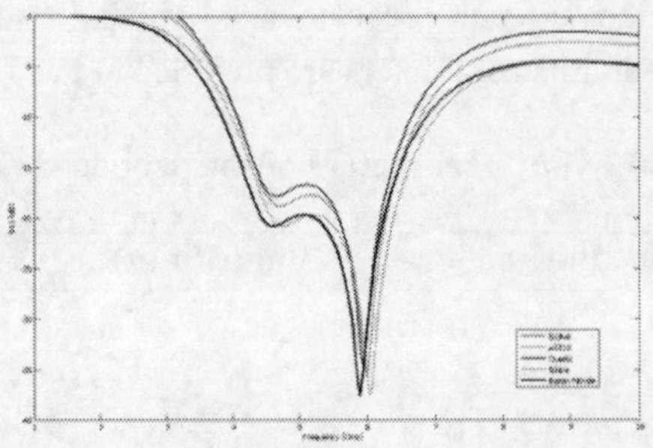

Figure 2. Return loss characteristics of CNT based Patch antenna for different substrates

Table 2. Variations in S11 characteristics for different substrates

Radiating Patch Materials	Substrate Materials				
	Si3N4	Al2O3	BN	Silica	Quartz
Copper	-26	-24	-19	-21	-23
CNT	-37	-35	-32	-32	-34
Graphene	-31	-29	-24	-26	-28

The voltage standing wave ratio (VSWR) measures the impedance mismatch between the antenna port and feed line connecting to it. The VSWR curves of proposed antenna structure with CNT on different substrates are plotted in Figure.3. The proposed antenna design attains minimum VSWR of 1.076 at the operating frequency of 2.45 GHz, which is well below the industrial standard.

3.2 Radiation characteristics

The strong and weak current distributions of the antenna are indicated in red and blue color. The ground plane of the antenna is etched to increase its electrical length. The stepped ground structure also alters the reverse current distribution which enhances the radiation pattern of the antenna.

The shape of radiation pattern is constant at far fields and hence radiation characteristics of the proposed antenna are evaluated in that region. E and H fields are perpendicular with each other and along with direction of propagation. Both E and H field strength decreases with increase in the distance. The antenna gain is always measured with respect to isotropic antenna. The gain characteristic of the final antenna structure is shown in Figure 4.

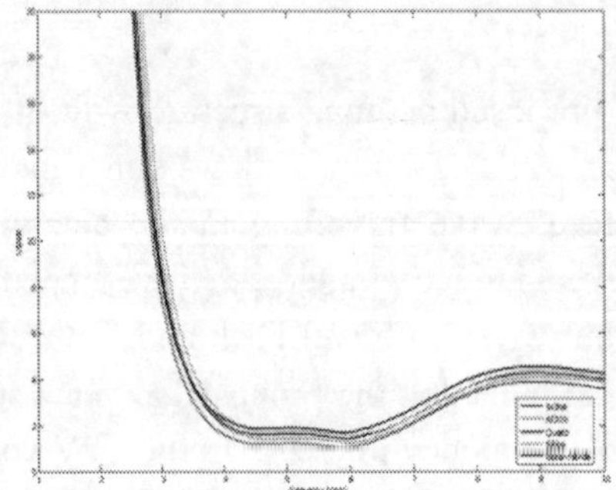

Figure 3. VSWR of CNT based Patch antenna for different substrates

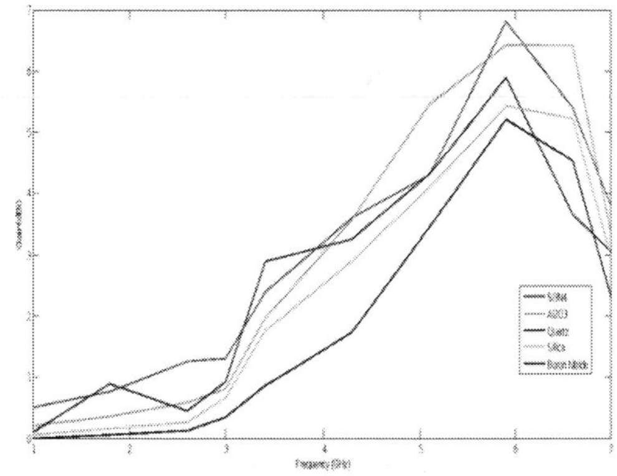

Figure 4. Gain of CNT based Patch antenna for different substrates

The radiation parameters for different substrates are highlighted in Table 3.

Table 3. Variations in radiation characteristics for different substrates

Patch Materials	Parameters	Substrates Si3N4	Al2O3	BN	Silica	Quartz
Copper	Gain	3.12	3.01	2.17	2.76	2.94
	Directivity	4.18	4.11	2.98	3.76	3.98
	Efficiency	76	74	72	72	73
CNT	Gain	6.87	6.42	5.21	5.42	5.89
	Directivity	7.67	7.53	6.98	7.08	7.10
	Efficiency	89	85	74	77	83
Graphene	Gain	5.68	5.14	4.62	4.04	4.91
	Directivity	6.94	6.47	6.32	5.73	6.45
	Efficiency	82	79	73	70	76

At resonant frequency, the radiation pattern of the proposed patch antenna is measured for particular angle. The radiation field pattern of E and H planes are examined at $\varphi = 0^0$ and $\theta = 90^0$ and its three dimensional perspective is shown in Figure 5.

IOP Conf. Series: Materials Science and Engineering **764** (2020) 012027 doi:10.1088/1757-899X/764/1/012027

Figure 5. 3D Radiation Pattern

When compared to other substrates, the CNT radiating patch on Si3N4 substrate provides good impedance and radiation characteristics, hence it is suitable to use for V2V communication.

4. Parameter study

The parametric study on antenna design layout is studied. The physical substrate thickness and ground length are varied to study the antenna performance.

4.1 Effect of Substrate thickness (h)

The effect of substrate thickness on return loss characteristics of the antenna is plotted in Figure 6. All printed antennas are constructed using substrate material which provides mechanical support to antenna and also it shows the variations in radiation characteristics. Hence, selection of appropriate substrate material is a key aspect in antenna design. Thick substrates with low dielectric constant increases the fringing fields for radiation, but substrate with high dielectric constant may also be used in special conditions. Hence, substrate thickness is varied and analyzed the antenna performance and desirable thickness is considered for further analysis.

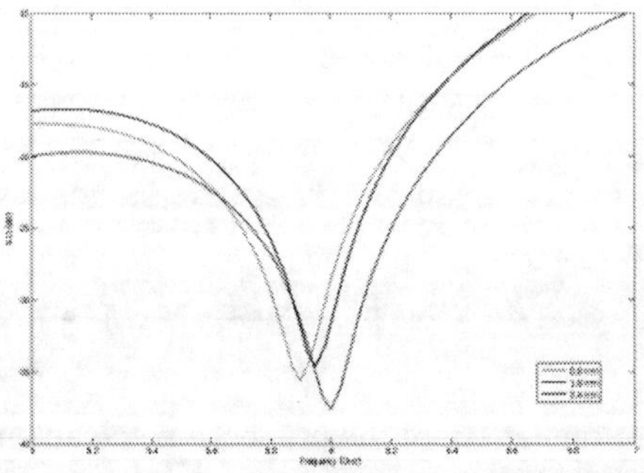

Figure 6. Return loss characteristics of CNT based Patch antenna for different substrates heights

4.2 Effect of Ground length (lg)

The effect of ground length on return loss characteristics of the antenna is plotted in Figure 7. The ground length is varied from 2.5 mm to 4mm, where its length greatly affects the input impedance of the antenna and cause impedance mismatch between the port and antenna. Hence most of the input power to reflect back to port and thus have increased effect of return loss.

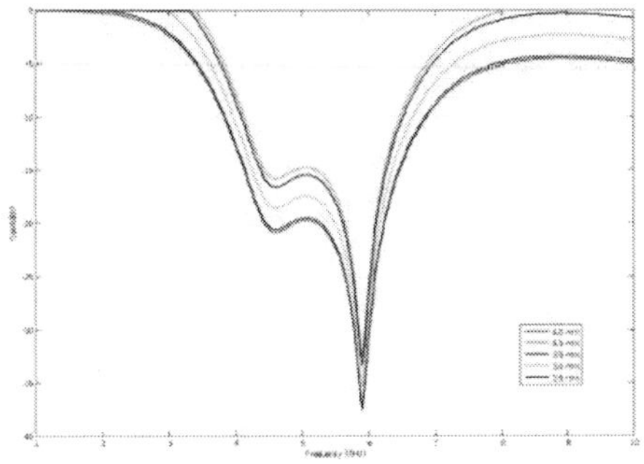

Figure 7. Return loss characteristics of CNT based Patch antenna for different ground heights

5. Conclusion

In this paper, performance of the microstrip patch antenna was studied for different radiating and substrate materials. The stepped ground structure was used to optimize impedance and radiation characteristics. A method of Moments based ADS simulation was performed to model the antenna for resonant frequency of 5.9 GHz. The simulation was performe for various novel materials which have high performance impact over V2V communication. Si3N4 with carbon nanotube radiator realized the minimum return loss of -37dB with VSWR of 1.027 close to the ideal case. It also achieves the maximum gain of 6.87 dBi with stable radiation characteristics. When compared to other materials, it also attains the maximum radiation efficiency of 89%. Hence, it is concluded that CNT based patch on Si3N4 substrate can be a suitable choice for high speed V2V communication in future which could ease traffic congestion, prevent injuries, save human lives and improve the environment. The modern vehicles have around 20 antennas in and around the vehicle body to offer various communication services and hence SAR exposure has to be evaluated for proposed antenna in future. As vehicles are more susceptible to damage, the future directions also include to undergo stress analysis for proposed communication device.

Acknowledgment

"The authors acknowledge the Management of Bannari Amman Institute of Technology, Sathyamangalam for providing high standard infrastructure and laboratory facilities to carry out this work."

6. References

[1] C.A Balanis, Antenna Theory Analysis and Design, John-Wiley and Sons Publications, 2nd edition, ISBN No. 978-81-265-1393-2, 1997

[2] A.B. Mutiyara, R. Refianti, and Rachmansyah, Design of Microstrip Antenna for Wireless Communication at 2.4 GHz, Journal of Theoretical and Applied Information Technology, vol. 33 no.2, pp. 184-192, Nov. 2011

[3] J.S. Moon and D. Gaskill, Graphene: Its Fundamentals to Future Applications, Microwave Theory and Techniques, IEEE Transactions on Microwave Theory and Techniques, vol. 59, no. 10, pp. 2702–2708, Oct. 2011

[4] Z. H. Ni, H. M. Wang, J. Kasim, H. M. Fan, T. Yu, Y. H. Wu, Y. P. Feng, and Z. X. Shen, Graphene Thickness Determination Using Reflection and Contrast Spectroscopy, Nano Letters, American Chemical Society, vol. 7, no. 9, pp. 2758-2763,

2007
[5] S. Chaudhary and A. Kumar, Use of Graphene as a Patch Material in Comparison to the Copper and other Carbon Nanomaterials, Intl. Journal of Emerging Technologies in Computational and Applied Sciences, vol. 3, issue 3, pp. 272-279, 2013
[6] Oliver Klemp, Performance consideration for automotive antenna equipment in vehicle to vehicle applications, URSI International Symposium on Electromagnetic Theory, 2010,doi: 10.1109/URSI-EMTS.2010.5637361
[7] Y.Ming Lin, K.A.Jenkins, D.B.Farmer,P.Avouris, "Oeration of Graphene devices at GigaHertzfrequecnies ", Nano Letters, vol.9, issue 1, pp. 422-426, 2009
[8] Jornet JM, AkyildizIF."Graphene -based nano -antennas for electromagnetic nanocommunications in the Terahertz Band",European Conference on Antennas and Propagation;Barcelona Spain, pp. 1–5,2010
[9] Li H, Xu C, Srivastava N, Banerjee K."Carbon nanomaterials for next-generation interconnects and passives: Physics,status and prospects", IEEE Transactions on Electronic Devices, vol.56, no.9, pp.799–821, 2009
[10] E.L Wolf, "Application of Graphene- A Overview", vol. I. Springer Publication,pp.1-3,2014
[11] Wu R. Jaang et.al, 'The Graphene nanoribbon terahertz based antenna on polyemid substrate", Journal of Optics, pp.5547–5548,2014
[12] Chen H M, Lin Y F, Chen C H, et al. "Miniature folded patch GPS antenna for vehicle communication devices", IEEE Transactions on Antennas and Propagation, vol.63, issue.5, pp. 1891–1898, 2015
[13] Klemp O, Thiel A. "Performance considerations for automotive antenna equipment in vehicle-to-vehicle communications",Proceedings of the 2010 URSI International Symposium on Electromagnetic Theory (EMTS ' 10),Aug 16 – 19, 2010, Berlin, Germany. Piscataway, NJ, USA: IEEE, pp.934 – 937,2010
[14] Westrick M, Almalkawi M, Devabhaktuni V, et al. "A low-profile, low-cost antenna system with improved gain for DSRC vehicle-to-vehicle communications", International Journal of RF and Microwave Computer-Aided Engineering,vol.23, issue.1, pp.111-117, 2013
[15] Varum T, Matos J N, Pinho P. "Printed antenna for on-board unit of DSRC

Heat Transfer in Square Porous Cavity Due to Radiation and Heat Generating Strip - Part I

N. Ameer Ahamad[1], Azeem[2], Maughal Ahmed Ali Baig[3*], Sarfaraz Kamangar[4], Irfan Anjum Badruddin[4], S. Lohith Reddy[3]

[1]Mathematics Department, Faculty of Science, University of Tabuk, P.O.BOX.741, Zip.71491, Saudi Arabia
[2]Department of Mathematics and Computer Science, Taylor's University, No. 1 Jalan Taylor's Subang Jaya - 47500, Selangor DarulEhsan, Malaysia
[3]Department of Mechanical Engineering, CMR Technical Campus, Hyderabad, India
[4]Department of Mechanical Engineering, College of Engineering, King Khalid University, PO Box 394, Abha 61421. Kingdom of Saudi Arabia

*Author for correspondence E-mail: mabaig13@gmail.com

Abstract. Investigation of heat transfer is presented due to a vertical heat generating strip inside the square porous cavity. The heat is assumed to be generated at isothermal rate at a strip inside the medium. The vertical walls of the cavity are cooled to lower temperature Tc. The two horizontal surfaces are maintained adiabatically so that heat does not cross the boundary. Finite element method is applied as a solution tool to solve the governing equations. The impact of various parameters such as heating strip length, Rayleigh number and Radiation parameter are presented in terms of temperature and streamline distribution inside the porous cavity.

1. Introduction

The impact of heat transfer in porous medium is quite distinctive due to various kinds of boundary conditions being applied to the physical domain [1-10]. The shape of the domain itself plays a crucial role in deciding the heat transfer characteristics inside any given porous medium. For instance, the heat transfer in square cavity [5] is substantially different from that of annular geometry [8-9] or vertical plates [4]. It is of paramount importance to the scientist and engineers working in the field of porous media to know the amount of heat that can be transferred from the porous domain to decide its suitability to a particular application. Thus there has been lot of effort being invested to understand the heat transfer in porous medium due to various boundary conditions. The present article investigates the heat transfer due to a small heat generating strip being placed inside the porous domain.

2. Analysis

Consider a square porous domain as shown in figure 1. The heat generating strip is present at the center of cavity placed vertically. The various boundary conditions applicable are shown in figure 1.

The governing equations for the problem can be given as

$$\frac{\partial u}{\partial x}+\frac{\partial v}{\partial y}=0 \quad (1)$$

$$\frac{\partial u}{\partial y}-\frac{\partial v}{\partial x}=-\frac{g\beta K}{\nu}\frac{\partial T}{\partial x} \quad (2)$$

$$u\frac{\partial T}{\partial x}+v\frac{\partial T}{\partial y}=\alpha\left(\frac{\partial^2 T}{\partial x^2}+\frac{\partial^2 T}{\partial y^2}\right)-\frac{1}{\rho C_p}\frac{\partial q_r}{\partial x} \quad (3)$$

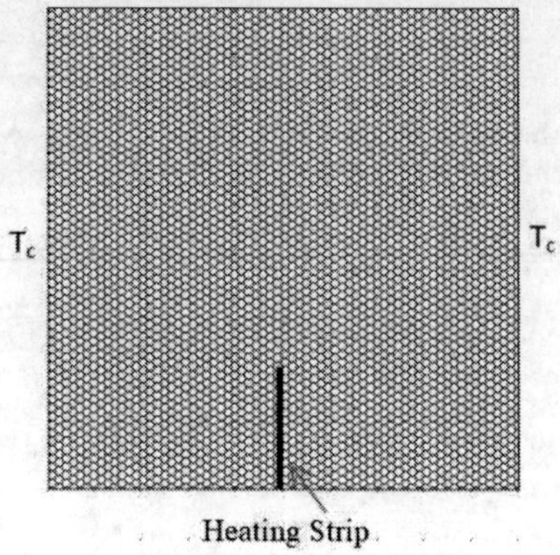

Figure 1: Porous annulus

3. Results and Discussion

The investigation is carried out for three parameters such as the height of the heating strip (SH), Rayleigh number (Ra) and Radiation parameter Rd. Figure 2 shows the heat transfer behavior when the strip height is varied as 17%, 33% and 50% of the cavity height at Ra=100 and Rd=0. It is seen that the isotherms and streamlines are symmetrically divided about the left and right side of the heating strip. The smaller size of the strip shows that the isotherms are placed further to each other. This indicates that the more of the porous medium is occupied with similar temperature. The streamlines clearly show that the strength of the fluid is less when the heating strip is smaller. The increase in the strip height pushes the isotherms closer to each other that reflects that the heat transfer increases with increase in the strip height. It is also noted that large amount of heat is dissipated from upper portion of the cold surfaces as compared to its lower portion which is very much obviousfrom the isotherm corresponding to 0.1. The fluid cell tilts towards the top corners when the height of the strip is increased. This shows that the fluid flows from the strip towards the upper corners and returns back by dissipating its thermal energy in that area.

Figure 3 shows the effect of changing Rayleigh number at SH=25% and Rd=0. The Rayleigh number is changed as Ra=10, Ra=100 and Ra=150. The low Rayleigh number is known to support conduction heat transfer due to lack of fluid movements. This is very much clear from figure 3a which shows smoothly distributed isotherms with the fluid movement restricted to lesser area in the porous domain. The increase in Rayleigh number completely changes the isothermal and streamlines distribution owing to increased fluid velocity which is reflected in terms of higher value of stream function.

The effect of changing the Radiation parameter is shown in figure 4. It is found that the isotherms are evenly distributed when radiation parameter is increased. It must be noted that the mathematical model used incorporates radiation effect as equivalent conduction thus the isotherms and streamlines shows that the conduction effect increased due to increase in Rd.

IOP Conf. Series: Materials Science and Engineering **764** (2020) 012028 doi:10.1088/1757-899X/764/1/012028

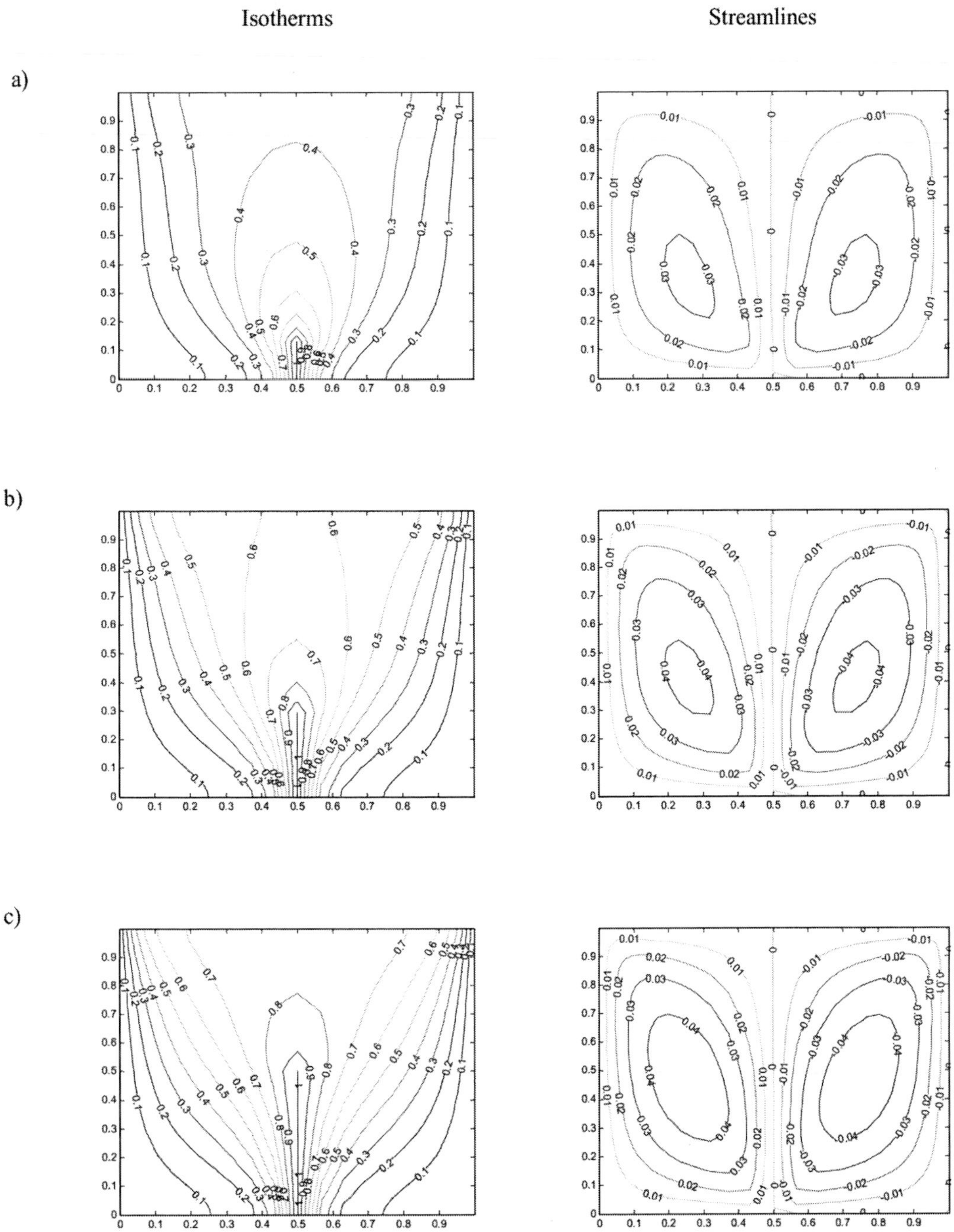

Figure 2: Effect of Heater Height (SH) a) SH = 17% b) SH = 33% c) SH = 50%
Ra = 100, Rd = 0.05

IOP Conf. Series: Materials Science and Engineering **764** (2020) 012028 doi:10.1088/1757-899X/764/1/012028

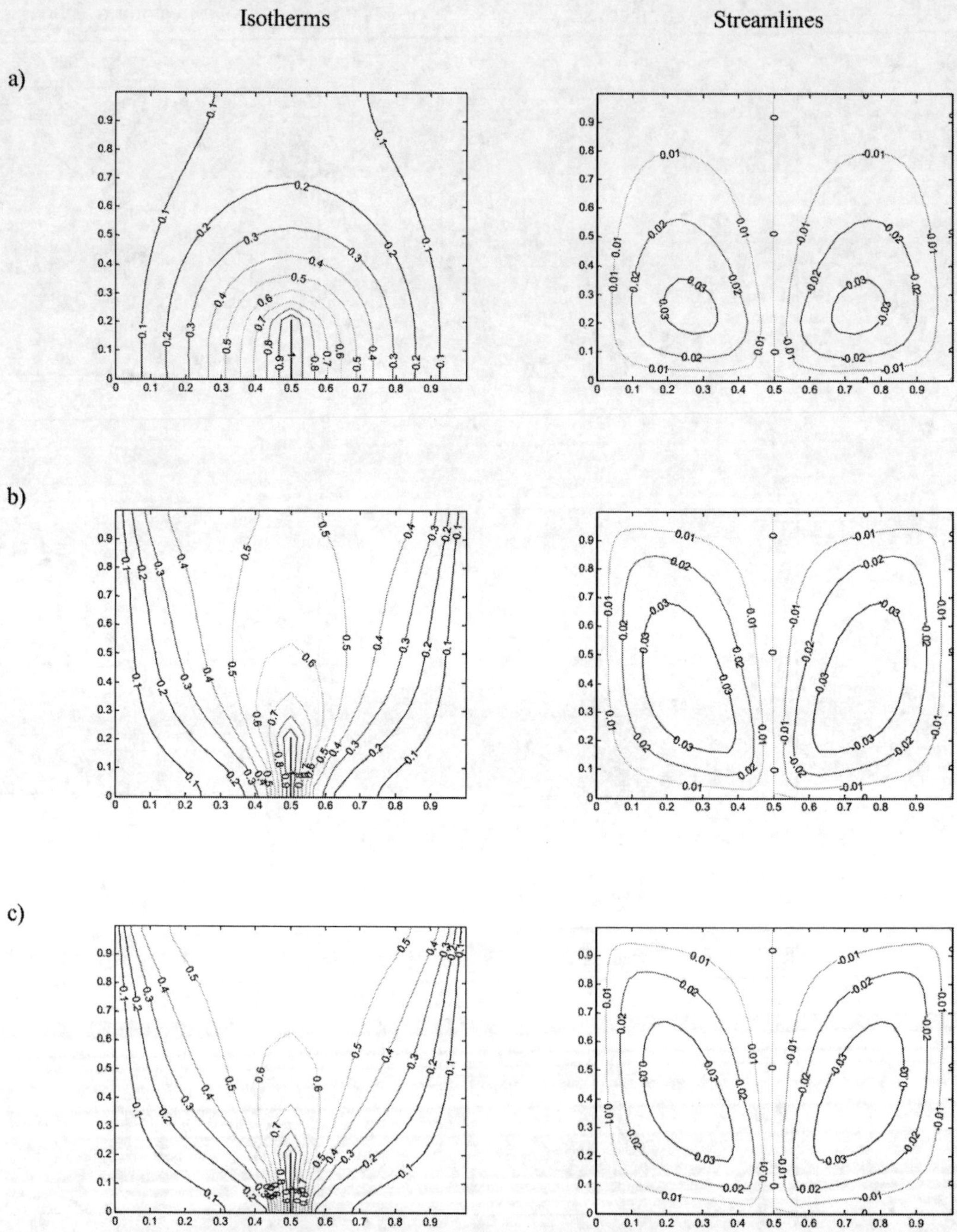

Figure 3: Effect of Rayleigh Number (Ra) a) Ra = 10 b) Ra = 100 c) Ra = 150
SH=25%, Rd=0.05

IOP Conf. Series: Materials Science and Engineering **764** (2020) 012028 doi:10.1088/1757-899X/764/1/012028

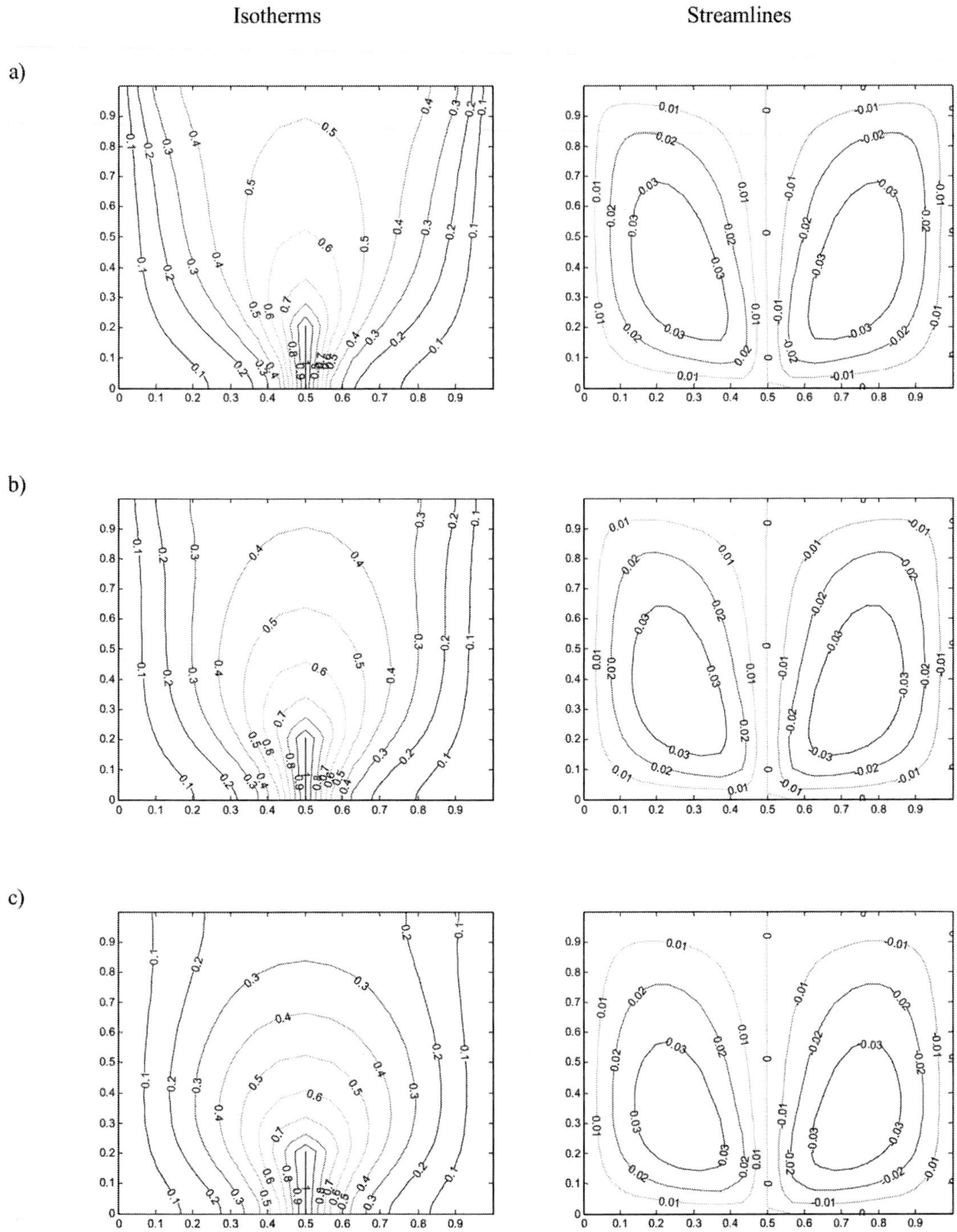

Figure 4: Effect of Radiation Parameter (Rd) a) Rd = 0.1 b) Rd = 0.5 c) Rd = 1
Ra = 100, SH = 25%

4. Conclusions

Heat Transfer in porous cavity due to a small heat generating strip inside the porous medium is analyzed with the help of FEM. It is noted that the

- Isothermal lines and streamlines are distributed symmetrically about the strip.
- Heat transfer increases with increase in strip length.
- Higher Rayleigh number increases heat transfer.

References

1. VafaiK ., Hand book of porous media, Marcel Dekker, New York, 2000.
2. PopI., InghamD.B., Convective heat transfer: Mathematical and computational modeling of viscous fluids and porous media, Pergamon, Oxford, 2001.
3. InghamD.B., Pop I.(Eds.),Transport phenomena in porous media, Pergamon, Oxford, 1998
4. Nield D. and BejanA., Convection in Porous Media, ed. 3rd. New York: Springer Verlag.,2006
5. I.A.Badruddin, Investigation of Heat Transfer in Irregular Porous Cavity Subjected to Various Boundary Conditions, International Journal of Numerical Methods for Heat and Fluid Flow, 29 (1), pp. 418-447 (2019). https://doi.org/10.1108/HFF-04-2018-0132
6. I.A. Badruddin., Heat and Mass Transfer with Soret/Dufour Effect in Irregular Porous Cavity, Journal of Thermophysics and Heat Transfer , 33 (3), pp. 647-662(2019), https://doi.org/10.2514/1.T5666.
7. S. K. Nath, V.V. Satyamurthy, Effect of aspect ratio and radius ratio on free convection heat transfer in a cylindrical annulus filled with porous media, HMT C16-85, proceedings of the 8th Nat. Heat and Mass Transfer Conf, India, 1985, pp. 189-193.
8. V. Prasad, F.A Kulacki, Natural convection in a vertical porous annulus, Int. J. Heat Mass Transfer 27 (1984) 207-219
9. A.P. Bassom, D.A.S. Rees, Free convection from a heated vertical cylinder embedded in a fluid-saturated porous medium, ActaMechanica 116 (1996) 139-151.
10. I.A.Badruddin, , Azeem, Mohd Yamani IdnaIdris, Nik-Ghazali. N., Salman Ahmed N.J, Abdullah A.A.A. Al-Rashed, Simplified Finite Element Algorithm to Solve Conjugate Heat and Mass Transfer in Porous Medium, International Journal of Numerical Methods for Heat and Fluid Flow, 27(11), pp. 2481-2507 (2017).

IOP Conf. Series: Materials Science and Engineering **764** (2020) 012029 doi:10.1088/1757-899X/764/1/012029

Effect of Manganese Doping on Dielectric Characteristics of Lead Zirconate Titanate of Different Zirconium/Titanium Ratios

Balgovind Tiwari[1], T. Babu[2], R.N.P. Choudhary[3]

[1] Physics Department, IIIT-RK Valley, RGUKT, A.P., India
[2] Physical Science Department, SCHS, Vempalli, A.P., India
[3] Physics Department, ITER, SOA-University, Bhubaneswar, Orissa, India
balgovindtiwari@gmail.com

Abstract. We are presenting in this work, the effect of Zr/Ti ratios on PZT ceramics modified by manganese. The studies include structural, microstructural, and dielectric properties of the samples. A solid state reaction method was used to prepare samples of PZT in different ratios of 65/35, 52/48, 35/65 modified with 0%, 2%, 6%, and 10% of manganese. Phase formation of compounds was checked by X-ray diffraction (XRD) technique. The surface morphology of the compounds was studied using scanning electron microscopy (SEM). Dielectric and electric characteristics show that Mn modification has increased the T_c and also, the values of ε_r whereas tangent loss is decreased.

1. Introduction

In lead zirconate titanate (PZT) material, the ratio of zirconium and titanium influences the way of behavior of material to a large extent and it's tailoring for different kind of applications. PZT being a versatile ferroelectric material, it is used extensively in the fabrication of capacitors, memories, sensors, electrooptic devices, actuators, etc. It is highly used in the above mentioned applications because of its high piezoelectric, ferroelectric, and dielectric properties [1-3]. Lead zirconate titanate (PZT) is formed by a combination of a ferroelectric $PbTiO_3$ ($T_c = 490\,^{o}C$) and anti-ferroelectric $PbZrO_3$ ($T_c = 230\,^{o}C$) as a solid solution [1]. PZT shows cubic structure above the Curie temperature, but shows either tetragonal or rhombohedral structure below Curie temperature depending on Zr/Ti ratio [7]. At room temperature, separation of tetragonal and rhombohedral phase of PZT with a boundary is called morphotropic phase boundary, at which both phases coexist in nature and the properties of PZT material will have extreme values [5]. Bulk and thin films of PZT has been modified with several types of modifiers by doping either at Pb site or Zr/Ti site to improve its dielectric, ferroelectric and piezoelectric properties [5].

From the literature review and to the best of our knowledge, we found that there are reports only on a particular Zr/Ti ratio. We rarely observed the effect of different Zr/Ti ratio studies for bulk PZT ceramics except the investigation of Zr/Ti ratio on dielectric and ferroelectric properties of PZT thin films [6], and effects of Zr/Ti ratio on structural, dielectric and piezoelectric properties of Mn- and (Mn, F)-doped lead zirconate titanate ceramics [7]. Hence, in this regard, we present the effect of changing Zr/Ti ratio on PZT. Moreover, most of the work on PZT is industry and application oriented.But in this work, we are trying to present a bigger picture of PZT with different Zr/Ti ratios such as 65/35, 52/48, 35/65 in terms of understanding through structure, morphology, and dielectric properties.

IOP Conf. Series: Materials Science and Engineering **764** (2020) 012029 doi:10.1088/1757-899X/764/1/012029

2. Experimental

We have used very high purity ingredients to prepare samples of $Pb(Zr_{0.65}$-$xMn_xTi0_{.35})O_3$ [$PZ_{65}T_{35}$],$Pb(Zr_{0.52-x}Mn_xTi_{0.48})O_3$ [$PZ_{52}T_{48}$], $Pb(Zr_{0.35-x}Mn_xTi0_{.65})O_3$ [$PZ_{35}T_{65}$] (x = 0, 0.02, 0.06, 0.10), by a high temperature solid state reaction method with appropriate stoichiometric calculations using oxides of lead, zirconium, titanium and manganese i.e PbO, ZrO_2, TiO_2 and MnO_2. The ingredients were mixed in an agate mortar-pestle for two hours. First hour in air atmosphere and then in acetone for next hour. The mixtures were then calcined in alumina crucibles. The process of grinding and calcination was repeated several times until the formation of compounds of desired structure was confirmed with final calcinations temperatures being 950 $^{\circ}$C for 2 h, 975°C for 3 h, 950°C for 4 h to $PZ_{65}T_{35}$, $PZ_{52}T_{48}$, $PZ_{35}T_{65}$ samples respectively. An excess of 3% of PbO was added to all the compositions to minimize Pb loss. The information regarding making of pellets can be found elsewhere [8]. The pellets of $PZ_{65}T_{35}$, $PZ_{52}T_{48}$, and $PZ_{35}T_{65}$ were sintered at an optimized temperature of 1100°C for 2 h, 1150°C for 4 h, 1100°C for 4 h respectively in an air atmosphere using alumina crucible. The information about the recording of dielectric and electrical characteristics is also reported elsewhere [8].

3. Results and Discussion

3.1. Structural and Microstructural Studies

From figures 1(a), 1(b), 1(c) the structural change in the material system was observed with change in the amount of Zr/Ti ratio as expected. The ratio 65/35 shows rhombohedral structure, the ratio 52/48 shows mixed phase of tetragonal and rhombohedral and the ratio 35/65 shows tetragonal phase. The XRD patterns of all ththree ratios show a sharp pattern with distinct diffraction peaks, indicating good homogeneity and crystallization of the samples. All the patterns exhibit most of the peaks corresponding to the perovskite phase. Substitution of Mn shows reduction in intensity of peaks in all the ratios as well as the appearance of some new peaks in the ratio 65/35. The reduction in intensity of peaks in the ratio 52/48 indicates a shift in phase from tetragonal to rhombohedral, whereas in 35/65, it shows reduction in tetragonality of the structure. Mn also induces a reduction in crystallite size in all the ratios.

Figure 1(a)

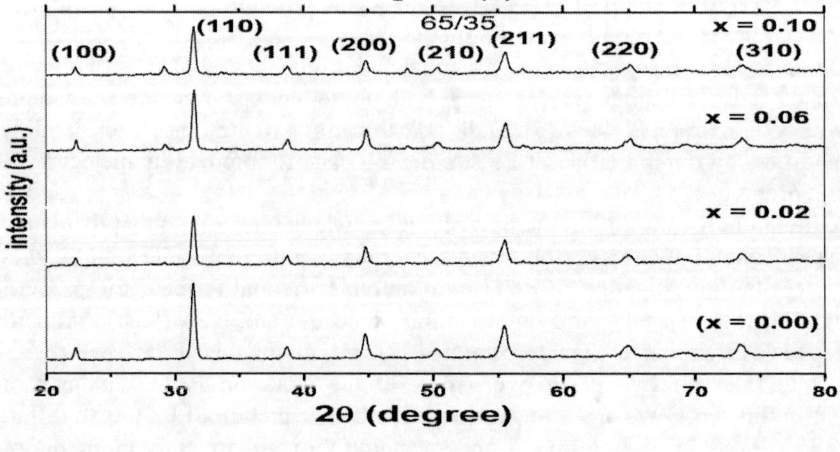

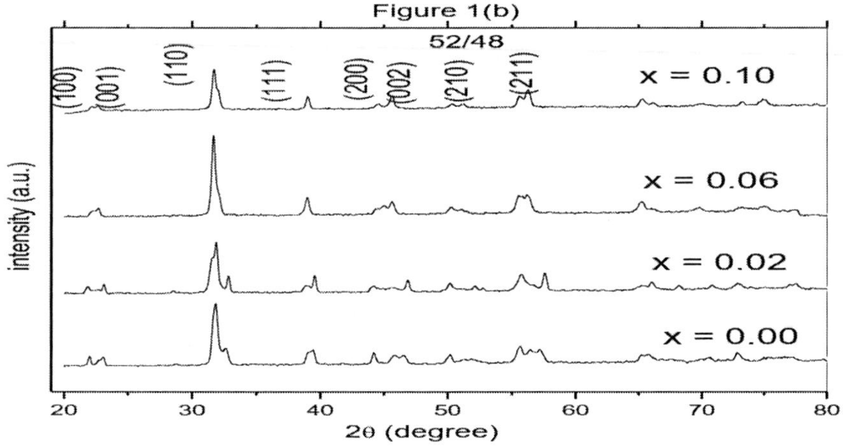

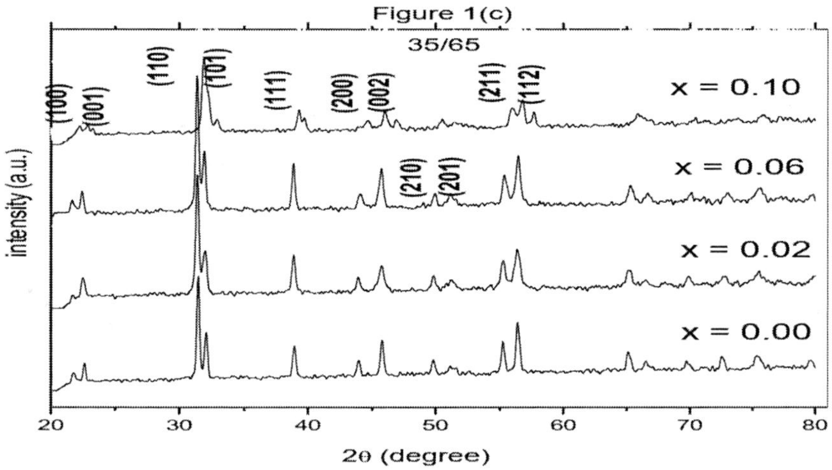

Figure 1: Comparison of XRD of PZT of 65/35, 52/48 and 35/65 ratios

The microstructures can be compared from SEM micrographs of figure 2. It was observed that almost all the grains were spherical in shape and density in distribution across the sample surface. Only in the ratio 52/48 some samples show some melting appearance. The grain size varies with Mn concentration. The grains of ratio 35/65 are the biggest and show highest decrease in grain size with increase in the amount of Mn modification. The range of grains size is 1.1 – 1.8μm for the ratio of 65/35, 1.4 – 1.7μm for the ratio of 52/48, and 1.3 – 3.3μm for the ratio of 35/65. The polycrystalline nature of all the samples can be seen from the shape, size and distribution of grains.

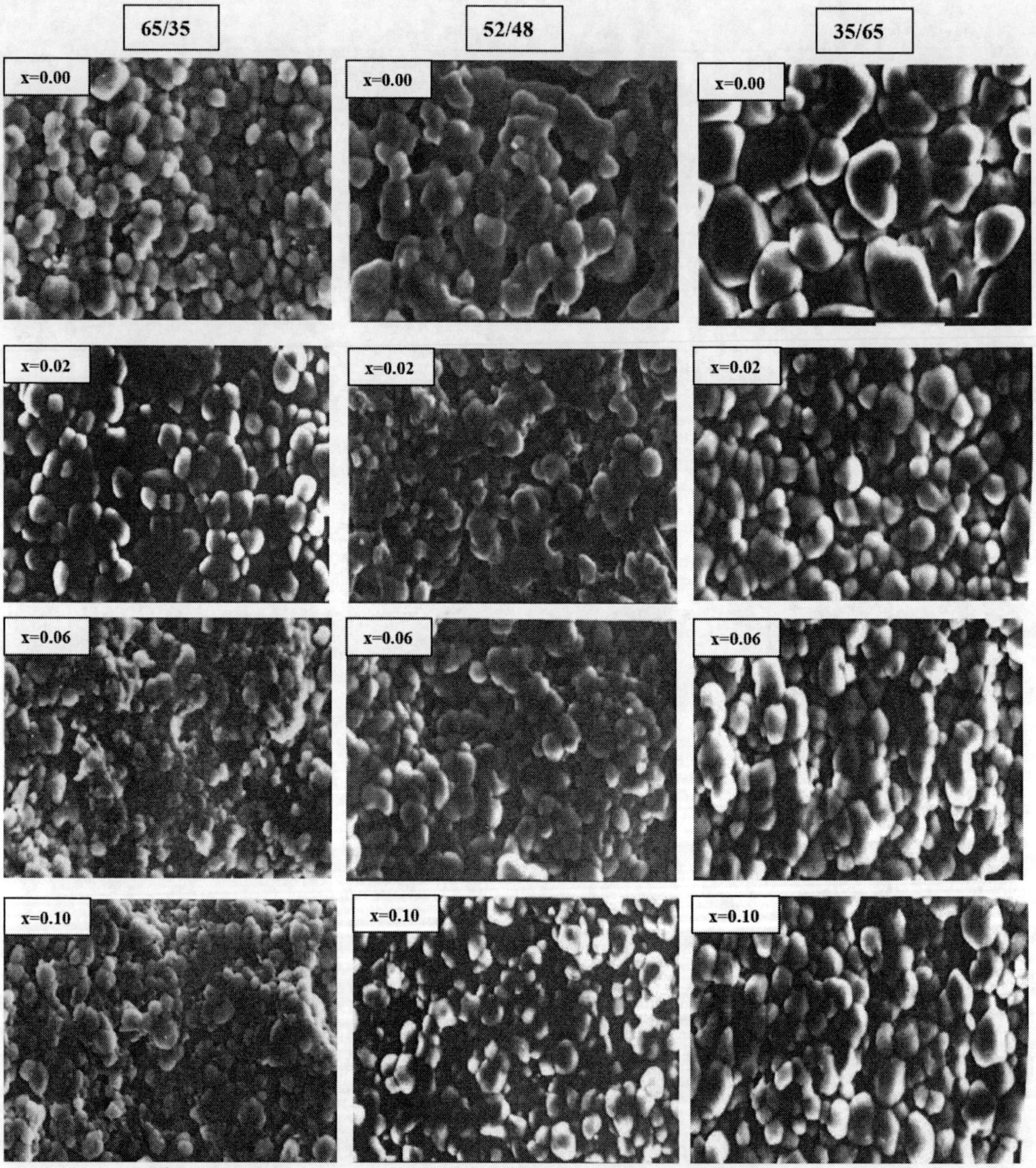

Figure 2: SEM micrographs of PZT of 65/35, 52/48 and 35/65 ratios

3.2. Dielectric Studies

Figure 3 shows the variation of ε_r with a temperature of $PZ_{1-x}T_x$ (x = 0.00, 0.02, 0.06, 0.10) for different Zr/Ti ratios. From Figure 3(a) we can see that the value of ε_r increases continuously on continuous increasing temperature to its maximum value (ε_{max}), and then show decrease. It indicates that a phase transition (from the ferroelectric to non-ferroelectric phase) is occurring at a specific temperature called the Curie temperature (T_c) [9].

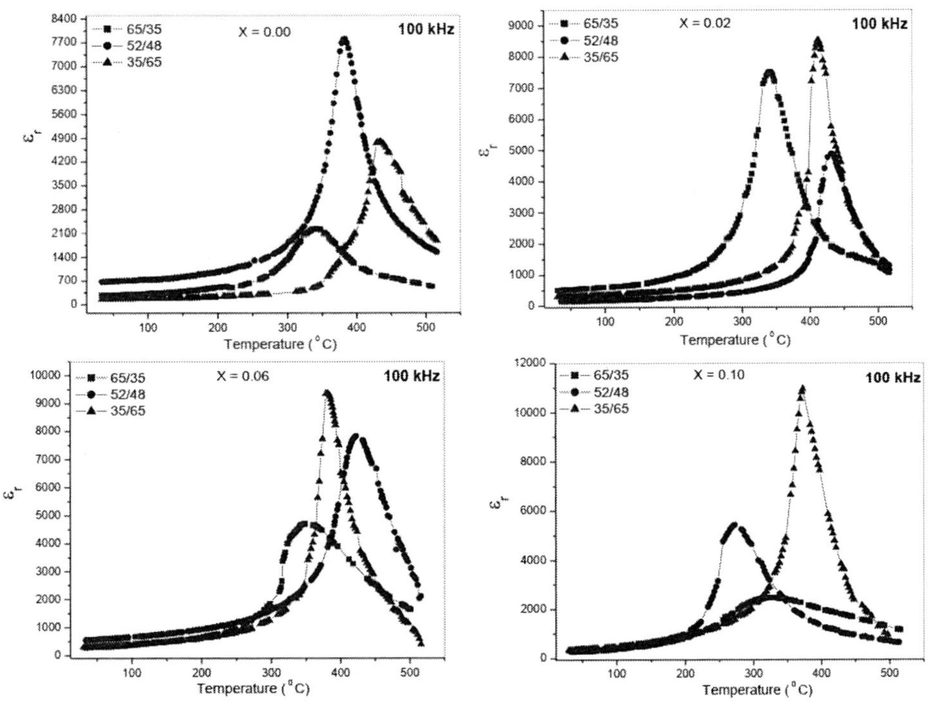

Figure 3: the variation of ε_r with a temperature of $PZ_{1-x}T_x$ (x = 0.00, 0.02, 0.06, 0.10) for different Zr/Ti ratios

The T_c changes with increase of Mn concentration. For ratio 35/65 T_c shows a decrease with increase in Mn concentration, whereas for other ratios, first it shows an increase and then T_c decreases. The value of ε_{max} of pure PZT(65/35 and 35/65) is less compared to Mn substituted samples. The value of ε_{max} shows a decreasing trend after initial an increase for 65/35 whereas for 35/65, ε_{max}increases with Mn concentration.

Figure 4 shows the variation of tanδ with a temperature of $PZ_{1-x}T_x$ (x = 0.00, 0.02, 0.06, 0.10) for different Zr/Ti ratios at 100 kHz. The value of tan δfor PZT is very less up to 400°C for all the ratios at every Mn concentration. And beyond these temperatures tanδ starts to increase rapidly upto 500°C. As can be seen that for all concentrations, all the three ratios show a decrease in tanδ, this indicates that introduction of Mn has resulted in decrease of loss in the PZMT samples. The increase of tanδ at higher temperatures is due to space charge polarization [10]. There is rise in the value of tanδ with increase in concentration of Mn at normal room temperature. For pure PZT and x = 0.02 natures of variation of tanδ is same for all ratios. Only for x = 0.06 and 0.10, the ratio 35/65 shows more loss.

Figure 5 shows the study ln $(\frac{1}{\varepsilon} - \frac{1}{\varepsilon max})$ and ln (T-T_c) of PZMT at 100 kHz for different Zr/Ti ratios. Diffusivity was calculated (values are given in figure also) with the help of graphs slopes. All the values of γ is between 1 (ideal value for a normal ferroelectric material) and 2 (ideal value for a ferroelectric material that is fully disorded).

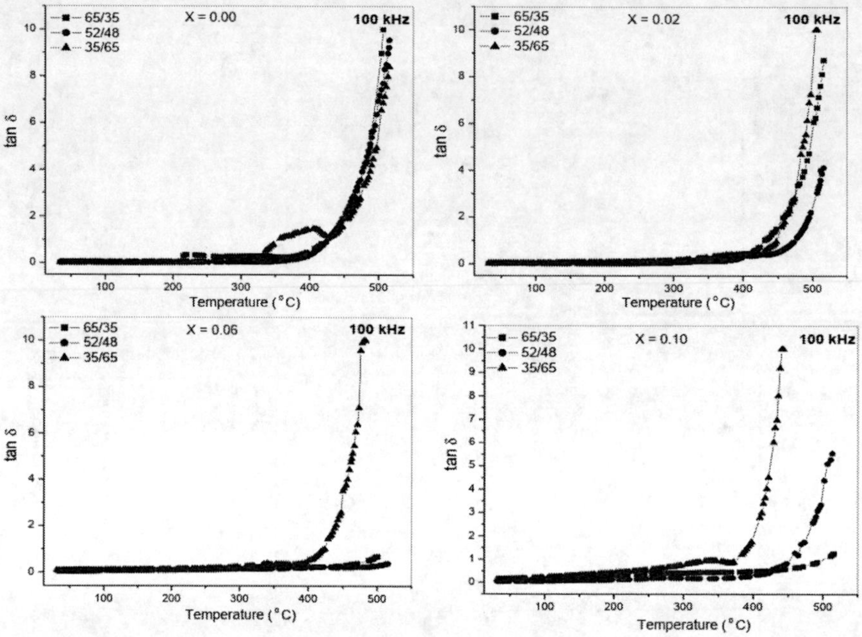

Figure 4: variation of tanδ with a temperature of $PZ_{1-x}T_x$ (x = 0.00, 0.02, 0.06, 0.10) for different Zr/Ti ratios at 100 kHz

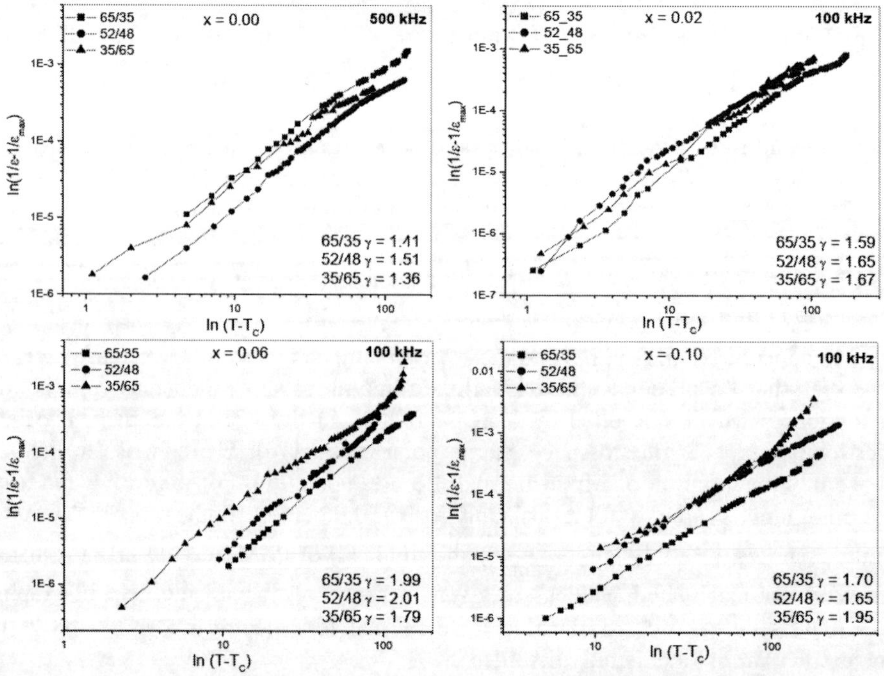

Figure 5: the study between ln $(\frac{1}{\varepsilon} - \frac{1}{\varepsilon max})$ and ln (T-T$_c$) of PZMT at 100 kHz for different Zr/Ti ratios

The materials exhibit diffuse phase transition confirmed with their value of exponent γ (>1). The diffuse phase transition is generally observed in the PZT material systems, which can be due the presence of non-equivalent position in the elementary cell. From the diffusivity values, it can be seen that γ value increases with the introduction of Mn and as the concentration of Mn increases the value of γ changes indicating that the disordering in the modified system changes. For ratio 35/65, γ shows an increasing trend with an increase in Mn concentration. The other two ratios show an increase up to x = 0.06 then decreases. The presence of broadened or diffused dielectric peaks is due to statistical composition fluctuations and these fluctuations have occurred because distinct cations have occupied the equivalent sites. Also the value of γ is in the range of 1 and 2. So it is concluded that the materials are disordered.

4. Conclusions

The comparative study of all the three ratios 65/35, 52/48 and 35/65, for all the concentrations of Mn was done (x = 0.00, 0.02, 0.06, 0.10). All the samples were prepared by a high-temperature solid-state reaction technique. The properties are summarized in table 1.

Table 1: Properties of PZT ceramics of different ratios

X →	0.00		0.02		0.06		0.10	
Ratios ↓	ε_r	T_c	ε_r	T_c	ε_r	T_c	ε_r	T_c
65/35	2232.0	341.5	7501.9	340.4	4705.0	348.8	2462.0	327.3
52/48	7808.7	383.5	4876.3	432.3	7826.8	422.8	5427.0	272.3
35/65	4792.6	433.5	8530.1	412.0	9354.5	381.2	10954.0	372.8

Very less in the structure has been observed in PZT due Mn modification but the size of particle and density on the surface was changed. The dielectric and electric properties of Mn modified PZT shows increase in the value of T_c and relative dielectric constant (ε_r), and decrease in the tan δ.

References

[1] Pulkit Sharma, SugatoHajra, SushrisangitaSahoo, P. K. Rout, R. N. P. Choudhary, Capacitive and resistive characteristics of gallium modified lead zirconate titanate, J Mater Sci: Mater Electron, 28 (2017), 12048–12055.

[2] Nguyen Dinh Tung Luan, Le Dai Vuong, Truong Van Chuong, Nguyen Truong Tho, Structure and Physical Properties of PZT-PMnN-PSNCeramics Near the Morphological Phase Boundary, Advances in Materials Science and Engineering, 2014 (2014), 821404 (8 pages).

[3] Sweta Sharma, S. C. Azad, Md. Soaib Khan, Rajiv Ranjan, B. Behera, R.N.P. Choudhary, Effect of Gd^{3+}-substitution on structural and electrical properties of lead zirconate titanate (PZT;Zr/Ti=55/45) ceramics, Journal of Metallurgy and Materials Science, 2, 56 (2014), 1-6.

[4] JANA PINTEA, ANA MARIA MOISIN, ALINA DUMITRU, EFFECT OF Nb DOPPING ON PIEZOELECTRIC PROPERTIES OF PZT TYPE CERAMICS, Romanian Reports in Physics, 56 (2004), 441-444.

[5] ZHENG WANG, WEI REN, JIANBAO REN, XIAOQING WU, PENG SHI, XIAOFENG CHEN, AND XI YAO, Ferroelectrics, 383(2009), 151-158.

[6] P. Khaenamkaew, S. Muensit, I.K. Bdikin, A.L. Kholkin, Effect of Zr/Ti ratio on the microstructure and ferroelectric properties of lead zirconatetitanate thin films, Materials Chemistry and Physics, 102 (2007), 159-164.

[7] Eric Boucher, Benoit Guiffard, Laurent Lebrun, Daniel Gutomar, effects of Zr/Ti ratio on structural, dielectric and piezoelectric properties of Mn- and (Mn, F)-doped lead zirconatetitanate ceramics, Ceramics International, 32 (2006), 479-485.

[8] B. Tiwari, R.N.P. Choudhary, Effect of Mn-substitution on structural and dielectric properties of $Pb(Zr_{0.65-x}Mn_xTi_{0.35})O_3$ ceramics, Solid State Sciences, 11 (2009), 219-223.

[9] BanarjiBehera, P.Nayak, R.N.P.Choudhary, Study of complex impedance spectroscopic properties of $LiBa_2Nb_5O_{15}$ ceramics, Mater.Chem.Phys.106 (2007)193–197.

[10] Patri, S.K., Synthesis and characterization of Bismuth layered multiferroic derivatives, Ph.D. Thesis, IIT Kharagpur, India (2009).

IOP Conf. Series: Materials Science and Engineering **764** (2020) 012030 doi:10.1088/1757-899X/764/1/012030

Heat Transfer in Square Porous Cavity Due to Radiation and Heat Generating Strip - Part II

N. Ameer Ahamad[1], Azeem[2], Maughal Ahmed Ali Baig[3*], Sarfaraz Kamangar[4], Irfan Anjum Badruddin[4], Jana Nagarjun[3]

[1]Mathematics Department, Faculty of Science, University of Tabuk, P.O.BOX.741, Zip.71491, Saudi Arabia
[2]Department of Mathematics and Computer Science, Taylor's University, No. 1 Jalan Taylor's Subang Jaya - 47500, Selangor DarulEhsan, Malaysia
[3]Department of Mechanical Engineering, CMR Technical Campus, Hyderabad, India
[4]Department of Mechanical Engineering, College of Engineering, King Khalid University, PO Box 394, Abha 61421. Kingdom of Saudi Arabia

*Author for correspondence E-mail: mabaig13@gmail.com

Abstract. The current article discusses the effect of a small heating strip placed inside the porous medium enclosed in a square cavity. The left vertical surface is maintained at hot temperature and right vertical surface remains at cold temperature that allows the heat to be dissipated outside the system. The other two surfaces of the cavity are maintained adiabatic. A comparison is drawn between the present work and the work where the left vertical wall is maintained at cold temperature. The different parameters such as length of the heating strip inside the domain, Rayleigh number and Radiation parameters are discussed with respect to the square porous domain.

1. Introduction

The heat transfer in porous medium can be found in plenty of devices that need to be designed either to dissipate more heat or to retain the heat depending upon the application requirement. This very requirement has motivated many scientists to study the heat transfer in porous medium that can be seen in number of articles published in this particular topic [1-11]. The variation of heat transfer across a porous domain depends on many parameters amongst which the geometry of the domain [5-9] and the boundary conditions [10-11] can be of paramount importance. This is the main reason that the heat transfer in porous domain has been studied to larger extent according to afore mentioned parameters. The current article is an effort to understand the heat transfer behaviour in a square porous domain which has two heat sources. The first heat source is the applied isothermal temperature at the left vertical surface and the second thermal source arises due to heat generated by a small strip inside the medium. The present article compares the heat transfer behaviour with that of the case when the left vertical surface is also maintained at cooling temperature.

2. Analysis

A square porous medium as shown in figure 1 is considered. The boundary conditions are depicted in figure 1.The governing equations for the problem can be given as

IOP Conf. Series: Materials Science and Engineering **764** (2020) 012030 doi:10.1088/1757-899X/764/1/012030

$$\frac{\partial u}{\partial x}+\frac{\partial v}{\partial y}=0 \tag{1}$$

$$\frac{\partial u}{\partial y}-\frac{\partial v}{\partial x}=-\frac{g\beta K}{\nu}\frac{\partial T}{\partial x} \tag{2}$$

$$u\frac{\partial T}{\partial x}+v\frac{\partial T}{\partial y}=\alpha\left(\frac{\partial^2 T}{\partial x^2}+\frac{\partial^2 T}{\partial y^2}\right)-\frac{1}{\rho C_p}\frac{\partial q_r}{\partial x} \tag{3}$$

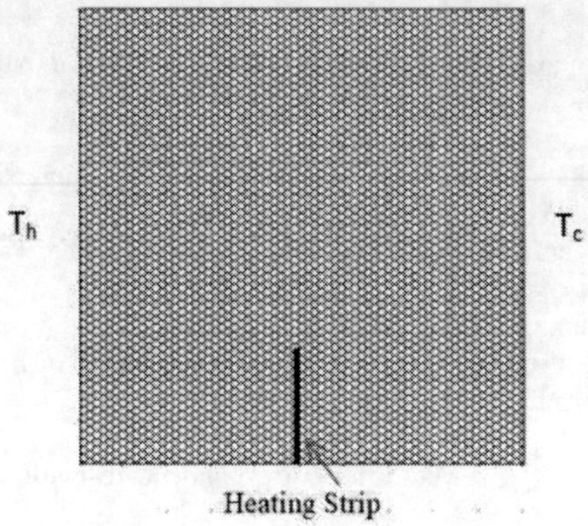

Figure 1: Porous annulus

3. Results and Discussion

Figure 2 shows the impact of varying the heat generating strip in three steps such as SH = 17%, 33% and 50% of cavity height. The heat distribution in this case is found to be unsymmetrical about the heating strip which is different from the case when the left vertical surface was maintained at cooling temperature Tc, as presented in Part I of this paper. It is seen that the heat content of the cavity towards the left of heating strip is lesser when the length of the strip is smaller. As the length of the strip increases the left half of the cavity accumulates more and more thermal energy which is reflected by higher temperature isothermal lines. It is further noted that almost half of the cavity is at equal temperature as that of left heating wall when strip length is increased to 50%. The thermal gradient at right side`of the strip increases with increase in the strip length that in turn should increase the heat transfer rate from strip to porous medium.

The effect of changing the Rayleigh number is shown in figure 3. Expectedly, the isotherms get distorted with increase in Rayleigh number and fluid strength increases. One notable difference between present results and the case of cooled left wall, as presented in Part I, is that all the isotherms touch the top surface in present case for Ra=10, whereas it was only one isotherm (0.1) which touches when left surface is also cooled. This shows that the substantial thermal variation exists at upper side of the cavity in present case. The increase in radiation parameter Rd pushes the isotherms towards the left surface as shown in figure 4.

IOP Conf. Series: Materials Science and Engineering **764** (2020) 012030 doi:10.1088/1757-899X/764/1/012030

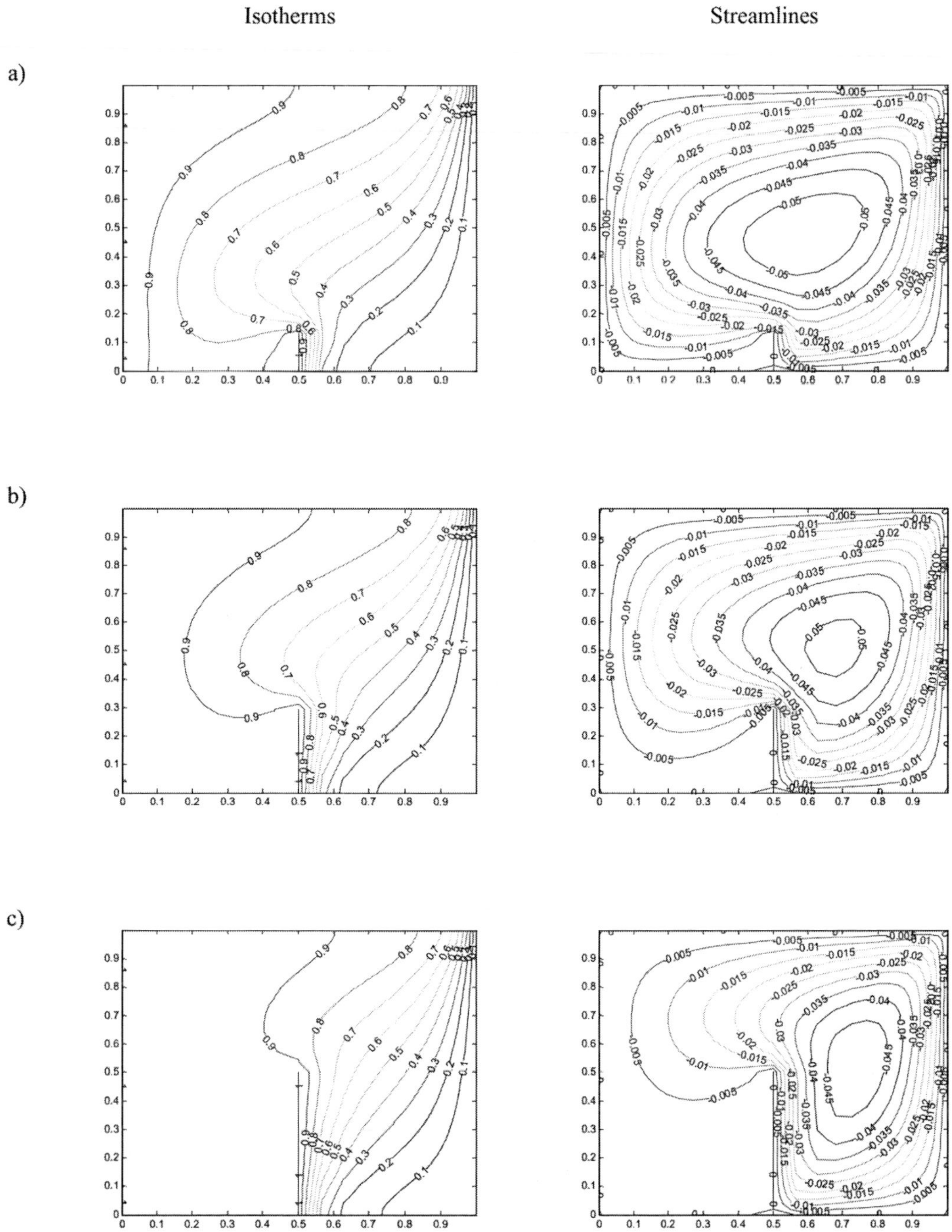

Figure 2: Effect of Heater Height (SH) a) SH = 17% b) SH = 33% c) SH = 50%, Ra = 100, Rd = 0.05

IOP Conf. Series: Materials Science and Engineering **764** (2020) 012030 doi:10.1088/1757-899X/764/1/012030

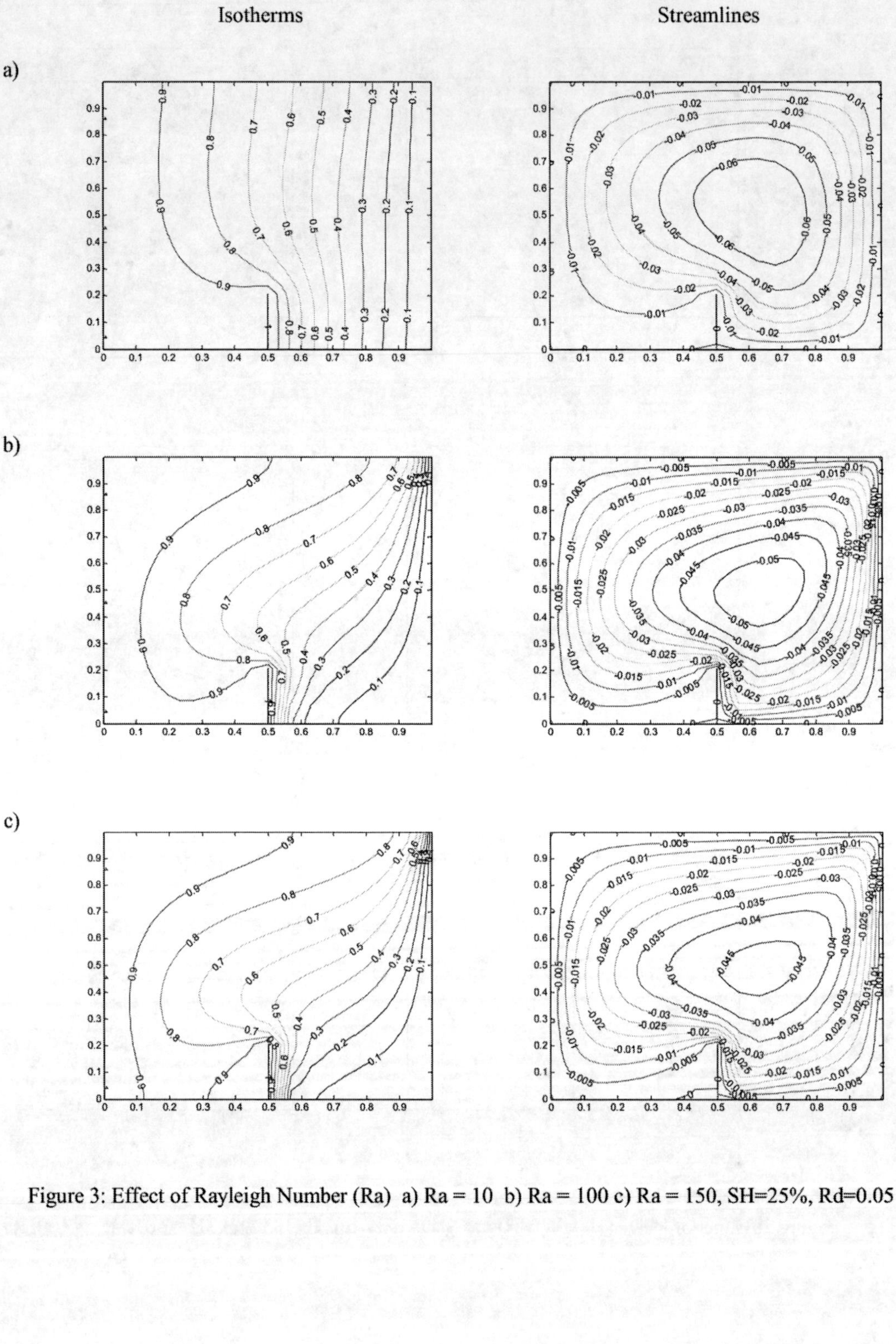

Figure 3: Effect of Rayleigh Number (Ra) a) Ra = 10 b) Ra = 100 c) Ra = 150, SH=25%, Rd=0.05

IOP Conf. Series: Materials Science and Engineering **764** (2020) 012030 doi:10.1088/1757-899X/764/1/012030

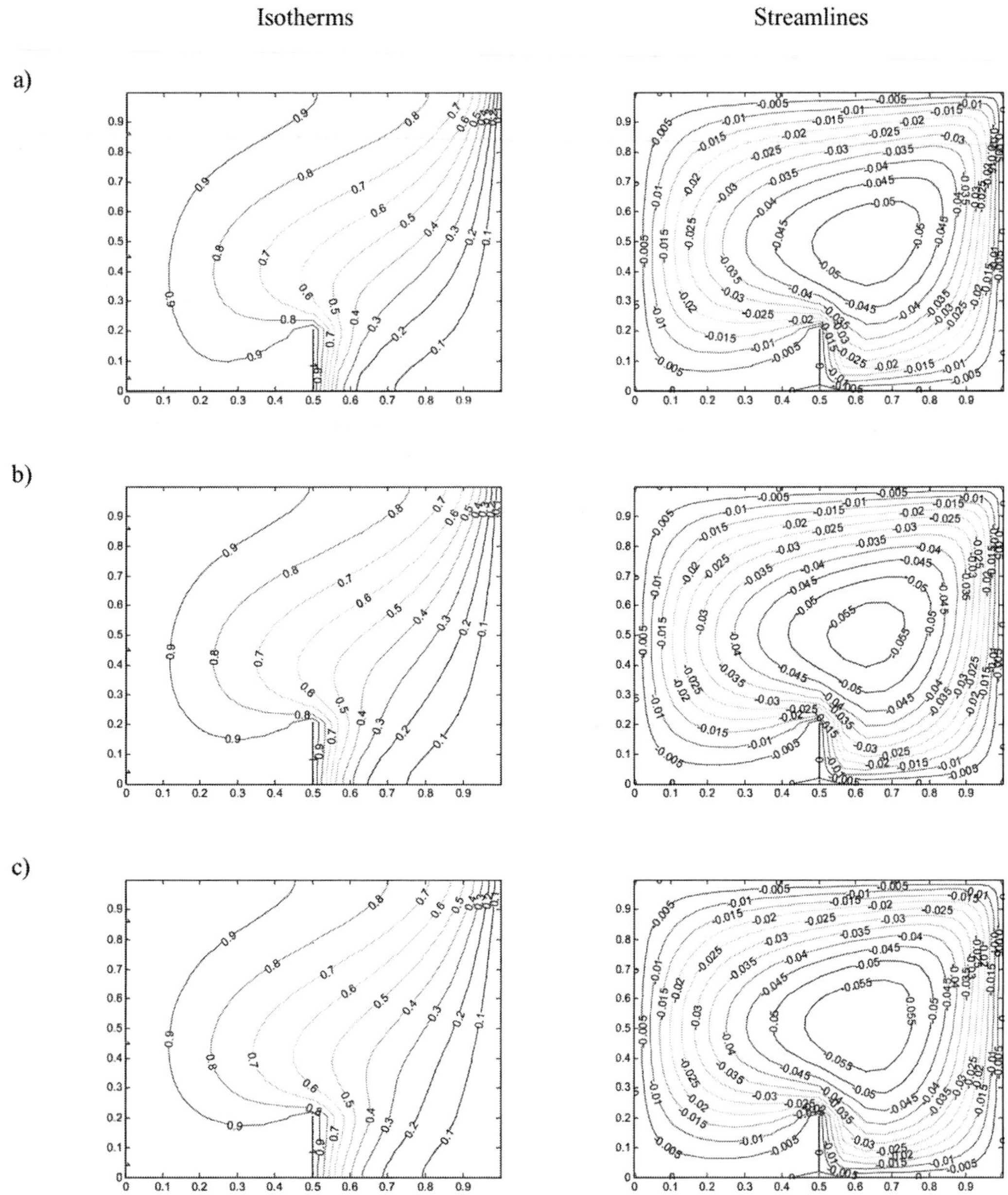

Figure 4: Effect of Radiation Parameter (Rd) a) Rd = 0.1 b) Rd = 0.5 c) Rd = 1, Ra = 100, SH = 25%

4. Conclusions

Heat transfer analysis of a square porous medium is carried out with a small heat generating strip being placed at the center and left vertical surface heated to hot temperature. It is noted that the

- Isotherms and streamlines follow unsymmetrical distribution in the cavity
- Increase in strip length brings higher thermal energy in the left part of the cavity.
- Increased radiation pushes the isotherms towards the left surface.

IOP Conf. Series: Materials Science and Engineering **764** (2020) 012030 doi:10.1088/1757-899X/764/1/012030

References

1. VafaiK ., Hand book of porous media, Marcel Dekker, New York, 2000.
2. PopI., InghamD.B., Convective heat transfer: Mathematical and computational modeling of viscous fluids and porous media, Pergamon, Oxford, 2001.
3. InghamD.B., Pop I.(Eds.),Transport phenomena in porous media, Pergamon, Oxford, 1998
4. Nield D. and BejanA., Convection in Porous Media, ed. 3rd. New York: Springer Verlag.,2006
5. Badruddin, I.A., Ahmed, N. J. S., Al-Rashed, A.A.A.A., Kanesan, J., Kamangar, S. and Khaleed, H.M.T. Analysis of Heat and Mass Transfer in a Vertical Annular Porous Cylinder Using FEM. Transport in Porous Media, 91(2), 697-715 (2012).
6. Azeem, I.A. Badruddin, N. Nik-Ghazali, Mohd Yamani Idnaldris, Ali E. Anqi, Salman Ahmed N.J , SarfarazKamangar , A. A.A.A. Al-Rashed, Effect of Size and Location of Solid on Conjugate Heat Transfer in Porous Cavity, Indian Journal of Pure & Applied Physics, 56 (10), pp. 792-801 (2018).
7. V. Prasad, F.A Kulacki, Natural convection in a vertical porous annulus, Int. J. Heat Mass Transfe 27 (1984) 207-219
8. Badruddin, I. A., Ahmed N. J. S, Al-Rashed, A. A. A., Nik-Ghazali, N., Jameel, M., Kamangar, S., Khaleed, H. M. T., and Yunus Khan, T .M., Conjugate Heat Transfer in an Annulus with Porous Medium Fixed Between Solids, Transport in Porous media, 109(3), 589-608 (2015).
9. Raptis.A., Radiation and free convection flow through a porous medium. Int. Commun. Heat Mass Transfer 25(2) (1998) 289-295.
10. Sarit K. Das, TanmayBasak, ,D.Jaya Krishna, Natural convection in a non-Darcy anisotropic porous cavity with a finite heat source at the bottom wall. Int. J. Therm. Sci 48(7) (2009) 1279-1293.
11. Amahmid.A., HasnaouiM., , M.Mamou, , P.Vasseur,Boundary layer flows in a vertical porous enclosure induced by opposing buoyancy forces. Int. J. Heat Mass Transfer. 42(19) (1999) 3599-3608.

Detection of skin Cancer using PIFA antenna

M.Leeban Moses [1*], **C.Raju** [1] and **Ajin R Nair** [1]

[1]Department of Electronics and Communication Engineering, Bannari Amman Institute of Technology, Sathyamangalam-638 401, India.
[*]Author for correspondence E-mail: leebanmoses@bitsathy.ac.in

Abstract. Due to the advancements in electromagnetic fields, it can be widely used in medical applications to detect various diseases. In this paper we present a novel idea to detect a biological skin cancer tissue by designing a Planar Inverted F antenna (PIFA) which is more commonly available in mobile devices. Our basic idea is to design PIFA antenna using CST software and thereby fabricating the antenna with one port and identifying its reflection parameters and VSWR ratio. The Biological skin tissue is modelled in CST software by using the dielectric properties of a normal tissue and a cancerous tissue. The antenna is excited by a source and radiated towards the normal and affected skin tissue. The permittivity, Conductivity, permeability and absorption power by the skin tissues are identified and compared with the affected tissues. Results showed that the Energy received from the normal and affected skin tissues gets varied and from this simulation results we can be able to conclude that weather a cancerous tissue is present in a human body.

1. INTRODUCTION

In the field of medical applications, cancer is the second leading cause of increase in mortality rate globally. Cancer or malignant neoplasm is a disease caused due to abnormal cell growth. Over 100 types of cancer affects humans. One among those is Skin cancer which is caused mainly due to exposure to ultra violet radiation from the sun. Basal cell carcinoma is the most common type of skin cancer that occurs on the sun exposed areas of the body, especially the head and neck which grows on the lowest layer of epidermis. Squamous cell carcinoma also occurs due to exposure to sun but then it grows deeper into the layer of the skin and spread to other parts of the body. These two types of carcinoma are called Melanoma cancers. Nonmelanoma cancers occur due to genetic defects or uncontrolled growth of skin. Skin cancer also has the danger of being spread to other parts of the body. But this can be prevented by early diagnosis and treatment.

The detection techniques of melanoma include Biopsy and certain Imaging techniques. All biopsy methods, except optical biopsy requires the removal of tissues from the affected or doubted area. Later the removed tissue undergoes microscopic examinations under chemicals and sometimes genetic tests to get nearly accurate results.

This procedure takes a long time and the main disadvantage is the removal of tissues. But some newer types of biopsies, such as Reflectance co focal microscopy (RCM), can be done without needing samples of skin. But this method isn't cheap or available in all hospitals and testing centers.

Imaging techniques are done by creating pictures by the use of X rays and other radioactive substances. Magnetic Resonance Imaging (MRI) can also be used for imaging the damaged tissue. Both the techniques are not preferable because of the ionized radiations.

To avoid the ionized radiations, microwave imaging technique has been developed which reduces the cost of operation and can be availed easily everywhere. The principal of operation of microwave imaging technology is based on the dielectric constant between the healthy tissues and cancer affected tissue. In order to identify the pathology, the difference in properties such as conductivity and permittivity between normal skin and malignant skin tissue are taken into account. But this technique cannot show a single cancer cell or even a few. Imaging technique requires millions of cancer cells to make a tumor big enough to be shown. Sometimes the cancer cells detected in imaging techniques turn out non-cancerous in biopsy.

In the literature, Optical fluorescence imaging or chemical indicator sensing technique has been adapted to determine the malignant tissue by using the optical waves. An optical signal is transmitted with different nanometer wavelength between the normal tissue and malignant tissue. The absorption, scattering and reflection ratios are compared in order to determine the affected tissues from the normal tissue [1], [2].Near infra-red(NIR) is the other technique implemented to enhance the scattering ratio by changing the wavelength requirements. The refractive index of the malignant tissue gets varied with the normal tissue. This exhibits Rayleigh scattering (weak scattering) and Mie scattering (strong scattering) in biological tissues. The results shown in [3], [4] dictates that terahertz wavelength(THZ) exhibits weak scattering compared with NIR. All the above methods in literature, depicts the imaging technique with non-ionizing radiations [7]. But it fails to identify the location and size of the tumor cells.

In today's world the use of antennas in medical applications is increasing because of its feasibility and low power transmission. In literature [5] an antenna with circular patch was tried to design to detect cancerous cells. The results were successful that they showed differences between malignant and normal tissues.

This paper presents a novel technique to design an antenna with 2.4GHZ frequency(ism band) to detect the presence of cancerous tissue. The rest of the paper is organized as follows. Section II describes the design of antenna in CST software and fabricates the antenna with 2.4 GHZ frequency and test the same antenna with Vector network analysis. Section III describes the study of different types of tissue with respect to the dielectric properties. The conductivity and permittivity of the normal tissue and malignant tissue are identified and stimulated in CST software. Section IV describes the transmission and receiving parameters of the normal and cancerous tissues.

2. DESIGN OF PIFA ANTENNA

This section describes the design of a planar inverted F antenna to detect the different types of tissues present in human body. In mobile phones there are 6 different types of antennas used for different application. The planar inverted F antenna is a type of antenna which is used in mobile phone for Wi-Fi and Bluetooth applications.

The antenna of designed for a frequency range of 5GHz. The ground plane is made up of copper by using the dimensions of {Ls/2 ,Ws/2, Tg} where Ls denotes the substrate length with 12 mm and Ws denotes the substrate width with 12 mm and Tg denotes the ground plane thickness with 0.2 mm. The substrate is made up of fr4 by using the dimensions {-Ls/2, -Ws/2, Tsub} where Ls denotes substrate length and Ws denotes the substrate width and Tsub denotes the substrate thickness. The antenna patch is made up of copper by using the dimensions of {-Lp/2, -Wp/2, Tp} where LP denote the length of the batch Wp denotes the width of the patch and t p denotes the thickness of the patch. On the patch the planar inverted F antenna is created by using different dimensions.

Figure 1. PIFA Antenna design

The planar inverted f antenna is designed for a single port at the edge of the first patch. The S parameter of an antenna is used to define the signal wave response of an electrical element at a given frequency. The S11 parameter of an antenna is used to find the input output relationship between the ports and estimates how much power is reflected from antenna and it is also called as reflection coefficient or return loss.

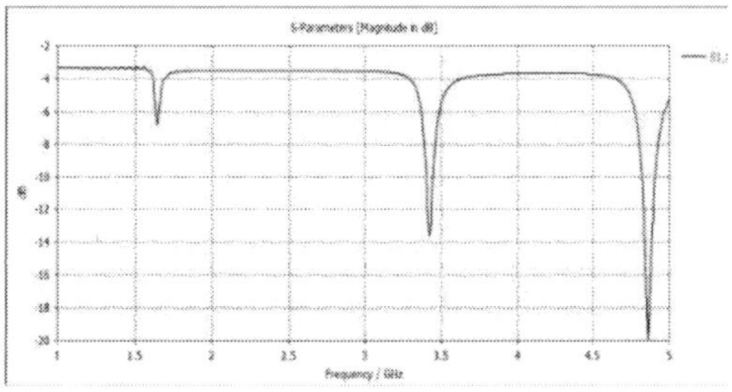

Figure 2. S_{11} parameter of PIFA antenna

When a power is transmitted to the port 1 of an antenna, the antenna gets radiated. All the power radiated from the antenna will not be transmitted instead some signals may be transmitted and few of the signals will be reflected back. This represents the imperfectness of an antenna and this can be calculated by using voltage standing wave ratio. The VSWR ratio describes how the antenna will is impedance matched to the radio.

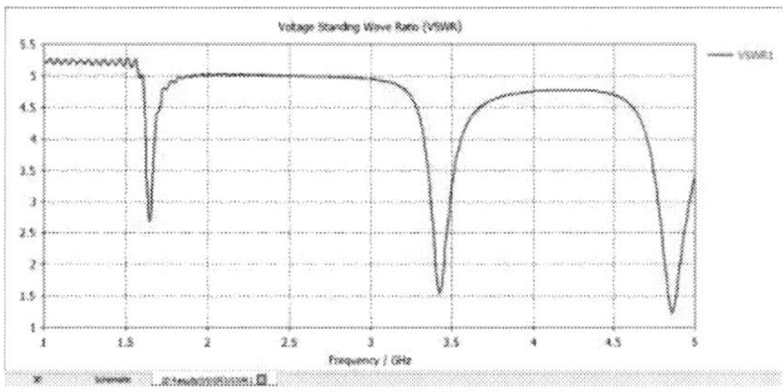

Figure 3. VSWR for PIFA antenna

The planar inverted f antenna is designed in CST software using the above specifications and it's simulated to obtain the reflection coefficient and voltage standing wave ratio. The designed antenna in the software is fabricated and tested in vector network analyzer to determine the return loss parameter and frequency of the specified antenna.

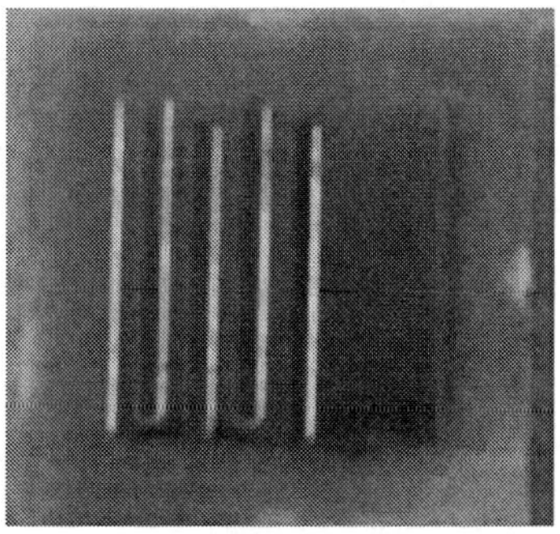

Figure 4. Fabricated PIFA antenna

IOP Conf. Series: Materials Science and Engineering **764** (2020) 012031 doi:10.1088/1757-899X/764/1/012031

3. MODELED PATHOLOGICAL TISSUE

Due to the advancements in electromagnetic spectrum, it's applied it in the field of medical sciences to detect various diseases. In this paper, electromagnetic signals are transmitted and received in order to detect the presence of skin cancer tissue in human body. This can be identified by studying the dielectric properties of the normal tissue as well as the cancerous tissue. The dielectric medium also depends upon the frequency transmission and it also various depending upon different frequencies. The normal tissue and the tumor tissue have different dielectric properties and these dielectric properties are studied by using the conductivity, permeability and permittivity of the tissues. In literature it has been show that the dielectric value of the normal skin is 38F/m and the dielectric value of tumor tissue is 50 F/m with the working frequency of about 1 to 5 GHz.

Figure 5. Modelled biological skin tissue

A rectangular skin model is developed in CST software by comparing the different dielectric properties of a normal and tumor tissue. The antenna is placed as a substrate here and a layer is grown above with dimension of 10mm which specifies the air medium. Above the air medium rectangular skin tissue is developed as shown in the figure. The green color in this figure represents the normal biological tissue with the dielectric properties of ε=1.0, $\nabla\varepsilon$=66.31, τ=7.585ps, α=0.063, σ=1.370s/m, ρ=1010.

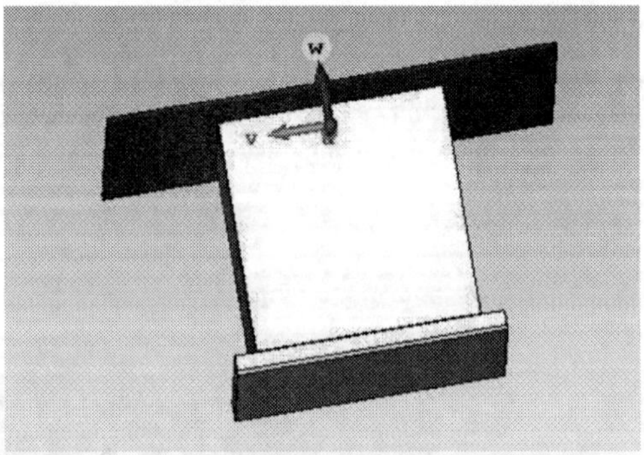

Figure 6. Modelled biological cancerous skin tissue

The above figure represents a rectangular skin model which is developed in CST software air by comparing the different dielectric properties of a tumor tissue. The antenna is placed as a substrate here and a layer is grown above with the dimension of 10 mm which specifies the air medium. About the r medium rectangular skin tissue is developed shown in the figure. The brown color in this figure represents the tumor biological tissue with the dielectric properties of ε=9.058, ∇ ε=51.31, d=10.34ps, α=0.022, σ=0.889s/m, ρ=863.

In order to define the dielectric properties of any material over the wide range of frequencies in CST simulation tool we need the definition of the permittivity of the medium and is denoted by first order and second order equations.

$$\varepsilon_r(\omega) = \varepsilon_\infty + \frac{\beta 0}{\alpha 0 + j\omega}$$

$$\varepsilon(\omega) = \varepsilon_\infty + \frac{\beta_0 + j\omega\beta_1}{\alpha_0 + j\omega\alpha_1 - \omega^2}$$

Where β_0, α_0, β_1, $\alpha_1$are the parameters that are derived from square fitting method in Single-Pole Cole-Cole dispersion model.

4. RESULTS

The antenna is fabricated in section II and a biological normal tissue and tumor tissue is developed in CST software in section III. The modelled biological tissue with different dielectric properties is simulated and the results are shown below.

The antenna is excited by a given signal source which is available in CST software. The radiated power from the antenna reaches the normal biological tissue with the given dielectric property and exhibits reflection, diffraction and scattering off the signal source. The reflected diffracted and scattered signals from the biological tissue reach the antenna with different time delays. Depending upon the reception of the strength of various signals with time delay permeability and permittivity off the particular skin can be estimated with different angles.

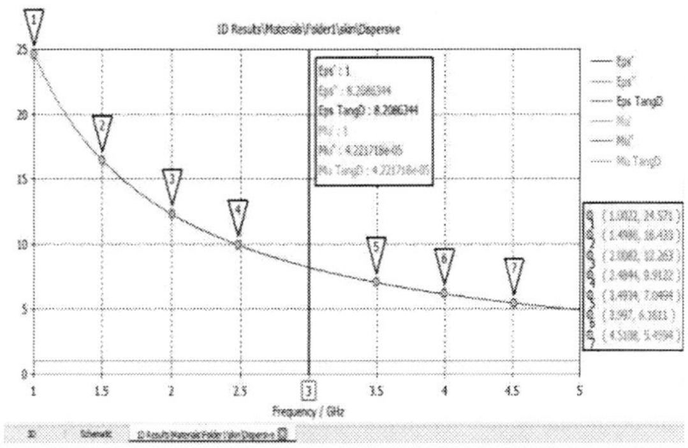

Figure7. Dispersive parameters of skin tissue

The results shown in figure 7 represent the different parameters such as permeability and permittivity for the normal biological tissue. For different frequency ranging from 1 to 5 GHz the energy of the signal has been estimated. The values of energy vary from 25 to 5 for different frequencies. From the estimated energy values the permittivity and permeability values can be estimated. Depending upon the different values of permittivity and permeability the dielectric property of a tissue can be identified.

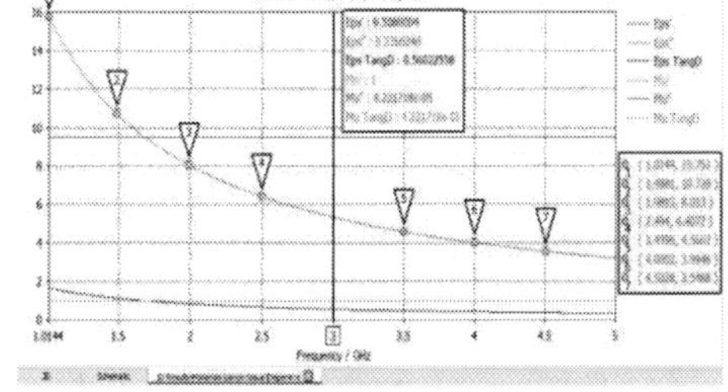

Figure 8. Dispersive parameters of cancerous skin tissue

IOP Conf. Series: Materials Science and Engineering **764** (2020) 012031 doi:10.1088/1757-899X/764/1/012031

The results are shown in figure 8 represents the different parameters such as permeability and permittivity for the cancerous biological tissue. for different frequencies ranging from 1 to 5GHz the energy of the signals has been estimated and it is found that the values of energy varies from 16 to 4 for the different frequency levels. Depending upon these values the dielectric property and of this tissue can be identified by calculating the permittivity and permeability. By these results we can come to a conclusion that this is a different type of tissue with different dielectric properties. So that cancerous tissue who can be present at this area.

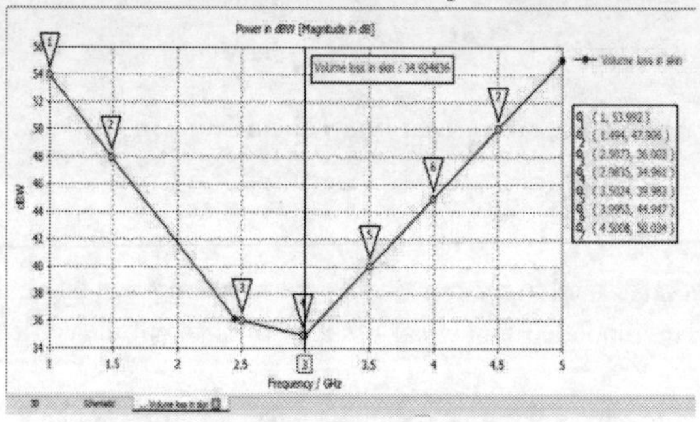

Figure 9. Volume loss in skin tissue

The figure 9 represents the volume loss in skin. The antenna is excited by a signal source with certain energy. The power of the signal is radiated through the antenna and received at a normal skin. As this tissue who has some dielectric property which makes the signal to reflect or scatter in different directions. Buy reflecting or scattering a signal in different directions the power of the signal main get degraded. This degradation in power will be received at the antenna at different angles with a proper time delay. By measuring this power off the signal we can estimate the loss of energy that has been occurred in the tissue. The simulation results show that the power degradation starts from 54 dbW to 34 dbW for different type of frequencies ranging from 1 to 5 GHz. According to the simulation the volume loss in the skin is found to be 34.92 dbW.

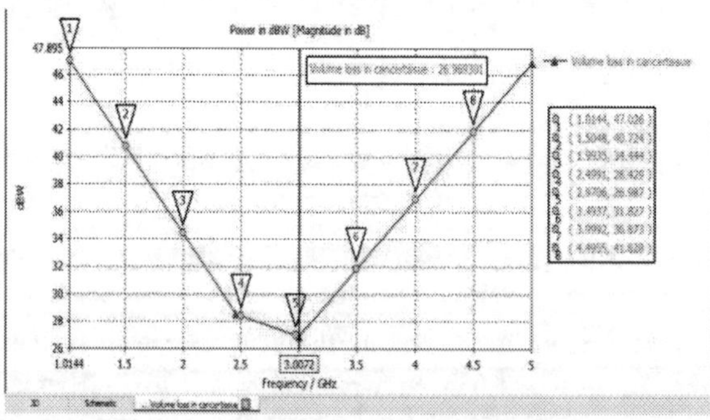

Figure 10. Volume loss in cancer skin tissue

Figure 10 represents the volume loss in cancerous tissue. The antenna is excited with a given amount of power by using a signal source and is transmitted via air medium to the affected skin. The affected skin has different dielectric properties than the normal skin it exhibits some different reflection and scattering parameters. Due to the difference in refraction and scattering parameters the energy of the signal get degraded more when compared to the normal tissue. This degradation in power is measured in the receiver antenna and it is noted that it the energy gets degraded from 47 dbW to 26 dbW for different frequencies ranging from 1 to 5 GHz. According to the simulation the volume loss in the cancer tissue is found to be 26.96.

IOP Conf. Series: Materials Science and Engineering **764** (2020) 012031 doi:10.1088/1757-899X/764/1/012031

5. Conclusion

This paper presents a technique to detect skin cancer by using mobile phones. In order to detect the skin cancer using mobile phones a planar inverted f antenna is designed using CST software with the frequency of 1 to 5 GHz. The s parameters and VSWR ratio of the antenna has been estimated. The dielectric properties of a normal skin tissue and a tumor skin tissue will be varied. This can be estimated by using the permittivity, permeability e and conductivity of the skin tissue. Biological skin tissues have been developed in CST software by finding the dielectric property of the normal skin tissue. This tissue has been grown above the antenna with the air medium located at 10 mm apart. The antenna has been excited by a signal source and power has been radiated from the antenna to the normal skin tissue. The reflected and scattered signals from the normal skin tissue are measured with different parameters such as its energy, permittivity, permeability, conductivity and volume loss in the tissue. A cancerous biological skin tissue that has been developed in CST software by finding the dielectric property of the affected tissue. The affected tissue has been grown above the antenna with the air medium located at 10 mm apart. The antenna is excited with the given signal source. The reflected and scattered signal from the cancerous tissue is measured with different parameters such as its energy, permittivity, permeability, conductivity and volume loss in the biological skin tissue.

We may observe that the energy received from a normal tissue and a cancerous tissue is different, the conductivity, permittivity and the permeability of a normal tissue and cancer tissue is different and also the volume loss in the biological tissue for a normal and abnormal tissue is different. This results shows that a mobile antenna can detect and normal tissue and a cancer tissue.

We may extend this work by selecting a different type of antenna with two port device's such that its reflection parameters and scattering parameters can be properly estimated. Also so this can be e tested by using original biological tissue from microbiological laboratories.

References

[1] V. Sharma et al., "Auto-fluorescence lifetime and light reflectance spectroscopy for breast cancer diagnosis: Potential tools for intraoperative margin detection," Biomed. Opt. Exp., vol. 3, no. 8, pp. 1825–1840, Jul. 9, 2012

[2] J. Q. Brown et al., "Optical assessment of tumor resection margins in the breast," IEEE J. Sel. Topics Quantum Electron., vol. 16, no. 3, pp. 530– 544, May/Jun. 2010

[3] P. Y. Han, G. C. Cho, and X.-C. Zhang, "Time-domain transillumination of biological tissues with terahertz pulses," Opt. Lett., vol. 25, no. 4, pp. 242–244, Feb. 15, 2000.

[4] G. J. Wilmink and J. E. Grundt, "Current state of research on biological effects of terahertz radiation," J. Infrared Milli-Terahertz Waves, vol. 32, pp. 1074–1122, 2011.

[5] Microstrip Antenna Design with Circular Patch for Skin Cancer Detection Rabia Top, Yaşar Ünlü, Seyfettin Sinan Gültekin, DilekUzer Advanced Electromagnetics, VOL. 8, NO. 2, MARCH 2019.

[6] Development Of Accurate UWB Dielectric Properties Dispersion At CST Simulation Tool For Modeling Microwave Interactions With Numerical Breast Phantoms, Ahmed Maher, Kaydar M. Quboa 2011 8th International Multi-Conference on Systems, Signals & Devices.

[7] Moses ML, Kaarthick B. Multi-objective cooperative swarm intelligence algorithm for uplink resource allocation in LTE-A networks. Trans Emerging Tel Tech. 2019;e3748. https://doi.org/10.1002/ett.3748

IOP Conf. Series: Materials Science and Engineering **764** (2020) 012032 doi:10.1088/1757-899X/764/1/012032

Analysis of effectiveness of power on refined numerical models of floating point arithmetic unit for biomedical applications

Dinesh Kumar J R[1,a], Ganesh Babu C[2,b], Balaji V R[1], Visvesvaran C[1]
[1]Department of Electronics & Communication Engineering, Sri Krishna College of Engineering & Technology, Coimbatore, Tamil Nadu, India
[2]Department of Electronics & Communication Engineering, Bannari Amman Institute of Technology, Sathyamangalam, Tamil Nadu ,India

Email: [a]dineshkumarjr@skcet.ac.in, [b]bits_babu@yahoo.co.in

Abstract:"The technology defined today might be redefined tomorrow"- from the quotes, in the modern era redefining the architecture with the optimization of area, delay and reduction of power is focused more than fixed models. This paper describes the floating point arithmetic unit and insists the importance of it, in real time signal processing. The present trends of portable devices for medical applications need occupancy of smaller area with long span time life of battery. The bio medical signals are converted into IEEE 754 format as BIF. Hence it is easily analyzed in FPGAR devices and also this work insists the importance of redefining the architecture with reduction of critical path delay. The FPU arithmetic unit has the blocks to perform the computations like addition, Multiplication and division operation. The main architecture is constructed using top level design with individual blocks defined as a part of it. The floating point unit defined with modern FPGAR (Reconfigurable FPGA) increase the speed of computation and enables the flexibility and adopted to the hardware reconfigurable models. A 32 bit representation of IEEE 754 results of FPU is stored in the data storage and this process optimize the critical path between blocks. The way its programmed consumes lower level of power consumption in the range of 50.4 nw and delay of 25.2 ns and also modern adder structure with 6T type was used to implement the intermediate blocks of multiplier in FPU block as part of computing partial products. By this process of FPU redefining, reduce the number of slice to 206 and increase the operating frequency of 46.12 MHz.

1. Introduction

In the modern era of digital world all the process of computation plays important role. The computations are done by the arithmetic unit of the processors [1]. The need for the accuracy in floating point computations are increased compared to Integer computation techniques [2]. The processors which is defined to perform the specific signal processing applications like speech signal processing[8], digital signal processing are expected to have maximum efficiency and more accurate with precision values. But the in Bio medical applications the real time data are much complex to analyze as this data's are having low bit variation and highly sensitive to external noises generated by the device. The error present in the signals will not define in precise way to identify the type of problem associated with the human body. Hence defining the biomedical system with precision and high accurate data plays a vital role in bio medical processors. The main idea is to refine the processors in which floating point operations is optimized in terms of power and area occupancy of the device. The arithmetic unit of the processors is defined to perform the computation in floating point or integer way of representation. The accuracy of the processing data is in depended of the data obtained and it is depends on the no of bits taken for its representation [4].

The computation of any process is done through the addition operations. Even though computing operation of multiplication can be done using repetitive addition. Hence the application of real time signals and processing of it through floating point unit should provide more accurate and précised results of computation. Mostly the solution required for the different problems associated with the real time signals needs to be regularized with high level computation accuracy [5]. A Floating point representation of number should be mentioned in IEEE 754 format. The IEEE 754 format representation is implemented in most of the modern processors and controllers and this format representation is classified as three blocks connected in cascaded manner to indicate sign-bit, exponent-bit and mantissa-bit with bit size of 1,8 and 23 bits and the total size of 4 bytes (32 Bits) as shown below Fig.1 [16].

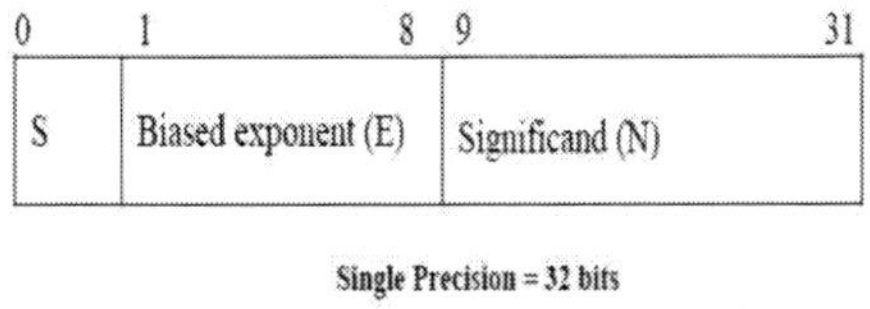

Figure 1. Single Precision IEE 754 format

The effective use of silicon chip area, power reduction and clock tree synthesis are major issues in focus to the embedded processors which are designed to operate with real time signals. The allocation of silicon area has significant impact on the power consumption level of the processor in integer computations, but the advancement in integrated circuit fabrication techniques narrated that the Integer operation ALU and FPU has same constraints with respect to the area, delay and power [12]. So the implementation of floating point computation was done with the help of FPGAs available in today's market with different algorithm. The different architecture of FPU can be simulated in the same FPGA using reconfigurable architecture [10]. The complexity of computation depends on the algorithm in programming levels of abstraction. The FPGA which is designed to implement the FPU operations can be redefined to develop the systems in the field of industrial control systems, Robotics, Digital computations and traffic controllers and biomedical [15] / DSP processors [8]. This paper is organized as sections of survey, computation methodology, result analysis and simulation results with conclusion in the following sections.

2. Related works and survey

The development of optimized model for floating point computation is always gets much priority, especially defining the processors associated with real time application and bio medical signal processing [6]. The various architectures are developed for forming the floated value computation, but the issues are related to either power or area. Importance of the architecture is defined in terms of blocks connected to it. Simple adder structure may not be used for floated value operations. The floating point unit should be perform the computation of real time signals. Initially the real time signals are dynamic and it should be represented to certain defined format. The IEEE 754 standard format is referred with single floating point precision with the bit size of 32 bits, where as the double precision method consumes more storage area (64 bits) [7]. The full adder circuit which performs bit wise addition is extended for the bit operation of 32bit. The blocks of floating point representation which defines in the previous sections, is taken into the part of computation as sign-bit, mantissa-bit and exponent bit. The sign-bit is calculated by the Ex- OR operation, the sign-bit is asserted to "0" for negative number and "1" for positive number and the remaining block of IEEE 754 format to be taken to compute the other progression based on the computation [9]. The two different forms of floating point representation are there, like DIF (Decimal Interchange Format) & BIF (Binary Interchange format) in VLSI, however the BIF format is referred mostly as ease of computation with 32 bits and also representing the binary number (1 & 0) is much easier then decimal representation. Hence the implementation of real time signals in BIF (IEEE 754) format reduce no of cycle of operation in terms of critical path delay computation.

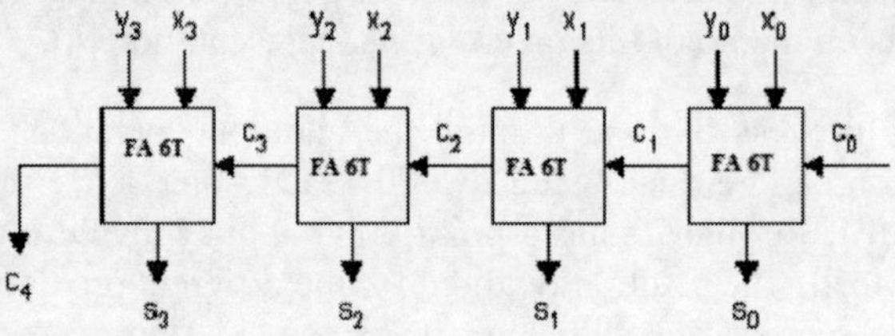

Figure 2. Full adder circuit designed using 6T Structure

The figure2 shows how the high speed adders (RPcY) are constructed using basic block of full adder with 6T as explained in [1]. The usage of this architecture is defined only for 4 bits of operation, so this structure is redefined to perform the computation of exponent bit(biased) of size 8 bits [3]. This structure is very useful as the number of transistor is reduced so the area of FPU can be reduced to the minimize level.

Figure 3 shows the another high speed adder (CSA) based computation for extracting the exponent bit [3]..The same part of the circuit can be used to redefine the multiplier block usually the multiplier is calculated by partial product (PP) calculations and this PP is stored in the floated buffer.

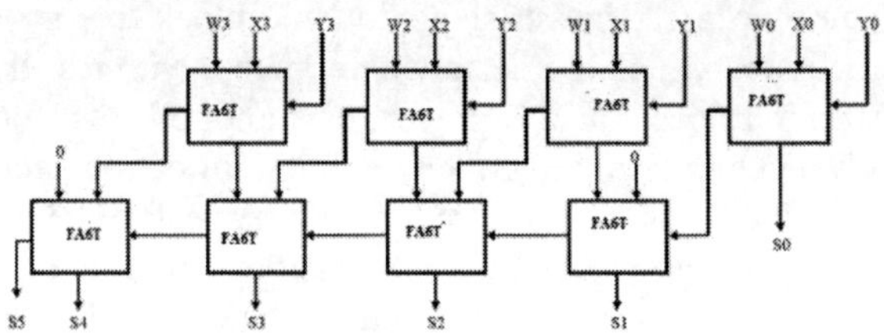

Figure 3. Carry save adder designed from FA of 6 T

The blocks of multiplier is redefining here using the high speed adder blocks and this ensures the reduction of delay and improves the speed of computation. The high speed operation of circuits ensures optimization on area and critical path delay. The biomedical signal [6] obtained from the human body is converted into 32 bits, by using various process like quantization and precision. After the conversion the corresponding DIF is converted to BIF, to perform the floating computations.

3. Computation methodology

3.1 Multiplication and Addition of Floated Values

The floating point computation is performed as specified in the flow graph. The operations involved in various steps. The initial process of operation is to get the numbers (1 & 2) to perform the desired operation [5]. The second step is to calculate the sign bit value using Ex-OR operation. Then check bit is used to decide the operation is addition or multiplication. This step is operated to identify the operation. If the selected operation is Addition of two floating number s then MSB polarity is to be identified, whether it is addition of two (+) ve number or(-)ve number or the mixed of the polarity. Suppose if (I) both the MSB are positive then calculate the difference between the exponent bit and conclude the operation by checking the difference (d) bit, if this bit is equals to zero then interchanging of the number is not permitted otherwise shift the number 2 by the no of bits founded as difference [14]. Then calculate the new value of is added with the Number 1 after the shift of number 2 by the d bits and then add the significant bits of both numbers. Check for the carry is generated, if the carry is found add that with the "1" found in the previous steps.

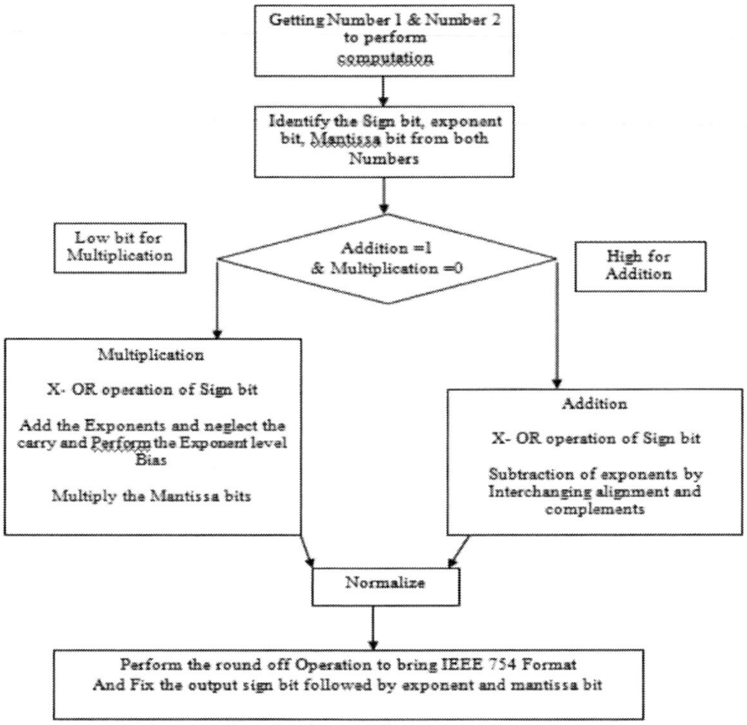

Figure 4. Floating point Addition/ Multiplication

Align the number by 32 bits of standard IEEE 754 format. If the numbers are differ then compute the 2's complement of number 2 and add them with the number 1. If the carry is generated, here it can be neglected and shift the result on left till the MSB becomes "1", also make a note on the number cycles it is taken to do this operation. Then subtract the exponent bit from the count calculated in the above steps. To make the both exponent bits are equal append "D" (Exponent – Count) no of zero to this. Align the numbering format to make 32 bits of standard IEEE 754 representation.
Similar to the steps the floating point multiplication is done by calculating the sign bit using EX-OR gate ad initial step. Then the exponent bits of both numbers were added and mantissa bit are multiplied. Generally the mantissa bit is represented in 23 bit. So the multiplication of mantissa might give total resultant bit of 46 bits [7]. Hence it requires the normalization process. The 46th bit is "1" then no need to shift the mantissa as its in the prescribed format, else the 47th bit is "1" then obtained mantissa has to be rotated to right and 1 bit is added with the exponent bit. The intermediate products obtained through the steps fix the decimal points when it is converted to decimal format representation. The overflow / underflow in the multiplier is not allowed since the FPU follows the standard form representation. The overflow is raised when the result is infinity, underflow is raised if the exponent bit is less than "0" [16]. So the compensation steps is done by adding bias or removing bias value depends on the value of exponent obtained from the previous step.

3.2 Division and Subtraction of Floated Values

The previous section discussed on how the floating point multiplication is performed. This section describes how the division is performed in floating point computations. All the basic computation of floating uses the same operation to calculate the sign value. Obtained the subtraction of two exponent bit and perform the division operation of mantissa of number 1 & 2 also perform the round off values and normalize the value of mantissa bit by shifting "1" to left or right. Overflow may occur if mantissa of number 2 is zero, ie, Divide by zero case will update the overflow operation. The floating point subtraction performed same as explained in floating additions. The floating point subtraction was done by performing addition of numbers 1 & the 2's complement of number 2.

IOP Conf. Series: Materials Science and Engineering **764** (2020) 012032 doi:10.1088/1757-899X/764/1/012032

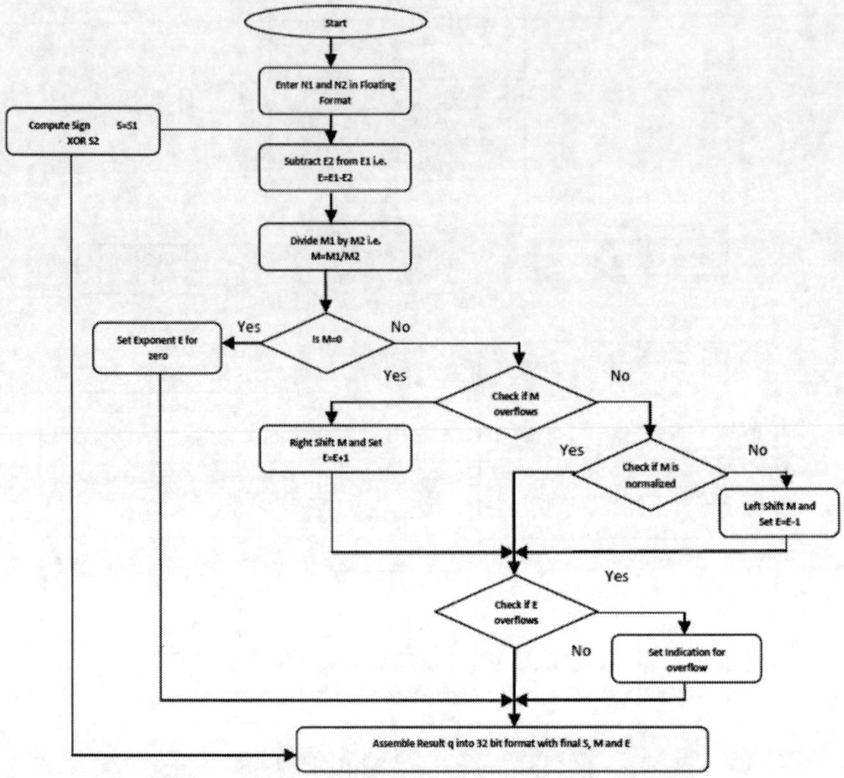

Figure 5. Floating Point Division

The developed architecture of Floating point computation methods like addition-multiplication & subtraction -divider were combined to refine the floating point arithmetic unit. This unit can consumes very less power compare to other structures since the optimization was performed on the silicon area allocation. The most critical path in building the final structure is supply distribution and clock synthesis along with critical path delay. The supply problem was compensated by constructing common rail at the core die area of the device [6]. The clock synthesis problem was nullified by performing de clone the clock gating element at the manufacture level. The critical path delay is reduced by connecting the register are closer and cutest the longest circuit path.

The floating point computation was defined in previous section. The need for floating point unit for real time signal processing applications depends on the power consumed and delay generated in critical path. In the redefined architecture the multiplier unit was constructed with the adder whose internal transistor count is less, 6 Transistor. The internal transistor count was reduced to improve the effectiveness of area of silicon die area and also the reduced no in operating transistor consumes less power, also the transistor was defined to operate with low operating voltage approximately 1.2V. The low transistor (6T) adder is replaced in the multiplier block of floating point unit to ensure the high speed in computation [1] and suppress the critical delay for generating the computed results [15]. The proposed FPU consists of floating point registers, Stack unit and buffers for accessing the data and moving the data's to the further computation purposes. The decoder, rounding and normalization unit are placed in the purpose of maintain no of bits to be equal to IEEE 754 standard format [7], after the computation which defines in the precious section, calculate the 32 bit format IEEE 754 format which was divided into three blocks of sign, mantissa and exponent bit. The blocks of FPU is designed initially and suing the top order architecture implementation method from the base to final architecture was constructed. Here the common bus designed to have the operation of 32bits. The architecture was defined in programming as the per the HDL description [17] and the optimization was done to reduce the critical path delay and less switching operation ensures the low power operation of the entire circuit.

IOP Conf. Series: Materials Science and Engineering **764** (2020) 012032 doi:10.1088/1757-899X/764/1/012032

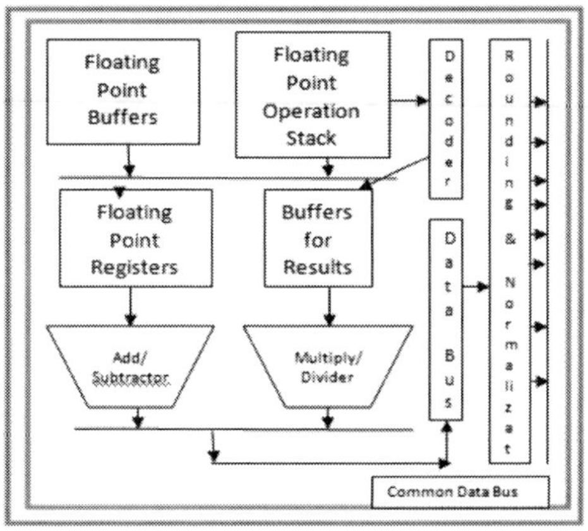

Figure 6. Floating Point Unit Architecture

4. Results & discussions

The proposed methodology flow graph was constructed using System Verilog (C) language. The code was simulated in Model sim at the initial level. The top level model and the combined structure was simulated in Xilinx 14 and the results obtained shows as this methodology provides high reduction of power in the range of 50.42 nW and also the delay by 25.2 ns. The critical path time delay is vary from 0.24us to 0.30us, and this variations was depends on the computation operation preferred by the programmer.

This architecture was implemented in Spartan VI processor which has the operating core voltage of 1.2V and supports the maximum I/O and adapted to 45nm Technology with high logic pin connections.

Table 1 : Delay & power comparison

Types of Unit	Delay (S)		Power (W)	
	**	##	**	##
FP Adder	25.9μ	18.7n	203m	34.42 n
FP subtractor	27.0μ	21.6n	203 m	40.12 n
FP multiplier	18.5μ	16.9n	24m	60.81 n
FP Divider	NA	16.2n	NA	60.36 n
FP Arithmetic Unit	NA	25.2 n	NA	50.42 n

** Existing Method ## Proposed Method

IOP Conf. Series: Materials Science and Engineering **764** (2020) 012032 doi:10.1088/1757-899X/764/1/012032

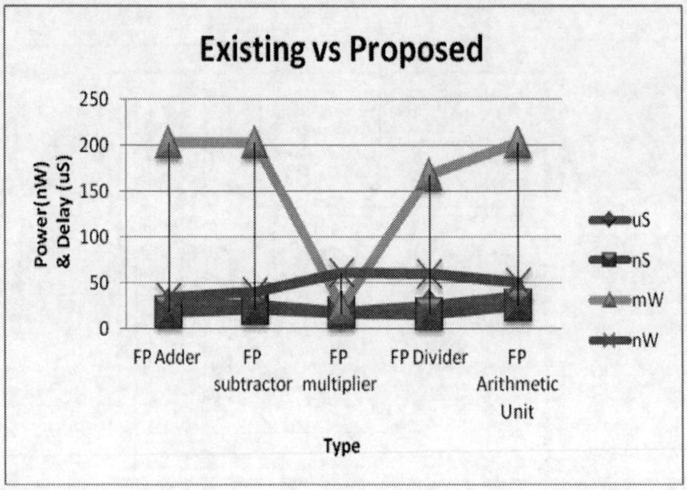

Figure 7. Comparison graph of existing system and proposed system

Table 2. Parameter of Refined multiplier in FPU

Adders Replaced in Multiplier Unit	Speed (MHz) **	Speed (MHz) ##	No of Slices **	No of Slices ##
Carry Save	44.72	52.4	776	542
Pipelined	151.2	94.4	126	154
FPU with Carry save adder based Multiplier	58.72	36.48	446	168
FPU with pipelined Multiplier	141.7	46.12	140	204

** Existing Method ## Proposed Method

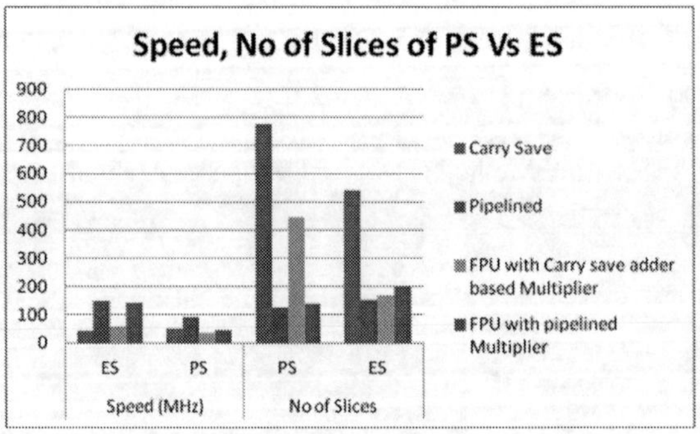

Figure 8. Comparison graph of refined FPU architecture

5. Conclusion

This paper explains the various steps involved in floating point computation like addition, multiplication and division. The process of computation and usage of optimization of programming leads to reduction of critical path delay and area. The field of bio medical needs much accuracy with the real time signal process. The FPU defined on FPGAR can be implemented to the DSP applications. Such that a way this system can be converted into single block of Simulink models in SCI lab or MAT lab tools. The FPGAR was used here so that the hardware can be reconfigured using programming techniques so that architecture was modified with less no of transistor count and less occupancy of area in silicon Die. Hence this FPU can be used in real time (bio signal) devices for the improvement of computation speed and reduce the other parameters of VLSI like delay, area and power. The bio medical devices which uses this FPU have high span of time in power with more portable devices

6. References

[1] Dinesh kumar J R, Ganesh babu C, Karth S P 2019 Performance Analysis of 16 bit Adders in high speed computing applications" IEEE –Processings of ICACE

[2] Mandal. S.K. , Smaranika Routi " Implementation of Low Power and High Speed Single Precession Floating Point Multiplier using Kogge-Stone Adder and Dadda Multiplier" IJVDCS, Sep2016

[3] Dinesh Kumar,C. Ganesh Babu "Survey on Consummation of Subsided Power and High Speed Adders For Mixed Signal Processing Applications" IJPAM, 2018.

[4] Sonali Mehtha, Balwinder Singh, Dilipkuamr "Performance Analysis of Floating Point MAC Unit" International journal of computer applications, Sep 2013

[5] Lathapriya V and Sivagurunathan P "Design And Implementation Of Low Power Floating Point Unit Using Reconfigurable Data Path: A Survey" IJTCSE, Feb 2016

[6] Dinesh Kumar, C. Ganesh Babu "Performance Investigation of Multiplier for Computing & Control Applications" in Jour of Adv Research in Dynamical & Control Systems, Vol. 10, 2018

[7] R. Prakash Rao " Implementation of The Standard Floating Point Mac Using IEEE 754 Floating Point Adder" Proceedings of the Second International Conference on Computing Methodologies and Communication (ICCMC 2018)

[8] Balaji, V R & Subramanian, S 2014, 'A Novel Speech Enhancement Approach Based On Modified DCT and Improved Pitch Synchronous Analysis', American Journal of Applied Sciences(SciencePub), vol.11, no.1, pp.24-37.

[9] Naresh Grover, M.K.Soni" Design of FPGA based 32-bit Floating Point Arithmetic Unit and verification of its VHDL code using MATLAB"Information Engineering and Electronic Business, 2014

[10] Michael Nachtigal, Himanshu Thapliyal and Nagarajan Ranganathan,, "Design of a Reversible Floating-Point Adder Architecture," proceeding of the 2011 11th IEEE International Conference on Nanotechnology Portland Marriott August 15-18, Portland, Oregon, USA, pp.451-456,2011.

[11] Yee Jern Chong, Sri Parameswaran, "Configurable multimode embedded Floating point unit for FPGA," IEEE Transactions on Very large scale integration (VLSI) Systems, vol. 19 no. 11, 2011

[12] Balaji, V R & Thangamonikha 2016, 'Implementation Of An Adaptive Fir Filter Using High Speed Distributed Arithmetic", Proceedings of International Journal of Advanced Research in Biology Engineering Science and Technology (IJARBEST), Vol. 2, No. 10, pp. 191-201.

[13] X. Wang and B. E. Nelson, "Tradeoffs of designing floating-point division and square root on virtex fpgas," in Proceedings of the IEEE Symposium on Field-Programmable Custom Computing Machines, (Napa Valley, CA), pp. 195–203, April 2003.

[14] KKK.Deergha Rao, PP.V. Muralikrishna "FPGA Implementation of 32 Bit Complex Floating Point MultipliernUsing Vedic Real Multipliers with Minimum Path Delay" 5th IEEE Uttar Pradesh Section International Conference on Electrical, Electronics and Computer Engineering,(UPCON), 2018

[15] Balaji V R, Sathiya Priya.J 2018 "Enhancement of Speech Signal Using Modified Binary Mask Based Algorithm for Vehicular Noise", Journal of Advanced Research in Dynamical & Control Systems, Vol. 10, 12-Special Issue, 2018

IOP Conf. Series: Materials Science and Engineering **764** (2020) 012032 doi:10.1088/1757-899X/764/1/012032

[16] Vladik Kreinovich and Siegfried Rump, "Towards Optimal Use of Multi-Precision Arithmetic: A Remark", published in Reliable Computing, 12:365-369, 2006.
[17] Bhaskar J. "A VHDL Primer" Pearson Printice Hall Publication Third Edition 2007.
[18] A. M. Shams, T. K. Darwish and M. A. Bayoumi "Permormance Analysis of Low Power 1-Bit CMOS full adder cells", IEEE Transaction on VLSI Systems, Vol. 10, Feb. 2002.
[19] Rakesh Chowdary Gutta " Design and Analysis of one bit Hybrid full Adder using pass transistor Logic" IJPAM 2017
[20] Kazi Nikhat Parvin , Md.Zakir Hussain , Md.Ali Ghazi Islam "High precision arithmetic unit for biomedical and radar system Applications" by IEEE NetACT .978-1-5090-6590-5/17/ ©2017 IEEE.

IOP Conf. Series: Materials Science and Engineering **764** (2020) 012033 doi:10.1088/1757-899X/764/1/012033

Study of machining parameters of micro ECM on SS304 steel with nickel coated copper tool

K Gnanasekaran[1], K Ramesh[1], S Jagadeesh [1], K Narayanamoorthy[1] and D Elil raja[1]

[1]Department of Mechanical Engineering, St. Joseph's Institute of Technology, OMR, Chennai, India.

Abstract. The aim of present study is to determine the optimum parameters on machining the material SS304 and nickel-coated copper tool by micro electrochemical machining. The coating process is done by an electroplating method. The parameters involved in the process are feed rate (0.5,0.6,0.7) duty ratio (0.6,0.7,0.8) voltage (16,17,18). Taguchi approach is used to find out the optimum machining parameter and the Minitab 17software is used to achieve better results. Copper coated tool life is improved when compared to the normal tool. The material removal rates (MRR), tapered angle, circularity, overcut are calculated by measuring the values on the video measuring machine.

1. Introduction

Micro Electrochemical machining is one of the most recent and useful machining processes. In this process, the electrolysis method is used to remove the metal from a workpiece. The raising of demand for producing micro slices and structures in most of the industries, and also with quick improvement in micro-electro-mechanical systems (MEMS), F.KLOCKE et al.[1] perform research on titanium - nickel-based alloys which are mainly used for aero-engine parts. Results show about the performance of material properties was analyzed by using a scanning electron microscope (SEM). Abhishek Tiwari et.al[2] researched on to minimize the MRR error by using the fuzzy logic system. the minimized error for MRR by using the input parameters is voltage, feed rate mm/sec, and gap mm. Minghuan Wang et.al [3] studied the application of drilling a micro-hole on electrochemical machining .the results show that multiple holes can be done by using a single electrode using a matrix array based on machining the hole quality is studied. Vladimir M. Volgin et al [4] perform research on modeling and analyzing the process on machining .the results help us to study the changes of characteristics in the worked surface and also it helps to find the morphological of the machined surface. J.M. Park et.al [5] study of electrochemical machining and its application on micro machined parts and to study about machining properties. The results show that the chemical reaction between the tool and the workpiece were verified based on the machining parameters and the properties were tested and burr formed during machining is minimized.

IOP Conf. Series: Materials Science and Engineering **764** (2020) 012033 doi:10.1088/1757-899X/764/1/012033

2. Materials and Experimentation.

The workpiece used in this project is SS304, because its easily available and also it has wide range of applications in industry./figure 1 shows the SS304 material and figure 2 shows the microstructure of SS304. SS304 Material is widely applicable for gears, Pinions, Shaft, Spindle, Cranes, Rolling mill.

Figure 1. SS304 Steel

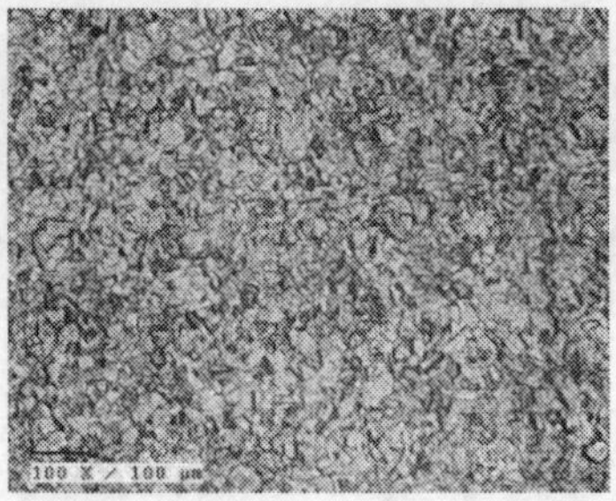

Figure 2. Micro structure of SS304

Table 1. Chemical composition of SS304

Element	Carbon	Manganese	Silicon	Sulphur	Nickel	Chromium	Phosphorous
Percentage	0.07	1.86	0.35	0.008	8.33	18.02	0.039

2.1 Selection of tool electrode

Copper is selected as a base tool for machining. Although, copper is widely used as electrode material in ECM due to it is high wear rate. The measurement of the electrode is 90 mm length and its inner diameter is 0.15mm and the outer diameter is 0.5mm. The hole section is generally made to provide the dielectric medium in the machining area for removal of burr also reduce the heat generation.

2.2 Selection of coating materials

The coating material increases the MRR and may reduce the tool wear rate. Here but we some improvement in the MRR and tool wear rate when compared with the normal tool, and also coating will help to increase the tool life. With the help of the above image, we can calculate the coating thickness, here base material was copper and coated material was a nickel. With the help of an electro less plating method nickel-coated above the copper. After coating the sample send through for SEM image, with help of SEM analysis calculate the thickness of the coating.

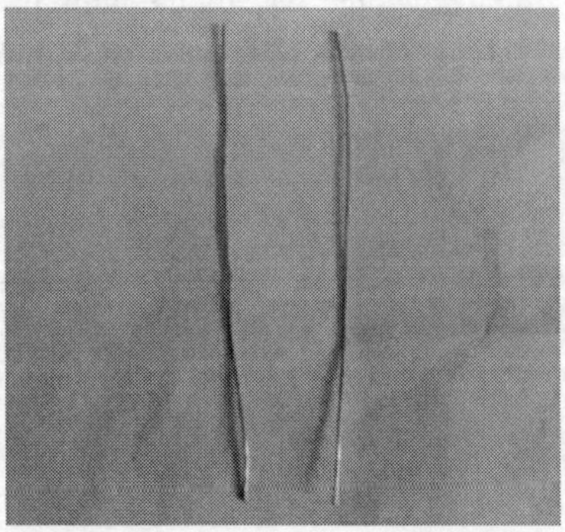

Figure 3. Nickel Coated tool image

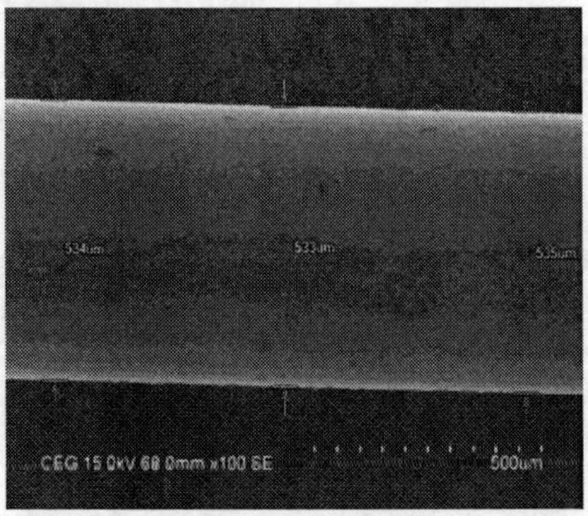

Figure 4. SEM image of Nickel Coated tool

IOP Conf. Series: Materials Science and Engineering **764** (2020) 012033 doi:10.1088/1757-899X/764/1/012033

2.3 Electrolyte

Figure 5. Sodium Nitrate

- **Symbol** : $NaNO_3$,
- **Molar mass** : 84.9947 g/mol
- **Melting point** : 308°C,
- **Density** : 2.26 g/cm^3
- **Boiling point** : 380°C

Table 2. Parameters and its levels

Machining parameters	Units	Level 1	Level 2	Level 3
Voltage	V	16	17	18
Feed	mm/min	0.5	0.6	0.7
Duty ratio	-	0.6	0.7	0.8

Figure 6. ECM Machine

3. Results & Discussion

3.1 Determination of optimal parameters

The results of the drilling process were tabulated in Table 3. "Smaller-the-better" quality characteristic was used to find the tapered angle, circularity& over cut and larger-the-best was used to find maximum MRR.

IOP Conf. Series: Materials Science and Engineering **764** (2020) 012033 doi:10.1088/1757-899X/764/1/012033

Table 3. Results for coated tool

S.No	Feed rate (mm/min)	Volt (V)	Duty ratio	MRR (mm^3/min)	Over cut of coating tool (mm)	Taper angle of coated tool	Circularity (mm^3/min)
1	0.5	16	0.6	0.0245	0.0904	4.460	1.060
2	0.6	16	0.7	0.1078	0.0739	4.118	1.025
3	0.7	16	0.8	0.1278	0.0659	5.710	1.039
4	0.5	17	0.6	01.599	0.0639	5.767	1.015
5	0.6	17	0.7	0.1487	0.0654	6.107	1.021
6	0.7	17	0.8	2.6560	0.1119	5.483	1.034
7	0.5	18	0.6	0.1350	0.1059	6.760	1.152
8	0.6	18	0.7	0.1383	0.0719	7.857	1.041
9	0.7	18	0.8	0.1459	0.1154	6.503	1.040

3.2 Signals to noise ratio

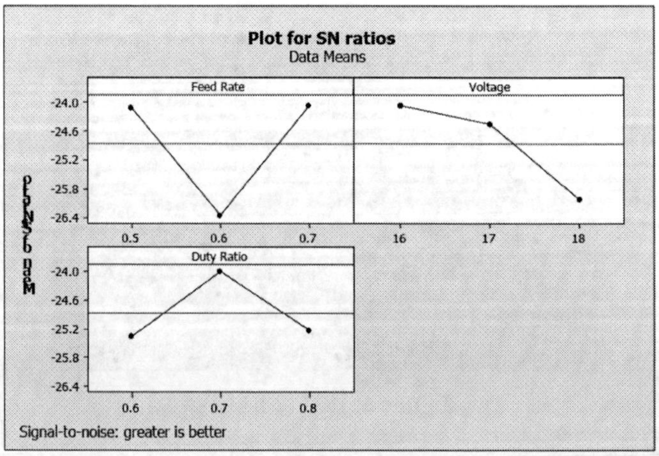

Figure 7. MRR

From figure 7 .The results shows that greater the better quality characteristic The S/N ratio is help to obtain the rank influencing parameters on MRR ,overcut ,tapered angle &circularity "larger-the-better" quality characteristic is help to find the optimum process parameters that affect the MRR.

Followed by feed rate is highly influenced by the voltage & duty ratio .the optimal parameter for MRR is duty ratio is 0.7, voltage is 16, feed rate is 0.7.

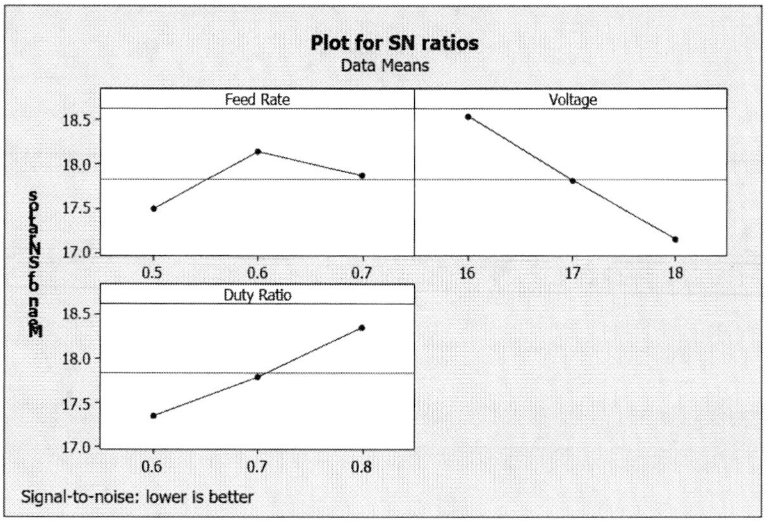

Figure 8. Over Cut

From figure 8. The results shows that lower the better quality characteristic The S/N ratio is help to find the rank of influencing parameters on the, overcut "lower-the-better" quality characteristic is used to find the optimal process parameters that affect the overcut . Followed by duty ratio is highly influenced by the voltage & feed rate the optimal parameter for overcut is duty ratio is 0.8,voltage is 16, feed rate is 0.6.

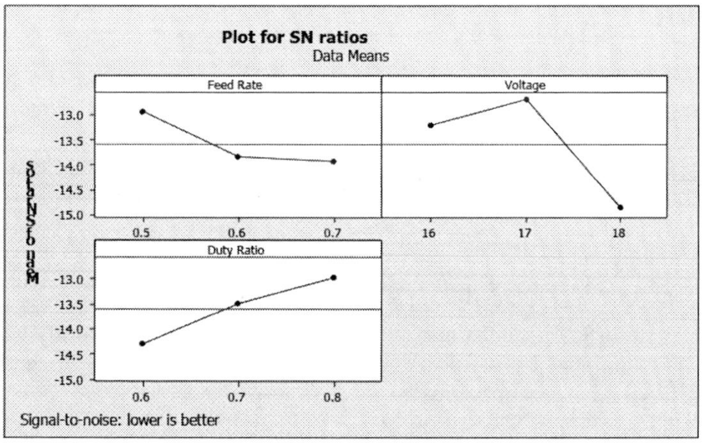

Figure 9. Circularity

From figure 9.The results shows that lower the better quality characteristic The S/N ratio is help to find the rank of influencing parameters on circularity "lower -the-better" quality characteristic is used to find optimal process parameters that affect the circularity . Followed by duty ratio is highly influenced by the voltage & feed rate. the optimal parameter for circularity is duty ratio is 0.8,voltage is 17, feed rate is 0.5.

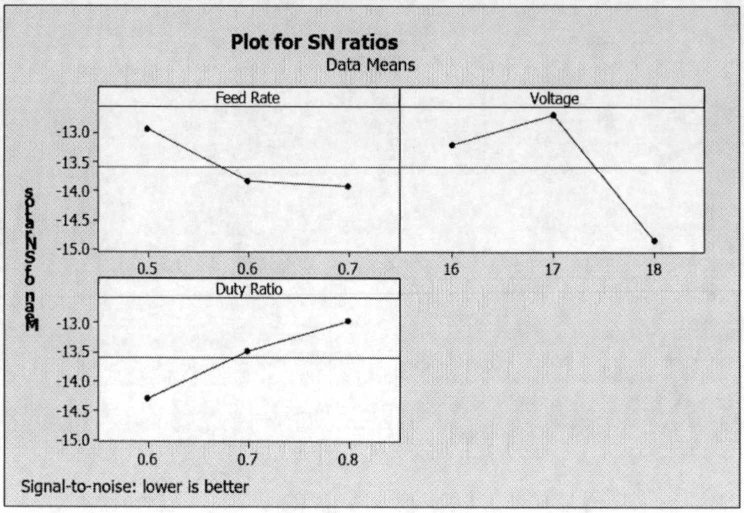

Figure 10. Tapered Angle

From figure 10.The results show that lower the better quality characteristic The S/N ratio is help to find the rank of influencing parameters on the, overcut "lower-the-better" quality characteristic is used to find optimal process parameters that affect the overcut. Followed by duty ratio is highly influenced by the voltage & feed rate the optimal parameter for overcutting is duty ratio is 0.8, voltage is 16, the feed rate is 0.6.

3.3 Confirmation test

The confirmation experiment is done with optimum process parameters. The predicted MRR was calculated by using a linear regression model. The linear regression equation obtained using MINITAB 17 is given below.

MRR = 0.166 - 0.0054 Feed Rate - 0.00631 Voltage + 0.0047 Duty Ratio (1)

The confirmation experiment is done with optimum process parameters. The predicted TAPERED ANGLE was calculated by using a linear regression model. The linear regression equation obtained using MINITAB 17 is given below.

TA = - 2.67 + 2.70 Feed Rate + 0.525 Voltage - 4.30 Duty Ratio (2)

The confirmation experiment is done with optimum process parameters. The predicted CIRCULARITY was calculated by using a linear regression model. The linear regression equation obtained using MINITAB 17 is given below.

CIRCULARITY = 1.58 - 0.005 Feed Rate - 0.0500 Voltage + 0.237 Duty Ratio (3)

The confirmation experiment is done with optimum process parameters. The predicted OVERCUT was calculated by using a linear regression model. The linear regression equation obtained using MINITAB 17 is given below.

OVER CUT = 0.0222 - 0.0267 Feed Rate + 0.0102 Voltage - 0.0717 Duty Ratio (4)

IOP Conf. Series: Materials Science and Engineering **764** (2020) 012033 doi:10.1088/1757-899X/764/1/012033

4. Conclusion

The Micro-ECM experimental work was conducted on SS304 steel workpiece using a Nickel coated Copper tool. The material removal rate, tapered angle, overcut, along with circularity is compared based on coated tool. In Micro-ECM the current is directly equivalent to the material removal rate.

- MRR is improved by the nickel coated electrode compare to the uncoated electrode. The nickel coated electrode produces high Material removal rate due to its low electrical conductivity.
- Life span of the tool is improved by the coated tool compare to the uncoated tool.

The optimal process parameter for the nickel coated copper tool MRR the feed rate is 0.7, voltage is 17, duty ratio is 0.8.followed by for tapered angle feed rate is 0.6,voltage is 18,duty ratio is0.7, for circularity feed rate is 0.5 voltage is 18 duty ratio is 0.6 and for overcut feed rate is 0.7 ,voltage is 18, duty ratio 0.8.

5. References

[1] Ferraris E, Reynaerts D and Lauwers B 2011 Micro ECM process investigation and comparison performance of AL_3O_2 and ZrO_2 based ceramic composite Journal of Manufacturing Technology **60** pp.235-238

[2] Ferraris E, Castiglioni V, Ceyssens F, Annoni M, Lauwer B and Reynaerts D 2013 ECM drilling of Ultra- high aspect ratio micro holes with insulated tools Journal of Manufacturing Technology **62** pp.191-194

[3] Rie Tanabe, Yoshiro Ito and Naotake Mohri 2011 Development of peeling tool for Micro EDM Journal of Manufacturing Technology **60** pp.227-330

[4] Tong hao, Li Yong, Zhan long and Li Baoquan 2013 Mechanism design and process control of micro ECM for drilling spray holes of diesel injector nozzles Journal of Precision Engineering **37** pp.213-221

IOP Conf. Series: Materials Science and Engineering **764** (2020) 012034 doi:10.1088/1757-899X/764/1/012034

Voice inspired entrenched arrangement for electrical utilizations

Vadivel.M[1], Sinivasan.M[1], Senthamilselvan.R.M[1], K.Rajalashmi[1]
Department of Electrical and Electronics Engineering, Bannari Amman Institute of Technology, Sathyamangalam, India – 638401

Abstract: The ultimate of this voice-controlled speech activated embedded system is to build a better companion for users. In this prototype was developed Remote control system for electrical appliance. This voice-controlled system will help in controlling different types of loads available. Voice controlled home appliances system that uses computers or mobile devices to control the appliances based on the command from mobile application. This application will enable a connection with any Arduino that involves a Bluetooth module. This technique of operating the home appliance is said to be safer one for the handicapped and aged peoples. Sprinkler controller which allows o connect to plant sensors and weather stations to help better manage of garden. This can be operate the appliances with less requirement of manpower.

Key words: Remote control system, Speech activated, Electrical appliances

1. Introduction

This paper explains the design and implementation of Arduino based Voice controlled embedded system for electrical appliance. Turn off the devices when not in use and to reduce the power consumption, which will reduce the e-bill. So voice control allow to control the appliances from the place [1-4]. Voice controlled home appliances refers to any devices, appliances or systems, which are connected to a common network that can be controlled remotely and independently [5-9]. This system are used for controlling indoor and outdoor lights, entertainment system and to control electrical and electronic appliances. This device can respond with speech and information is shown in the driver display.

2. Block diagram

The Arduino UNO acts as an integrated computer system and controls all subsystem activities. It is interfaced with a Bluetooth module, relay, and voice recognition Arduino application. The Arduino monitor the status of all devices regularly. If user voice is recognized, Bluetooth module will receive the user voice command and send the command Arduino UNO. The command matches with the code, there will be the output for that command.

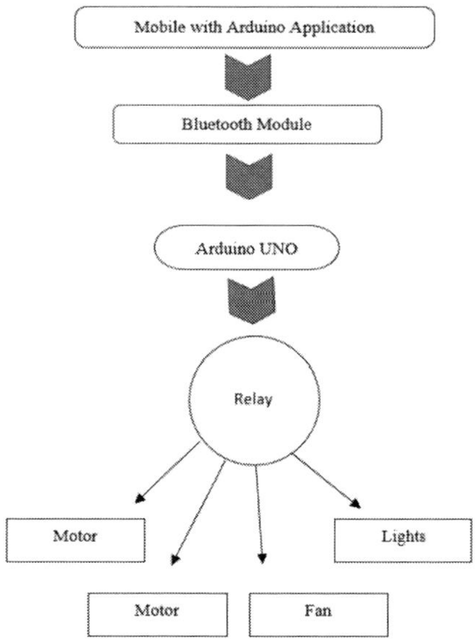

Figure 1. Block diagram of proposed system

Figure 1 shows the conceptual system block diagram.

3. Circuit diagram

The proposed system used the following components

1. Arduino UNO
2. HC-06 Bluetooth Module
3. Arduino Bluetooth Voice Controller (mobile app)
4. Electrical appliances
5. Relay

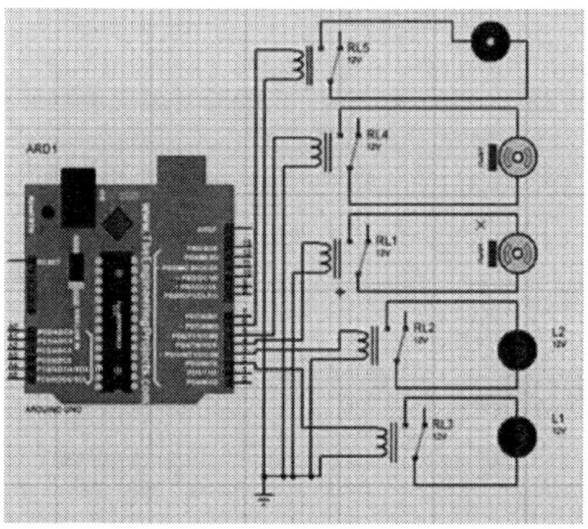

Figure 2. Circuit diagram of proposed system

IOP Conf. Series: Materials Science and Engineering **764** (2020) 012034 doi:10.1088/1757-899X/764/1/012034

The supply voltage of '5V' is given to Arduino via cables or adapter. Once, the Bluetooth module is on by supply given to the Arduino, switch on the Bluetooth in the mobile having Arduino voice controller application. In addition, pair with the corresponding Bluetooth source by giving the default password ("1234"). Tap the voice recorder option in the application in the mobile. Command the corresponding command to operate the home appliances. The command is sensed by the Bluetooth module, which is given as input control to the Arduino board. The output of Arduino is based on the input signal to the Arduino by the Bluetooth module. The output electrical signal is used to control the relay, which is an electronic switch. This is connected to the electrical appliances. Based on the voice command, the particular pin in the Arduino will generate the output electrical signal. This enable the relay to be act as switch. So that particular devices will be operated.

4. Flow chart

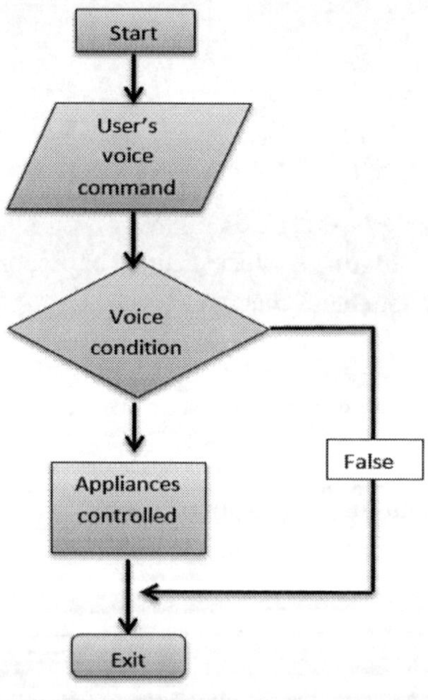

Figure 3. Flow chart of the proposed system

Figure 3 shows the flow chart of the proposed system. When the program is started, it will get the user voice command as input. Then it will check for the correct match. If the command is matched with the condition, body of the particular if statement is executed. That is, appliances is operated by user's voice.

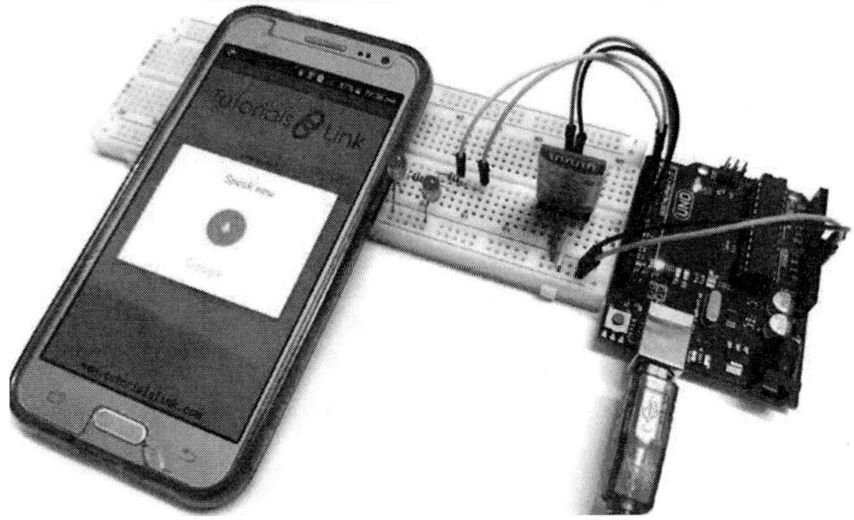

Figure 4. Hardware module of the proposed system

Figure 4 shows the hardware module of the proposed system.

5. Result & discussion

When the user's command is said the particular devices connected to the system will turn on or turn off.

Figure 5. Sprinkler controller using proposed system

IOP Conf. Series: Materials Science and Engineering **764** (2020) 012034 doi:10.1088/1757-899X/764/1/012034

Table 1. Voice command

VOICE COMMAND	PIN ENABLED	OUTPUT	
Turn on all light	Pin – 2,3	Appliance connected to pin 2,3 will be turned on.	Light 1 & Light 2 ON
Turn on a1	Pin – 2	Appliance connected to pin 2 will be turned on.	Light 1 ON
Turn on a2	Pin – 3	Appliance connected to pin 3 will be turned on.	Light 2 ON
Turn on all fans	Pin – 4,5	Appliance connected to pin 4,5 will be turned on.	Fan1 & Fan 2 ON
Turn on b1	Pin – 4	Appliance connected to pin 4 will be turned on.	Light 1 ON
Turn on b2	Pin – 5	Appliance connected to pin 5 will be turned on.	Light 2 ON
Turn on c1	Pin – 6	Appliance connected to pin 6 will be turned on.	Motor ON
Turn off all light	Pin – 2,3	Appliance connected to pin 2,3 will be turned off.	Light 1 & Light 2 OFF
Turn off a1	Pin – 2	Appliance connected to pin 2 will be turned off.	Light 1 OFF
Turn off a2	Pin – 3	Appliance connected to pin 3 will be turned off.	Light 2 OFF
Turn off all fans	Pin – 4,5	Appliance connected to pin 4,5 will be turned off.	Fan1 & Fan 2 OFF
Turn off b1	Pin – 4	Appliance connected to pin 4 will be turned off.	Fan1 OFF
Turn off b2	Pin – 5	Appliance connected to pin 5 will be turned off.	Fan 2 OFF
Turn off c1	Pin – 6	Appliance connected to pin 6 will be turned off.	Motor OFF
Turn off all	Pin – 2,3,4,5,6	All the appliance will be turned off.	Light 1, Light 2 , Fan1, Fan 2 , Motor OFF

Table 1 show the voice command to control the electrical appliances of fan, light and motor.

The command get executed the corresponding pin of the Arduino will give the output, while other will not. To control the devices using voice control, connect the devices to suitable pin of Arduino for that particular command. This allow the user to operate that appliance for that command only, if not some other appliance will be operated.

5.1Advantages

Provides safety to us from electric short circuit while using the conventional switches to operate loads.

- The cost is said to be minimum and it is considered as reliable one.
- By using this type of systems, we can save much time to operate home appliances from anywhere.

- It will enable the handicapped and old peoples to operate the electrical instruments easily.
- The manpower required for the operation of instruments is greatly reduced.

5.2 Disadvantages

- Limited to shorter distance only.
- Mostly voice recognition software will not put your words accurately on the input screen.
- If not maintain properly it may go wrong.

6. Conclusion

This voice controlled home control system will help in controlling different types of loads available. The major pros are man power reduction, cost efficient. This automation can be applied in various automation applications such as home automation, automatize military, industry automation, health care, transport...etc.

References

[1] Gerhart J 1999 Home automation and wiring. Mcgraw-hill. (ISBN 0070246742)
[2] Mann W C (Ed) 2005 Smart technology for aging, disability, and independence: The state of the science. John Wiley & Sons.
[3] Nave R May 2014 Hyper physics Microphones Retrieved 2014 May
[4] Yerrapragada C & Fisher P S 1993 Voice controlled smart house. In International Conference on Consumer Electronics 1993 Digest of Technical Papers ICCE (pp 154-155) IEEE
[5] Mohamed Abd, El-LatifMowad, Ahmed Fathy and Ahmed Hafez Smart Home Automated Control System Using Android Application and Microcontroller International Journal of Scientific & Engineering Research Volume 5 Issue 5 May-2014 ISSN 22295518
[6] Ming Yan and Hao Shi Smart Living Using Bluetooth Based Android Smartphone International Journal of Wireless & Mobile Networks (IJWMN) Vol 5 No 1 February 2013 DOI : 10.5121/ijwmn.2013.5105 65
[7] Chakradhar B, Krishnavenil S and Naresh D 2013 Bluetooth Based Home Automation and Security System Using ARM9 International Journal of Engineering Trends and Technology (IJETT) Vol 4 Issue 9 pp 4053-4058
[8] B Yukesekkaya, A A Kayalar, M B Tosun, M K Ozcan, and A Z Alkar A GSM, Internet and Speech Controlled WirelessInteractive Home Automation System IEEE Transactions on Consumer Electronics vol 52 pp. 837-843 August 2006
[9] D Brunelli, M Maggiorotti, L Benini, and F L Bellifemine Analysis of Audio Streaming Capapbility of Zigbee Networks EWSN 2008 2008 LNCS 4913 pp 189-204

IOP Conf. Series: Materials Science and Engineering **764** (2020) 012035 doi:10.1088/1757-899X/764/1/012035

Triphibian - an urban future transportation system

Prakash Eswaran[1,a], Balamurugan Tamilmani[2,b], Rarity Karna D[3], Mohamed Arif A[4], Susheel R N[5], Saravana Kumar B[6] and Ramachandra Raju K [7]

[1]Department of Mechanical Engineering, Bannari Amman Institute of Technology, Sathyamangalam

[5]Department of Electrical and Communication Engineering, Bannari Amman Institute of Technology, Sathyamangalam

[a]prakashe@bitsathy.ac.in, [b]balamurugantamilmani@outlook.com

Abstract. The future requires a new concept of saving the energy besides to save the nature for maintain the sustainable resources. A single vehicle can transform into a car, a helicopter, a boat and a submarine by utilizing the parts which taken from used parts. This concept vehicle can be used as a multi utility platform which encompasses latest technologies like image processing and artificial intelligence for motion control. The concept of connected vehicle and handless driving with electric driven is introduced for road ways which helps in reduction in size of the vehicle. Self-driving feature is powered by nexus of electric and solar energy. Proposed concept supports to rescue people in endangered areas and a better alternative for public transports.

Keywords: Multipurpose Vehicle, Save Energy, Waste Utilization, Image processing and Artificial Intelligence, Self-driving and Public transport.

1. Introduction

In modern days the surging consciousness of pollution and energy shortage crises, sectors in automobiles and motorcycles are no longer the best for transportation.[1]As the price of petroleum products raising now-a-days, there comes a need for cheaper and more efficient form of transport. In addition, saving energy in order to determine the problem of remnant fuel depletion is becoming increasingly important. Electrical vehicle technology as a step towards fulfilling these goals. An amphibious aircraft or amphibian is an aircraft that can take off on both land and water. In the underwater the vehicle moves as like recent submarine due to the arrangement of fins inspired from the bio mimics like shark fish. The Mono wheel is the weightless device that help the users for travel in local areas [2]. The design vehicle conceptually as well as aesthetically made with better simulation work using Ansys for testing in land, water and air. According to different modes of transformation, this vehicle is transformed into Car, Boat, Submarine, and Helicopter. In the air the vehicle moves as like the helicopter but the systematic arrangements are totally unique [3] . From the stable telescopic setup of a vehicle promotes the up thrust force for upward movement. In the water the vehicle moves as like a boat, the main parts are airbags for floating in the water the propeller along with turbine helps for direction control and speed control the telescopic system come from the side

IOP Conf. Series: Materials Science and Engineering **764** (2020) 012035 doi:10.1088/1757-899X/764/1/012035

surface of the vehicle helps to land the setup to perform those operations. Example, In urban areas like Chennai, Bangalore, Mumbai and Kolkata the usage of automobiles are very higher in these areas the Triphibian make a big change.

2. Existing design

In the era there are lot of vehicles even some special automobiles are design and developed .But they are either possible in land or water or air ways alone. Mixing of all these are very rare cases. The below figure are mixing of some two modes and single mode like land and air ways, air and water ways, or water ways alone in Figure 2.3 . Classic landing craft in figure.2.2 are not amphibious vehicles as they do not offer any real land transportation at all, although they are part of amphibious warfare. Ground effect vehicles, such as ekranoplans, will likely crash on any but the flattest of landmasses so are also not considered to be amphibious vehicles. The Harlingue monowheel in April 1914 in figure.2.1.The proportions of the propeller look even more extreme in real life- it doesn't look like a standard aeroplane propeller, it expected to produce thrust at a lower forward speed. The round thing at the rear is a roller- when stationary the machine rested on that and the side skids. Once it starts moving the driver would lean forward to raise the roller from the ground. Those vehicles facing problems on cost, efficiency, usability, ergonomics, less user friendly more and more not accessible by all. The Triphibian is sketched and modelled such a way that possible to well and good at land, air, and water and even in under water.

Figure 2.1

Figure 2.2

Figure 2.3

3. Proposed design

Figure 3.1. shows the Isometric view of the Triphibian. The materials are sourced specifically with varying environment conditions and mechanical strengths like abrasion, torsion and payload .The sustainable consideration that 70% these materials for assembly are directly taken and reused from the stores of old as well as scrab areas and rest are also possible with largest recycling technology. The examples, the propeller from the helicopter, the electrical system like lights ,battery, switches even PDM from the damaged super cars with some add-ons in autonomous control with smart python coding are implemented.

The normal Lander which is running in the roadways as like every vehicle. In case of some natural calamities like sudden flood or disaster with appropriate water trapped in a city at the time the Triphibian is transforming by the alarming to driver and the airbag comes out with the telescopic arrangement from the sideways and perform the swimming action. The steering system is carried out with the design of propellers inside the airbags. The vehicle also able to submerged by sucking out the air and utilize the fins for proceed inside the water through high viscosity of the liquid.

IOP Conf. Series: Materials Science and Engineering **764** (2020) 012035 doi:10.1088/1757-899X/764/1/012035

Other than submerged driving while in top surface the vehicle have to drive smoothly by using the designed air bag out using joystick operation at driver cabin shown in figure.3.2.Based on the surrounding environment the structure modification occur.

While in need of flying, the concept of heliport setup is modeled with beautiful propellers at the side and rear portion. In normal mode the vehicle comes to close setup of propeller at side and rear, two tyre make the point contact with road promotes the land operation. The controller module is embedded with extend reality system make the hologram view for the front, back and side view is processed by image processing technology. The hand/ leg operating arrangement is designed in better ergonomic way.

Figure 3.1 Isometric view

Figure 3.2 Cockpit view

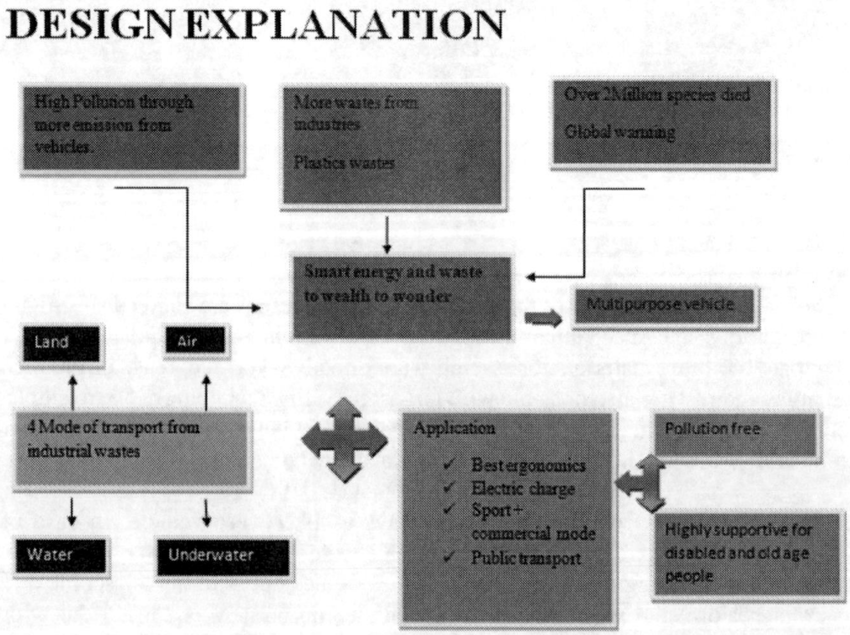

Figure 3.1.1 Concept of the product

IOP Conf. Series: Materials Science and Engineering **764** (2020) 012035 doi:10.1088/1757-899X/764/1/012035

3.1. Product development

The flowchart in figure.3.1.1 shows that the concept of the product from initial sketch to product development and applications. Automobile industries get benefitted from pollution and able to quick recharge facility with Li-Ion battery setup below the driver and passenger cabin. The materials using or fabricated not requires any special machinery. It is already available in the market more and more the used parts from truck, buses, helicopter, and ships through 3R(Reduce Reuse and Recycle) policies .It is employed by reduce the production of materials and reusage the materials from automobiles thereby recycle the entire process to get environmentally friendly vehicle.

Table 1. Bill of materials

S.No.	PARTICULARS	MATERIAL	QUANTITY
1	SPHERE	FIBER GLASS	1
2	WINDOW	FIBRE GLASS	2
3	CHASIS	STEEL ROD	1
4	TYRE	HARD RUBBER	(2)(2)
5	WING PROPELLER	AL-MG ALLOY	2
6	WING FRAME	POLYMER	2
7	TELESCOPIC SUSPENSION	CHROME ALLOY	2
8	TAIL PROPELLER	AL-SS ALLOY	2
9	TAIL FRAME	AL-SS ALLOY	2
10	MAIN TAIL	STAINLESS STEEL	1
11	HEAD LIGHT	LED	2
12	FINS	SHEET METAL	2
13	SEAT	SPONGE(POLYURETHANE FOAM)	4
14	LCD SCREEN	LIQUID CRYSTAL	1
15	JOYSTICK & PEDALS	PLASTIC	2
16	SWITCH	BUTTON	10
17	MOTOR	DC & BRUSHLESS(TITANIUM)	(2)(2)(2)
18	GEAR	CAST IRON	6
19	DRIVEN SHAFT	STEEL ROD	2
20	BATTERY	ULTRA CAPACITOR(LI-ION)	10000

4. Dimensional details

The multiutility vehicle consists of outer body made up of fiber reinforced polymer and airbag ,propeller, telescopic system requires high elastic material for closing and opening of the side support for flying as well as movement in water. The tyres which are attached with rim by force fit with spherical body at the mid plane of the vehicle thereby support the entire vehicle to withstand the overall weight and also helps to balancing the vehicle .The LED (Light Emitting Diode) and headlights designed during dark, warning and under the water conditions, then the passengers and driver seats can be changed according to the requirement, transparent solar panel and circuit connections which are interconnected with one another.

The transformation of the vehicle are given below:

4.1. Construction details of car

Software used - CAD/CAS-2D Sketch, 3D model, simulation, rendering, flow simulation, electric works, visualization all done using SolidWorks.

IOP Conf. Series: Materials Science and Engineering **764** (2020) 012035 doi:10.1088/1757-899X/764/1/012035

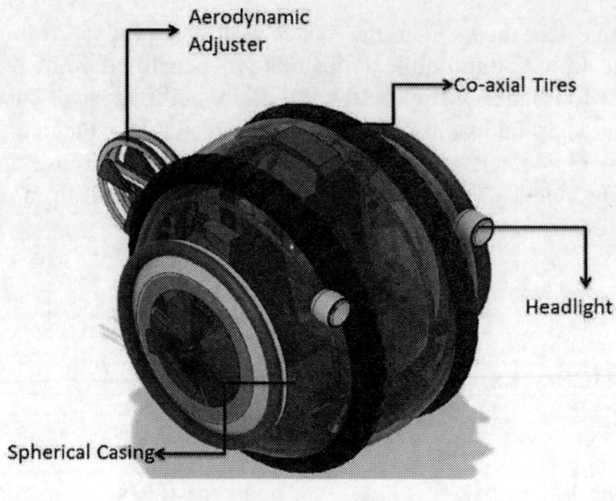

Figure 4.1.2 Front view

For the land motion consists of two number of coaxial tyres for moving forward and backward even in 360degree shown at Figure .4.1.2. This movements can controlled by either hand lever or pedal arrangement. The vehicle balancing is controlled by hover ring attached at the back (Air dynamic adjuster).

4.2. Construction details during float

In The water the vehicle moves as like a boat, the main parts are airbags for floating in the water the propeller along with turbine helps for direction control and speed control the telescopic system come from the side surface of the vehicle helps to land the setup to perform those operations as shown in Figure .4.2.1.

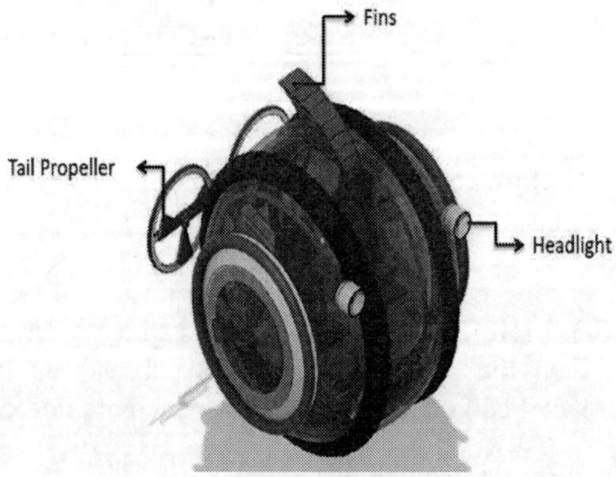

Figure 4.2.1 Isometric View when submerged

IOP Conf. Series: Materials Science and Engineering 764 (2020) 012035 doi:10.1088/1757-899X/764/1/012035

4.3. Construction details during submerging

Figure .4.3.1 Backview

Figure .4.3.2 Side view

In the underwater the vehicle moves as like recent submarine due to the arrangement of fins inspired from the bio mimics like shark fish in Figure .4.3.2. While moving in the underwater the total weight of the vehicle is balanced and controlled by propeller which gives the thrust for moving .This is shown the two rings located at the Figure 4.3.1.

5. Construction details during fly

In the air the vehicle moves as like the helicopter but the systematic arrangements are totally unique. From the stable telescopic setup arise from the sides of a vehicle promotes the up thrust force for upward movement shown in Figure .5.1.

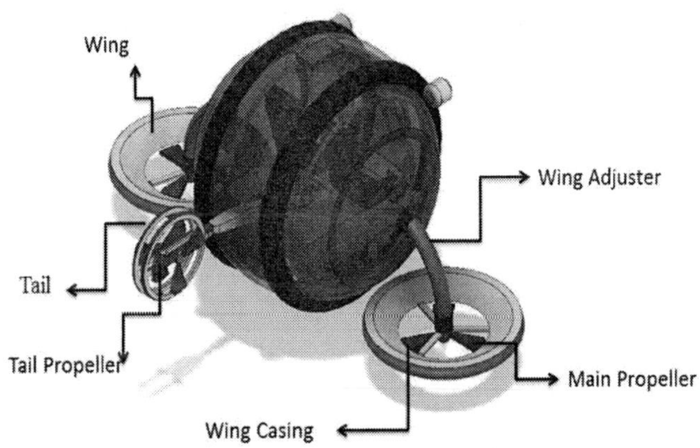

Figure 5.1 Isometric View during float

6. Result and discussion

The result of the proposed solution provides the alternative solution for driving the vehicle at their wish not to worry in any cause at flood, or any situations. It make more advantage in the field of military areas to rescue the people from endangered areas. The seating capacity can be increased and can be used like public

IOP Conf. Series: Materials Science and Engineering **764** (2020) 012035 doi:10.1088/1757-899X/764/1/012035

transport with smart mobilty systems. It add the further advantage of the outer body is made with the solar powered glass able to absorb the solar energy thereby saving it and can be used for the instant or later usage.

7. Conclusion

Thus the three mode of transportation vise land, air and water even in underwater is satisfied by saving energy in eco-friendly manner as well as using the best transmission with electrical system. The design criteria of time efficient cost, by saving the fuel and pollution free are defined. While consider for usability the vehicle can be used for across the year not seasonally, here the ultimate way of mounting and dismantling is cleared. Significantly, passenger safety is more considered here. In traffic as well as parking, the dimension of the vehicle is optimized. It is completely new medium of transportation with versatility.

8. Future scope

The future scope of utilizing the all energy from the renewable sources thereby reducing the fuel power and cost to make everyone accessible. The technology of artificial intelligence and image processing is going to be interlinked with Extend reality to experience the customer satisfaction before buy this can test the ride and feel the comfort with free of cost.

9. References

[1] Anup M Gawande, Akshay P Mali 2016 Amphibious Vehicle International Research Journal of Engineering and Technology Vol 3 Issue 10
[2] ADM Klunder, Prewitt R 1945 Submarine technology with undersea warfare Basic Factors of Helicopter Design SAE Technical Paper 450215
[3] Xiaolin Xie, Feng Gao, Chuan Huang, Wen Zeng Design and development of a new transformable wheel used in amphibious all-terrain vehicles (A-ATV) Journal of Terramechanics Vol 69 February 2017 Pages 45-61
[4] A P Khode, K Senthilkumar, Bhushan S Patil, Nagaraj Kulkarni, M W Trikande Shape 2017 Optimization And Weight Reduction Of Seat Structure For Wheeled Armoured Amphibious Combat Vehicle Materials Today: Proceedings Vol 4 Issue 2 pp 1917-1926
[5] Sebnem Helvacioglu, Ismail Hakki Helvacioglu, Burak Tuncer Improving the river crossing capability of an amphibious vehicle Ocean Engineering, Vol 38, Issues 17–18 2011 pp 2201-2207

IOP Conf. Series: Materials Science and Engineering **764** (2020) 012036 doi:10.1088/1757-899X/764/1/012036

Product design and development of googles and hand stick for blind people

D Rajith Karnal[1], R Rajiev[1], S Saravanan[1]

[1]Dept. of Mechanical Engineering, Bannari Amman Institute of Technology Sathyamangalam, 638401, Tamilnadu

Abstract. Project vision to world focuses on the field of medical science, which paves a way to the next level of medical technology. The theme of the vision to world is to bring back the vision to the blind people, with the help of this gadget. It is a blind helper device. The Gadget is engaged with UV emitter and receiver, sound senor, LED, circuit board, spectacles. It can be used by all kinds of people. It doesn't need much knowledge. The device works by detecting the signals outdoor through the UV signals and that signals were converted into electric signals through which the sound and light signals were produced. The sound signal is used as an indicator to the blind and the light signal is used as an indicator to the normal people. It creates a mutual understanding with the blind and the normal people. This helps to rectify the visual problem of blind people. The gadget is in the shape of normal spectacles so it would be easier to carry everywhere. The origin of this device is bionic eye; the concept of bionic eye is to fix an electrode in the eyes. But in order to rectify that this project vision plays a major role in serving blind. The main objective is to give vision to the blind with artificial means without any side effects. Keywords. Visually Impaired, Goggles, Hand Stick, UV Sensor, Buzzer, And Micro-Controller.

1. Introduction

In the modern world, electronics plays a major role in all fields. Some of the fields like medical, space, science, and so on. This project holds on to the field of medical science. This is completely operated with electrical device. The project mainly focuses on the struggles of blind people. It holds a good hope for the blind people to lead a normal life. It is just a simple project but with more effective means. The technique used in this device is not that difficult, so that it could be handled by any engineers. The cost to manufacture the device is very much low when compared to other devices. Hence it is more affordable to the people of all criteria. This is just a small step taken to help the blind but in future it would be more powerful with new designs and technologies included in it. To help the visually impaired person several steps were taken. The following were the part of steps explained. When obstacle is ahead over the visually impaired people, automatically [1] the device helps to avoid the upcoming obstacles. Blind used to travel to different places from their native places, the travel guide [2] will help the blind people to move on to the exact location. The blind cannot read the text in front of them, as a text recognizing technology [3] will be very much helpful for the blind people to interact with the texts over the surroundings. The blind will suffer a lot inside their houses; hence to help the blind to walk in safely without hitting over the wall an indoor blind assistant [4] plays a major role in helping them. To make the gadget easily wearable [5], all the electronic components are attached over a belt for easy wearable. The neuro science paved the way for the development of brain waves that help the blind people to get back the vision with the help of new [6] neurological researches. To identify the person speaking with the blind people, the 2D and 3D face recognition system is useful. By identifying the depth of the face [7] the person's identity is verified and is known to the blind with the help of the device. An electronic aided device [8] greatly helps the visually impaired to move themselves safe over the surroundings. They were guided with the help of smart glasses [9] with ultrasonic sensors for identifying the obstacles. The blind device holds some sensors with the help of the sensor [10] the blind is guided to the direction. To rectify the problems faced by the blind peoples, therefore to make their life peaceful. This project has the power to change the view of society about the blind peoples. In order to give a helping hand to the blind, this is designed to show a way for blind to lead a better life.

2. Existing design

The principle mechanism of bionic eye figure 1 is to generate artificial visuals, by operating the nervous system by doing surgery. An electric plate is implanted inside the retina of the sensory system. It acts as a mediator between the surrounding and to the brain. This is the mechanism used in abroad for the welfare of blind people. As it mainly involves in the surgery process, more advanced technology would be required additionally. Some time it also becomes failure. The principle mechanism used by the eye stick figure 2 is to detect the stairs case while walking. It is capable of identifying the distance of the upstairs and sends the data according to that. This mainly helps only for climbing purpose and not for other purposes, for the blind people. The sensor used in it is expensive. Currently available devices were bionic eye, eye Stick, etc. Every device always has some drawbacks among huge benefits. The draw backs of the bionic eye are implanting an electrode inside the retina of eye, so that the cost of the implanting is very much high. The surgery done in the retina of eyes will cause some side effects in eyes. The drawbacks of eye stick are, it can't detect the water stokes on road sides and also drainages. This makes them to stumble when they step on these waters stokes or drain, they get fell down.

Figure 1. Bionic Eye

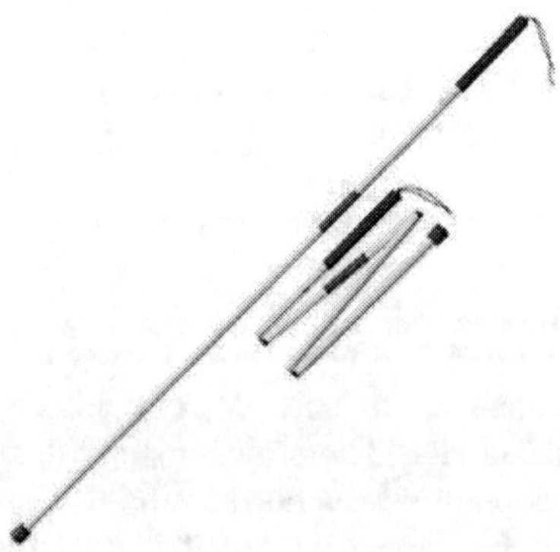

Figure 2. Eye Stick

3. Proposed design

The device V gear we have implied the mind signals which carries the digital signal produced by sensors and digital gadgets. Through this we can make the blind to get real visuals by artificial means. Additional

IOP Conf. Series: Materials Science and Engineering **764** (2020) 012036 doi:10.1088/1757-899X/764/1/012036

features like detection of water depths, detecting human bodies. In the device we have used several principles to enhance the quality of artificial vision helper. Doppler Effect takes a major role in the detection of humans or any object movements around the blind. Figure 3 photoelectric Effect is also used along with the Doppler Effect, which carries the digital signals. Our device holds the main unique of transferring signals from an artificial external device to natural internal device (brain) without causing any damage to the system. This is why it gives a unique identity among other all products. We have the concept of transmitting and converting one signal into other signal in to the brain, without doing any surgery. Also the stick figure 4 helps mainly in guidance of the blind people. That's the main highlight of this device. The component table 1 details are used for proper fabrication of the product. This concept is not an existing one. It is a new thought of innovation for blind people. It provides a solution, which other devices failed to give. It provides vision to blind people without any surgery or side effects. This solution shows difference among other existing devices.

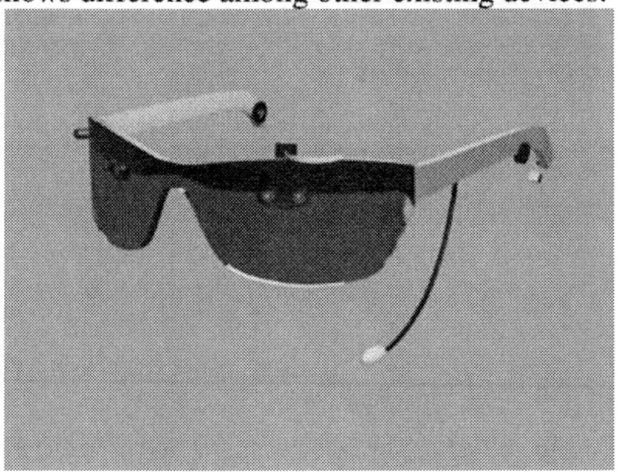

Figure 3. Google Design

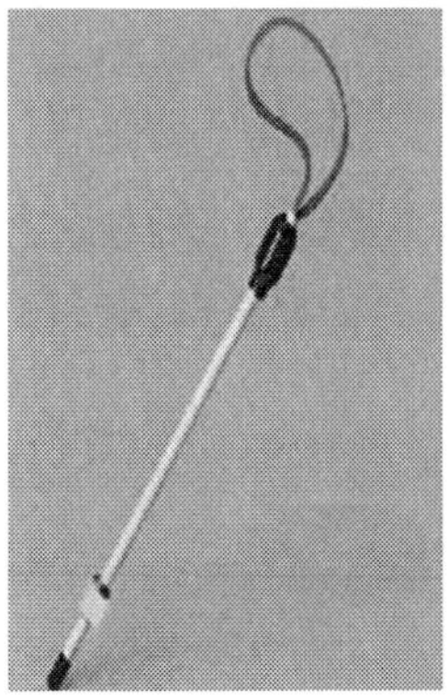

Figure 4. Hand Stick Design

3.1. Product development

3.2. Design of goggles

The main principle of this project is echo location; this is what bats use to catch prey. The design of the project is made on a simple spectacles design, which is most frequently used by all the people who have defects in their eye. Figure 5 shows that goggle holds all the circuit boards and other accessories. In the Figure 6 detects the object to a particular distance so that the objects at long distance will not interrupt the UV signals, hence forth there won't be any interference in the signals while transmitting and receiving. The source of ultraviolet radiation emitter and receiver is placed over the front face of goggle so that it covers the whole 180' angle, which is equal to the human eye.

Figure 5. Goggle

Figure 6. Ultraviolet emitter and receiver

When the circuit is on the source starts to emit UV signals when it strikes the objects in front, the signals were retreated and falls on the receiver, thereby it receives the signal which is transmitted from the object. These signals were converted in to electric signals and were feed to a Figure 6 Shows splits up the signals

IOP Conf. Series: Materials Science and Engineering **764** (2020) 012036 doi:10.1088/1757-899X/764/1/012036

according to the program. These signals make the work function of the sound generator and light emitting diode (LED).

Figure 7. Mother board

Figure 8. Sound Producer

The signal detected by the source from the object produces different frequency at particular distances. Figure 7 produces the sound at different tones with high pitch and low pitch. That helps in analyzing the distance of the object from the blind people. It produces a beep sound that will not make irritation to the blind. So, there is no defect while sound is produced. Figure 8 starts to blink when the blind is close to some obstacles; this indication of light helps normal people to understand that the person is blind and that would be easy for normal people to give way to the blind. The sound is also produced at the same time when the light starts to blink.

Figure 9. Light Emitting Diode

Figure 10. Power Source

Therefore, both the process takes place at a time and there won't be any mistakes by the surroundings. The main problem arises now only, the Figure 9 for the device must be rechargeable, must with stand for long duration time period. So that initially power bang is used at this stage but we can alternate a better source for the power distribution. But for better cost and for easy recharging power bang is used. The overall methods were taken from simple science, which is known wide. This project is a combination of all signals and sources that are simple and known.

IOP Conf. Series: Materials Science and Engineering **764** (2020) 012036 doi:10.1088/1757-899X/764/1/012036

Table 1. Bill of materials

S. No	PARTICULARS	SPECIFICATIONS	USES
1	ARDUINO NANO	Circuit Board, Microprocessor, and Resistor, Capacitor.	Acts as a mother board.
2	UV SENSOR	Circuit board, UV HC-SR04, IC chip.	Detects external specimens.
3	LED	Light Emitting Diode, Tungsten Filament.	Produces Light signal for identification
4	BUZZER	Copper Plates, Magnet.	Produces Sound signal for identification
5	COIL MACHINE	Copper Wires, Foam Sheet.	Indicates water level in pits.
6	VIBRATIONAL MOTOR	Magnet, Coil windings.	Generates vibration for identification.
7	PIPES	Poly Vinyl Chloride (PVC)	Holds the particulars of stick
8	SPECTACLES	Glass Frame, Lens.	Holds the particulars of Goggles.
9	BATTERY	Li – ion electrolyte	Acts as a Power supply.

3.3. Design of hand stick

The hand stick is mainly used for the detection of obstacles at the ground level, such as troughs, craters, water pits, lakes, etc. It uses vibration to indicate these obstacles, where a coil mechanism is used for the indication of water depth or flow.

Figure 11. Coil Machine

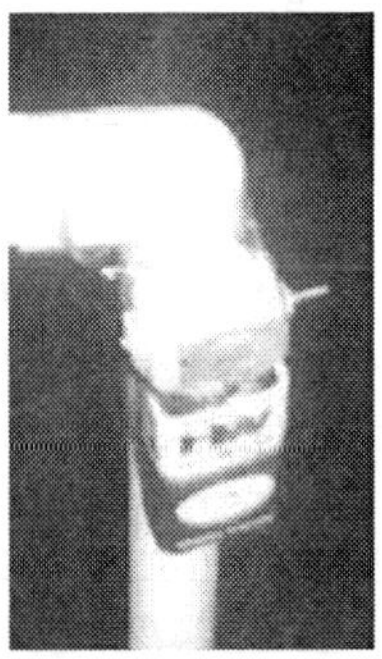

Figure 12. Vibrational Motor

The Figure 10 has a simple mechanism in it, in which two coils are placed using a foam sheet at a certain distance. One coil is fixed rigid, and the next coil is placed in a floating sheet. The two coils were connected in an incomplete circuit. When the stick enters the water, the floating sheet gets floated due to the density of sheet is less than the density of water. At that time the incomplete circuit gets completed and the power transmission takes place. The Figure 11 gives the indication to the blind that carries the stick with him. This works with the principle of electromagnetic induction, the repulsion between the magnet and the coil windings makes the vibration to be produced.

IOP Conf. Series: Materials Science and Engineering **764** (2020) 012036 doi:10.1088/1757-899X/764/1/012036

4. Result and discussion

The result of the project would be positive if it reaches the blind people figure 12 at right time and right place with all affordable commitments. The drawbacks present in this project would be rectified in the next stage of this project vision because the stages of this project won't stop until it gives a solution which will rectify all the problems of blind.

Figure 13. Complete Model

5. Conclusion

Here by I conclude that this is the start of a new era. This invention is just an initial level this will make the world turn around one day. This device will help the blind to get relief from all their struggles and sufferings. Compared to all other inventions in the field of creating a solution to solve the problem of blinds, this project would probably be the best of all. Even though all other projects were in advanced level they have much more side effects than this. Some inventions are pretty much expensive and some were too risky like operating the eye system and placing some electronic devices in to it, at times it may become very worst and would lead to the loss of eye. Though our device may have some disadvantages that are not harmful to the people and also to the environment, the risk factor is zero percent in this project. This device has the capacity to withstand at all stages and will get succeed in all ways. We have been still working in this field to improve its rate of efficiency and to make a real artificial visual device.

6. Future scope

In future the technology will be more advanced so that the technical facilities will be used wider and deeper. In future the mechanism, the UV signals which were received from the object by the ultraviolet radiation emitter and receiver is send to the mother board were these signals are transferred as electric impulses, so that those generated impulse were directly transmitted to the nerves system which is located at the back of the head. These signals which is transmitted to the nerves and enters to the eye system where is provides black and white image of the objects in the present world, detected by the UV signal. This type can be made colorful by using photocell to capture image as it is and transfer those signals into the mother board and then it enters in to the nerves of eyes and the brain will receive the actual image of the outer world like the same as natural eye, the rate of electric impulse passed is very much low so that it won't be harmful for the people.

7. References

[1] Surapol Vorapatratorn and Kowit Nambunmee, An obstacle warning device for the totally blind, Journal of Assistive, Rehabilitative & Therapeutic Technologies, 2014; 2.

[2] Ram Tirlangi, Ch. Ravi Sankar, Electronic Travel Aid for Visually Impaired People based on Computer Vision and Sensor Nodes using Raspberry Pi, Indian Journal of Science & Technology, 2016; 47.

[3] Tejaswani G, Afroz B, and SunithaS, A Text Recognizing Device for Visually Impaired, International Journal of Engineering & Computer Science, 2018; 7.

[4] Bing Li, Juan Pablo Muñoz,Xuejian Rong, Qingtian Chen, Jizhong Xiao, Yingli Tian, and Aries Arditi, Vision-Based Mobile Indoor Assistive Navigation Aid for Blind People, IEEE Explore, Volume: 18 Issue: 3.

[5] Eklas Hossain, Raisuddin Khan, and Ahad Ali, Design and data analysis for a belt-for-blind for visual impaired people, Int. J. Advanced Mechatronic Systems, 2011; 3: 315-328.

[6] Monica Gori, Giulia Cappagli, Alessia Tonelli, Gabriel Baud-Bovy, and Sara Finocchietti, Devices for visually impaired people: High technological devices with low user acceptance and no adaptability for children, Neuroscience & Biobehavioral Reviews, 2016; 69:79-88.

[7] Andrea, F. A., Michele, N., Daniel, Riccio, and Gabriele, Sabatino. 2D and 3D Face Recognition: A Survey, IEEE Computer Vision, Vol. 3; 345.

[8] Kailas Patil, Qaidjohar Jawadwala, and Felix Che Shu, Design and Construction of Electronic Aid for Visually Impaired People, IEEE Transactions on Human-Machine Systems, 2018; 48.

[9] Jinqiang Bai, Shiguo Lian, Zhaoxiang Liu, Kai Wang, and Dijun Liu, Smart guiding glassesfor visually impaired people in indoor environment, IEEE Transactions on Consumer Electronics, 2017; 63:3.

[10] Wafa Elmannai and Khaled Elleithy, Sensor-Based Assistive Devices for Visually-Impaired People: Current Status, Challenges, and Future Directions, Sensors Basel, 2017; 17:565.

IOP Conf. Series: Materials Science and Engineering **764** (2020) 012037 doi:10.1088/1757-899X/764/1/012037

Performance Analysis of Medium Altitude Low-Cost Autonomous Quadcopter

D Lakshmanan[1*], P Saravanan[2], P Vadivelu[1], D Nivitha[1], Yaswanth M S[1]

[1]Department of Aeronautical Engineering, Bannari Amman Institute of Technology, Sathyamangalam

[2]Department of Mechanical Engineering, Mahendra Engineering College, Namakkal

*Corresponding Author: lakshmanand@bitsathy.ac.in

Abstract. In modern engineering, the rapid growth in the development of unmanned aerial vehicles which deserves the major space in the area of intelligence and surveillance operations. Unlike the conventional aircraft, the number of rotors mounted in an autonomous body that determines the propulsive efficiency as well as the maneuvering capabilities of unmanned aerial vehicles. The main objective of this research reveals that the low cost and an autonomous aerial vehicle with four rotors configuration design to achieve the long endurance flight limit. The developed Quadcopter flying model integrated with Arduino autopilot module which effectively controls the various maneuvering actions during the flight. This paper also reveals the entire hardware structure of the designed Quadcopter and its flight dynamic response characteristics. The remotely operated Quadcopter model was subjected to the flight test and the performance parameters have recorded. The flight test was conducted in an open environment in which stability parameters has been observed during onboard. Pitch angle, the Rotational speed of four rotors and the various attitudes of Quadcopter were changed accordingly to maintain the steady level flight. Findings are plotted with respect to the flight endurance. **Keywords:** - Quadcopter, Maneuvering, Attitude, Flight test

1. Introduction

The special abbreviation of "DRONE" that is expressed as Dynamic Remotely Operated Navigation Equipment and which are trending the aviation market in recent times. Many aerial vehicles are booming in the aviation sector with various configurations. In general, the aviation hunters will always stick with the reliability, cost-effectiveness and the performance of Drones. In a Quadcopter Drone, the directed upwards rotors are constructed with square shaped frames with equally in the distance which are a reference to the center of mass of the body.(1) In general, the angular velocities acting in the rotors are mainly used to control the trajectory of the Quadcopter model.(2) As like conventional unmanned Aerial Vehicles, Quadcopter can perform search and rescue, inspections, traffic monitoring, aerial photography, terrain monitoring and several other applications. Recent days Quadcopter has gained more responses from military sectors and research scholars.(3) In the modern aerial vehicle industry, most of the rotor based autonomous vehicles are implemented with Integrated Modular Avionics package and Global Positioning System for the effective control of its flight profile in a 3-Dimensional region. In this research paper, a genuine practical approach is proposed to build and design of a prototype Quadcopter which is low cost and minimum power consuming avionics bay for the efficient endurance.

1.1 Structure of Quadcopter

The Quadcopter structure is assumed as rigid body configuration with rotors on each frame which produce angular velocities with respect to its clockwise and anticlockwise rotational effects.(4) The angular positions which determine the attitude of Quadcopter in an inertial reference frame. Pitch angle is always expressing the rotation of Quadcopter structure with respect to Y-axis. Rotational moment around the X-axis is explicated with Roll angle and the moment about Z-axis stated based on Yaw angle. (5) In general, Quadcopter has six degrees of freedom as three rotational and three translational motions which are developed from four independent rotors. The coupling effect of translational as well as rotational motion will provide the betterment in achieving six degrees of freedom during the flight test.

2. QUADCOPTER HARDWARE MODULE

The Quadcopter hardware model integration process was performed at the Flight-Sim Lab, at Bannari Amman Institute of Technology, Erode. The overall Quadcopter structure was built around the gross weight of 0.9 kg which can carry the payload capacity of 0.15gram. The designed model aims to operate at the maximum voltage of 12v and that can be sufficient for all four motors which were mounted on the Quad frame. Low cost and lightweight hardware have been utilized to achieve better thrust to weight ratio. (6) The entire Quadcopter structure was designed in consideration with reliability as well as low power consumption criterion to maintain the maximum endurance limit. The total cost of this Quadcopter was estimated at around $246. The total breakdown of the hardware components integrated with Quadcopter is mentioned in the Table. (1)

Table 1. Quadcopter components and Cost Estimation

S.No	Name of the Hardware Components	Price ($)
1	Rotobotix Frame F450 Integrated PCB version	11.16
2	Propeller pairs named REES52 1405	3.66
3	Electronic Speed Controller M30-A	8.66
4	Lithium-Polymer 3000 mAh (milliamp-hour), 12volt	33.86
5	Ardupilot 2.8 mega flight controller	45.28
6	GPS Telemetry	43.86
7	BLDC: 4xMotors(1400volt)/Accessories	38.01
8	Six channel: Transmitter plus Receiver	49.71
9	4xServo MotorSG90	8.77
10	3-Axis: Accelerometer and Gyroscope Module	2.68

2.1 Speed controller and the motor configuration

The small, low cost, four lightweight Brushless Direct Current motors are arranged in the "X" shaped Quadcopter frame. (7)The multicolored identified motor consists of fourteen motor windings with 13 stator arms which can be operated at 1400 Kv. (8)The lower kilovolt DC motors provide very high torque value even at lower rotational speed to push the propeller blades towards maximum thrust. The all four motors together were able to produce 1.5 kg of thrust to lift and create momentum about the Quadcopter model. The four motors are driving the fixed pitch propeller via the Electronic Speed Controller connected with Microcontroller unit. The reinforced carbon-fiber based propeller pairs named REES52 1405 were used to produce maximum rotational speed during the operation. The Electronic Speed Controller M30-A operated at the frequency 50 Hz was integrated and that will transfer the sufficient amount of electric energy to enable all the electronic components mounted in the Quadcopter structure.

2.2 Energy storage device

The operating level of remotely operated unmanned aerial vehicles is purely depending on the energy storage device. If the efficiency of the battery is not capable of storing and discharging the sufficient power, then the endurance level of an unmanned aerial vehicle will be unsatisfactory. This research paper reveals about the performance of 3 cell-series Lithium-Polymer 3000 mAh (milliamp-hour), 12v with the instantaneous discharge of 55A battery source was implemented in the Quadcopter Model. The major advantages of using Li-Po batteries are always lighter, smaller and lifelong as well as the maximum charging capacity when compared with Ni-Ca, Ni-MH, and Lead acid batteries. The designed Quadcopter model carries the Li-Po based battery with 184 gram of weight.(9)

2.3 Panel of Quadcopter Avionics

Heart of this Quadcopter structure was strongly built enough to control the trajectory with the help of reliable avionics panel. The Integrated Modular Unit consists of Ardupilot 2.8 mega flight controller with Global positioning system (GPS) tracker, 3-axis ADXL-345 series accelerometers, 3- Axis L3G4200D Gyroscopes with 16-bit resolution, receiver, and transmitter. The new version of Ardupilot APM 2.8 is capable of performing autopilot mode of operation in which altitude and coordinated can be controlled. The APM 2.8 mega microcontroller is one of the reliable open source autopilot system adopted with the latest sensor technology. (10)This GPS has an accuracy of 5 meters in position measurement which uses 45 milliamps at 6.5 volts and that is enabled with a Wide Area Augmentation System. The Radio Frequency enabled six channel receivers (433 MHz operating frequency) were integrated into flight control avionics unit.

When the input signal is transmitted from the ground controlled transmitter unit, the receiver reads the signal and sends it to the microcontroller. In this six-channel operation, the first four channels are uses as roll, pitch, and yaw and throttle control. The fifth channel is used to trigger the switch control to initiate the programming sequences of Quadcopter flight. The sixth channel is used for safe control of the ground handling process and finally, that will turn-off the closed loop operation.

Figure. 1. Assembled Low-Cost Quadcopter

2.4 Processing with Arduino Microcontroller

Arduino 2.8 mega microcontroller is basically inexpensive and powerful in nature also simple to perform programming sequences and provides tremendous benefits as compared to other Microcontrollers. (11)Arduino 2.8 mega operates at 16 MHz clock speed and input/output pins that can be used to integrate Electronic Speed Controller, Input Receiver, Global Positioning System, Storage Device Port, slots for connecting personal computer and other peripherals. The input receiver's signals are mapped with Electronic speed controller output.

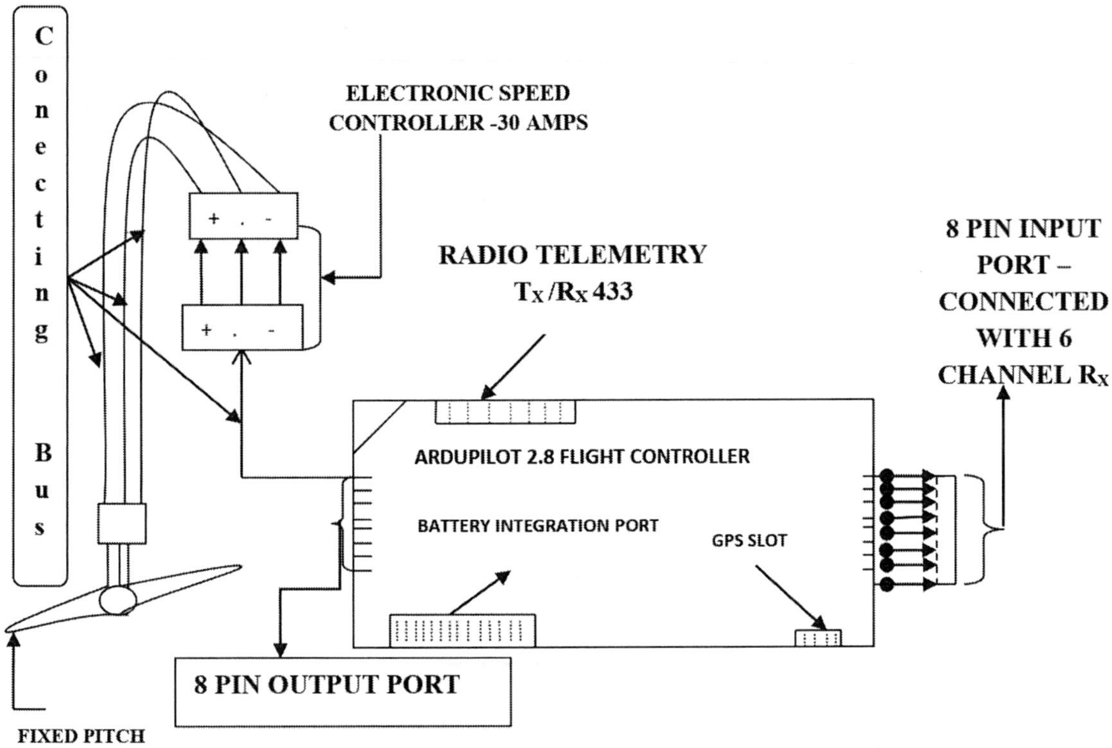

Figure.2. Representation of Quadcopter hardware assembly

Therefore the special output signals are sent from the Arduino board to Electronic Speed Controller which controls the motor speed. Measured data from Global Position System or Photography/Video or attitude information from gyroscope were stored in the Microstorage device mounted in the Arduino 2.8 autopilot module.

3. Dynamics of Quadcopter – "X" Shaped configuration

The four rotors mounted on the Cross symbol ("X") shaped Quadcopter produces angular velocity due to the rotational effect of propellers. (12)According to the Law of Physics, if two propellers are spinning clockwise direction, then the other two adjacent propellers will be spinning in a counterclockwise direction. With this phenomenon, the torque produced on the four rotors can be balanced when all propellers rotate at the same rate. The torque produced in electric motors can be expressed as

$$\tau = K_t(I_i - I_o) \tag{1}$$

Where τ is the motor Torque, I_i - is the input current supplied to all the motors, I_o - is the supplied current when no load on the motors and K_t is the proportional torque constant.

Considering voltage drop V across each motor, Motor resistance R_m, V_∞ is the angular velocity of the motor and also K_v is the proportional constant of electromagnetic force due to the propeller rotational speed.

Power produced from the motor can be represented as

$$P = IV = \frac{(\tau + K_t I_o) K_v \omega}{K_t} \quad (2)$$

Further simplifying the Quadcopter rotor dynamics model and assuming that $K_t I_o$ is very much less than Torque produced on the motors. The power produced from the model can be defined as

$$P \approx \frac{\tau K_v \omega}{K_t} \quad (3)$$

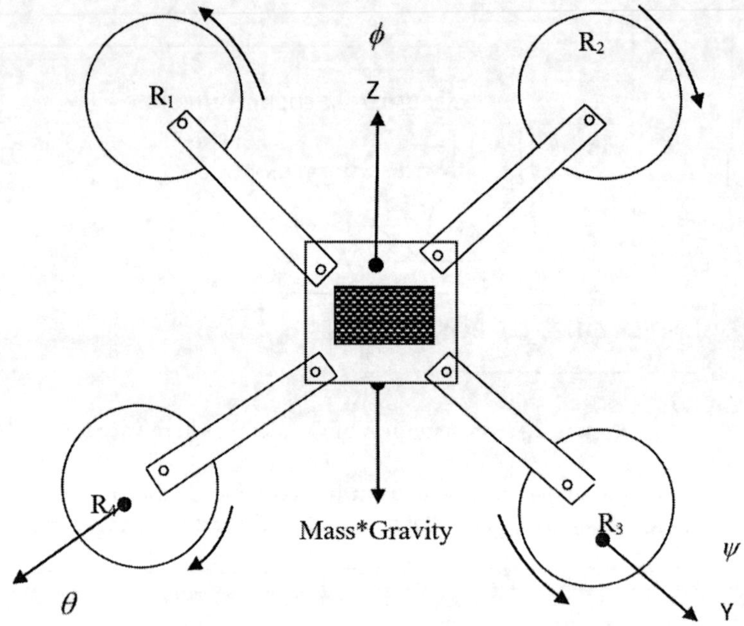

Figure.3. Outline of “X” type Quad-Model

When the Quadcopter model is assumed to be operated at low speed and maintaining constant thrust as well as floating same attitude, this maneuvering is said to “Hovering”. Most of the Quadcopter models are flown at such attitude to obtain the exact aerial photography or terrain monitoring applications in real-time applications.

Considering the free stream velocity V_∞ is zero (air surrounding the Quadcopter is zero), the expression of hovering velocity V_h of designed Quadcopter can be determined.

According to the Momentum theory, the equation of hovering velocity of a Quadcopter model is framed as

$$V_h = \sqrt{\frac{T}{2\rho A}} \quad (4)$$

ρ is the density of the surrounding air and A is the area swept out by the rotor.

Relating the hovering velocity which influences the power required to maintain a steady attitude and that can be represented as

$$P \equiv \frac{\tau K_v \omega}{K_t} = \frac{T^{\frac{3}{2}}}{\sqrt{2\rho A}} \tag{5}$$

In general, Torque is a kind of twisting force that used to rotate the propeller and also provides thrust to move the Quadcopter to and fro. Here each rotor creates some amount of torque about the directional axis and that leads to contribute sufficient angular acceleration as well as the phenomenon to overcome frictional drag forces around the Quadcopter model. As per principle of aerodynamics, the frictional drag force component can be written as

$$D = \frac{1}{2} S v^2 C_d \rho \tag{6}$$

Here D is the Drag force, S is the reference area (Cross-section of Propeller), C_d is the coefficient of Drag. The effect of frictional drag force which creates some amount of opposite torque on the Quadcopter inertial frame during the flight. Due to this frictional drag, the undesirable torque on the Quadcopter model can be indicated as

$$T_D = R\frac{1}{2} S v^2 C_d \rho = R\frac{1}{2} S (R\omega)^2 C_d \rho \tag{7}$$

Where R is the Propeller Radius, from the above equation, it was observed that the torque due to drag component is always proportional to the square of angular velocity.

$$T_D \alpha \omega^2$$

Considering the all four rotors in the Quadcopter model, the total torque on the directional axis (Z-axis) has expressed with respect to the rotational directions of propeller blades. Total torque is the summation of all the torques from the propeller

$$T_{total} = c(\omega_1^2 - \omega_2^2 + \omega_3^2 - \omega_4^2) \tag{8}$$

Where c is the aerodynamic coefficient. If the propeller is spinning in a clockwise direction, it considered as positive and the counterclockwise rotation of propeller is represented as negative. In this research paper, the designed Quadcopter model which consists of two equally distributed uniform rods crossed at the center point of the structure arranged diagonally with four motors. According to the equation of motion, the acceleration of the Quadcopter is mainly due to the thrust, gravity and linear frictional effects.(13)

By considering the position of Quadcopter is "o", g is the acceleration due to gravity, Drag force D, Thrust Component T_c, Then the linear motion is represented in the following form

$$\dot{p}a = D + RT_c + \begin{bmatrix} 0 \\ 0 \\ -mg \end{bmatrix} \tag{9}$$

IOP Conf. Series: Materials Science and Engineering **764** (2020) 012037 doi:10.1088/1757-899X/764/1/012037

Therefore the rotational equation of motion of the designed rigid Quadcopter model can be represented from the Euler`s equations of for rigid body dynamics with respect to the X, Y and Z axis.

$$\omega_r = \begin{bmatrix} \tau_\alpha I_{xx}^{-1} \\ \tau_\beta I_{yy}^{-1} \\ \tau_\gamma I_{zz}^{-1} \end{bmatrix} \tag{10}$$

Here ω_r is the rotational motion expresses with respect to maneuvering, α is the roll torque, β is the pitch torque and γ is the yaw torque.

3.1 Quadcopter flight test at open environment

The Quadcopter flight test was conducted in the open environment around the radius of 200 meters. From the flight test, it was observed that the designed Quadcopter was able to carry 0.15gram of weight and continuously flown up-to 30 minutes of endurance. During the flight test, the major manoeuvring parameters such as pitch angle, roll angle, and yaw angle were changed as well as controlled with respect to the trajectory. Also observed the power consumptions of each component to accelerate the Quadcopter in a 3-Dimensional frame.

Figure. 4. Quadcopter Flight Test conducted at Bannari Amman High Ground

The power requirement at every of each stage of flight test and throttle was adjusted from the zero to maximum to pull the Quadcopter with the payload. The discharging level of the attached Li-Po battery was observed in connection with the endurance limit.

4. Result and Discussions

Observation from the flight data, it clearly shows that the influences of vibration on the Quadcopter which severely affects the stability. When the initial throttle was activated from the ground-based transmitter controller, there itself the motors start working and by the way, the four propellers were spinning to produce power. Due to this, vibrations were acting on the frame and that will cause a little flutter effect during the flight. Though manned aircraft/Rotor-based fixed-wing unmanned aerial vehicles were able to control during the gliding operation, but form the flight test, it was clearly monitored that the Rotary wing Quadcopter were difficult to control when there is a discontinuity in transmitting power and operating signals.

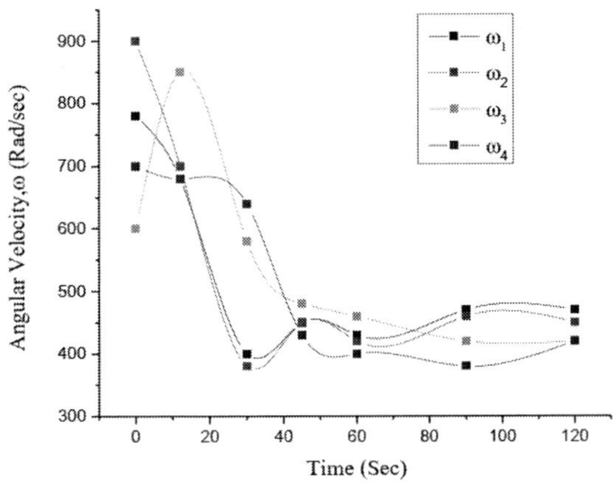

Figure. 5. Quadcopter angular velocities with respect to the reference trajectory

The variation in Roll angle, Pitch angle, and Yaw angle has been observed during the maneuvering process and the dynamic response of the designed Quadcopter model was explicated based on the angular displacement/angular velocity.

Pitch angle θ is initially increased to maximum and maintained around 1.9 degrees (with little oscillation) to achieve the steady level condition during the flight test. Pitch angles have recorded when increasing/decreasing the front and back rotor blades rotation, Roll angles were observed by changing the rotation of left and right rotors and also yaw angles have observed with changing the rotation rate of the counterclockwise set of rotors.

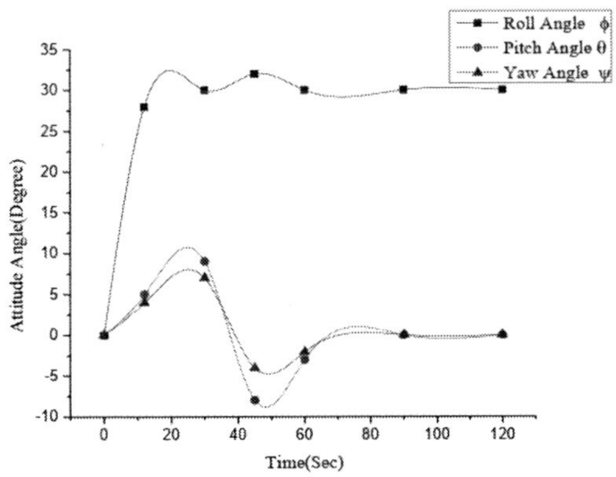

Figure. 6. Measured attitude values from the Quadcopter following a maneuver

The attitudes values are measured from the onboard 3-axis accelerometers during the flight test. After a few seconds of flight, Quadcopter settled with a balanced attitude and that has been clearly executed in the above graphical presentation.

The mentioned graphical representation in Fig.7.denotes the variations of time versus motors RPM (Rotation per Minute) leads to produce sufficient thrust to take Quadcopter for the further level of attaining range and endurance.

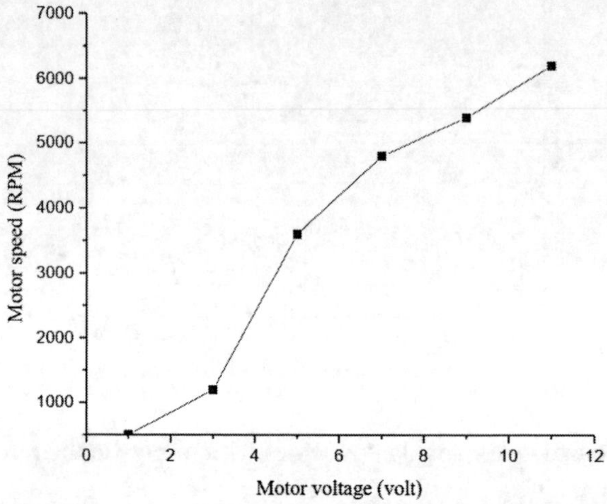

Figure 7. The response of Motor speed with supplied voltage

Here the designed Quadcopter model has the gross weight of 900 grams. According to the basic law of physics, nearly 1.8kilo-gram of thrust is required here to lift this flying model. Each rotor has the tendency to generate at-least .45 kilogram of thrust during the operation. It was clearly identified that when four motors were attaining the constant speed of 6200 RPM, then it is maintained throughout the flight test which is more sufficient to produce required thrust with the accelerating voltage of 11v.

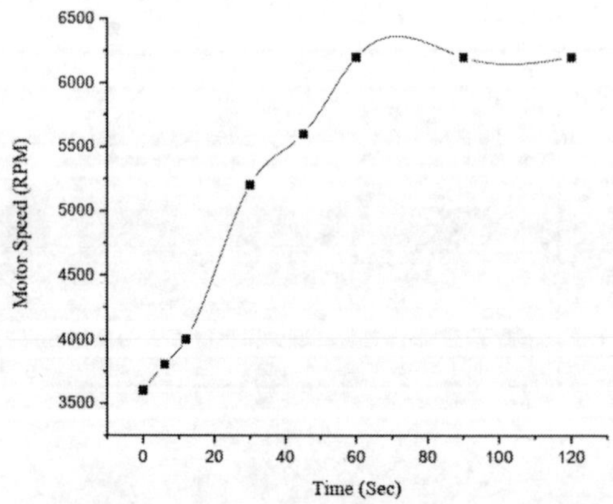

Figure. 8. The response of Motor speed with supplied voltage

Here in this design, fixed pitch propellers were used and only by changing the Motor RPM the variation in thrust with respect to time has been plotted. Also, the variation in Motor RPM versus voltage levels has been projected for the convenient flight.

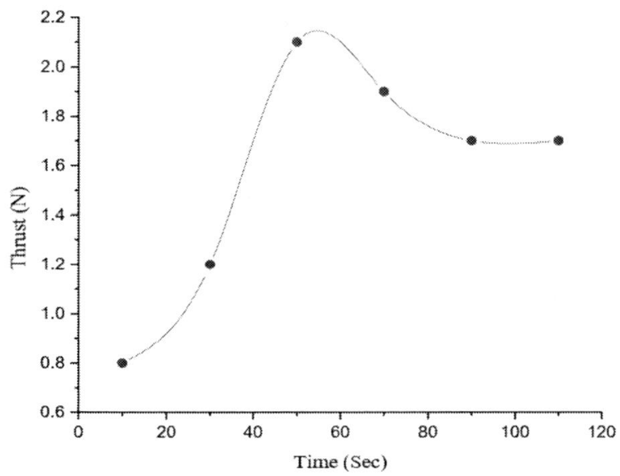

Figure. 9. Variation in Thrust Level

It was captured that the kinetic energy of motor-propeller is always more during the Quadcopter hovering and therefore the kinetic energy has been converted into thrust based on the increases in propeller rpm. The constant coordinates at the end of the plot which expresses the variation in thrust when the motor speed maintained constantly with respect to time.

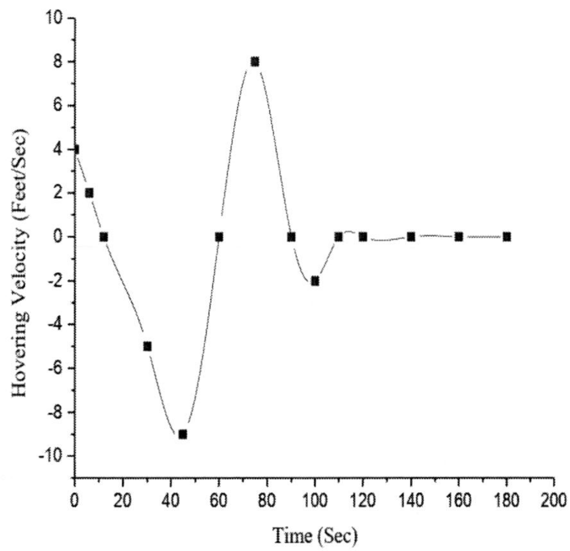

Figure. 10. Quadcopter Hovering with Time interval

IOP Conf. Series: Materials Science and Engineering **764** (2020) 012037 doi:10.1088/1757-899X/764/1/012037

The Hovering velocity of Quadcopter with respect to time has been graphically represented. It was represented that the displacement response for a fixed pitch propeller reaches the excellent accuracy when the battery power is varied from zero to maximum. The angular velocities produced on the fours rotors have been recorded.

5. Conclusion

In this research paper, an introduction about the Quadcopter is exemplified followed by the physical structure. Highly reliable, reusable kind of low-cost hardware specifications has been suggested in this research paper to build an autonomous Quadcopter. The Quadcopter test-bed was prepared and programmed using Arduino 2.8 mega flight controller. It was observed form the flight test that the endurance level rotary wing Quadcopter depends on the availability of battery power as well as the stability responses of its components. Later the dynamic responses of Quadcopter were also represented with flight test analysis. The designed Quadcopter model achieved the endurance limit with the support of avionics panel unit. A small 0.15gram of a small and 360-degree movable camera was fitted on the frame and photographs were been taken during the flight test. Variation in thrust, propeller RPM, angular velocity, endurance time, pitch, roll, and yaw angle was determined from the flight test.

References

[1] J M Alias, A R Anandhan and A. M. Menon 2018 Design Of Different Controllers In Quadcopter *IOSR J. Eng.* **61** 72

[2] N. Promkajin and M. Parnichkun 2018 Development of a robust attitude control for non identical rotor quadrotors using sliding mode control *Int. J. Adv. Robot. Syst.* **15**

[3] D C Patel, G S Gabhawala, A K Kapadia, N H Desai, and S M Sheth 2015 Design of Quadcopter in Reconnaissance Design of Quadcopter in Reconnaissance. *International Conference on Innovations in Automation and Mechatronics Engineering.*

[4] M Thomas, A T Albin, C Joseph, and A K Mathew 2016 Design and Analysis of a Quadcopter Using Catia *Int. J. Sci. Eng. Res.* **140** 156

[5] M. Khan 2014 Quadcopter Flight Dynamics, *Int. J. Sci. Technol. Res.* **130** 135

[6] J. Eker 1999 Design and Implementation of a Quadcopter *Int. J. Res. Electron. Commun. Technol.* (vol.4) **5** 9

[7] G Ostojić, S Stankovski, B Tejić, N Đukić, and S. Tegeltija 2015 Design, Control and Application of Quadcopter *Int. J. Ind. Eng. Manag.* (vol. 6) **43** 48

[8] P Gadiya, A Patheria, K Chib, and A Gore 2017 Design, Analysis, and Fabrication of QuadCopter for Emergency Medical Services using GPS *Int. J. Res. Mech. Eng. Technol* (vol.7) **65** 68

[9] L Y Sørensen, L T Jacobsen, and J P Hansen 2017 Low cost and flexible UAV deployment of sensors *J. Sensors* (vol.17) **1**13

[10] G Ononiwu, A Okoye, J Onojo, and N Onuekwusi 2016 Design and Implementation of a Real-Time Wireless Quadcopter for Rescue Operations *Am. J. Eng.* (Vol.59) **2320**

[11] U Hasnain and N Ahmed 2015 Design Parameters of Indigenously Developed Quadcopter for Area Surveillance *Students Research Papers Conference* (Vol.54) **266** 269

[12] E Balasubramanian and R Vasantharaj 2013 Dynamic Modeling and Control of Quad Rotor *Int. J.Engg (*vol. 5) **63** 69

[13] M W Mueller and R D Andrea 2014 Stability and control of a quadrocopter despite the complete loss of one, two, or three propellers *IEEE International Conference on Robotics and Automation* **45** 52

IOP Conf. Series: Materials Science and Engineering **764** (2020) 012038 doi:10.1088/1757-899X/764/1/012038

Speed control of three phase induction motors using PowerFlex 525 compact AC drive

Lakshmi K, Kumaresan K*, Sridhar C and Shylesh K

Department of Electronics and Instrumentation Engineering, Bannari Amman Institute of Technology, Coimbatore-638 401, India.

*Author for correspondence E-mail: mailforkumares@gmail.com

Abstract. Most of the control systems have to make things move and that generally involves motors. Lifting, blowers, conveyors, pumping, robotics, and fans – pretty much everything uses a motor of some kind. All-purpose single or three-phase motors are huge for simple on/off systems and Inverter-duty motors are particularly designed for operation with variable frequency drives (VFD). They convert electrical energy into mechanical energy, which produces a mechanical rotary action that does some form of work. Alternating current motors are the common types of motors used in commercial and industrial applications. The fundamentals of drive programming and configuration with the PowerFlex 525 compact Alternating current drive, the CompactLogix PLC and the Studio 5000 Logix Designer software is used for configuring these drives. A characteristic that provides complete parameter groups for all the range applications used for the PowerFlex 525 AC Drive. The PowerFlex drive is used with ControlLogix, CompactLogix controllers and different Allen-Bradley products. Point-to-point positioning mode options used for the accurate motor stop at positioning in that system to save money and time during system operation, development, and maintenance. The addition of Allen-Bradley motor control devices into the Logix control devices helps to reduce programming time, ease startup, commissioning and streamline diagnostics.

1. Introduction

The VFD (Variable Frequency Drive) is used to reveal and manipulate parameters of an induction motor by using variable frequency and voltage to manage the velocity of induction motor. Change in the frequency is applied to regulate the speed of induction motor using the formula N= (120*f)/P where f= frequency of supply p=pole [1]. Change in the frequency gives magnetizing flux among air gaps of induction motor. But the induction motors are meant to on its characteristics. PowerFlex 525 drive is a modern, modular layout to help rapid and easy installation and configuration. This powerflex525 Drive of modular drives offers embedded Ethernet/IP connections, USB programming, and better protection capabilities. PowerFlex 525 drives, which suggest a power rating of 0.4 kW to 22 kW (0.5 to.30 Hp) with global voltage classes of 100v to 600Vand they are used in wide for packages which includes conveyors, fans, pumps, and mixers [2].

The PowerFlex 520-Series drive conveys an imaginative structure that is incredibly flexible and can contain frameworks going from independent machines to basic framework incorporation. By combining a many range of motor control option, energy savings, communications and standard safety features in a cost-effective drive, the PowerFlex 520-Series drive is used for a wide range of applications [3]. It improves the work execution and efficiency by taking benefits of the accompanying key highlights offered in a PowerFlex 520-Series drive. Implanted Ethernet/IP port

IOP Conf. Series: Materials Science and Engineering **764** (2020) 012038 doi:10.1088/1757-899X/764/1/012038

permits simple design, control, and gathering of information over the system. Double port Ethernet/IP alternative card used to Device Level Ring (DLR), giving issue tolerant network to most choose drive accessibility [4]. Control Performance on Sensor less Vector Control (SVC), range of motor control alternatives, Volts per hertz control (V/Hz), Closed-loop speed vector control, Permanent Magnet motor control and Variety of Positioning Control, including, point to stop mode is utilized to stop the motor in a consistent position with no encoder criticism and shut circle input with a discretionary encoder gadget. Basic PID usefulness upgrades more application adaptability and Power Flex 525 drives comprise of two PID circles and Point-to-point position. Stop mode alternatives are utilized to stop the motor at any position [5].

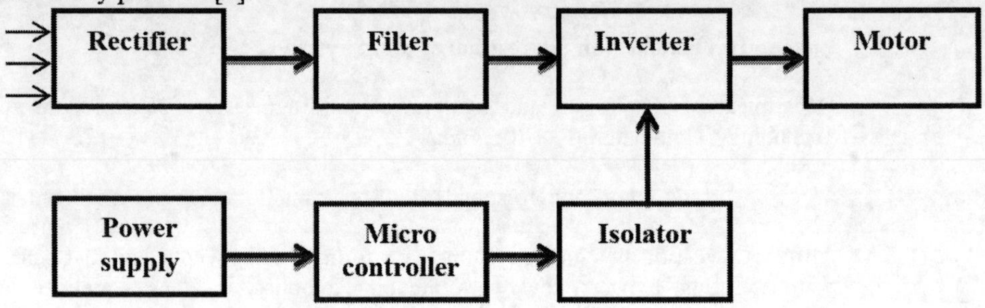

Figure 1. Block diagram of VFD.

2. Methodology

2.1. AB Compact Logix PLC

Programmable logic controller is an industrial PC used for the automation of numerous electro-mechanical methods in industries which have been ruggedized and modified for the control of overall manufacturing processes. Hydromantic Division of the General Motors Corporation proposed the first criteria for the programmable controller in 1968 [6]. Their foremost purpose changed into to get rid of the high charges and continuous protection. It is associated with inflexible, relay-controlled systems he Compact Logic 5370 L1 controllers that combine the power of the architecture with the easy utilize of POINT I/O modules in a compact, inexpensive package [7]. Machine developers and end users can take advantage of an Integrated Architecture system with which requires low rotate development and I/O point tallies and backing for Coordinated Motion over Ethernet/IP for extended adaptability. Expels the requirement for lithium batteries utilized for inherent vitality stockpiling incorporate 2-GB Secure Digital (SD) card for high speed program, restore and a tiny structure constituent for enlarged field. Open access capability of this PLC support for Modbus TCP like gadgets, for example, printers, barcode reader and server.

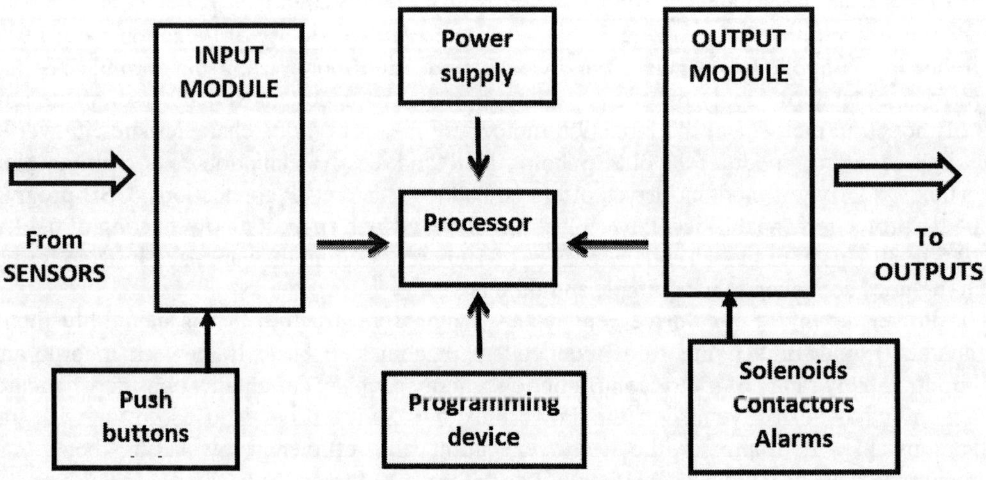

Figure 2. Components of PLC.

Expanding on the adaptability of the AB Logic families of controllers, Programmable Automation Controllers (PACs) are intended to meet on the expansion for an unrivalled exhibition controller in a conservative, moderate bundle. These types of controller also support Point IO configuration. It reduces wired connections between controllers to field devices [8].

2.2. Configure the embedded Ethernet/IP adapter

The PowerFlex 525 Embedded Ethernet/IP system adapter requires a network IP address to function on an Ethernet/IP network. The main benefit of this method is to drive PLC system with no wire connection and connect to Ethernet/IP. There are two techniques for configuring the embedded Ethernet/IP adapter IP deal with BOOTP Server – Use BOOTP server if we like enhanced for controlling the IP addresses of the drive with a server. In the BOOTP server the following IP Address, subnet mask, and gateway address are set to enable its function [9]. If it is set ON in the network addresses manually with parameters, it is necessary to set C128 [Enable EN Address] to 'framework'.

Figure 3. Hardware setup PLC and VFD.

2.3. Configuring drive parameters

- The keypad operation in the device to the Reset Drive to Defaults section to construct the following Communication group parameter changes:

  C129 – Enter IP AddressCfg 1 = 192
  C130 – Enter IP AddressCfg 2 = 168
  C131 – Enter IP Address Cfg 3 = 0
  C132 – Enter IP Address Cfg 4 = 22
  C133 – Enter Subnet Address Cfg 1 = 255
  C134 – Enter Subnet Address Cfg 2 = 255
  C135 – Enter Subnet Address Cfg 3 = 255
- After setting the IP address for Drive and Compact Logix PLC two devices are added with programming software using Ethernet/IP communication protocol. The dominant general for the layer of many industrial protocols together with Ethernet, Ethernet TCP/IP, Modbus TCP/IP and Profibus.
- In Compact Logic controller it is necessary to ensure whether it is NOT in Run Mode. If move the switch to PROG (bottom/down) position at the time of program uploading, the RUN LED on the controller is turned OFF, then REM position goes to middle at a time.
- Using RSLinx classic configuring the ETHERNET\IP address Driver for compact logic controller and PowerFlex 525 Drive. RSLinx classic is also used to know the current status of

IOP Conf. Series: Materials Science and Engineering **764** (2020) 012038 doi:10.1088/1757-899X/764/1/012038

communications between PLC and PC. Select Ethernet\IP driver for PLC and Drive not require for wire these two controllers and it Require Ethernet cable

- Configure the input and output address to assign input address to bush button local: 1:I.Data.0 to Data.3 and select corresponding output parameters form VFD outputs.
- AB compact logic process for these and it has changed with 525 firmware over the years.

The last one I used was: Rung 1: XIC (your Motor start bit) - XIC (Motor1:I.Ready)-OTE (Motor1:O.Start) Rung 2: XIO (your Motor start bit) - XIC (Motor1:I.Active) - OTE (Motor 1:O.Stop) Rung 3: XIO (Motor Reverse Bit)

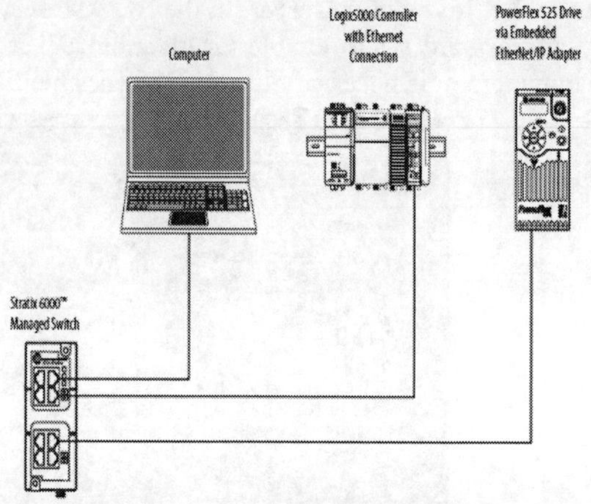

Figure 4. Hardware connection of PLC drive.

2.4. PowerFlex 525 drive add-on profile with Studio 5000

A configuring drives like a timed, side-by-side assessment, Integrated Drive Profiles in Studio 5000 Logix Designer and minimize drive system improvement time by as much as 70% compared to predictable design. Studio 5000 Logix Designer software is integral to drive diagnostics, faults, alarms, and event information. It also to gather all data with respect to a drive, its peripherals, different programming segments, and PC working system Drive to Tech maintain Wizard can be run. Studio 5000 Logix Designer software flashed many drive parameters and updated. Having a distinct repository of drive configuration data to the controller scheme file, it reduces downtime by speeding drive configuration. "Add-On Profiles "are Integrated Drive Profiles, independent of exacting releases of Studio 5000 Logix Designer software and many are compatible to work with past versions of the programming software as well, helping to prevent obsolescence of the controller when new drives are connected to it. Click on the tab named "Drive". The Add-On Profile for the PowerFlex 525 drive in the Logic Designer software provides a general appear and think to the Connected Components Workbench software. This studio5000provides the same ability to download, upload a program, view, and compare drive parameters, as well as access the Wizards. Another characteristic of Logic Designer is to enable Automatic Device Configuration (ADC) in the drive. To expand Ethernet Tree, write click add the PowerFlex 525-EENET Drive and set default parameters values.

The "Parameters" tag is clicked on. It is easy to use the Connected Components, Workbench software and working go online with the drive, analyzing the parameters with the Parameter groups, Custom View groups, AppView, and standard default application parameters like as Conveyor, Mixer, compressor, centrifugal pump and positioning.

IOP Conf. Series: Materials Science and Engineering **764** (2020) 012038 doi:10.1088/1757-899X/764/1/012038

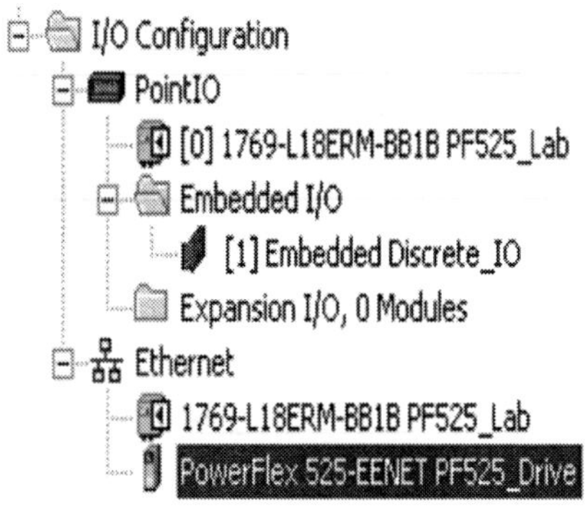

Figure 5. Add-on profiles with studio 5000 logic designer.

The PowerFlex 525 Module arrangement window gives the capacity to sort out up to four terms of Input Data and four words of Output Data to be acknowledged throughout the embedded Ethernet link. Drive status and control information will be communicated by default data's, Status Information "Drive Status" statement contains the drive status bit in sequence, such as Fault, Ready, and at Reference. Open the drive Module Properties window by double-clicking on the "PowerFlex 525-EENET PF525_Drive" node in the I/O design folder. Click the "Download" button, Expand the "AB_ETHIP-1" Ethernet\ IP driver and select the "192.168.0.22, PowerFlex 525" node.

The "Output Frequency" expression contains speed input data, which demonstrates the genuine working frequency (Hz) of the PowerFlex drive. Control Information on "Logic Command" expression contains the drive control Digital bit information, like Start, Stop, Reverse, Forward, and Clear Faults.

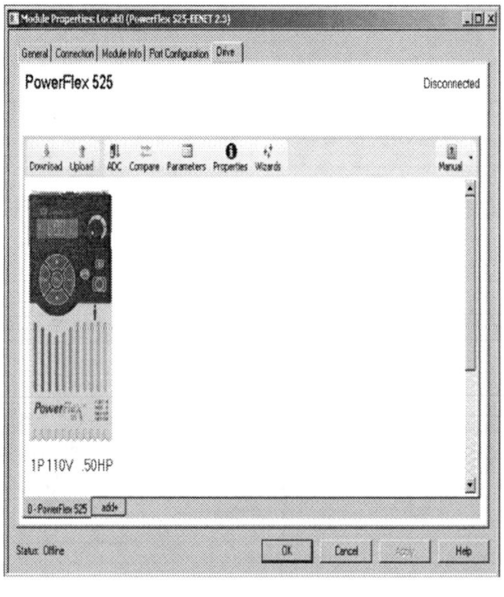

Figure 6. Drive module properties and parameters.

The "Frequency Command" statement contains the Input reference frequency value for the drive to run the motor. Drive parameters are pointed to Data links. Our drive I/O get assembly is dynamic and enables the developer to pick and pick the favoured parameters to impart as system I/O [10]. In another one to storing the port configuration data in the drive through controller and manages the communications to the drive with proper tag-based addressing. The "Controller Tags" double-click on icon under the "Controller Power_flex1.ACD" folder and change the values.

IOP Conf. Series: Materials Science and Engineering **764** (2020) 012038 doi:10.1088/1757-899X/764/1/012038

3. Program for speed control of an induction motor

The speed control of induction motor is done with low electrical power factor. PowerFlex 525 point to point speed control uses encoder device and it is used to set stop mode value as 0 in Frequency Command register. Frequency of induction motor is 50 HZ and speed of induction motor is 1500 rpm at no load condition.

Synchronous Speed Ns=120f/P Where, N is the speed of a rotor of induction motor, Ns is the synchronous speed, S is the slip.

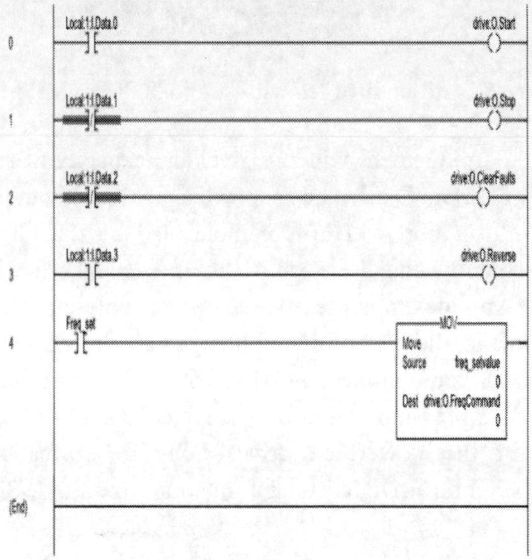

Figure 7. Ladder diagram for VFD control.

The three phase induction motors is used for two types of speed control method and Changes the frequency of motor speed with variable load.

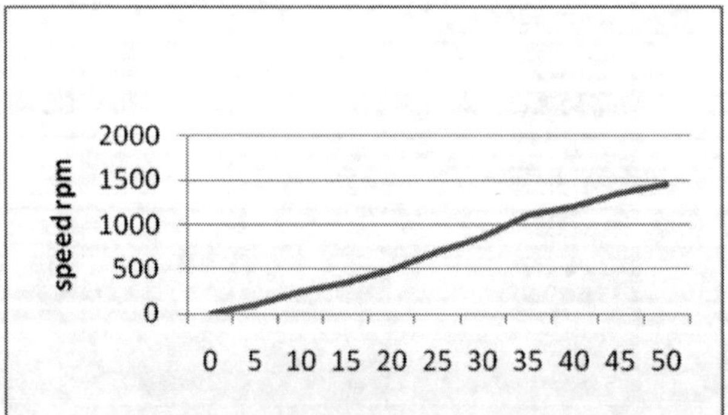

Figure 8. Speed vs. frequency charts.

4. Conclusion

The objective of this paper is to control the three-phase induction motor varying with motor input frequency and also to control the motor in forward and reverse direction using PowerFlex525 and Compact Logic PLC with studio 5000 using Ethernet\IP. This Ethernet\IP implementation is used to control the speed of induction Motor automatically with low cost and high speed transmission protocol. Comparable sort of work is possible with the assistance of HMI, SCADA, etc. Further restorative measures can also be done while structuring such frameworks. Thus, the speed control strategy in this paper is controlled by utilizing Ethernet\IP technique.

References

[1] Mahesh Kumar K M and B Ramachandra, "Speed Control of Three Phase Induction Motor Using PLC under Open and Closed Loop Condition", International Journal of Engineering Research and Application, Vol. 7(1), pp.34-39, 2017.

[2] S.Sneha, S.Radhika, "Controlling and Protection of Three Phase Induction Motor Using PLC", International Journal for Research in Applied Science & Engineering Technology, Vol.4(11), 2016.

[3] Yasar Birbir, H.Selcuk Nogay, "Design and Implementation of PLC-Based Monitoring Control System for Three-Phase Induction Motors Fed by PWM Inverter", International journal of systems applications, engineering & development, Vol.2(3), 2008.

[4] Loannides, M.G., "Design and implementation of PLC-based monitoring control system for induction motor", IEEE Transactions on Energy Conversions, Vol. 19(3), pp.469–476, 2004.

[5] S. Takiyar and B. K. Chauhan, "Hybrid Method for Customized Control of Induction Motor", International Journal of Computer and Electrical Engineering, Vol.5(4), 2013.

[6] Pradip M. Ambore, M. S. Badmera "PLC & SCADA based Condition Monitoring of Three Phase Induction Motor", International Journal of Innovative Research In Computer and Communication, Vol.4(6), pp.11574-11580, 2016.

[7] Sujith John Mathew, B Hemalatha, "PLC Based Induction Motor Starting and Protection", International Journal of Engineering Research and General Science, Vol.3(2), 2015.

[8] Ms. Bhagyashree S. Bhosale, Mr. Mohit V. Burad, Mr. Ajay B. Chavan, R N Baji, "Automation and protection of induction motor by using PLC", International Research Journal of Engineering and Technology, Vol.5(4), 2018.

[9] Ravi Masand, Deepika Jadwani, Abhishek Sahu, "PLC and SCADA Based Fault Diagnosis of Induction Motor", International Journal of Digital Application & Contemporary Research, Vol.2(6), pp.1-6, 2014.

[10] M. Peltola, "Slip of Ac Induction Motors And How to Minimize It", ABB Drives Press Releases Technical Paper, ABB, New Berlin, pp.1–7, 2003.

IOP Conf. Series: Materials Science and Engineering **764** (2020) 012039 doi:10.1088/1757-899X/764/1/012039

Certain investigation on feature extraction in dorsal hand vein image

K S Vairavel[1,a], R Nevetha[1,b], S Mekala[1,c]
Department of Electronics and Instrumentation Engineering, Bannari Amman Institute of Technology, Sathyamangalam.
Email: [a]eievairam@gmail.com, [b]neve.rajagopal@gmail.com, [c]mekala62665@gmail.com

Abstract. Biometrics refers to people being known by their fingerprints or attributes. This research takes into account veins on the back of the hand called the dorsal vein. The primary target of this work is to build up an individual acknowledgment framework dependent available dorsal vein design with high acknowledgment rate. Attributes such as LBP (Local Binary Pattern), HOG (Histogram of Oriented Gradients) and WLD (Weber Local Descriptor) are extracted and the technique of Chisquare, Cityblock, Euclidean and Minikowski is used for recognition speeds.The quality assessment is conducted in terms of accuracy of category. Hand vein biometrics offer higher security, simple to get the hand vein picture and extremely difficult to fashion the information contrasted with progressively settled biometric check techniques.

1. Introduction

1.1 Image

An image is a square pixel (picture components) cluster or lattice masterminded in segments and columns [1]. A picture can be portrayed as a two-dimensional flag that contains force and shading data sorted out along a x and y spatial hub [2].

1.2 Digital image processing

Computerized picture preparing incorporates forms whose data sources and yields are picturesand envelops forms that remove characteristics from pictures and furthermore incorporates the acknowledgment of individual items [3].

1.3 Biometrics

There many methods to authenticate an individual as shown in the Figure 1.1. The simplest form of authentication is to enter a password or PIN (knowledge) [4].

1.3.1 Types of biometrics

Biometrics is the investigation of consequently perceiving people by methods for inalienably one of a kind physical or social attributes. Biometrics has been used mostly in the computer and physical security industry, where robust authentication methods used to control access to physical locations or computer systems were needed [5]. Biometrics to be used as feature for authentication is shown in the Figure 1.

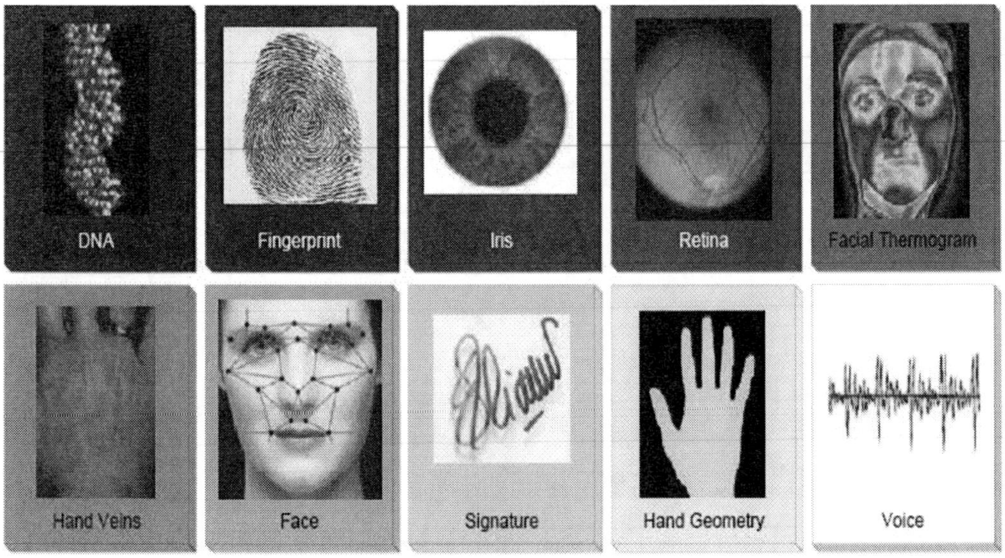

Figure 1. Biometrics Types

As an issue of first significance, it ought to be across the board, with each individual having the given trademark. Next, there ought to be certain immutability intrinsic to the component, i.e., the purposeful segments should remain commonly steady over a period of time [6].

1.4 Dorsal hand vein

Dorsal hand vein pattern recognition system is a very hygienic contactless device and removes any apprehension between people about coming into contact with something that has already been touched by other people as shown in Figure 2 As near-infrared light is used to acquire the image it is non-harmful to the humans [7].

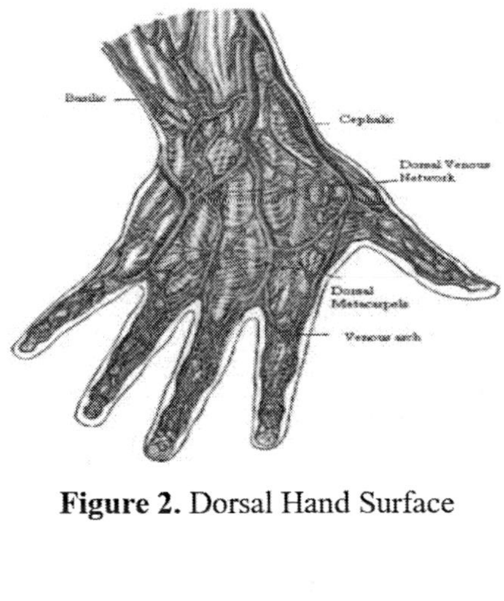

Figure 2. Dorsal Hand Surface

2. Proposed system

Bio metric plays a vital role in authentication. The system merges with the idea that for authentication purposes the dorsal vein image is used. Here are the algorithms as shown in Figure 3.

IOP Conf. Series: Materials Science and Engineering **764** (2020) 012039 doi:10.1088/1757-899X/764/1/012039

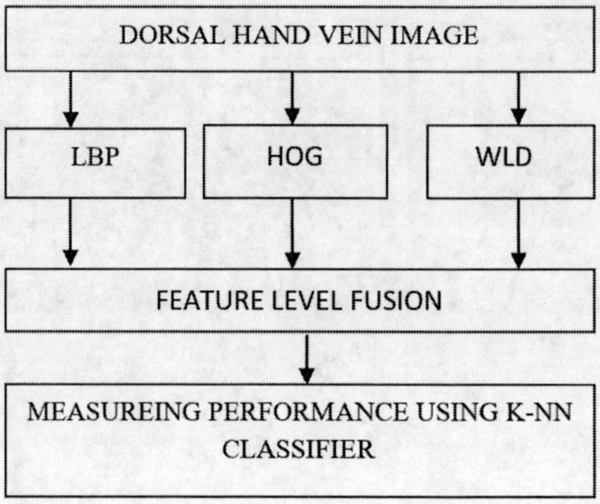

Figure 3. Proposed System of Dorsal Vein Biometric

2.1 Roi extraction

The highlights of vein designs extricated from a similar district in various dorsal hand vein pictures are thought about for acknowledgment step [8]. The removed locale is known as the area of intrigue (ROI). The procured picture is first changed over to a gray scale picture and after that its ROI separated utilizing default editing capacity from MATLAB [9].

2.2 Local binary pattern

The basic idea for working up the LBP manager was to represent twodimensional surfaces by two proportional measures: adjacent spatial models and separate diminishing scale [10].

LBP provides a strong surface descriptor with an invariant turn, and Figure 4 shows a diagram of the basic LBP head. For each pixel in an image, it differentiates its view and all the adjacent pixel values [12].

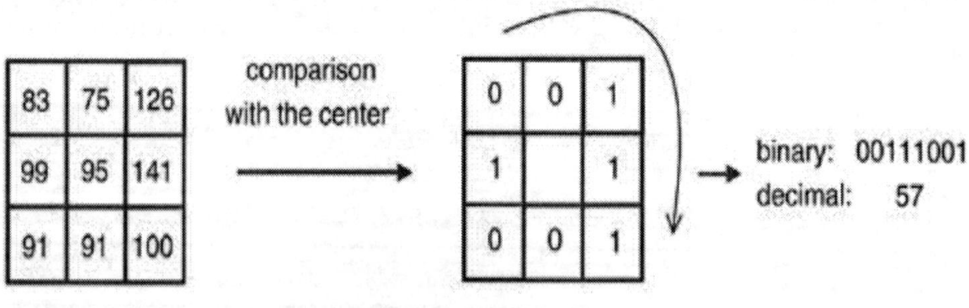

Figure 4. Example of Operator Basic LBP

The LBP administrator can be stretched out to neighborhoods of various sizes. The pixel esteems can be interjected for round communities to allow any sweep and number of pixels in the areas [13].

2. 3 Weber local descriptor

Find out that the chief figuring the power contrasts of the current pixel is really Laplacian head for the critical differential excitation element against its neighbours. Since the chairman of Laplacian is prone to uproar, in this procedure substitute with LoG (Laplacian of Gaussian). As a prominent descriptor of the area, LBP was essentially highly equivalent, computationally efficient and invariant to changes in monotonic dim measurement.

2.4 Histogram of oriented gradients

HOG are highlight descriptors used in PC vision and image processing for the end goal of object recognition. The approaches include restricted parts of an image's angle introduction events [14].This technique is like edge implementation histograms, scale-invariant feature shift descriptors, and shape settings, but it differs in that it is processed in a thick network of continuously divided neighborhood cells and is used for more precise discernment of neighborhood standardization. HOG descriptors rely on neighborhood objects appearance and shape within a picture that can be represented by the dispersion of inclinations of force or edge directions. The descriptor of the HOG retains a few primary focuses over other methodologies of descriptors.

3. RESULTS AND DISSCUSSION

3.1 Histogram of oriented gradients

The dorsal vein identification method using Oriented Gradient Histogram is performed for verification purposes with the K-NN classifier. The obtained result is from MATLAB simulation. Figure 5 shows the characteristic extraction of the Histogram from the dorsal vein illustration. Table 1 displays the chi-square, city block, euclidean and minikowski simulated values.

Table 1. Tabulation of HOG Left

Distance Measure		**Chi-square**	**City Block**	**Euclidean**	**Minkowski**
K fold cross validation	**Testing Images**	**Recognition Rate(%)**			
K = 1	204	95.5882353	90.19607843	85.29411765	83.33333333
K = 2	204	98.0392157	97.05882353	93.62745098	91.66666667
K = 3	204	97.0588235	96.07843137	94.11764706	92.15686275
K = 4	204	98.0392157	98.03921569	97.05882353	95.58823529
K = 5	204	98.0392157	96.56862745	92.15686275	91.66666667
Average Recognition Rate		97.35294118	95.58823529	92.45098039	90.88235294

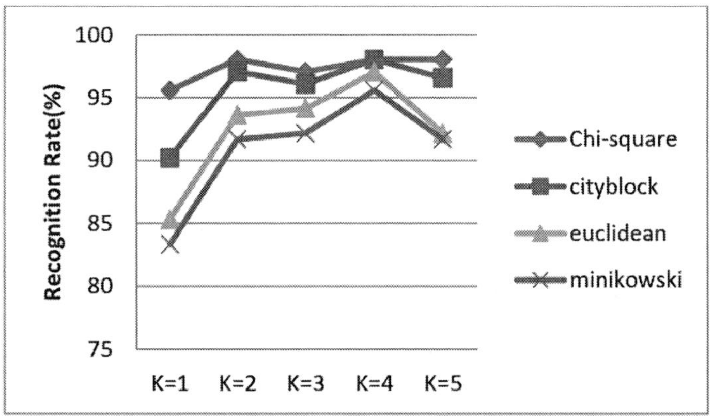

Figure 5. Graphs of HOG Left

The HOG identification rate left for chi-square is 97%, city block is 95%, Euclidean 92% and minikowksi 90%.The chi-square includes in the Oriented Gradient Histogram with high recognition values.

With the K-NN classifier, the validation goal is accomplished using the Oriented Gradient Histogram. The rate of identification of chi-square, city block, euclidean and minikowski as shown in Table 2.The graph is drawn for recognition rate as shown in the Figure 6.

Table 2. Tabulation of HOG Right

Distance Measure		Chi-square	City Block	Euclidean	Minkowski
K fold cross validation	**Testing Images**	**Recognition Rate(%)**			
K = 1	204	95.0980392	92.15686275	88.23529412	87.25490196
K = 2	204	99.0196078	97.54901961	97.05882353	94.11764706
K = 3	204	99.5098039	99.50980392	97.54901961	96.56862745
K = 4	204	98.5294118	98.52941176	96.07843137	95.58823529
K = 5	204	99.5098039	98.03921569	94.11764706	94.11764706
Average Recognition Rate		98.33333333	97.15686275	94.60784314	93.52941176

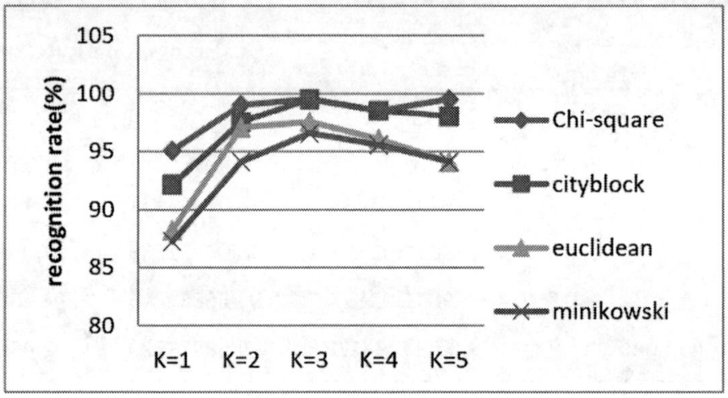

Figure 6. Graph of HOG Right

The recognition rate in HOG right for chi-square is 98%, cityblock is 97%, Euclidean is 94% and minikowksi is 93%. The chi-square consists of high recognition rate in Histogram of Oriented Gradient.

3.2 Local binary pattern

The validation objective is achieved with the K-NN classifier using Local Binary Template. Table 3 shows the recognition frequency of local binary pattern chi-square, cityblock, euclidean, and minikowski.The graph is drawn for recognition rate as shown in the Figure 7.

Table 3. Tabulation of LBP Right

Distance Measure		Chi-square	City Block	Euclidean	Minkowski
K fold cross validation	**Testing Images**	**Recognition Rate(%)**			
K = 1	204	87.254902	83.82352941	75.98039216	68.62745098
K = 2	204	92.1568627	91.66666667	86.2745098	82.35294118
K = 3	204	95.5882353	93.62745098	88.7254902	86.76470588
K = 4	204	93.1372549	94.11764706	89.21568627	84.80392157
K = 5	204	93.1372549	91.17647059	83.33333333	75.98039216
Average Recognition Rate		92.25490196	90.88235294	84.70588235	79.70588235

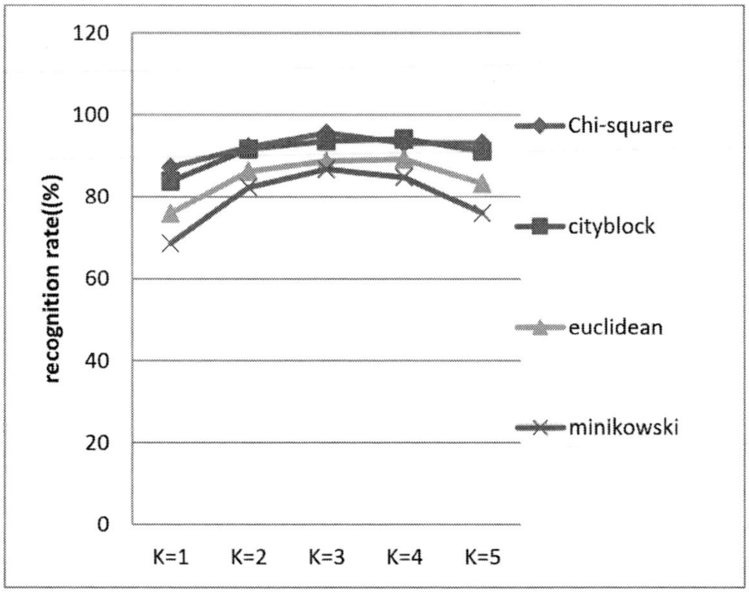

Figure 7. Graph of LBP Right

The recognition rate in LBP right for chi-square is 92%,cityblock is 90% , Euclidean is 84% and minikowksi is 79%. The chi-square consist of high recognition rate in Local Binary Pattern.

The validation purpose is performed with the K-NN classifier using the Local Binary Pattern. Table 4 reports the recognition frequency of local binary pattern chi-square, cityblock, euclidean, and minikowski. The graph is drawn for recognition rate as shown in the Figure 8.

Table 4. Tabulation of LBP Left

Distance Measure		**Chi-square**	**City Block**	**Euclidean**	**Minkowski**
K fold cross validation	**Testing Images**	**Recognition Rate(%)**			
K = 1	204	81.8627451	77.45098039	65.19607843	55.88235294
K = 2	204	92.1568627	90.19607843	85.29411765	79.90196078
K = 3	204	92.6470588	91.17647059	84.80392157	80.39215686
K = 4	204	96.0784314	92.64705882	87.25490196	83.33333333
K = 5	204	90.6862745	85.29411765	79.41176471	74.50980392
Average Recognition Rate		90.68627451	87.35294118	80.39215686	74.80392157

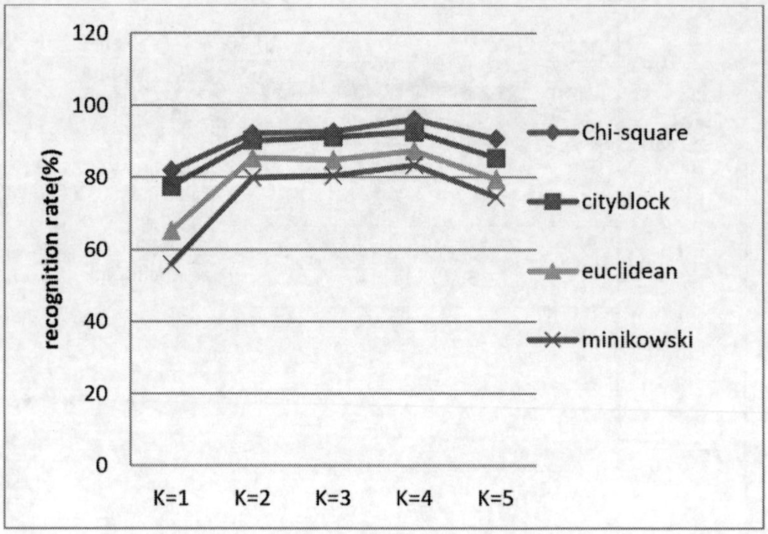

Figure 8. Graph of LBP Left

The recognition rate in LBP left for chi-square is 90%, cityblock is 87%, Euclidean is 80% and minikowksi is 74%. The chi-square consists of Local Binary Pattern's high recognition frequency.

3.3 Weber local descriptor

The validation purpose is performed with the K-NN classifier using Weber Local Descriptor. The chi-square, city-block, euclidean, and weber local descriptor minikowski recognition level as shown in Table 5.The graph is drawn for recognition rate as shown in the Figure 9.

Table 5. Tabulation of WLD Left

Distance Measure		**Chi-square**	**City Block**	**Euclidean**	**Minkowski**
K fold cross validation	**Testing Images**	**Recognition Rate(%)**			
K = 1	204	54.4117647	55.39215686	47.54901961	41.66666667
K = 2	204	65.6862745	68.62745098	59.3137254	52.45098039
K = 3	204	70.0980392	73.52941176	63.7254902	56.8627451
K = 4	204	61.2745098	68.1372549	56.37254902	51.96078431
K = 5	204	63.7254902	70.09803922	56.37254902	51.47058824
Average Recognition Rate		63.03921569	67.15686275	56.6666666	50.88235294

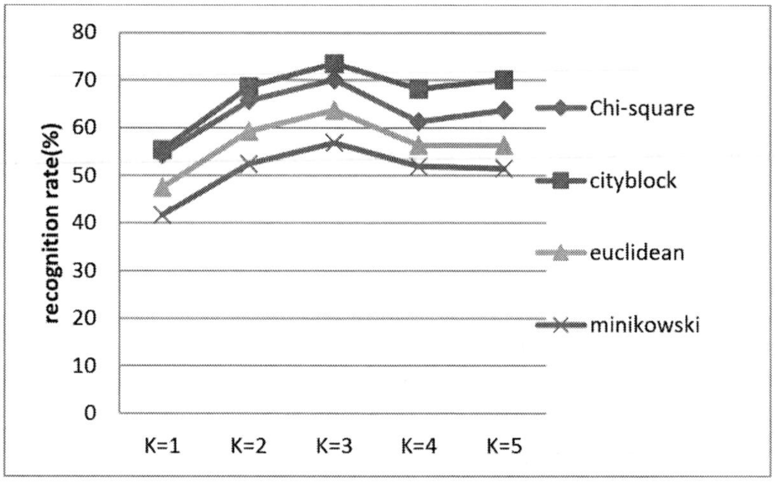

Figure 9. Graph of WLD Left

The recognition rate in WLD left for chi-square is 63%, cityblock is 67%, Euclidean is 56% and minikowksi is 50%. The Cityblock consist of high recognition rate in Weber Local Descriptor.

The validation objective with the K-NN classifier is achieved using Weber Local Descriptor. The recognition level of the local weber descriptor chi-square, city-block, euclidean and minikowski as shown in Table 6.The graph is drawn for recognition rate as shown in the Figure 10.

Table 6. Tabulation of WLD Right

Distance Measure		**Chi-square**	**City Block**	**Euclidean**	**Minkowski**
K fold cross validation	**Testing Images**	**Recognition Rate(%)**			
K = 1	204	60.2941176	64.21568627	55.88235294	47.54901961
K = 2	204	67.1568627	68.62745098	59.80392157	52.94117647
K = 3	204	69.1176471	70.09803922	62.25490196	57.84313725
K = 4	204	68.1372549	71.56862745	62.74509804	55.39215686
K = 5	204	58.8235294	63.23529412	51.96078431	46.07843137
Average Recognition Rate		64.70588235	67.54901961	58.52941176	51.96078431

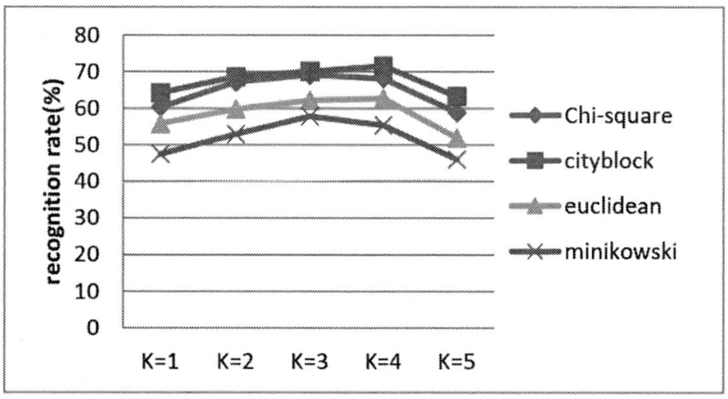

Figure 10. Graph of WLD Right

IOP Conf. Series: Materials Science and Engineering **764** (2020) 012039 doi:10.1088/1757-899X/764/1/012039

The recognition rate in WLD right for chi-square is 64%, cityblock is 67%, Euclidean is 58% and minikowksi is 51%. The Cityblock consists of Weber Local Descriptor's high recognition level.

4. Conclusion

Dorsal vein is eliminated using local binary pattern (LBP), local weber descriptor(WLD) and K-NN classifier.. The MATLAB simulation evaluated the function level fusion for each of its recognition. While the WLD, LBP and HOG are contrasted, in cityblock technique The WLD has a low recognition rate, the LBP has a high recognition level in chi-square technology and the chi-square technology has high recognition rate. The LBP is successful n comparing WLD and HOG and has a high recognition level. The proposed system increases the biometric system's trustworthiness and also enhances the system's overall performance.

5. Future work

The future work will concentrate with the filters on the endowment of the process. The reliability is improved by applying the filters to the proposed work and can make sure and contribute to the output.

References

[1] T Tanaka and N Kubo 2004 Biometric authentication by hand vein patterns Proc. SICE Annu. Conf., Yokohama, Japan, pp. 249–253.
[2] Gopal A, Smriti Srivastavaa, Saurabh Bhardwaj B, Sandeep Bhargavaa, 2016 Fusion of palm-phalanges print with palmprint and dorsal hand vein vol 47 pp.12–209.
[3] Sanchit Ramalho M, Correia P L, Soares, L.D, 2011 Biometric identification through palm and dorsal hand vein patterns International Conference on Computer as a Tool (EUROCON) IEEE
[4] Khan M, Khan N M 2011 Dorsal hand vein biometric using Independent Component Analysis (ICA) International Conference on Internet Technology and Secured Transactions (ICITST) IEEE
[5] C L Lin and K C Fan 2004 Biometric verification using thermal images of palm-dorsa vein patterns IEEE Trans Circuits & Sys For Video Technology vol 14 pp199 – 213
[6] Bogdan Belean, Mihaela Streza, Septimiu Crisian, Simian Emerich 2017 Dorsal hand vein pattern analysis and neural networks for biometric authentication vol 26 pp-305-314
[7] Z Honarpisheh & K Faez Biometric 2013 Identification by Clustering the Dorsal Hand Vein Patterns using the Firefly Algorithm International Journal of Electrical and Computer Engineering (IJECE) vol 3 pp 30-41
[8] Ajay Kumar, David Zhang 2006 Personal Recognition Using Hand Shape and Texture Vol 15
[9] M Rajalakshmi, V Ganapathy, R Rengaraj 2017 Palm-dorsal vein pattern authentication using convoluted neural network (cnn) Vol 116 pp 525-532.
[10] Fransisca J Pantah, Jayanti yusmah sari, Amil A Ilham and Ingrid nutanio 2018 Multispectral dorsal hand vein recognition based on local line binary pattern Vol 11
[11] M Heikkil, M Pietikinen, and C Schmid 2006 Description of interest regions with center-symmetric local binary pattern Proceedings of 5th Indian Conference of Computer Vision Graphics and Image Processing vol 4338 pp 58–69
[12] R Nathiya, B Shanmugapriya, G Sivaseema, J Priyadharshini 2015 Dorsal fingerprint patterns using lbp and gabor for identifying human identities Vol 3 Issue 3, pp 236-243
[13] A Petpon & S Srisuk 2009 Face recognition with local line binary pattern In Image and Graphics ICIG'09 5th International Conference pp 533-539
[14] Mostafa A Ahmad, Ahmed H Ismail, Nadir Omer 2018 An Accurate Multi-Biometric Personal Identification Model using Histogram of Oriented Gradients (HOG) Vol 9

IOP Conf. Series: Materials Science and Engineering **764** (2020) 012040 doi:10.1088/1757-899X/764/1/012040

EFFECT OF SILICON CARBIDE IN MECHANICAL PROPERTIES OF ALUMINIUM ALLOY BASED METAL MATRIX COMPOSITES

B Siddharthan[1], R Rajiev[2], S Saravanan[2], T K Naveen[3]
[1]Assistant Professor, Department of Mechatronics, Bannari Amman Institute of Technology, Sathyamangalam.
[2]Assistant Professor, Department of Mechanical Engineering, Bannari Amman Institute of Technology, Sathyamangalam.
[3]Assistant Professor, Department of Mechanical Engineering, Sri Krishna College of Technology, Coimbatore.
Email: [a]siddharthan@bitsathy.ac.in

Abstract. Metal Matrix Composites are a class of composites with evaluated properties that get another perspective of material structure and are generally utilized in the aviation and automobile industry. These materials have the properties of both the base matrix material as well as the reinforced materials. These metal matrix composites have low residual stress and thermal residual stresses. Due to this concept, the materials posses enhanced adhesive bonding between the matrix and reinforcements. Aluminium alloy with silicon carbide is applicable where hardness and toughness are the requirements. Metal Matrix Composites are designed in order to have the combined properties of both metals and ceramics. The specimens are fabricated for various proportions of the materials. In the fabrication, it begins with Aluminium alloy and Silicon carbide. The specimen is evaluated for diverse mechanical tests, and the experimental result is the pure aluminium undergoes more wear when compared to the reinforced specimen.

1. Introduction

The word composite generally refers to a material system which constitutes two materials that is a metal and a ceramic distributed evenly throughout the cross section. On the basis of physical and chemical nature of the matrix material, Composite materials are classified as metal matrix composites, ceramic matrix composites and polymer matrix composites. In composite material, the individual components retain their individual characteristics and to reveal only their positive outcomes and not their drawbacks, in order to acquire an improved material.

Metal Matrix Composite (MMC) is composed of a metallic metals mostly magnesium, aluminium, Titanium, cobalt and copper and Ceramics (Carbides and Oxides) or metallic (Tungsten, Molybdenum) phase. When three materials are identified in a composite, it is known as hybrid metal matrix composite.

Metal Matrix Composites were made of dispersed reinforcement material into base metal matrix. The reinforcement material is coated in order to stay away from chemical reaction with the matrix material. The matrix is the single block of material into which the reinforcement is dispersed and is continuous throughout the surface. The matrix material is usually a lighter metallic material like Aluminium or Titanium or Magnesium to provide complete support to the reinforcement.

The reinforcement materials were surrounded with the matrix, which is used to alter the mechanical and electrical properties. The reinforcement material is dispersed neither continuous nor discontinuous. Aluminium alloy metal matrix composite (AMMC) offers designers with many included advantages, as they are especially appropriate for applications requiring improved strength, good structural stiffness, and dimensional constancy and lightweight. MMCs provide enhanced properties over monolithic alloy. In Aluminum alloy metal matrix composite (AMMC) the major constituent is aluminum or aluminum alloy, which form porous surface and is term as matrix layer. The other constituents embedded in this aluminum /aluminum alloy porous matrix layer and serve as reinforcement, which is usually a nonmetallic and commonly ceramics such as Silicon Carbide or Aluminium Oxide. Properties of AMMCs can be varied by varying the composition of constituents and their volume fractions.

2. Literature Review

The effect of the addition of Silicon Carbide reinforcement to aluminium on microstructure and wear properties of Al 5083-SiC by adding the SiC 2%, 3% and 5wt.% through stir casting is experimented by [1]. By adding 5% wt. of Silicon Carbide particles to aluminium alloy the maximum hardness value, maximum ultimate tensile strength and tribological properties were improved due to a fair distribution of Silicon Carbide particles in aluminium matrix. Hybrid metal matrix composite was fabricated by [2] by reinforcing the Silicon carbide and Titanium diboride on aluminium and the results shows that addition of reinforcement upto 15% wtreveles the reduction of the hardness value and its tensile test results in reduction of strength to 50-60% is by TiB_2 but the addition of Titanium diboride increases the wear resistance of the material, maximum addition of TiB_2 is limited to 2.5% because 5% lead to porosity and affect the hardness value. [3] fabricated a MMC by stir casting method and investigated the mechanical properties by hybrid reinforcing Aluminium 8011 matrix with boron carbide (B_4C) and alumina particles and summed up that with increase in weight percentage of reinforcement particles increases the mechanical strength while the composite with 90% Al + 5%B_4C+5% Al_2O_3 have greater mechanical properties than the other samples prepared.
Long-term naturally aging die cast of Aluminium-Silicon-Copper alloys is studied by [4] in the presence of corrosive environment, the material increases over 10 year period and variation in microstructure affecting the physical properties. The cause for modification in microstructure is mainly of higher casting pressure and quick cooling process. But in Al alloy heavy walls, the change in microstructure not occur even after natural aging. [5] Proposed an investigation on the effect of the addition of Silicon Carbide (SiC) to the Aluminium alloy upto 60% SiC on the composites. The porosity of the composites is reduced and microscopically it's found to be homogeneous in distribution. [6] evaluate the mechanical properties of hybrid reinforced Al alloy 7075 reinforced with SiC and Al_2O_3 metal matrix composites in the proportions of 3%,5%,7% and 9% weight of Al_2O_3 and 2.% weight of SiC by stir casting method and concludes that ductility alone decreases while hardness and tensile increases with their addition and the sample containing the highest percentage of Al_2O_3 exhibit high tensile strength and hardness with a minimum % of elongation.

The investigation conducted by [7] on the mechanical properties of particulate reinforced aluminium and Silicon Carbide MMC's by stir casting route and stirring the MMC slurry in partial molten state helps to disperse ceramic particles into matrix material properly while holding temperature, stirring speed and the size and position of impeller are the important factors to be considered while casting of these composites since they have an crash on the mechanical properties and the wear rate tends to decrease with the increase in weight percentage of SiC and these materials can be used in high elevated temperatures, for better wear resistance and corrosion resistance. [8] fabricated Aluminium-Silicon Carbide composites with weight percentage varying from 5% to 30% weight of SiC using powder metallurgy process and its various tests results were measured. Alloying of composites for 16 hours of milling resulted in uniform powder structure and the quality of the final product depends upon the initial compact. It resulted in the increased hardness, compressive strength, density by the reinforcement of contents from 5% to 30% weight of SiC. [9] Investigated the aluminium alloy metal matrix composite by stir casting method for various proportions of boron carbide(B_4C) and silicon carbide (SiC), and concluded that 96% Al, 2% SiC and 2%B_4C has a higher strength due to higher quantity of carbides in composite and for different casting method its mechanical properties can be found. An overview of Aluminium matrix composite of the effect on addition of reinforcements in aluminium alloy with different mechanical properties like a tensile test, strain, hardnessand wear are discussed by [10]. It concludes that SiC reinforced have high wear resistance and compressive strength compared to other reinforcements like Al_2O_3 and B_4C in MMC.

3. Materials and Sample Preparation

Aluminum and its combinations have outstanding properties like density, good plasticity, great versatility and flexibility and great wear resistance. They find broad applications in aircrafts, astronautics, automotives and train fields. Neverthless, low hardness and poor shock resistance bring about their restricted application in heavy duty conditions. Similar to all composites, aluminium based composites were not a solitary material, but rather a group of materials whose strength, flexural, density, stiffness, electrical and heat carrying properties can be customised. The proportions of matrix material, reinforced material, shape and size, the method of reinforcement and the method of manufacturing can be varied inorder to enhance the required properties. Despite the varieties, be that as it may, Aluminium composites offer incredible thermal conductivity, high shear strength, magnificent wear resistance, high-temperature activity and non-combustibility.

The chemical composition of Al 8011 alloy consists of various materials as listed in Table 1.

Table 1. Chemical Composition of Al 8011 alloy

Material	Weight %
Fe	1
Si	0.9
Mn	0.3
Zn	0.1
Cu	0.1
Ti	0.08

IOP Conf. Series: Materials Science and Engineering **764** (2020) 012040 doi:10.1088/1757-899X/764/1/012040

Mg	0.02
Al	97.5

Silicon carbide (SiC) is made up of tetrahedra of carbon at the core and shared silicon atoms with tough bonding in crystal lattice. This results in a very hard and sturdy material. SiC is not affected by any acids or alkalis up to around 900 °C. When exposed to air, SiC reacts with air and form a Silicon oxide coating at 1300 °C and is able to be utilized till 1650 °C. The low thermal expansion and high thermal conductivity gives this material an outstanding thermal shock resistant quality. The strength of Silicon Carbide ceramics vary with grain boundary impurities. SiC particles reinforced Aluminium, are one of the generally known composites due to their predominant mechanical properties.

4. Sample Preparation

Table 2 shows the samples with different weight percent proportions of 8011 Al alloy and SiCthat arecasted.

Table 2. Weight Percent of Samples Prepared

Samples	**Aluminium**	**Silicon carbide**
1.	100%	-
2.	90%	10%
3.	80%	20%
4.	70%	30%

Total weight of each sample = 145.8 g.

MMC can be a manufacture by utilizing a few procedures which can be a solid, liquid and vapour phase. Stir casting (Liquid state) procedures were generally used to fabricate AMMCs. In stir casting technique, MMCs are manufactured by dispersing the preheated reinforcement material into high temperature molten matrix material by applying stirring activity and pouring the combined material into the die and it is made solidified as shown in Figure 1. To manufacture the enormous number of MMC parts, stir casting is the most efficient way of fabrication in fluid state.

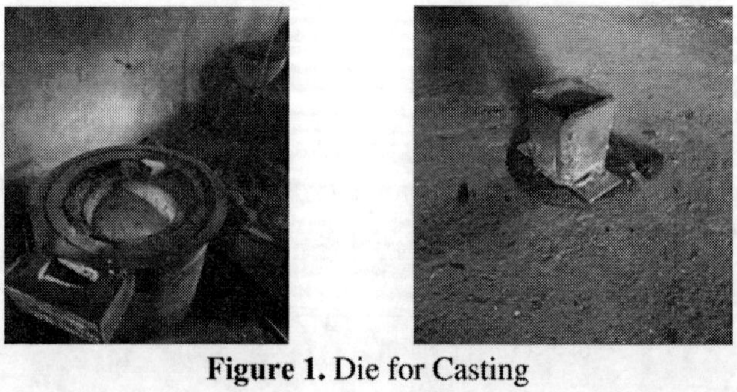

Figure 1. Die for Casting

5. Experimental Method

The prepared composite samples are investigated for their mechanical properties and are compared with the pure aluminium.

5.1 Tensile Test

Tensile test is an basic test for evaluation of mechanical property in which the sample is subjected to a tensile load till it fails. It results in the direct measurement of ultimate strength, elongation and reduction in area of the sample. As per the ASTM B557 standards, the prepared samples as shown in Figure 2 were subjected to tensile test and the results were noted.

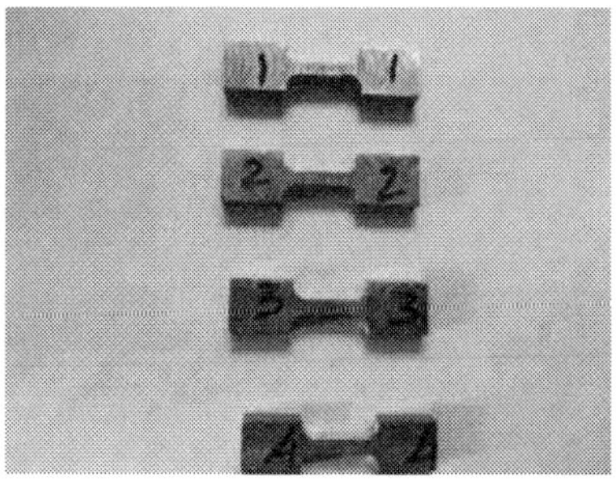

Figure 2. Samples for Tensile Test

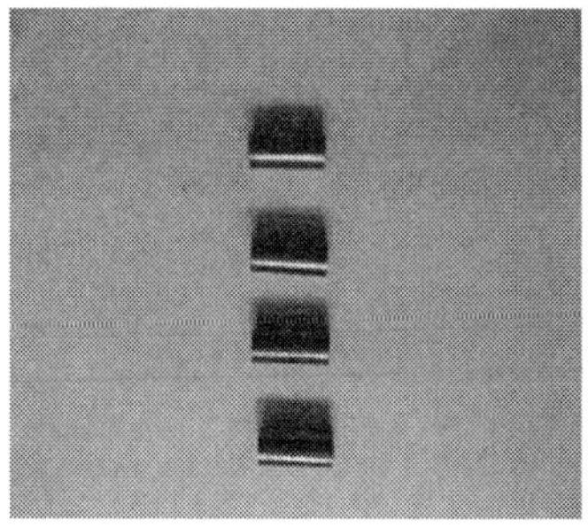

Figure 3. Wear Test Samples

5.2 Hardness Test

Hardness is the opposition offered by the material for contained deformation. This term can be applied for indentation or scratching or bending. Hardness measurement is done in Vickers Hardness Testing Machine. The microhardness test was carried in Micro Vickers Hardness Tester which has a testing load series of 10 gms to 1 kg load and testing scale used is HV. The prepared samples were subjected to a load of 0.5 kg with reside time 10 sec in three locations of each sample

5.3 Wear Test

Wear is associated to interactions between the sliding or rolling surfaces specify the subtraction and deformation of material on contact surface as a result of sliding or rolling actions. Wear test was carried in Pin on disc for ASTM G99 standard with the samples shown in Figure 3.

6. Results and Discussion

The experimentation on the prepared samples are conducted and the results are discussed as follows:

6.1 Tensile Test

Figure 4 shows that the casted pure Al 8011 alloy has the maximum tensile strength of 142.8 MPa. When adding 10% Silicon Carbide to Al 8011 alloy, the tensile strength of the 90% Al 8011-10% SiC metal matrix composite improves to 146.13 MPa as shown in Figure 5. This shows that there is a slight improvement in the tensile strength. In the same manner, Figure 6 shows the ultimate tensile strength of 20% SiC to 80% Al alloy particles as 155.63 MPa. The 70% Al 8011-30% SiC metal matrix composites have the maximum tensile strength of 162.61 MPa when compared to the remaining composites and pure alloy as shown Figure 7, due to the uniform reinforcement of the Silicon Carbide particles, this 70% Al 8011-30% SiC has the highest tensile strength.

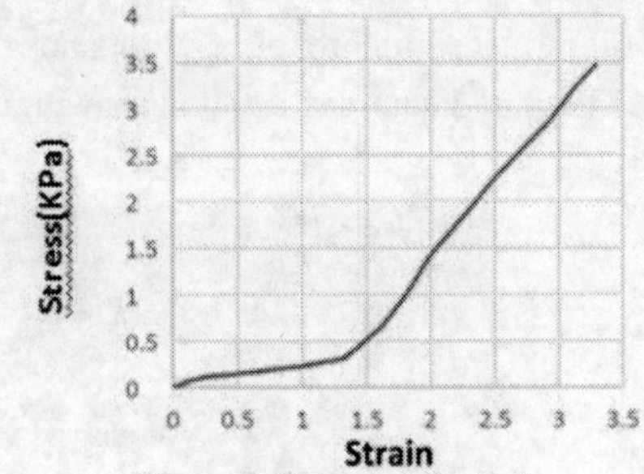

Figure 4. Stress vs Strain for Pure 8011 Aluminium

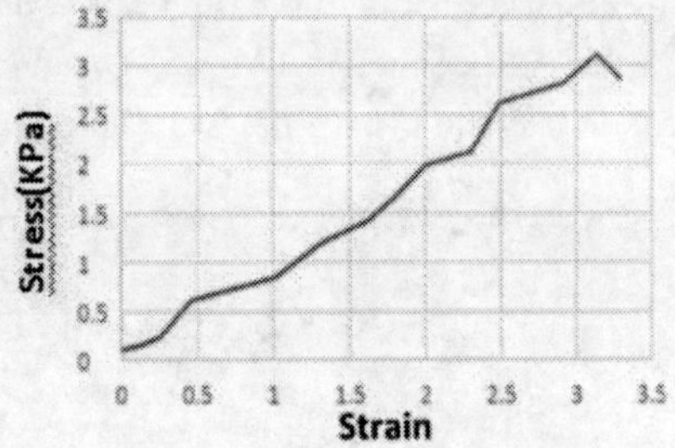

Figure 5. Stress vs Strain for 90%Al+10%SiC

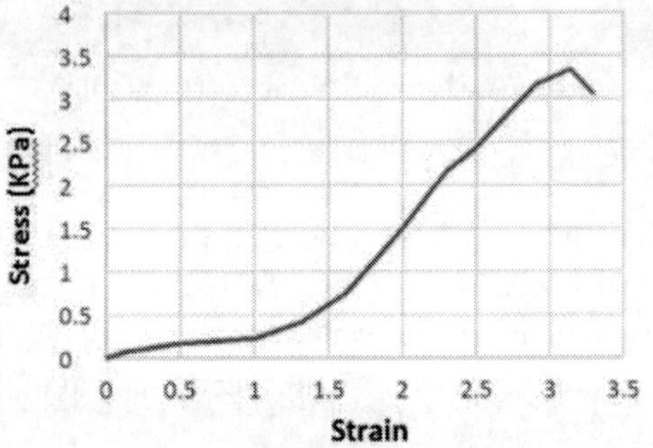

Figure 6. Stress vs Strain for 80%Al+20%SiC

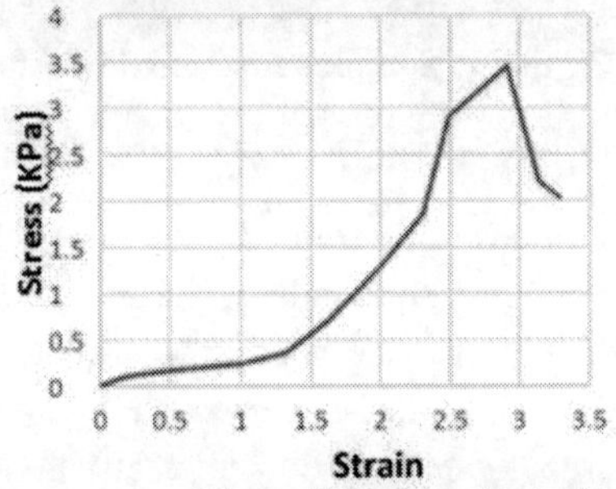

Figure 7. Stress vs Strain for 70%Al+30%SiC

6.2 *Hardness Test*

The microhardness test was carried in Micro Vickers Hardness Tester which has a testing load range of 10 gms to 1 kg load and testing scale used is HV. The prepared samples have undergone a load of 0.5 kg with reside time 10 sec in three locations of each sample. Figure 8 indicates the mean value of micro hardness was enhanced with the accumulation of weight percentage of ceramic (SiC) particle into the aluminium matrix.

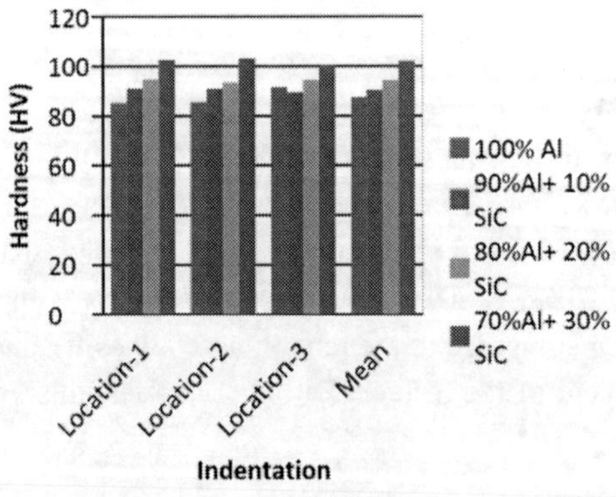

Figure 8. Hardness values of samples

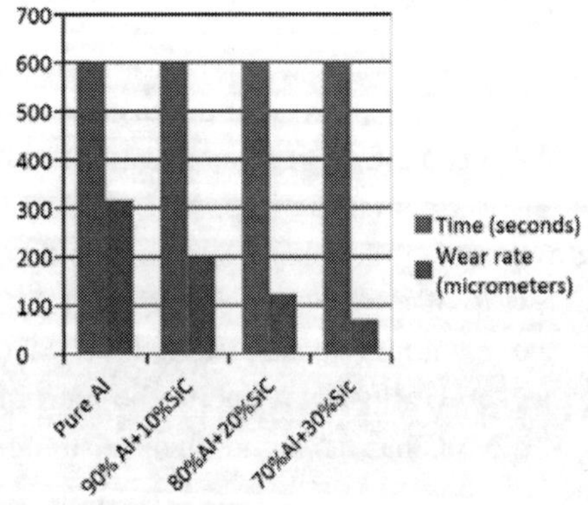

Figure 9. Wear rate

6.3 *Wear test*

Tribological properties of the produced samples were evaluated by conducting test on Pin and Disc set up. A load of 9.81 N was applied for a period of 600 seconds throughout the experiment for all the samples.

Casted Pure Al has a wear rate of 318 micrometers when subjected to a load of 9.81 N for a period of 600 seconds with a coefficient of friction of 0.320 as shown in Figure 5.5. The wear rate of 200 micrometers was drastically reduced as shown in Figure 5.6 for 10%SiC + 90% Al from the Pure Al alloy. Again with 20 wt.% of SiC and 80 wt.%Al8011 alloy a wear rate of 123 micrometers is observed Figure 5.7. The minimum wear rate of 71 micrometers was observed in Figure 5.8 for 30 wt.%SiC and 70 wt.% Al8011 alloy composite material. The wear resistance enhances by increasing the proportion of SiC to Aluminium matrix.

7. Conclusion

Aluminium alloy based Silicon Carbide Metal Matrix Composite have successfully fabricated by die casting. The addition of silicon carbide particles to the pure metal enhances the mechanical properties tensile strength, hardness and wear resistance of the composites. From this experimental study, it is noted that the higher the percentage of reinforced SiC, the higher is the tensile strength and hardness value with lower wear rate is observed. The sample with 70% Al + 30% SiC has the tensile strength of 162.61 MPa and the Vickers hardness of 103.1 H.V with the wear rate of 71 micrometers is observed.

REFERENCES

[1] Jaya Prasad Vanam, Chiranjeevi R, Santosh Kumar R, VenkataRamanan VSN and Sunny Kumar A 2018 Effect of SiC on Mechanical, Microstructure and Tribological Properties of Alumium MMC processes by Stir Casting, IOP Conference Series: Materials Science and Engineering **455** 1-9

[2] Johny James S, Venkatesan K, Kuppan P and Ramanujam R 2014 Hybrid Aluminium Metal Matrix Composite Reinforced with SiC and TiB2 Procedia Engineering Vol 97 pp 1018-1026

[3] Kandan R, Dhinesh Kumar, Sudharsanam M, Venkadesan A and Badrinath, R 2014 Investigation of Mechanical Properties on Newly Formulated Hybrid Composite Aluminium 8011 Reinforced with Boron Carbide and Aluminium Oxide by Stir Casting Method International Journal For Scientific Research And Development Vol 5 pp 962-965

[4] Montasser S and Tahat 2010 Alumnium-silicon alloy and its composites reinforces by silicon carbide particles Microelectronics International Vol 27 pp 21-24

[5] Okayasu M, Sakai K and Takasu, S 2011, Material properties of long term naturally aged die-cast Al-Si-Cu alloys International Journal of Cast Metal Research Vol.24 pp 286-298

[6] Pugalenthi P, Jayaram M and Natarajan A 2015 Evaluation Of Mechanical Properties of Aluminium Alloy 7075 Reinforced With SiC And Al2O3 Hybrid Metal Matrix Composites Applied Mechanics and Materials vol 76 pp 246-251

[7] Rajesh Purohit, Rana RS, and Verma CS 2012 Fabrication of Al-SiCComposites through Powder Metallurgy Process and Testing of Properties International Journal of Engineering Research and Applications,vol 2 pp 420-437

[8] Rajesh Agnihotri and SantoshDagar 2017 Mechanical Properties Of Al-Sic Metal Matrix Composite Fabricated by Stir Casting Route Research in Medical and Engineering Sciences vol 2 no 5 pp 1-6

[9] Subramanya Reddy P, Kesavan R and VijayaRamnath, B 2017 Investigation of Mechanical Properties of Aluminium 6061-Silicon Carbide, Boron Carbide Metal Matrix Composite Springer Science Business Media Dordrecht vol 5 pp 253-258

[10] VijayaRamanth B, Elanchezhian C, Annamali RM, Aravind S and Vignesh, V 2014 Aluminium Metal Matrix Composites – A Review Advanced Materials Science vol 38 pp 55-60

Comparison of geometrical accuracy and surface finish of cam profile generated by wire-EDM and CNC milling machine

C. Veera ajay[1*], B. A. Pradeep[2], C. Boopathi[3], R. K. Sanjeev[3], V. Meganathan[3]
[1]Department of Mechanical Engineering, National Engineering College, Kovilpatti, Tamilnadu, India
[2]Department of Production Engineering, PSG College of Technology, Coimbatore, Tamilnadu, India
[3]Department of Mechanical Engineering, Bannari Amman Institute of Technology, Sathyamangalam, Tamilnadu, India

*ajayveera2@gmail.com

Abstract. CAM is a machining element that has a curved groove or curved outline used for converting rotary motion into the form of linear motion or vice versa. The radial cam is a type of cam which having rotating plate or disc with an outer circumference shaped for producing a linear movement to a follower which is held against it. The quality of the final machined component surface depends on the accuracy and surface finish, position, orientation and run-out. The main aim of this paper is to machine a CAM profiles with CNC milling and wire cut EDM process and comparing the surface finish and accuracy of the machined surface with surface roughness tester and coordinate measuring machine (CMM).

Keywords: Cam profiles, CNC Milling, Wire-EDM, Surface roughness, Geometrical accuracy

1. Introduction

The invention of numerical control (NC) and computerized numerical control (CNC) machine tools are enhanced great value to automation and flexibility in product manufacturing and machining of complex contours. In general cam profiles are designed by geometric curvatures by the points traced follower motion as well as through a set of locus of points. Here, the cam profile is designed using CREO software and manufactured using CNC milling centre and wire EDM process. Wire electrical discharge machining also called as wire-EDM is one of the non-conventional machining process that is used widely for die manufacturing. A CMM is a device used for computing the physical geometrical characteristics of a machined component. Here the accuracy of machined cam profiles is checked using co-ordinate measuring machine. A good quality surface will increases the corrosion and wear resistance, fatigue strength of the workpiece.

Numerous studies about the CAM profiles for continuity Bezier curves based on various parameters and construction of CAM profiles plates on CNC machine tools with NC code generation are performed. The effects of input parameters on response such as surface roughness using wire EDM is investigated and the deviation error is checked with coordinate measuring machine (CMM) hence these are the manufacturing techniques developed during recent years. Inspite of these studies, new research efforts are essential to clarify the effect of various factors involved in the process.

IOP Conf. Series: Materials Science and Engineering **764** (2020) 012041 doi:10.1088/1757-899X/764/1/012041

Hetal et al. [1] concluded that end continuity conditions were applied for checking the tool path variations produced during machining. There is no tool path difference observed in parametric programmed machining or CAM system but the marks of machining commonly referred to as "manufacturing signature" that happens due to selection of manufacturing process. The final generated cam profile is checked using CMM. Kanlayasiri et al. [2] studied the effects of input variables on the surface roughness of DC53 die steel using wire-EDM process. In this work, the input process parameters such as pulse-on time, pulse-peak current, pulse-off time, and wire tension are selected. Analysis of variance method was used for finding the input process variables that affecting response such as surface roughness. The final component is manufactured using wire-EDM process and the surface roughness value is measured using surface roughness tester. Kang et al. [3] proposed a new system that was applied to marine engine cams. Only minimum time (5 mins) is required for the measurement but the old coordinate measuring machine would take over 1 hour for the measurement even with a skilled operator.

Andrzej werner [4] described a method for improving the accuracy of machined components manufactured using CNC machine and wire electrical discharge machines. After completing the machining process, coordinate measurements were performed. Ahmari et al. [5] used the Design of Experiment methods for optimizing the process parameters of surface reconstruction for freeform surfaces using reverse engineering technology. The input process parameters are number of points in percentage, noise reduction, triangle counts in percentage, and Sampling whereas the response parameters are the computational time, surface accuracy and the required computer memory space. The authors concluded that reverse engineered surfaces final accuracy depends on number of input points of point cloud data, number of triangles in polygon model, and reduction of noise.

2. Methodology and Material used

A work material such as aluminum AA6063 bar of dimension 200 mm x 150mm x 25mm was selected for the wire EDM machining and CNC machining. The design of radial disc cam profile is developed in CAM software and NC code is generated for the machining purpose in CNC milling and wire-EDM machine. The dimension of the CAM product is shown in Figure 1.

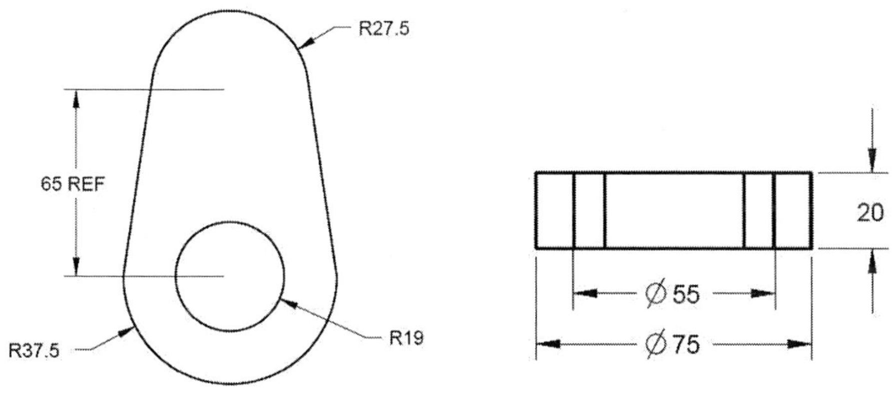

Figure 1. Dimension of the CAM product

The Figure 2 shows the Wire cut Electric Discharge Machining (WEDM) machine. This process can be used for cutting the plates having thickness upto 300mm. Electrode made of brass wire are generally used. The wire EDM parameters used for machined cam profile is shown in Table 1.

Table 1. Wire EDM Parameters

Parameters	Range
Voltage (v)	16
Pulse on time (µs)	1
Pulse off time (µs)	10
Wire speed (m/min)	5
Servo voltage(v)	45

Figure 2. Wire cut - Electrical Discharge Machine

The Figure 3 shows the machined cam profile on CNC machine. The material is surface finished before machining. LV 45 CNC milling machine is used for experimentation. The end milling cutter of diameter 15mm is used for machining process and for drilling operation the drill bit of diameter 12mm is used. The CNC milling parameters used for machined cam profile is shown in Table 2.

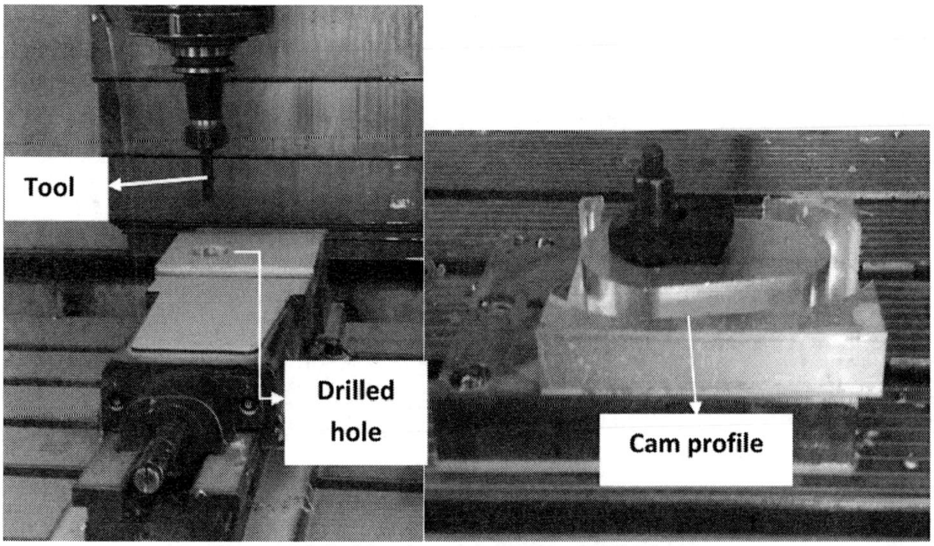

Figure 3. Machined cam profile in CNC Machine.

Table 2.CNC Milling Parameters

Drilling operation parameter	Outer profile parameter
Roughing operation	Roughing operation
Feed = 1200 mm/min Speed = 2200 rpm Depth of cut = 20 mm	Feed = 1850 mm/min Speed = 1850 rpm Depth of cut = 20mm
Finishing operation	Finishing operation
Feed = 1200 mm/min Speed = 3200 rpm Depth of cut = 20mm	Feed = 650 mm/min Speed = 3000 rpm Depth of cut = 20mm

The machined cam profiles are checked for surface roughness and accuracy in surface roughness tester and coordinate measuring machine respectively. The surface roughness tester is used to measure the surface roughness of the machined CAM surface. A Coordinate Measuring Machine (CMM) is used for measuring the physical geometrical characteristics of a CAM profile. This machine may be controlled manually by an operator or it may be controlled by computer. The measuring range of the CMM used is 500 mm × 400 mm × 400 mm in x, y, z axis respectively. The CMM used for measuring wall angle has a following specification: resolution: 0.0001 mm (0.1μm), drive speed: 8 – 300 mm/s (CNC mode), workpiece max height: 545 mm, workpiece max mass: 180 kg, air supply pressure: 0.4 MPa, probe material used: Ruby.

3. Results and description

3.1 Calculation of Material removal rate in wire EDM process

$$MRR = K \times H \times F_R \times \rho$$

IOP Conf. Series: Materials Science and Engineering **764** (2020) 012041 doi:10.1088/1757-899X/764/1/012041

Where, K is the cutting width in mm
L is the thickness of the material in mm.
F_R is the cutting speed mm/min.
Density of the workpiece g/mm

$$MRR = 25 \times 25 \times 0.5 \times (2.7 \times 10^{-3})$$
$$= 0.625 \text{ g/min}$$

3.2 Surface roughness comparison of cam profiles

The surface roughness is measured for machined CAM profile at different location such as top surface, botttom surface, outer profile and inner circle. The solid view with indication of different surface is shown in Figure 4. The surface roughness value for the machined CAM profile in Wire EDM and CNC Milling machine at different location is shown in Table 3.

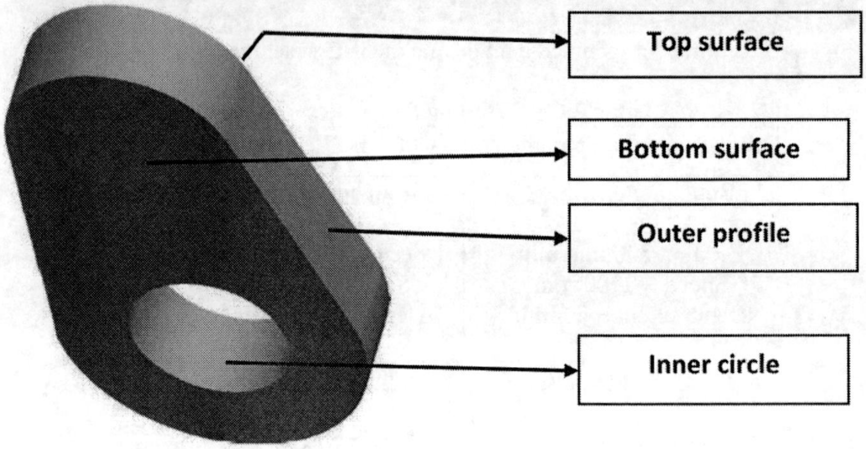

Figure 4. Solid view of cam profile

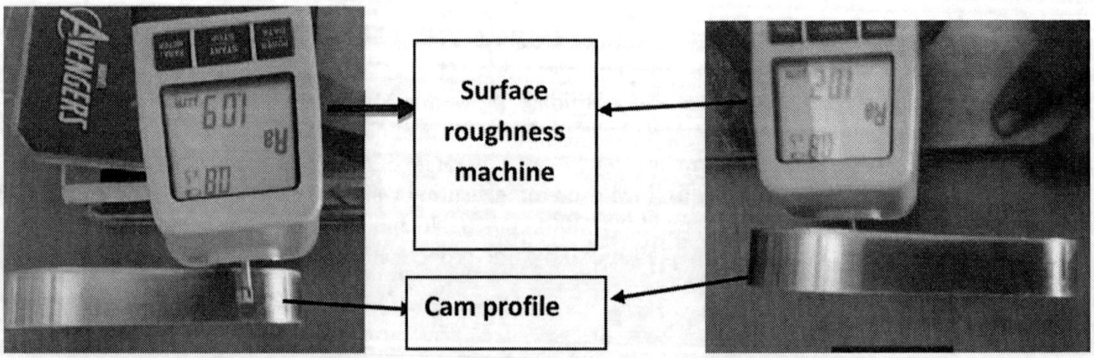

Figure 5. Surface measurement of outer profile and inner hole

The surface roughness measurement gives 0.21µm and 0.33µm variation in inner circle and outer surface of CAM profiles which shows that CAM 1(CNC milling) has higher surface finish than CAM 2 (wire EDM).

Thus from the Table 3, it is concluded that surface finishis better in CAM 1 (CNC milling) than CAM 2 (wire EDM). From the Figure 6, it is understood that CAM profile 1 (CNC milling) have lower surface roughness value compared toCAM profile 2 (wire EDM).

Table 3. Surface Roughness Measurement

S.No	SURFACE ROUGHNESS MEASUREMENT		
	Surface	CAM 1 (CNC milling)	CAM 2 (wire -EDM)
1	TOP	0.74 μm	0.80 μm
2	BOTTOM	0.76 μm	0.85 μm
3	INNER CIRCLE	1.02 μm	1.23 μm
4	OUTER PROFILE	1.09 μm	1.42 μm

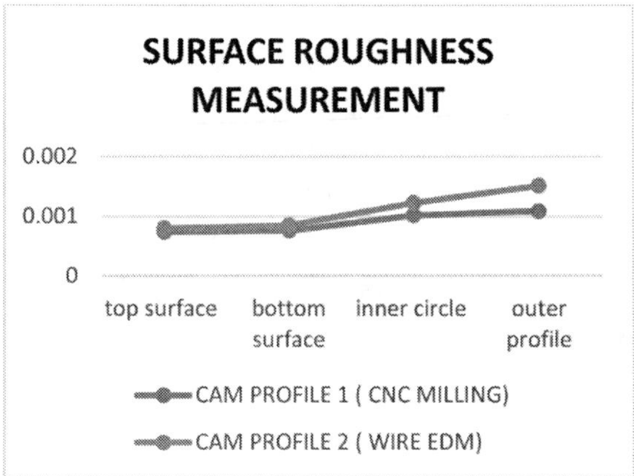

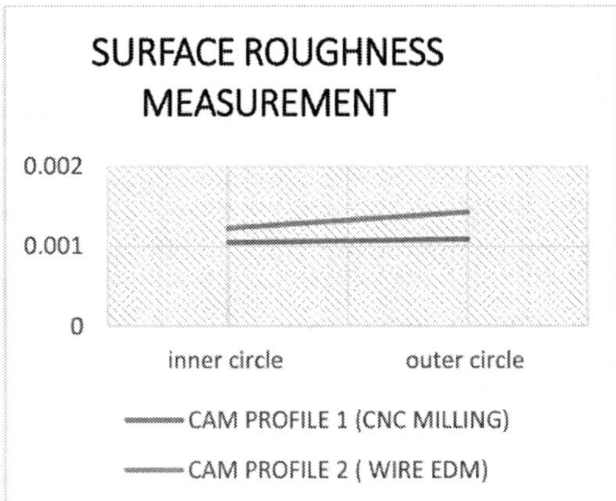

Figure 6. Surface roughness measurement

3.3 Geometric Accuracy Measurement

CAM profiles machined from CNC milling and wire EDM process are compared for geometrical accuracy. Figure 8 shows the depth and circularity measurement for the drilled hole in cam profile using CMM. Figure 9 shows the comparison of geometrical accuracy of the cam profiles machined from CNC milling and wire EDM process. Cam profile machined from wire-EDM is more accurate than cam profile machined from CNC milling (Only slight accuracy varies).

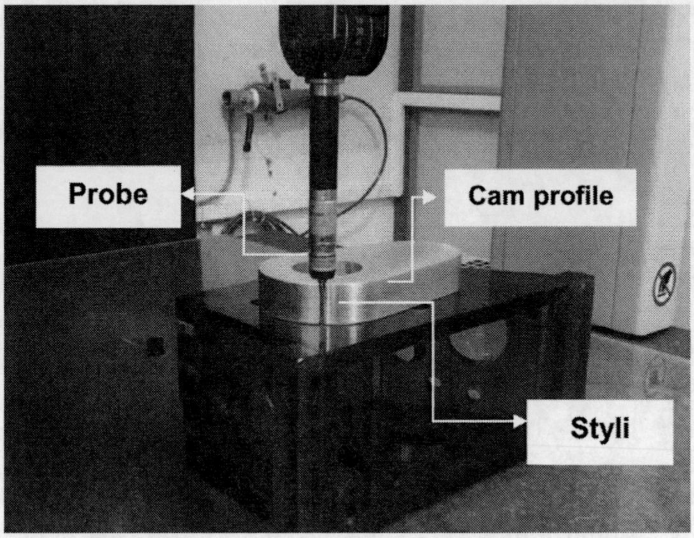

Figure 7. Coordinate Measuring Machine (CMM)

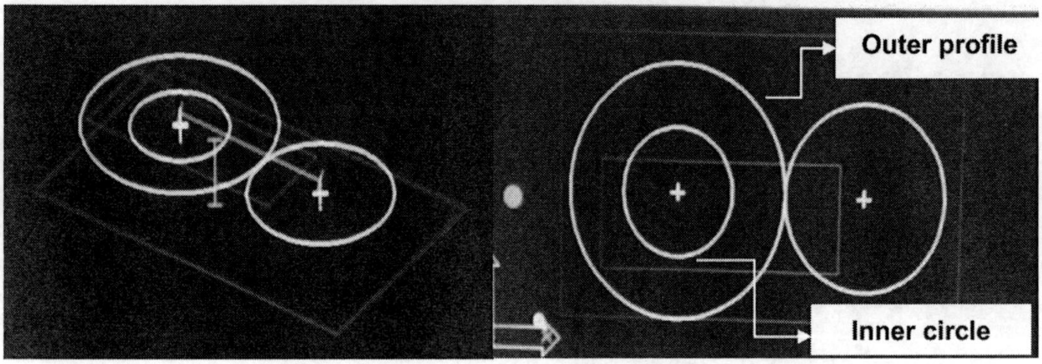

Figure 8.CMM – Depth and Circularity measurement.

4. Conclusion

Cam profile made of aluminum alloy AA6063 is machined with CNC milling and wire cut EDM process. The cam profile machined using CNC milling and wire EDM machine, which is compared to ensure the effective processing in the view of surface finish by surface roughness tester and also the surface accuracy using coordinate measuring machine (CMM).

a) The CAM profiles are machined from CNC milling with varying speed, feed and depth of cut as well as from wire EDM.
b) The surface roughness measurement gives 0.21μm and 0.33μm variation in inner circle and outer surface of CAM profiles which shows that CAM 1 (CNC milling) has higher surface finish than CAM 2 (wire EDM).
c) The geometrical accuracy measurement was performed using CMM where Cam profile machined from wire-EDM has more accurate than cam profile machined from CNC milling.

IOP Conf. Series: Materials Science and Engineering **764** (2020) 012041 doi:10.1088/1757-899X/764/1/012041

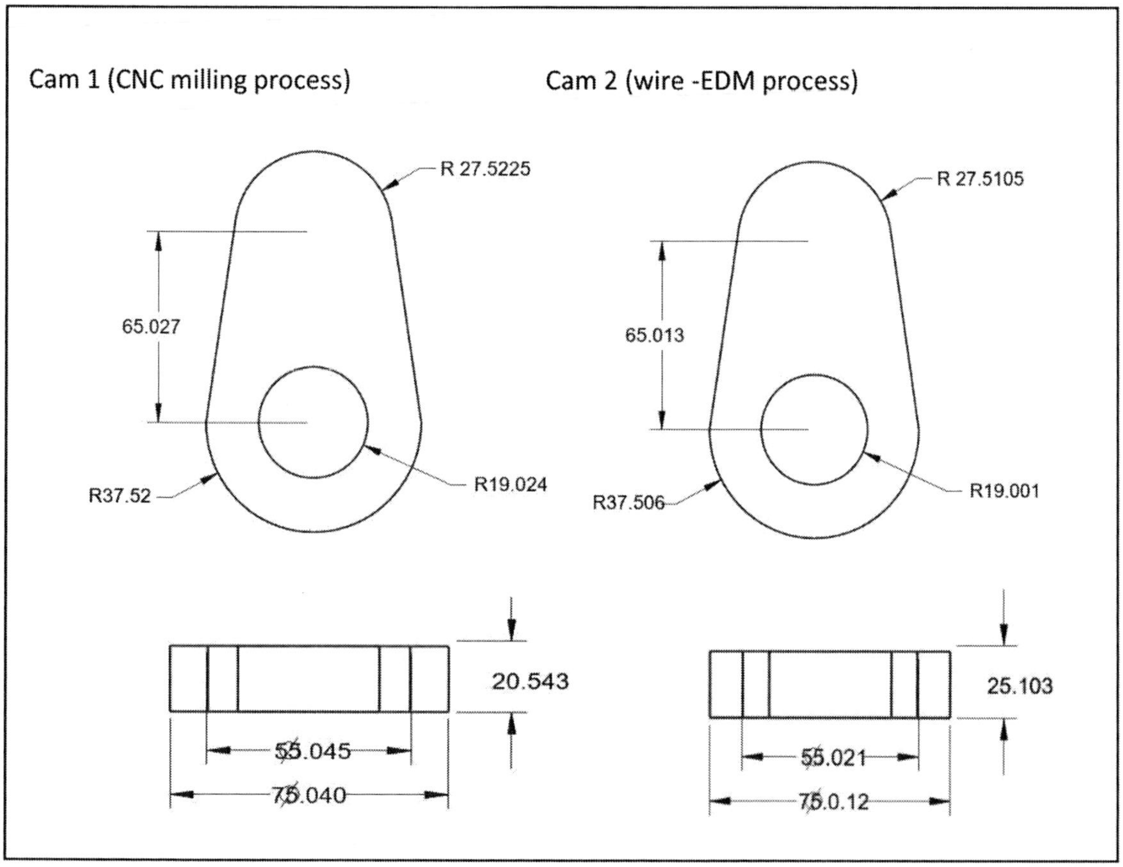

Figure 9. Geometrical comparison of two cam profiles

Reference

[1] Hetal N. Fitter, Akash B. Pandey, Jitendra M. Mistry, D. Divyang 2014 Patel comparison of parametrically machining with CAM machining for C0 continuity Bezier curves based on various parameters. Procedia Engineering 97:1167 – 1177.

[2] K. Kanlayasiri, S. Boonmung 2007 Effects of wire-EDM machining variables on surface roughness of newly developed DC 53 die steel: Design of experiments and regression model. Journal of Materials Processing Technology192-193:459-464.

[3] Jae-Gwan Kang, Sangbo Han 2010 Measurement and evaluation of form deviation error of disk cam with an exclusively built profile-measuring machine. Robotics and Computer-Integrated Manufacturing 26(6):686-693.

[4] Andrzej Werner 2016 Method for enhanced accuracy in machining curvilinear profiles on wire-cut electrical discharge machines. Precision Engineering44:75-80.

[5] A.M.A. Al-Ahmari, Javed Aalam 2015 Optimizing parameters of freeform surface reconstruction using CMM.Measurements 64:17-28.

IOP Conf. Series: Materials Science and Engineering **764** (2020) 012042 doi:10.1088/1757-899X/764/1/012042

The Result of Nano-Additives in Diesel Engine using Diesel-E Waste Plastics Pyrolysis Oil Fuel Blends: A Review

S. Deepankumar[1*], K. L. Senthil Kumar[2], L. S. Gokul[3], B. Saravanan[4]

[1,4]Department of Automobile Engineering, Bannari Amman Institute of Technology Sathyamangalam, India.
[2]Department of Mechatronics Engineering, Bannari Amman Institute of Technology Sathyamangalam,India.
[3]Department of Mechanical Engineering, Bannari Amman Institute of Technology Sathyamangalam,India.
[*]Author for correspondence E-mail: deepankumar@bitsathy.ac.in

Abstract. Primary pollutants IC engines are nitrogen oxide (NOx) and particulate matter (PM). For most recent years, researchers are rigorously working to discover the alternative fuel which will cut emission characteristics without compromising the performance. There are two methods in general to reduce emission in CI Engines. One way is using catalytic converter and another is to get better performance of Diesel engine using nano fuel additive. Adding fuel additive is one of the best ways to improve the performance of CI engines. Many researchers have contributed in this area and this paper consolidates the outcomes of different tests done on the CI engine utilizing nano fuel additives with pyrolysis oil in diesel fuel. It is understood that majority of the researchers report performance enhancement and reduction in emission characteristics with dosing of nano particles additives in diesel and biodiesel. The result of several nano particles on the enrichment in the performance characteristics and drop in emission of CI engine tested with diesel-biodiesel fuel blends are discussed.

1.Introduction

With the huge applications of plastics, the making of plastics is growing worldwide in leaps and bounds. Wood, metals and ceramics have been substituted by plastics as the quality of human improved with the discovery of plastics. From the start of development of plastics, growth of plastic waste gives increase to environmental problems. It may take billions of years for plastic to degrade in nature. Recycling and reuse of plastics is found to be the most favorable solution to administer the plastic waste. Chemical recycling conforms to the principles of sustainable development in future [1] and energy calamity, which has gain huge awareness by researchers. The challenges of E-waste management and increasing fuel energy crisis can be addressed simultaneously with the making of fuel from E-Waste plastics. Several researchers showed that the fuel formed from E-Waste plastics have properties like to that of petroleum fuels [2-9]. This review is to analyze the diesel engine's performance, emission and combustion characteristics with E-waste plastic oil blends as fuel produced by pyrolysis with nano fuel additive.

1.1 Pyrolysis mechanism

Proximate analysis is typically adopted to measure the chemical properties of E-plastics. The main elements in plastics are carbon, moisture content, volatile matter and ash content as the liquid oil yield depends on the assumed elements [10]. Thermo chemical action and Catalytic conversion are the generally used methods for manufacture of fuel from E-plastics [11]. In thermo chemical treatment, large polymers in E-plastics are split into smaller hydrocarbons in an oxygen free background at high-minded temperatures. Whereas, in catalytic conversion, a catalyst is used to modify the reaction mechanism which reduces the Pyrolysis temperature and selectivity [12,13].

1.2. Thermo chemical treatment

In pyrolysis process, the lighter molecular weight polymers require higher temperatures for degradation [14]. Parameters like degradation time, temperature, type of reactor and residence time govern the residue of plastic pyrolysis procedure. As such, the hydrocarbons having boiling point around 350C and 1850C are used as gasoline and that of 1850C and 2900C are used as diesel [15, 16]. At the temperature range of 6500C – 7000C, the weight % of oil yield and wide supply of hydrocarbon are obtained [17]. In fact, at the temperatures between 3700C and 4200C, in the incidence of oxygen, five liquid fraction having hydrocarbons involving C_4H_8 and $C_{28}H_{58}$ were obtained [18]. Nitrogen, when used for pyrolysis of mixed plastic E-wastes resulted in privileged concentration of aromatics and alkanes [19]. The Pyrolysis of E-waste electrical and electronic equipment at an elevated temperature of 8000C has a yield of 30.4 – 45.5% of gas [20].

1.3 Catalytic conversion

In the thermal cracking catalyst helps in growing the reaction rate along with reduction in pyrolysis temperature. This improves the quality of fuel, increase selectivity and residence time [21]. Nano crystalline zeolites like HZSM-5, Hβ and HMOR generally use catalysts [22]. On the outset, more amount of catalyst compact the yield of plastic oil [23]. Hydrocarbon in the plastic oil yield was improved from 52% to 60% by using activated carbon bed containing Fe in the incidence of hydrogen [24]. Catalysts HZSM-5 and SiO_2-Al_2O_3 when used in mixture resulted in the final product with better octane rating [25].

2. Engine Performance and Emission Analysis

In order to balance the energy required and substitute of fossil fuels at least to some extent, the development of alternative energy sources is in evitable. And also, global warming and E-waste management policies have enforced the utilization of alternative fuels on engines.

Experiments were conducted by Ioannis et al [26] on a CI engine fuelled with blends of waste plastic pyrolysis oil in proportions of 25%, 50%, 75%, 90% and 100% (v/v%). The Brake thermal efficiency (BTE) of WPO blends was found to be lesser than diesel. The upper blending ratios of WPO resulted in higher in net heat release rate, cylinder pressure, HC, CO , NOx, and CO_2 emissions. Paul Daniel et al [27] used blends of 10%, 20%, 30% and 50% (v/v%) and exhaust gas recirculation of 5%, 10%, 15% and 20%. BSFC increased by 2.3%, 4.6%, 16.3% and 22.6% by EGR of 5%, 10%, 15% and 20% respectively. PPO helped in NOx reduction. However, HC, CO and CO_2 emissions were increased with increase in PPO composition and EGR rate.

Plastic oil produced from crude landfills was tested as fuel on diesel engine by Adun et al [28]. The diesel equivalent fraction (LFPL3) was produced at a temperature range between 2500C and 3500C. They reported that BSFC by 12.5%, EGT by 11.6%, HC by 12.5%, NOx by 45.4% and CO by 5.4% with LFPL3. Investigation was carried out by Ravi Shankar et al [29] on a CRDI engine with plastic Pyrolysis oil blends in proportions of 10%, 20% and 30% on mass basis. The calorific value of the blend decreased with the increase in quantity of plastic oil. The brake thermal efficiency significantly decreased by 12%, 31% and 48% for plastic oil-diesel fuel blends of 10%, 20% and 30%

respectively. NOx emission was found to be 15%, 22% and 23% higher for 10%, 20% and 30% blends respectively compared to diesel under full load conditions. Sumit and Rohit [30] used semi batch type pyrolisis reactor to produce plastic oil from waste plastics and polyethylene. The Pyrolysis temperature was maintained between 5000C and 7000C. Plastic oil-diesel blends in proportions of 5%, 10%, 15% and 20% were used as fuels on diesel engine. They reported that the catalyst plays an important role in the thermal cracking. As such no catalyst was used in their process of production of plastic oil. Due to low oxidation stability, plastic oil been kept for some hours had gum out.

The addition of oxygenated additive, Di ethyl ether (DEE) reduced the emission levels with plastic oil-diesel blends [31]. Experimental investigations were carried out on VCR engine at compression ratios of 12, 16 and 20 with P2.5, P7.5, P12.5 and P100 blends along with 2.5% volume of DEE by volume. It was found that addition of DEE reduces the viscosity of fuel blends and increases the rate of reaction. Alcohols are highly volatile than DEE and when added to bio-diesels, reduce harmful emissions from diesel engine. Damodharan et al [32] synthesized oil from waste plastics in a laboratory scale batch reactor. Blends of diesel, waste plastic oil and n-Butanol like D50-WPO40-B10, D50-WPO30-B20 and D50-WPO20-B30 were used as fuel on a diesel engine. D50-WPO20-B30 delivered better performance than diesel. Smoke opacity of neat WPO was higher than neat diesel and blends of WPO. With addition of n-Butanol to blends, the smoke emission diminished. This may be due to the improved availability of oxygen and lower carbon content. On the summary, NOx, CO and HC emissions increased with increase in WPO composition.

Devraj et al [33] conducted experiments using diesel, WPPO(100% waste plastic oil), WD05(WPO blended with 5%DEE) and WD10(WPO blended with 10%DEE) on a diesel engine and found that BSFC for diesel was lower than WPPO by 12%. With addition of DEE to blends, HC and NOx emissions were higher than baseline fuel diesel. Under full load conditions, CO emission decreased by 10.5% when compared between WD05 and WD10. This may be due to availability of additional amount of oxygen in DEE. CeylaGungor et al [34] produced oil from waste polyethylene using sodium aluminum silicate catalyst in a thermal reactor. Experimental study was conducted on a 4 cylinder diesel engine with blends of WPE5, WPE10, WPE15 and WPE20 at varying speeds and the results were compared with diesel fuel. Result revolved that Power output was increased by 1.6% and CO emission was decreased by 20.63% for WPE5 blend. Hamzah et al [35] studied the performance of diesel engine using Waste Plastic Disposal Fuel (WPDF) and diesel as fuels at varying speeds and constant load on a single cylinder diesel engine. Reduction in brake power and torque was observed for WPDF at all speeds than diesel. This was attributed to lower cetane number of WPDF which is 5.5% lower than that of diesel. The peak combustion pressure at low speed (1200 rpm) and high speed (2400 rpm) for WPDF is inferior to diesel. Kaimal et al [36-39] synthesized plastic oil in a reactor at a temperature range of 3500C – 4000C with silica catalyst and obtained a yield of 80% by weight of plastic waste. Experiments were conducted with various blends of diesel-plastic oil such as PO, PO05, PO50, PO75 and diesel – plastic – Di ethyl ether like PD5, PD10 and PD15. They reported that the combustion of PO varies from 370CA to 460CA under full load conditions.

Emulsification of waste plastic oil with water was carried out by Senthil and Sankaranarayanan [40]. The emulsions like PW10 (Waste plastic oil 88% + Span 80-1% + Water 10% + Tween 1%), PW20 (Waste plastic oil 78% + Span 80-1% + Water 20% + Tween 1%) and PW30 (Waste plastic oil 68% + Span 80-1% + Water 30% + Tween 1%), PO (WPO 100%) and diesel were used as fuels. They reported that the NOx emissions for emulsions reduced greatly by 32%. Out of which PW30 recorded least NOx emissions under full load conditions. Rajan et al [41] used WPO diesel blends in proportions of 10%, 20% and 30% (v/v%) of WPO. The characterization of plastic oil revealed that the viscosity of WPO 100 is lower than that of diesel.

Plastic oil was produced by Kaoline catalyst from waste polypropylene at a temperature range of 680C to 3460C which is equivalent to diesel. Blends like, 10% B WPO, 20% B WPO, 30% B WPO, 40% B WPO and 50% B WPO were tested on a diesel engine. Brake thermal efficiency was greater than diesel for all blends more than80% load. The exhaust gas temperature and NOx emissions were originate to be higher for blends at all loads. This may be due to the improper penetration of WPO blends deeper into the CI engine combustion chamber. Also, the unsaturated hydrocarbons present in the fuel (WPO) might have lead to higher HC and CO emissions. It was also reported that the vibration of engine had occurred for 50B WPO at full load [42]. Mani et al [43-46] discovered most of the paths of utilizing WPO in neat and blend form. Tests were conducted at advance and retard injection timing using blends like WPO, WPO10, WPO30, WPO50, WPO70 and varying exhaust gas recirculation (EGR) rate as 10%EGR and 20%EGR. Characterization of WPO revealed that 75% of liquid hydrocarbon to be obtained on mass basis from E-plastic wastes. They observed that for WPO without EGR, heat discharge rate increased by 24.5% at max load. Marginal difference in BTE is found with 20% EGR. Also, the exhaust gas temperature, smoke, NOx and CO2 concentration decreased marginally. Table 1 inclines the summary of previous studies that confirmed the use of waste plastic oil and its mixtures as fuels and Table 2 is the list of properties of E-waste plastic oil and its mixtures from previous studies[48-54].

Table I : Table of previous Investigations

Ref No.	Blend	EM	BTE	BSFC	NOx	HC	CO	CO2	HRR	Smoke	EGT	Cylinder Pressure	Remarks
[26]	PPO		↓	↑	↑	↑	=	↑	↑		↑	↑	A blend between 60% PPO at 80-90% loadis optimized
	PPO25		↓	↑	↑	↑	=	↑	↑		↑	↑	
	PPO50		↓	↑	↑	↑	↑	↑	↑		↑	↑	
	PPO75		↓	↑	↑	↑	↑	↑	↑		↑	↑	
	PPO90		↓	↑	↑	↑	↑	↑	↑		↑	↑	
	PPO100		↓	↑	↑	↑	↑	↑	↑		↑	↑	
[27]	PPO 10			↑	↓								5% and 10% EGR can be particular as good value
	PPO 20			↑	↓								
	PPO 30			↑	↓								
	PPO 50			↑	↓								
		NO EGR		↑		↑	↑						
		5% EGR		↑		↑	↑						
		10 % EG R		↑		↑	↑						
		15% EGR		↑		↑	↑						
		20 % EG R		↑		↑	↑						
[28]	LFLP3		↑	↓	↓	↓	↓	↓			↓		LFLP3 is improved fuel over Diesel
[29]	PO 10		↓	↑	↑		↓	↑			↑		PO used in pure with Slight modifications to the engine
	PO 20		↓	↑	↑		↓	↑			↑		
	PO 30		↓	↑	↑		↓	↑			↑		
[30]	WPPO 5		↑	↓	↑	↓	↑	↓					20% blend is Optimized
	WPPO 10		↓	↑	↑	↓	↑	↓					
	WPPO 15		↓	↑	↑	↓	↑	↓					
	WPPO 20		↓	↑	↑	↓	=	↓					
[31]	P 2.5		=	↑	↑	=	↑	↓		↑			WPO can be mixed with diesel and DEE used as improver.
	P 7.5		↓	↑	↑	↑	↑	↓		↑			
	P 12.5		↓	↑	↑	↑	↑	↓		↑			
	P 100		↓	↑	↑	↑	↑	↓		↑			
		CR 12	=	↑	↑	↑	↑	↓					
		CR 16	↓	↑	↑	↑	↑	↓					
		CR 20	↓	↑	↑	↑	↑	↓					

IOP Conf. Series: Materials Science and Engineering **764** (2020) 012042 doi:10.1088/1757-899X/764/1/012042

[32]	D50WPO40B 10		=	↑	↓	↑				↓			D50WPO20B30 is Optimized ,n-Butanolisgood asimprover
	D50WPO30B 20		↓	↑	↓	↑				↓			
	D50WPO20B 30		↓	↑	=	↑				↓			
	WPO		↓	↑	↓	↑				↑			
[33]	WPPO		↓	↑	↑	↑	↑	↑		↑		↑	Adding of DEE is encouraging
	WD 05		↓	↑	↓	↑	↑	↓		↓		↓	
	WD 10		↑	↑	↓	↑	↑	↓		↓		↓	
[34]	WPE 5				↑		↑	↓					WPE 5 mixture is the good blend
	CPE 5				↑		↑	↓					
[35]	WPDF										↑	↓	100% WPO as base fuel
[36]	PO		↓	↑					↑			↑	Engine to be adept of run using 100% PO
	RME		↓	↑					↑			↑	
[37]	PO 25		↓						↑			↑	PO 25 shows improved results
	PO 50		↓						↑			↑	
	PO 75		↓						↑			↑	
	PO 100		↓						↑			↑	
[38]	PO 25		↓	↑	↑	↑	=			↑			25 % blend is an useful replacement for diesel
	PO 50		↓	↑	↑	↑	↑			↑			
	PO 75		↓	↑	↑	↑	↑			↑			
	PO		↓	↑	↑	↑	↑			↑			
[39]	PD 5		↓		↑	↑	↑		↑				Adding of DEE is a workable key without enginemodificat ions
	PD 10		↓		↑	↑	↑		↑				
	PD 15		↓		↑	↑	↑		↑				
	PO		↓		↑	↑	↑		↑				
[40]	WPO		↓		↑	↑			↑	↑	↑		PW 30 has improved performance.
	PW 10		↑		↓	↑			↑	↓	↓		
	PW 20		↑		↓	↑			↑	↓	↓		
	PW 30		↑		↓	↑			=	↓	↓		
[41]	WPF		↑	↓	↓	↑	↓	↓			↓		Performance of WPF10D90 is like to diesel
	WPF10D90		↑	↑	↓	↓	↓	=			↓		
	WPF20D80		↓	↑	↓	↓	↓	↑			↓		
	WPF30D70		↓	↑	↑	↓	↓	↑			↓		
[42]	10%B WPO		↓	↓	↑	↑	↑			↑	=		Improved performance up to 30% blend. BTE of blends is closely same as diesel
	20%B WPO		↓	↑	↑	↑	↑			↑	↑		
	30%B WPO		↓	↑	↑	↑	↑			↑	↑		
	40%B WPO		↓	↑	↑	↑	↑			↑	↑		
	50%B WPO		↓	↑	↑	↑	↑			↑	↑		
[43]	WPO		↓		↑	↑	↑	↑	↑	↓	↑	↓	WPO displays upper BTE up to 75% graded power
[44]	WPO	Retarded injection timing	↑	↓	↓	↓	↓	↓	↓	↓	↓	↓	Retardation solid NOx emission by 30%
[45]	WPO	NO EGR	↑		↑	↓	↑	↓	↑	↓	↑	↑	EGR reduces NOx emission
		10% EGR	↑		↑	↓	↑	↓	↓	↑	↑	↑	
		20% EGR	↑		↑	↓	↑	↓	↓	↑	↓	↑	
[46]	WPO 10		↓		↑	↑	↑	↓	↓	↑	↑	↑	WPO results inimprovedperfor mancethan blend of WPO anddiesel
	WPO 30		↓		↑	↑	↑	↓	↓	↑	↑	↑	
	WPO 50		↓		↑	↑	↑	↓	↓	↑	↑	↑	
	WPO 70		↓		↑	↑	↑	↓	↓	↑	↑	↑	
	WPO		↓		↑	↑	↑	↓	↑	↑	↑	↑	

↓ = Decreased; ↑ = Increased

Table II. Properties of E-waste plastic oil and its blends

Ref No	Blend	Density (kg/L)	Kinematic Viscosity (Cst)	Flash point (^{0}C)	Calorific Value (Mj/kg)	Cetane No.
[26]	PPO	0.9813	1.918	13	38.3	-
	Diesel	0.8398	2.62	59.5	42.9	-
[27]	WPO	0.8423	-	28	40.05	
	Diesel	0.8377	-	66	44.8	-
[28]	LFLP3	0.802	2.085	50	46	65
	Diesel	0.834	3.162	75	45.5	60
[30]	WPPO	0.78	3.8	100	39.76	-
	Diesel	0.79	1-4.11	52-96	44.8	-
[31]	P 2.5	0.831	1.568	49	44.414	42.51
	P 7.5	0.824	1.79	48	44.091	43.03
	P 12.5	0.819	1.92	47	43.868	43.61
	P 100	0.790	2.52	42	43.34	51
	DEE	0.714	0.23	-40	33.85	126
	Diesel	0.84	2.0	50	44.8	42
[32]	Diesel	0.838	3.8	70	41.82	54
	WPO	0.813	2.16	38	40.35	51
	n-Butanol	0.810	2.2	36	34	-
	D50WPO40B10	0.825	3.0	36	40.45	48.6
	D50WPO30B20	0.8249	3.1	36	39.815	44.7
	D50WPO20B30	0.8246	3.2	36	39.18	40.8
[33]	Diesel	0.84	-	50	46.5	40-45
	WPPO	0.798	-	42	45.216	51
	DEE	0.713	-	45	33.9	126
	WD05	0.794	-	43	44.65	-
	WD10	0.754	-	43.5	44.084	-
[34]	Diesel	0.833	2.52	58.5	10790 (kcal/kg)	54.58
	WPE 100	0.788	2.326	<100	10867 (kcal/kg)	43.751
	WPE 5	0.831	2.49	58.5	10800 (kcal/kg)	52.58
	WPE 10	0.828	2.48	59	10807 (kcal/kg)	51.63
	WPE 15	0.826	2.43	59.5	10812 (kcal/kg)	50.37
	WPE 20	0.824	2.4	61	10818 (kcal/kg)	49.62
[35]	Diesel	0.8416	3.05	84	42.49	68.2
	WPDF	0.7710	2.149	72	34.72	64.4
[36]	Diesel	0.84	2.15	45	43.5	54
	PO	0.83	2.64	40	44.2	50
[39]	Diesel	0.84	2.15	45	43.5	54
	PO	0.83	2.64	40	44.2	50
	DEE	0.714	0.22	-45	33.87	125
	PD5	0.83	2.52	40	43.167	-
	PD10	0.827	2.4	40	42.834	-
	PD15	0.821	2.37	39	42.501	-
[41]	WPF10D90	0.804	0.616 (Poise)	65	44.87	46
	WPF20D80	0.832	0.052 (Poise)	66	44.39	46
	WPF30D70	0.815	0.064 (Poise)	69	45.58	47
	WPF100	0.78	0.0457 (Poise)	72	42.90	44
	Diesel	0.80	0.61 (Poise)	61	45.35	49
[43]	WPO	0.835	2.52	42	44.34	51

IOP Conf. Series: Materials Science and Engineering **764** (2020) 012042 doi:10.1088/1757-899X/764/1/012042

	Diesel	0.84	2.0	50	46.5	55

3. Conclusions

The necessity for exploration of other alternate has raised because of quicker run-down of diesel fuels and natural consequences. The oil from E-waste plastic has start to become the alternative fuel whereby the controlling of the same is to be spoken mentioning the difficulty in energy management with respect to E-waste. In the research constrained review the combustion, emission and performance characteristics of E-waste plastic oil-diesel blends on internal-combustion engine using diesel as a fuel. But, there are some issues like kinetics of oil from E-waste plastic, improvising properties of combustion and engine alterations which shall be thought-about for exhaustive exploration of IC engine combustion consuming the blend of oil from E-waste plastic with nano fuel additive.

References

1. SL. Wong, N. Ngadi, TAT. Abdullah, IM. Inuwa, Current state and future prospects of plastic waste as source of fuel – a review, Renewable and sustainable energy review,50, 1167-1180,2015.
2. AK. Panda, RK. Singh, Mishra DK, Thermolysis of waste plastic to liquid fuel – a suitable method for plastic waste management and manufacture of value added products – a world prospective, Renewable and sustainable energy review,14(1), 233-248,2010.
3. Buekens AG, Huang H, Catalytic plastics cracking for recovery of gasoline range hydrocarbons from municipal plastic wastes, ResourConservRecycl, 1998; 23(3);163-181.
4. Miskolczi N, Bartha L, Deak G, Jover B, Thermal degradation of municipal plastic waste for production of fuel like hydrocarbon, Polymer degrade stab, 2004; 86(2);357-366.
5. Jan MR, Shah J, Gulab H, Catalytic degradation of waste high density polyethylene into fuel products using BaCO3 as a catalyst, Fuel processing technology, 2010; 91;1428-37.
6. Jung SH, Cho MH, Kong BS, Kim JS, Pyrolysis of a fraction of waste poly propylene and polyethylene for the recovery of BTX aromatics using a fluidized bed reactor, Fuel processing technology, 2010; 91;277-84.
7. Miskolczi N, Bartha L, Deak G, Thermal degradation of polyethylene and polystyrene from the packaging industry over different catalysts into fuel like feed stocks, Polymer degrade stab, 2006; 91(3);517-26.
8. Shah J, Jan MR, Mabood F, Jabeen F, Catalytic Pyrolysis of LDPE leads to valuable resource recovery and reduction of waste problems, Energy conversion manag, 2010; 51;2791-801.
9. Akminsky W, Schlesselmann B, Simon CM, Thermal degradation of mixed plastic waste to aromatic and gas, Polm degrade stab, 1996; 53;189-97.
10. Hong SJ, Oh SC, Lee HP, Kim HT, Yoo KO, A study on the pyrolisis characteristics of poly vinyl chloride, J Korean inst chem. Eng, 1999; 37;515-21.
11. Passamonti FJ, Sedran U, Recycling of waste plastics into fuels-LDPE conversion of FCC, Applcatal B environ, 2012; 125;499-506.
12. Al Salem SM, Lettieri P, Baeyens J, Recycling and recovery routes of plastic solid waste – A review, Waste manag, 2009; 29;2625-43.
13. Shah SH, Khan ZM, Raja IA, Low temperature conversion of plastic waste into light hydrocarbons, J hazard matter, 2010; 179;15-20.
14. Gao F, Pyrolysis of waste plastic into fuels, University of Canterbury,2010.
15. Sharma BK, Moser BR, Vermillion KE, Dell KM, Rajagopalan N, Production characterization and fuel properties of alternative diesel fuel from Pyrolysis of waste plastic

grocery bags, Fuel process technol, 2014; 122;79-90.
16. Encinar JM, Gonzalez JF, Pyrolysis of synthetic polymers and plastic wastes kinetic study, Fuel process technol, 2008; 89(7);678-86.
17. Agvado J, Serrano DP, Escola JM, Fuels from waste plastics by thermal and catalytic processes – a review, Indeng chem. Res, 2008; 47(21);7982-92.
18. Sarker M, Rashid MM, Molla M, Waste plastic converting into hydrocarbon fuel material, J Environ scieng, 2011; 5; 446- 452.
19. Ding F, Xiong L, Luo C, Zhang H, Chen X, Kinetic study of low temperature conversion of plastic mixtures to valueadded products, J anal appl Pyrolysis, 2012; 94;83-90.
20. Alston SM, Clark AD, Arnold JC, Stein BK, Environmental impact of Pyrolysis of mixed WEEE plastics part 1- experimental Pyrolysis data, Environ scitechnol, 2011; 45(21);9380-5.
21. Manos G, garforth A, Dwyer J, catalytic degradation of high density polyethylene over different zeolite structures, Indeng chem. Res, 2000,1198-202.
22. Sharuddin SDA, Faisal A, daud WMAW, Aroua MK, A review on Pyrolysis of plastic wastes, Energy conversion manag, 2016; 115;308-326.
23. Chika M, Jude A, paul TW, Thermal degradation of real world waste plastics and simulated mixed plastics in a two stage Pyrolysis catalysis reactor for fuel production, Energy fuels, 2015; 29(4),2601-2609.
24. Scott DS, Czernik SR, Piskorz J, Radlein DSAG, Fast Pyrolysis of plastic wastes, Energy fuels, 1990; 4(4);407-11.
25. Vemichi Y, Nakamora J, Itoh T, Sugioka M, Conversion of polyethylene into gasoline range fuels by two stage catalytic degradation using silica alumina and HZSM-5 Zeolite, Indengchem Res, 1998; 38(2);385-90.
26. IoannisKalargaris, Guohong Tian, Sai Gu, Combustion performance and emission analysis of a DI engine using plastic Pyrolysis oil, Fuel processing technol, 2017; 15;108-115.
27. Vijay Kumar K, Paul Daniel M, Durga Prasad B, Ravi Kumar P, Performance and emission characteristics of diesel engine operated on plastic Pyrolysis oil with exhaust gas recirculation, International journal of ambient energy, 2017; 38(3); 295- 299.
28. ChumsuntiSantaweesukSompop, AdunJanyalertadun, Fuel production performance and emission of a CI engine using waste plastic oil, World journal of engineering, 2017; 14(2);114-120.
29. Divakar Shetty AS, Ravi Shankar Shukla, Antony AJ, Performance and emission characteristics of CRDI engine working on plastic oil, Indian journal of science and technology, 2016; 9(45);1-6.
30. Sumit Bhat, Rohit Singh Lather, Production of oil form waste plastics and polythene using pyrolysis and its utilization in compression ignition engine, Indian journal of science and technology, 2016, 9(48),1-16.
31. Jayabal S, Thirumal P, Anantha Kumar S, Investigation on performance emission and combustion characteristics of variable compression engine fuelled with diesel waste plastic oil blends, J Brazsocmecheng, 2017; 39(1);19-28.
32. Damodharan D, Sathiyagnanam AP, Rana D, Rajesh Kumar B, Saravanan S, Extraction and characterization of waste plastic oil with the effect of n-Butanol addition on the performance and emissions of a DI diesel engine fueled with WPO diesel blends, Energy conversion and management, 2017; 131;117-126.
33. Devraj J, Robinson Y, Ganapathi P, Experimental investigation of performance emission

and combustion characteristics of waste plastic Pyrolysis oil blended with di ethyl ether used as fuel for diesel engine, Energy, 2015; 85;304-309.

34. CeylaGungor, Hasan Serin, Mustafa Ozcanli, SelahattinSerin, Kadir Aydin, Engine performance and emission characteristics of plastic oil produced from waste polyethylene and its blends with diesel fuel, International journal of green energy, 2015; 12(1);98-105.
35. MohdHerzwanHamzah, Abdul Adam Abdullah, AgungSudrajad, NurAtiqahramlan, NurFauziahJaharudin, Performance of diesel engine operating with waste plastic disposal fuel, Applied mechanics and materials, 2014; vols 465- 466,423-427.
36. Kaimal VK, Vijayabalan P, A detailed investigation of the combustion characteristics of a DI diesel engine fuelled with plastic oil and rice bran methyl ester, Journal of the energy institute, 2015; 90(2);324-330
37. Kaimal VK, Vijayabalan P, A detailed study of combustion characteristics of a DI diesel engine using waste plastic oil and its blends, Energy conversion and management, 2015; 105;951-956.
38. Kaimal VK, Vijayabalan P, A study on synthesis of energy fuel from waste plastic and assessment of its potential as an alternative fuel for diesel engines, Waste management, 2016; 51;91-96.
39. Kaimal VK, Vijayabalan P, An investigation on the effects of using DEE additive in a DI diesel engine fuelled with waste plastic oil, Fuel; 2016; 180;90-96.
40. Senthil Kumar P, Sankaranarayanan G, Investigation on environmental factors of waste plastics into oil and its emulsion to control the emission in DI diesel engine, Ecotoxicology and environmental safety, 2016; 134;440-444.
41. Rajankumar, Mishra MK, Singh SK, Arbind Kumar, Experimental evaluation of waste plastic oil and its blend on a single cylinder diesel engine, Journal of mechanical science and technology, 2016; 30(10);4781-4798.
42. Achyut Panda K, Murugan S, Singh RK, Performance and emission characteristics of diesel fuel produced from waste plastic oil obtained by catalytic Pyrolysis of waste polypropylene, Energy sources Part A- recovery utilization and environmental effects, 2016; 38(4);577-585.
43. Mani M, Subash C, Nagarajan G, Performance emission and combustion characteristics of a DI diesel engine using waste plastic oil, Applied thermal engineering, 2009; 29;2738-2744.
44. Mani M, Nagarajan G, Influence of injection timing on performance emission and combustion characteristics of a DI diesel engine running on waste plastic oil, Energy, 2009; 34;1617-1623.
45. Mani M, Nagarajan G, Sampath S, An experimental investigation on a DI diesel engine using waste plastic oil with exhaust gas recirculation, Fuel, 2010; 89;1826-1832.
46. Mani M, Nagarajan G, Sampath S, characterization and effect of using waste plastic oil and diesel fuel blends in compression ignition engine, 2011; 36;212-219.
47. P. Sushma, Waste plastic oil as an alternative fuel for diesel engine – A Review,2018:455; 1-7.
48. S.Deepankumar,B.Saravanan,J.Balaji,R.Gobinath,Experimental Investigation of Performance and Emission Characteristics of Diesel-Bio Diesel (CSOME) - Ethanol-Diethyl Ether Blends in CI Engine" in Journal of Thermal Energy Systems Volume 2 Issue 3. 148-155
49. GovindasamyRajamurugan, Sadhasivam Deepankumar, Anbalagan Ramakrishnan,

PrabuKrishnasamy, Dinesh Dhanabalan , Corrosion Characteristics on Friction Stir Welding of Dissimilar AA2014/AA6061 Alloy for Automobile Application,2019-28-0063

50. S. Deepankumar, M. Boopathi& J Balaji S. Balachandran, R. Gobinath, "Experimental Analysis Of Performance And Emission Characteristics Of Single Cylinder Direct Injection Diesel Engine Using Algae As A Biodiesel And Barium Oxide As A Nano-Fuel",International Journal of Mechanical and Production Engineering Research and Development (IJMPERD)Volume 9 Issue 1.35-42
51. S. Deepankumar,T.Raja,A. Tamilselvan,R. Gobinath,"Enhancing The Four Stroke Ci Engine Performance And Reducing Emission By Preheating And Oxidation Process Using Biodiesel", International Journal of Mechanical and Production Engineering Research and Development (IJMPERD), Volume 8Issue 7.230-240
52. S. Deepankumar,S. Jegadheeswaran& R. Thangavel "Experimental Investigation On Emission Analysis Of Multi-Cylinder Direct Injection Diesel Engine Using Manifold Injection Of Ethanol", International Journal of Mechanical and Production Engineering Research and Development (IJMPERD), Volume 8 Issue 7.704-710.
53. S Deepankumar, R Gobinath, S Balachandran, M Boopathi,"Experimental Investigation of Performance and Emission Characteristics of Diesel-Bio Diesel (CSOME) with Nano Additive Blends in CI Engine" ,AdvAutomobEng, Volume 7Issue 176,2.
54. S Deepankumar, R Gobinath, S Balachandran "Experimental Investigation of Performance and Emission Characteristics Of Diesel-Bio Diesel (CSOME)-Ethanol-Diethyl Ether Blends In CI Engine", in Journal of Thermal Energy Systems Volume 2 Issue 3. 148-155

IOP Conf. Series: Materials Science and Engineering **764** (2020) 012043 doi:10.1088/1757-899X/764/1/012043

Experimental study on concrete by partial replacement of fine aggregate by textile effluent treatment plant sludge

Loganayagan.S[1,a] Rajkumar G[1,b], Pavthra A[1,c], Poonkundran M[1,d]
Department of civil engineering, Bannari Amman institute of technology, Sathyamangalam-638401. Tamil Nadu, India
Email: [a]loganayagans@bitsathy.ac.in, [b]rajkumar.ce16@bitsathy.ac.in, [c]pavithra.ce17@bitsathy.ac.in, [d]poonkundran.ce17@bitsathy.ac.in.

Abstract. Chemical sludge produced by textile industries poses a high threat to the environment causing pollution in all stages of its treatment and disposal. In this study sludge from textile dying unit was collected. It has been partially replaced for fine aggregate in concrete up to 20% replacement to find out the possibility of reuse of this hazardous waste in build materials. Manufactured sand made by grinding angular aggregates was used as fine aggregate. Study shows that the samples failed to meet the desired compressive strength required for structural applications. The reason for failure in strength development is to be further experimented by individual analysis of sludge components.

Keywords: chemical sludge, pollution, hazardous waste, fine aggregate, concrete

1. Introduction

Developing industrialization is creating various items and furthermore heaps of waste. The vast majority of this waste makes danger of peril people and the earth. According to USEPA, "Perilous waste will be squander that is unsafe or conceivably hurtful to our wellbeing or nature. The level of peril may shift notwithstanding the amount of the waste produced. By nature, unsafe waste muddles the procedure of gathering, dealing with, treatment and transfer, and, obviously costly and hazard included. Run of the mill case of such squanders incorporate substantial metals, for example, Chromium, Mercury, Nickel, Cyanide, and so forth and oil and oil loaded squanders with poisonous metals. Material industry is one of the most established ventures in India and has a noteworthy nearness in the Indian economy, adding to about 14% of assembling esteem expansion and 1/third of the India's gross fare income. According to, yearly risky waste created is assessed to be around 6.23 million tons. Out of this, 49.55 % is recyclable, 6.67 % incinerable and remaining 43.78 % is expendable in verified landfills. Twelve conditions of the nation (Maharashtra, Gujarat, Andhra Pradesh, Tamil Nadu, Odisha, Madhya Pradesh, Assam, Uttar Pradesh, West Bengal, Kerala, Karnataka and Rajasthan) represent 97 % of the aggregate squander age. In Tamil Nadu, huge amounts of compound slop is created for example 55.76 tons/day from Balotra CETP, 130.58 tons/day from Pali CETP, 184.60 tons/day from Tirupur CETP and 90.40 tons/day from Karur CETP are created also, are lying in CETP premises anticipating transfer to landfill. Mindfulness has been made for the sheltered administration of unsafe squander. Be that as it may, it isn't yet placed by and by in the greater part of the conditions of India because of different issues. Notwithstanding appropriation of most costly and propelled strategies, the resultant built up could conceivably be dangerous yet it must be dealt with again through a reasonable innovation. One likewise needs to work toward creating different environment– well-disposed and powerful financial innovations for the administration of dangerous waste in India. Hardening/Stabilization (S/S) is an innovation utilized for treating mechanical strong squanders containing poisonous constituents to keep their disintegration and discharge to nature. It had been utilized for a considerable length of time as a last treatment venture before the transfer of both radioactive and synthetic dangerous waste. A few examiners have endeavoured S/S innovation for the treatment of various kinds of modern squanders, for the most part overwhelming metals containing inorganic/natural slope. The interest for development materials is dependably heightening because of expanding urbanization. In this way, usage of waste materials in development is the need of hour. It isn't just sets aside extra cash yet additionally diminishes load on the waning common assets. A few specialists have endeavoured to reuse squander materials into different kinds of development materials, for example, obstructs from sewage muck utilizing fly fiery debris, lime and gypsum what's more, Arsenic and Iron slop in block making, sewage and slime powder in block and tile making, and ooze cinder for light weight totals and.Few writing of portrayal and reuse of compound slime from material CETPs in India are accessible. Still the endeavours are on to discover a reasonable financial and condition agreeable answer for the safe reuse of chemical sludge. We have attempted the experiment with the partial replacement of sand with sludge material.

2. Materials

The textile sludge material for study purpose was obtained from Sri Mahalakshmi textile and dying unit, Asokapuram locality, Erode district, Tamil Nadu, India. The sludge was collected at sludge storage facility, where sludge is dried to a certain extent prior to storage and is transported for treatment and disposal at periodic intervals. Water samples were also obtained from the effluent treatment plant at three different stages (ETP outlet, RO permeate, sludge mix prior coagulation) for testing.

The sample sludge was dried at 100°C for 24 hours in a hot air oven and the moisture content obtained was 36%. The sludge was further dried in room temperature and was ground to fine particles by manual grinding by trowel. The sludge particles passing 300 micron sieve and retained on 75 micron sieve was taken for study. The sludge that passed 75 micron sieve was not involved in study purpose due to inconvenience in handling at dry state.

Ordinary Portland cement – 53 grade confirming to IS 12600-2009 is used. OPC grade-53 shows 53N/mm^2 after 28 days of curing in fresh water. Standard cubes of dimension 150mm x 150mm x 150mm is used for determination of strength with partial replacement of fine aggregate with certain percentages of sludge material and different curing periods.

Manufactured sand (M sand) is used as a substitute for natural river sand in this study. Particles passing through 4.75 mm sieve are used. Crushed angular aggregates of size 20mm and 10mm showing specific gravity 2.79 and 2.74 respectively was used in the study. Water for curing and mixing confirming to IS 456 was used.

3. Methodology

Specific gravity of the fine aggregate and the sludge which is used in the study was determined by standard pycnometer. The sludge was analysed by Tamil Nadu pollution control board, District Environmental Laboratory, Tiruppur. Standard cubes of 150 mm x 150 mm 150 mm dimension was casted for M20 concrete to determine strength characters with different percentages of sludge for 7 days and 28 days of curing. The standard cubes were made 5% to 20% sludge as partial replacement for fine aggregate. Cubes with different sludge percentages were casted and tested at periodic intervals of curing. The compression test was conducted in a compression testing machine as per IS standards.

For various sludge percentages two samples were made and tested for compressive strength at curing periods of 7 days and 28 days respectively and the average strength of the cubes were obtained.

4. Results

4.1. Characreristics of Sludge

The grade distribution for sludge sample shows that the maximum percentage of sludge particles was retained on 2.36mm sieve. The moisture content of the sludge particles was about 36%, whichwas determined by drying in a hot air oven for 24 hours.

Table 1. Grade size distribution of textile effluent sludge

S.No	IS Sieve No.	Weight retained	Percentage weight retained	Cumulative percentage weight passing
1	4.75	0.036	3.6	96.4
2	2.36	0.196	19.6	76.8
3	1.18	0.157	15.7	61.1
4	0.6	0.137	13.7	47.4
5	0.3	0.152	15.2	32.2
6	0.15	0.157	15.7	16.5
7	0.075	0.149	14.9	1.6
8	Pan	0.016	1.6	0

Table 2. Properties of textile sludge

S.No	Parameters	Sample LAB code	ETP outlet 111	RO permeate 112	Untreated sample 113
1.	pH	(Number)	7.56	6.95	8.82
2.	Total suspended solids	(mg/l)	284	4	280
3.	Total dissolved solids	(mg/l)	13008	368	71010
4.	Chloride(as Cl)	(mg/l)	4615	125	28842
5.	Sulphate(as SO4)	(mg/l)	1325	7	10142
6.	BOD 3 days at 27degree celsius	(mg/l)	10	4	54x
7.	COD	(mg/l)	384	16	71010
8.	Oil and Grease	(mg/l)	1	<1	2
9.	TKN	(mg/l)	18.48	<5	6.72
10.	Sulphide	(mg/l)	<1	<1	<1
11.	Phenolic compound	(mg/l)	<0.0005	<0.0005	<0.0005
12.	Percent sodium	(%)	97	65	92

The sludge was obtained from different points namely ETP outlet, RO permeate, and untreated sludge sample. The properties shows that the ETP outlet sample shows increased values of dissolved solids, chlorides, oil and grease and phenolic compounds making the sample non portable. The RO permeate shows improved characters under limits of standards making it portable. Untreated sample is highly contaminated with values more than prescribed limits for portable water making it non portable. The standards of portable waste as indicated my BIS standards are pH (6.5 – 8.5), dissolved solids (500 mg/l), chlorides (250 mg/l. max), and phenolic compounds (0.001 mg/l. max). the RO permeate alone resides within the limits of standards which makes it most probable for portable consumption. ETP outlet and untreated sludge sample are not suitable for portable consumption.

4.2. Characteristics of M-Sand

The crushed form of angular aggregates was tested for specific gravity by pycnometer shows a specific gravity of 2.49. The grade size distribution of the manufactured sand (M-sand) used in the study purpose showed allowable characters to be used in concrete casting and testing process.

Table 3. Grade size distribution of M-sand

S.No	IS Sieve No	Weight retained	Percentage weight retained	Cumulative percentage weight passing
1	4.75	0.006	0.6	99.4
2	2.36	0.129	12.9	86.5
3	1.18	0.259	25.9	60.6
4	0.6	0.122	12.2	48.4
5	0.3	0.262	26.2	22.2
6	0.15	0.156	15.6	6.6
7	0.075	0.049	4.9	1.7
8	Pan	0.017	1.7	0

4.3. Compressive Strength and Structural Applications

The compressive strength observed by testing in a compression testing machine at 7 days and 28 days respectively shows that the 5% replacement samples showed values with minor deviations from nonreplaced samples. The samples with 10%, 15% replacement shows greater deviation and 20% replacement sample shows poor strength development characters. The samples did not achieve desired strength characters at 28days of curing making it unsuitable for structural applications in building components. The compressive strength result shows that average compressive strength decreases with the increase in the percentage replacement of sludge in the concrete.

Table 4. Compressive strength results

S.No	Percentage replacement %	Comp strength 7 days (N/mm^2)	Comp strength 28 days (N/mm^2)
1	0	19.21	22.46
2	5	17.88	19.56
3	10	9.325	13.44
4	15	8.66	11.66
5	20	3.23	9.66

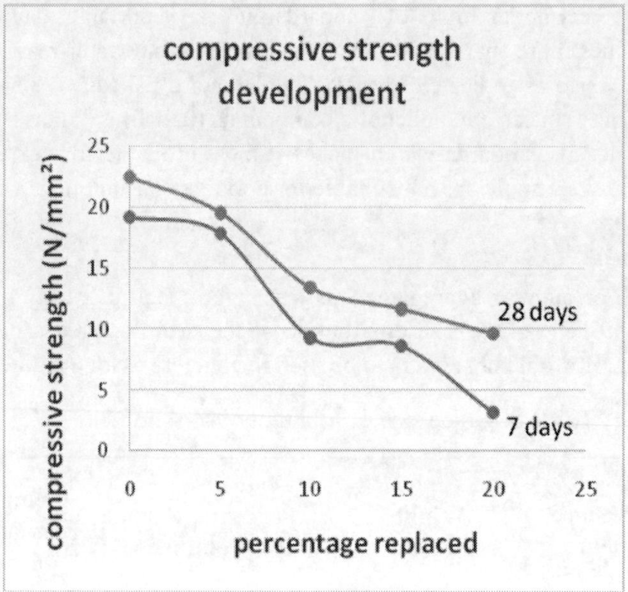

Figure 1. Compressive strength development

5. CONCLUSION

From this study it is possible to conclude that partial replacement of textile effluent sludge for fine aggregate in concrete is not possible and restricts the usage of this industrial waste as a component in concrete. The sludge material gets finer than cement particles which increases its volume and water demand during mixing which reduces the strength characteristics of the building. Addition of this sludge also delays the setting of concrete. Detailed study on causes for failure in strength gain in concrete due to individual chemical components of sludge is to be carried out as next step of research.

6. Acknowledgement

The authors of this paper are grateful to Sri Mahalakshmi textile and dying unit, Asokapuram locality, Erode district, Tamil Nadu, India for providing the samples and assistance in testing of sludge for various properties. The authors are also grateful to Department environmental laboratory, Tiruppur for analysing the sludge samples as per standard methods and Concrete lab officials of Bannari Amman Institute of technology, Sathyamangalam for support in casting and testing specimens as per standards.

7. REFERENCES

[1] Ansari A A and Thakur B D 1999 Waste Management in the Processing Unit of a Textile Industry Environmental Pollution Control Journal, 4(8): 30-36.
[2] Azeez P A 2001 Environmental Implications of Untreated Effluents from Bleaching and Dyeing Workshop on Eco-friendly Technology for Waste Minimization in Textile Industry 7-8
[3] Anil Laul 2002 Interview on Sustainable Architecture Practice Down to Earth 11(2) 43
[4] Bal A S 1999 Waste Management for Textile Industry Indian Journal of Environmental Health, 41(4) 264-265
[5] Chandrasekaran K 2001 Studies on Management of Sludge from Hosiery Knitwear dyeing Wastewater Treatment PlantsWorkshop on Eco-friendly Technology for Waste Minimization in Textile Industry 61-62

IOP Conf. Series: Materials Science and Engineering **764** (2020) 012043 doi:10.1088/1757-899X/764/1/012043

[6] Chaudhary S K 2002 Utilization of Industrial Waste Products and Flyash in Building Materials Civil Engineering and Construction Review 18 (2)
[7] Balasubramanian J, Sabumon P C, John U Lazar and Ilangovan R 2006 Reuse of textile effluent plant sludge in building materials Waste Management vol 26 pp 22-28
[8] Baskar R, Meera Sheriff Begum K M and Sundaram S 2006 Characterization and reuse of textile effluent plant waste sludge in clay bricks Journal of the University of Chennai Technology and Metallurgy vol 4 2006 473-478
[9] Hema Patel and Sunnel Pandey 2009 Exploring the reuse of potential of chemical sludge from textile waste treatment plants in India - A hazardous waste American Journal of Environmental Sciences 5 (1) pp106-110
[10] Hilary Nath 2006 Sludge Bricks Development Reach Journal, brandix inspired solutions Issue 3 pp 6
[11] Jayakumar P T, Ram Kumar V R, Suresh Babu R & Mahesh Kumar M 2013 Experimental Investigation on Paver Blocks using Steel Slag as Partial Replacement of Aggregate and sludge as partial Replacement of Cement International Journal of Science and Research vol 2 pp 87 – 91
[12] Mehdi Ahmadi, Nikoo Bayati, Ali Akbar Babaei & Pari Teymouri 2013 Sludge Characterization of a Petrochemical Wastewater Treatment Plant Iranian journal of health sciences vol 2 pp 11-13
[13] Palanivelu K, Rajkumar R 2001 Characterization and Leachability studies on textile effluent treatment plant sludge Environmental Pollution Control Journal
[14] Ramesh Kumar M, & K Saravanan 2009 Recycling of Woven Fabric Dyeing Wastewater Practiced in Perundurai Common Effluent Treatment Plant CCSE Journal 2009 vol 3 (4) pp 146
[15] Raghunathan T, Gopalsamy P, Elangovan R 2010 Study on Strength of Concrete with ETP Sludge from Dyeing Industry International Journal of Civil And Structural Engineering vol 1 pp 379-389

IOP Conf. Series: Materials Science and Engineering **764** (2020) 012044 doi:10.1088/1757-899X/764/1/012044

Influence of Water Ageing on Mechanical Performance of Glass Fiber Reinforced Polyester (GFRP) nanocomposites

S M Vinu Kumar[1*], V Hariprasad[2], K L Senthil kumar[1]
[1*]Department of Mechanical Engineering, Bannari Amman Institute of Technology, Sathy, Tamilnadu, India
[2] Department of Aeronautical Engineering, Bannari Amman Institute of Technology, Sathy, Tamilnadu, India,
Corresponding Author: vinukmr1988@gmail.com

Abstract. The current paper discusses about the effect of water ageing on mechanical properties of unfilled and nano-SiC filled Glass/Polyester (G-P) composites. G-P and its nanocomposites were synthesized using hand layup technique followed by compression moulding curing process. Kinetics of water absorption undergoing in each composites has been studied. Results showed that addition of nanofiller in pure G-P composites reduces the water uptake after attending its equilibrium level. Mechanical properties of the moisture absorbed sample exhibited low strength compared to dry or unexposed composites. In overall, amongst the fabricated composites, 4.5 wt.% nano-SiC Filled G-P composites showed better mechanical properties.

1. Introduction

Fiber Reinforced Polymer based composites are in huge demands for making aerospace components, automotive parts, satellite vehicles parts, marine parts and so on and so forth whenever there is a demand for replacing the heavy metallic material. The main mottos of synthesizing composites are to impart high strength (being low weight materials) and offer outstanding performance during its service as a component. Most of the composites which are in present services are generally made up of synthetic fiber reinforced types. Though its usage has become threat to the environment still it is best choice for applications, since yet to develop alternate natural fiber reinforced composites which could give best performance as synthetic one. Then It doesn't mean synthetic fiber reinforced composites (SFRP) are not having any issues. SFRP composites are facing environmental challenges, particularly when they are out to moisture, hydrothermal, corrosion, saline condition, UV with humidity. According to Huang, et al [1] when glass fiber reinforced composites exposed to moisture absorption for long duration water molecules will tends to enter the voids existing between fiber and matrix phase and deteriorate the mechanical performance of the composites. This happens due to poor interfacial adhesion occurred between primary and secondary phase and also due to hydrophilic group present in synthetic fiber especially glass fiber. Yan et al [2] concluded that mechanical performance of the flax fabric/epoxy composites degraded water, seawater and alkaline ageing. Tual et al [3] developed the carbon/epoxy composites for turbine blade. Their work inferred that mechanical properties of the composites are reduced during sea water ageing. Exhaustive studies revealed that pores/voids in the laminates can be reduced by incorporating filler/particles into the matrix phase. These fillers are either chemically bonded or physically tangled to polymer phase in micro and nano form. Most of the fillers

IOP Conf. Series: Materials Science and Engineering **764** (2020) 012044 doi:10.1088/1757-899X/764/1/012044

used for fabricating polymer hybrid composites are Inorganic in nature due to ease of availability and low manufacturing cost.
From the exhaustive literature survey it is found that glass fabric reinforced Polymer composites may be appropriate choice for making marine and low weight structures compared to other conventional/SFRP materials. Although there are plenty of works reported on Mechanical properties on Glass/Polyester composites, to our capacity no work has been reported on the effect of sea water ageing on mechanical performance of Nano-SiC filled Glass/Polyester composites. Thus the objective of this paper is to understand the influence of the water ageing on Tensile, flexural, Hardness, and Impact properties and also sea water diffusivity of nano-SiC filled Glass/Polyester composites. Finally, performance of the nanocomposites has been evaluated and compared with that of neat/Unfilled Glass-Polyester composites.

2. Materials and Method

2.1. Materials

In the present investigation unidirectional E-Glass fibers were purchased from the Local vendor, Chennai, Tamilnadu India. Nanofiller-SiC were procured from Universal Carborundum India Pvt Ltd, Kochi, Kerala, India. Unsaturated Polyester resin and its accelerator and catalyst were supplied by Vasa Vibala Resin Pvt Ltd, Chennai, Tamilnadu India.

2.2. Preparations of composite Laminates

Pre-calculated amount of polyester resin is mixed with 1.5 % MEKP catalyst as prescribed by vendors. SiC powder was dried in a muffle furnace at 150° C for about 1 hr before mixing it with the resin. Later 1.5 % Cobalt Naphthenate accelerator added to Resin and Filler mixture to initiate the curing process.

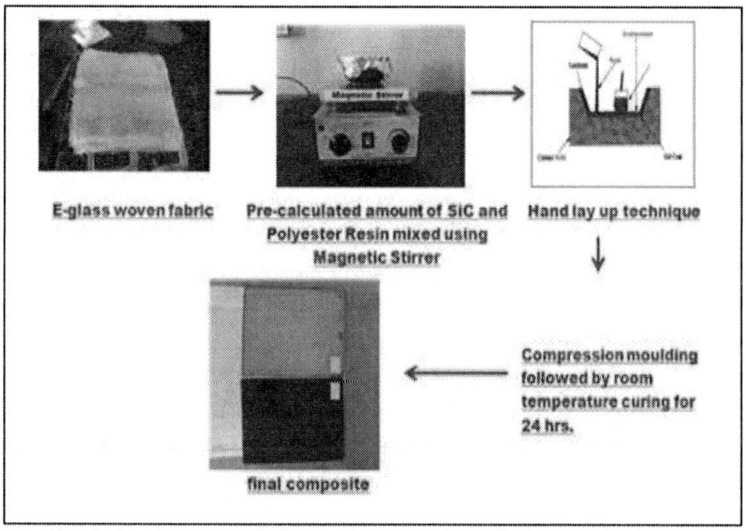

Figure.1 Process of Fabrication of unfilled and Nano-SiC filled Glass/Polyester (G-P) composites

Steps followed in fabrication of the hybrid laminates is shown in the figure.1 At an outset Resin and Filler mixture impregnates each layer of E-glass fabric layer (of size 220×220 mm) by laying down on the surface of the mold one above the other until desired thickness is achieved. Metallic hard roller and brush facilitates degassing and ensures the uniform distribution of the resin. Each fabricated laminates were cured under pressure of 35 bar using hydraulic press for 24 hrs. Further laminates are removed from the mold and cured for 48 hr at room temperature before use of the composites. All laminates for the mechanical tests were prepared from the eight layer of the E-glass fabric except the hardness test for which 16 layers of fabric were used. All laminates of the hybrid composites were processed at a

IOP Conf. Series: Materials Science and Engineering **764** (2020) 012044 doi:10.1088/1757-899X/764/1/012044

weight fraction of 60% (±1.5). Thickness of Laminates maintained uniform thickness of 3.2 mm using a spacer. Finally, neat and SiC-Filled G-P composites are successfully fabricated and composition of the composites has been detailed in Table 1.

Table 1. Composition of Unfilled and SiC filled G-P nanocomposites

Sl. No	Composite Designation	Quantity of Constituents (wt %)		
		E-Glass Fabric	Polyester matrix	Nano-SiC filler
1	G-P	60	40	-
2	1.5 G-P	65	38.5	1.5
3	3.0 G-P	65	37	3.0
4	4.5 G-P	65	35.5	4.5

2.3. Water Absorption test

Water absorption test were carried out on fabricated G-P based composites in accordance to ASTM D 570-98. At least three test specimens were taken from each composite having a dimension of length 6.4 cm and width 1.2 cm. specimens were Immersed in the sea water aqueous solution whose pH=8 was maintained at room temperature. Before immersing the specimen, its weight has to be determined using high precision digital weighing balance. After an Interval of 24 hours, Specimen should be taken off the sea water bath and weight has to be measured by wiping out of the surfaces of specimen using tissue papers. This process should be continued till weight gained by composites attains saturation levels.

Percentage in Weight gain is calculated by using following equation (1)

$$W(t) = \frac{W_x - W_O}{W_O} \times 100 \ldots\ldots\ldots\ldots\ldots (1)$$

Where W_x is the Mass gained by the specimen in regular interval of time during ageing and W_o is the initial mass before ageing. W(t) is the moisture gain in percentage.

When composites immersed for several period of time, weight changes in composites with respect to time was plotted to study effect of moisture on composite system. Diffusion coefficient is calculated using initial slope obtained from the plot drawn between moisture gain and square root of time using Equation (2). Where 'm' is the slope calculated from the graph and't' is the thickness of the sample[4].

$$D = \pi\left(\frac{t^2 m^2}{16\, W_\infty^2}\right) \ldots\ldots\ldots\ldots (2)$$

Besides, sorption coefficient which is equilibrium sorption of the penetrant and it is calculated using equation (3). Where W_∞ and W_t is the percentage of water uptake at equilibrium stage and at time't'[4-7].

$$S = \frac{W_\infty}{W_t} \ldots\ldots\ldots\ldots\ldots\ldots\ldots (3)$$

Net effect of the sorption and diffusion coefficient can be determined using Permeability coefficient using equation (4)[4-7]

$$P = D \times S \ldots\ldots\ldots\ldots\ldots\ldots\ldots (4)$$

3. Characterization of the composites

3.1. Barcol Hardness Test

The Hardness of the neat and Nano-SiC filled G-P composites determined as per ASTM D2583.

IOP Conf. Series: Materials Science and Engineering **764** (2020) 012044 doi:10.1088/1757-899X/764/1/012044

The specimens were positioned and loads had applied over it using hard Indenters. Nearly 10 indentions are made and its average value have been computed to assign final hardness number to the individual composites

3.2. Tensile Strength and Flexural Strength

Tensile test was carried out as per ASTM D-638 in universal testing machine (capacity 100kN). Specimen dimension for tensile test is shown in the Figure 2. During the test, machine's cross head speed of 4 mm/min was maintained. In order to evaluate flexural property, three-point bending test was carried as per ASTM D790 in universal testing machine. In this test effective length of 58 mm and cross head speed of 2 mm/min have been selected. Dimension of the specimen for conducting bending test is shown in the Figure 3.

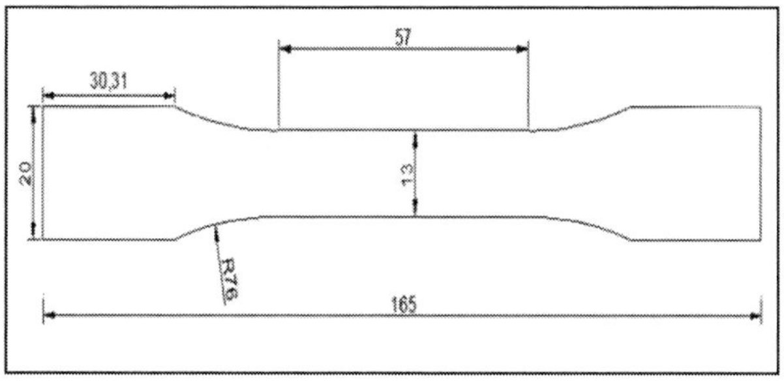

Figure. 2 Dimension of the Tensile specimen prepared as per ASTM D-638

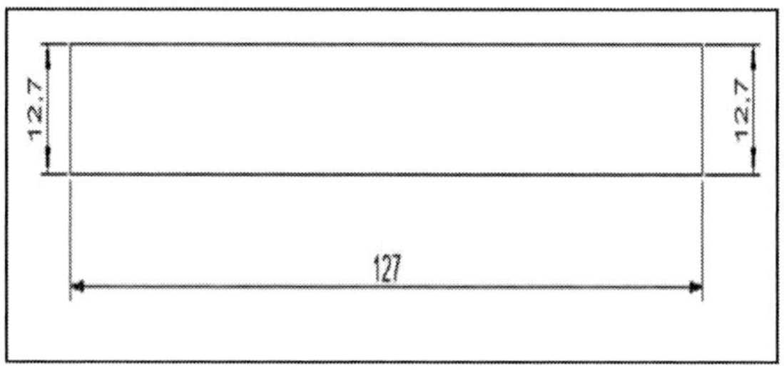

Figure. 3 Dimension of the Fexural speciemn prepared as per ASTM D-790

3.3. Impact strength:

Toughness of the composite determined using Impact test. Izod impact test was employed under ASTM D256. Specimen should be cut in accordance to standard as shown in Figure 4. The striking velocity of pendulum used in the test is 4 m/sec.

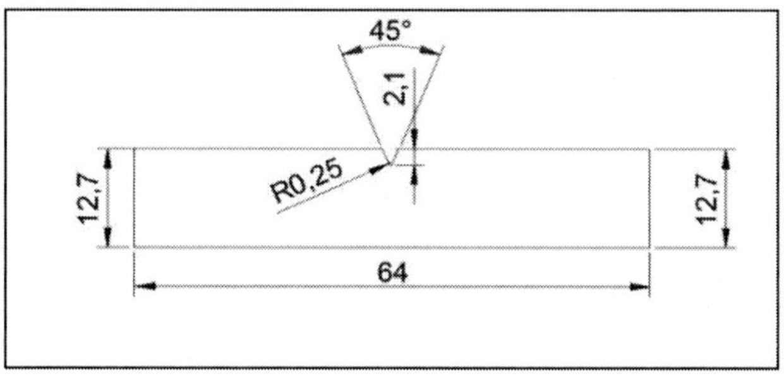

Figure. 4 Dimension of the Impact (Izod) specimen prepared as per ASTM D-256.

4. Results and Discussion

4.1. Water absorption Behavior of the Neat and Nano-SiC filled G-P hybrid composites:

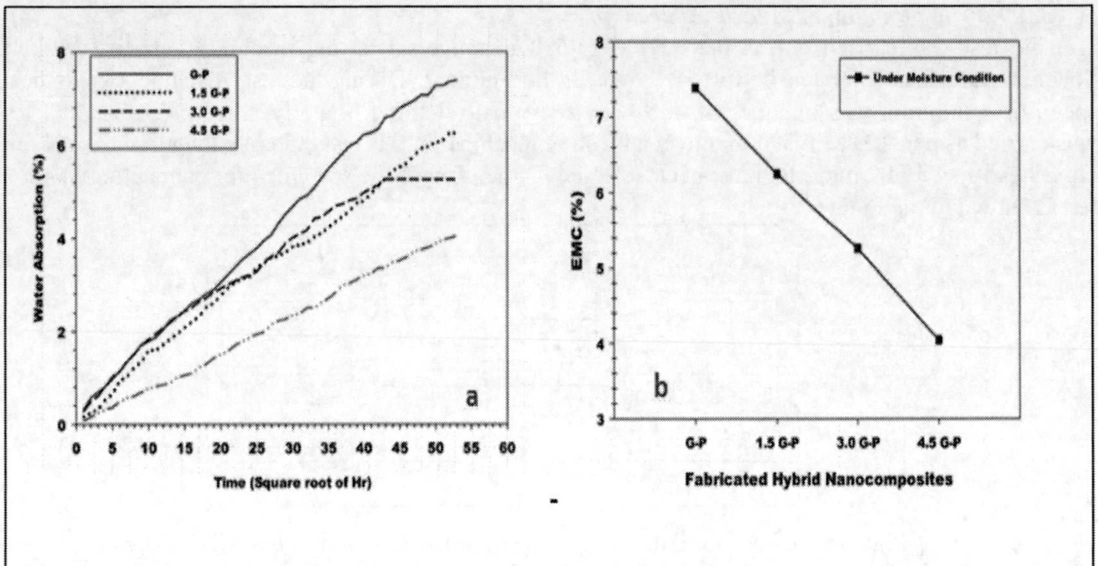

Figure. 5(a) Water absorption gain in percentage against time (square root of hour) (b) EMC curve obtained different composition of fabricated composites.

Water absorption behavior of G-P composites at various loading of nano-SiC particles is shown in the Figure 5(a). As the ageing time of composites in sea water increases, moisture uptake by the composites also increases up to saturation point. Beyond this point composites are no longer in the position to absorb any moisture or water molecules. This can be witnessed in the present work and is depicted in the Figure 5(a). From the plot it is interesting to observe that, as the nano filler addition increased from 1.5wt.% to 4.5wt.%, water absorption capacity decreases. From the plot 5(b) it is noted that EMC (Equilibrium moisture content) of composites reduced as the nano filler loading increases. This is attributed to strong interfacial adhesion imparting by the filler to matrix phase which in turn improves the fiber matrix region there by curtailing down the void content [5]. According to Anjum et al [8] voids can pave the way for the water molecule to enter through capillary action there by increase the water absorption rate. Therefore, it is essential to make sure that void in the laminate is least. Furthermore, moisture uptake attains the equilibrium state after initial take off, so this pattern is said to have followed Fickian Diffusion process [6].

Table.2 Sorption, Diffusion and permeability coefficient of the unfilled and Nano-SiC filled G-P composites.

Fabricated hybrid nanocomposites	Sorption Coefficient (S)	Diffusion Coefficient ($\times 10^{-4}$ mm^2/sec)	Permeability Coefficient (P) ($\times 10^{-4}$ mm^2/sec)
G-P	0.04315	1.28845	0.431592
1.5 G-P	0.0272	4.75667	3.691173
3 G-P	0.0558	2.82833	2.675603
4.5 G-P	0.0236	0.447917	0.032344

There are three important factors under Fickian Diffusion process responsible for moisture absorption in composites and they are diffusion coefficient (D), Sorption coefficient (S) and permeability coefficient (P) and these factors are calculated using Equations 2, 3, and 4 respectively. Calculated value has been presented in the Table 2. Diffusion coefficient (D) is one of the most essential parameter considered in the Fickian Diffusion Model. It shows about the ability of the water molecule penetrating the composites materials. Diffusion coefficient increases when initial nanoparticle loading increased to 1.5wt.% and decreases when filler loading increases further from 3 wt.% to 4.5 wt.%. Lowest diffusion coefficient was found for 4.5 G-P nanocomposites because filler improves the interfacial adhesion between matrix and itself. Apart from that, filler also reduces the gap/void existing between Fiber-Matrix thereby reduces the velocity of water molecule entering the laminates [5]. Thus water absorption of 4.5 G-P nanocomposites is low as compared to other fabricated composites. Another important factors considered in kinetics of water diffusion is sorption coefficient and permeability coefficient. Sorption coefficient of 4.5 G-P nanocomposites showed lowest value compared to other composites since minimum void in the composite resist the absorption of water molecules. Permeability coefficient of the nano hybrid composites behave little differently. It follows the same trend of Diffusion coefficient except by the 3 G-P nanocomposites. Lowest Permeability coefficient exhibited by the 4.5wt.% Filled nano-SiC/G-P composites.

4.2. Effect of Water Absorption on Hardness

Figure 6 presents the hardness of the various fabricated hybrid composites when they are subjected to different environmental condition. It is observed that, as the wt.% addition of nano-SiC particle increases, the hardness of the hybrid composites also increases. Same trend has been observed for the moisture exposed sample. Highest hardness number had exhibited by 4.5 G-P nanocomposites in both dry and moisture environments.

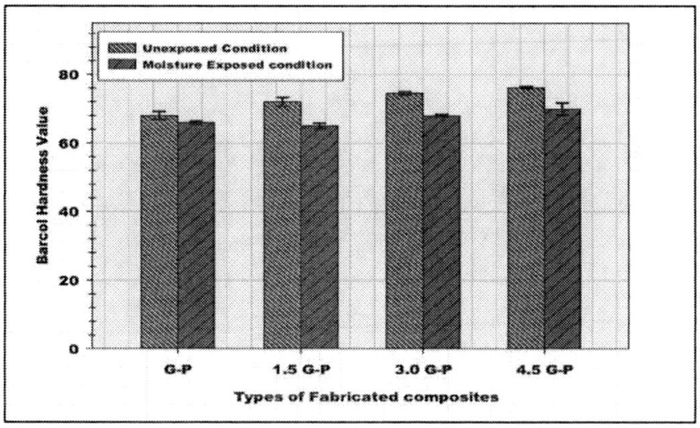

Figure 6 Barcol Hardness of the Unfilled and Nano-SiC filled G-P composites at dry and moisture condition

4.3. Effect of Water absorption on Tensile Strength

Figure 7 shows the tensile strength of the various fabricated hybrid composites under dry and moisture condition. It is seen in the figure that 25.95% improvement in the tensile strength was noted when 4.5wt.% of nano SiC filler added into pure G-P composites. Addition of nanofiller forms the chemical bonding and as well as physical tangling with polyester and it further fills the gap between fiber-matrix which leads to overall strong interfacial bonding among the reinforcements[9]. When the composites are subjected to tensile load, matrix phase receive the load and transfer the same to glass fiber through nano particles. Thus when composites exhibit better tensile property it means there exist a strong bonding between the reinforcements. Further when the composites exposed to moisture condition, it follows the same trend as seen in dry condition. Moisture exposed sample showed less strength compared to non-exposed sample because water molecules penetrated inside the composites allows the glass fiber to swell along with matrix phase and weaken/damage the fiber-matrix interface hence strength of the composites reduces[10]

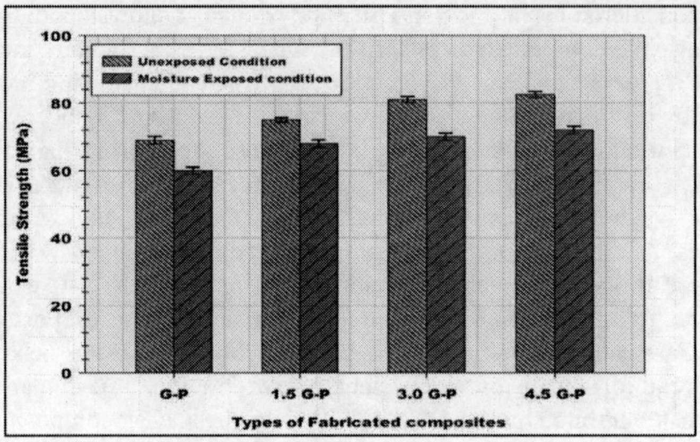

Figure. 7 Tensile Strength of the Unfilled and Nano-SiC filled G-P composites at dry and moisture condition.

4.4. Effect of water absorption on Flexural Strength:

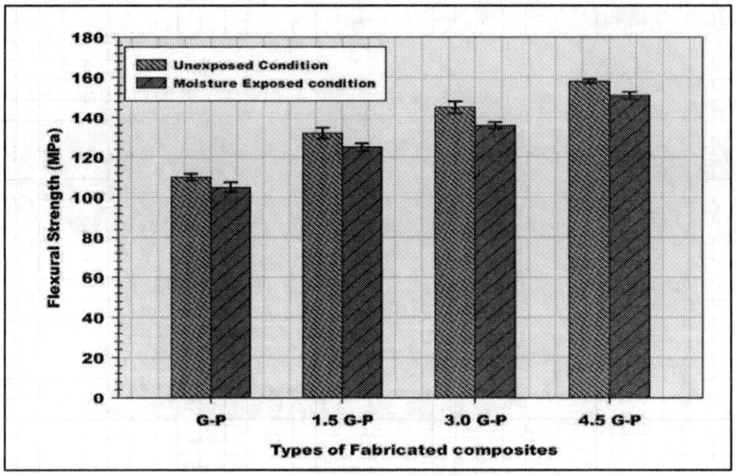

Figure. 8 Flexural Strength of the Unfilled and Nano-SiC filled G-P composites at dry and moisture condition.

Flexural strength of the unfilled and nano-SiC filled G-P composites under dry and moisture exposed condition has been presented in the Figure 8. It is observed that addition of nano filler into the G-P composites increases the bending strength of the composites. Unfilled G-P composite showed 110 MPa and for the same, when nano-SiC filler was added, bending rose to 158 MPa, that means 34.9 % improvement in the flexural strength was noted. Flexural strength is purely depending upon the fiber property since it is load bearing member of the laminates. When the composites are subjected to flexural test then bottom surface of the material experiences compression stress and top surface experiences tensile stress[11, 12]. Therefore, samples should have strong fiber –matrix interface to resist the bending load. From the plot it can be seen that 4.5 G-P nanocomposites has shown better flexural strength which means fiber-matrix interface as result of filler's chemical bonding imparts strong bonding and also it reduces void [11]. Further from the plot it can also be seen that, moisture exposed samples showed less strength compared to unexposed samples however it follows the same trend as observed in the dry sample. Difference in the flexural strength of the exposed composites with dry samples is very marginal. Moreover, when composites exposed to water immersion, moisture uptake increases. When void percentage in the sample increase, strength of the composite decreases because water molecule will get trapped between these voids causes the matrix and fiber to swell thereby weakens the interface badly, thus failure of the fiber in the composites takes at low loading during mechanical test[13]. Therefore, it has to ensure during fabrication that, voids should be minimal. However, it is impracticable to avoid the void formation while synthesizing thermoset based polymer composites. Figure 9 shows voids are predominantly observed in the 1.5 G-P nanocomposites. Flexural and tensile strength are fiber sensitive and water molecule diffused inside the composites further deteriorates the matrix-fiber interface which leads to early debonding of fiber phase. Water absorbed by synthetic fiber such as glass fiber and carbon fiber is less as compared to natural fiber however water molecules penetrating inside the matrix network accelerates the matrix cracking for pre-mature failure thus flexural and Tensile strength decreases[14, 15].

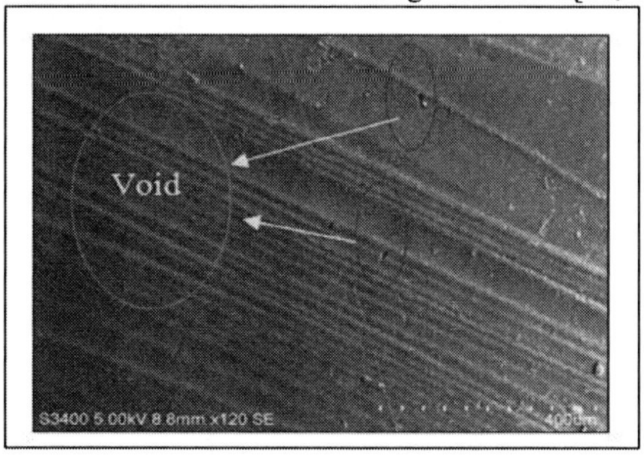

Figure 9 Voids formation in 1.5 wt.% nano-SiC filled G-P composites

4.5. Effect of water absorption on Impact strength:

Impact strength of unfilled and nano-SiC Filled G-P composites is shown in the Figure 10. Impact strength of 4.5 G-P nanocomposites exhibited 14.6 % improvement compared to unfilled G-P composites. This may be due to the uniform distribution on nano-SiC particles in polyester resin which transfer the load to the fiber via matrix phase. During water immersion, sample absorbs water molecules which accelerates interface degradation of fiber-Matrix-filler eventually leads to reduction in impact strength of the hybrid composites. It is interesting to see the effect of moisture on 4.5 G-P nanocomposites. Only 1.34% decreased in the impact strength was noted compared to unexposed sample because water molecule trapped in the gap existing between matrix and fiber also resist the impact energy to produce high impact values [11, 12].

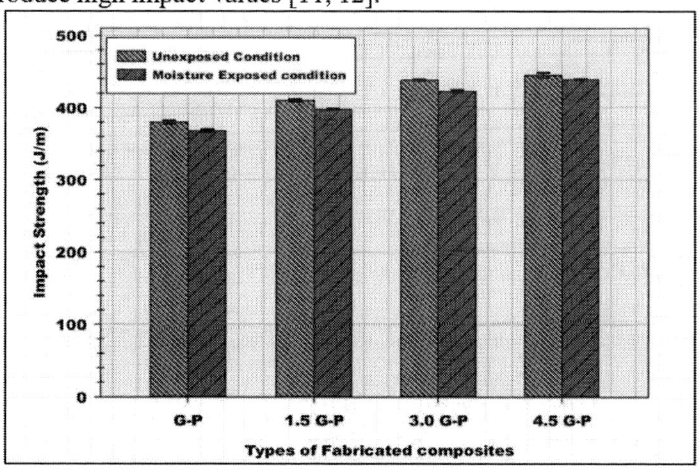

Figure 10. Impact Strength of the Unfilled and Nano-SiC filled G-P composites at dry and moisture condition.

IOP Conf. Series: Materials Science and Engineering **764** (2020) 012044 doi:10.1088/1757-899X/764/1/012044

Conclusions:

Mositure absorption property of the Glass fiber reinforced polyester composites with and without nanofiller SiC has been studied. From the Characterization studies following conclusions has been drawn :

1. Moisture abosption studies showed that reinforcement of nanofiller SiC in to pure G-P composites offers resistance to the moisture abosprtion. Amongst the fabricated composites, 4.5 wt.% nano-SiC filled composites exhibited low moisture uptake.
2. Mositure absorption endured by the unfilled and Nano-SiC filled G-P composites follows Fickian Diffusion Behavior. Additon of Nano-SiC fillers reduces the EMC from 7.5% to 4.1 % in sea water ageing condition.
3. Tensile strength, Flexural strength, Impact strength of the unexposed composites showed better strength compared to water absorption samples.

References:

[1] C. Huang, S. Fu, Y. Zhang, B. Lauke, L. Li, L. Ye, Cryogenic properties of SiO2/epoxy nanocomposites, Cryogenics 45(6) (2005) 450-454.
[2] L. Yan, N. Chouw, Effect of water, seawater and alkaline solution ageing on mechanical properties of flax fabric/epoxy composites used for civil engineering applications, Construction and Building Materials 99 (2015) 118-127.
[3] N. Tual, N. Carrere, P. Davies, T. Bonnemains, E. Lolive, Characterization of sea water ageing effects on mechanical properties of carbon/epoxy composites for tidal turbine blades, Composites Part A: Applied Science and Manufacturing 78 (2015) 380-389.
[4] M. Gupta, R. Srivastava, Mechanical, thermal and water absorption properties of hybrid sisal/jute fiber reinforced polymer composite, (2016).
[5] H. Jena, A.K. Pradhan, M.K. Pandit, Studies on water absorption behaviour of bamboo–epoxy composite filled with cenosphere, Journal of Reinforced Plastics and Composites 33(11) (2014) 1059-1068.
[6] H. Jena, M.K. Pandit, A.K. Pradhan, Effect of cenosphere on mechanical properties of bamboo–epoxy composites, Journal of Reinforced Plastics and Composites 32(11) (2013) 794-801.
[7] S.B. Harogoppad, T.M. Aminabhavi, Diffusion and sorption of organic liquids through polymer membranes. 5. Neoprene, styrene-butadiene-rubber, ethylene-propylene-diene terpolymer, and natural rubber versus hydrocarbons (C8-C16), Macromolecules 24(9) (1991) 2598-2605.
[8] N. Anjum, B. Suresha, S.L.A. Prasad, Influence of Water Ageing on Mechanical Properties of CaCO3 Filler Filled Epoxy Resin and Sansevieria/Carbon Fiber Reinforced Composites, (2019).
[9] Y. Hu, G. Du, N. Chen, A novel approach for Al2O3/epoxy composites with high strength and thermal conductivity, Composites Science and Technology 124 (2016) 36-43.
[10] A. Pandian, M. Vairavan, J. Thangaiah, W. Jappes, M. Uthayakumar, Effect of moisture absorption behavior on mechanical properties of basalt fibre reinforced polymer matrix composites, Journal of composites 2014 (2014).
[11] S. M. Vinu Kumar B. Suresha, G. Rajamurugan, A. Megalingam, Mechanical and abrasive wear behavior of cenosphere filled carbon reinforced epoxy composites using Taguchi-Grey relational analysis, Materials Research Express 6(1) (2018) 015307.
[12] K.S. Ahmed, V. Mallinatha, S. Amith, Effect of ceramic fillers on mechanical properties of woven jute fabric reinforced epoxy composites, Journal of Reinforced Plastics and Composites 30(15) (2011) 1315-1326.
[13] J. Yao, G. Ziegmann, Water absorption behavior and its influence on properties of GRP pipe, Journal of composite materials 41(8) (2007) 993-1008.
[14] L. Bian, J. Xiao, J. Zeng, S. Xing, Effects of seawater immersion on water absorption and mechanical properties of GFRP composites, Journal of Composite Materials 46(25) (2012) 3151-3162.

[15] T. Gentry, L. Bank, A. Barkatt, L. Prian, Accelerated test methods to determine the long-term behavior of composite highway structures subject to environmental loading, Journal of Composites, Technology and Research 20(1) (1998) 38-50.

IOP Conf. Series: Materials Science and Engineering **764** (2020) 012045 doi:10.1088/1757-899X/764/1/012045

Filter Design for High Frequency Applications

T.Perarasi[1*], R.Gayathri[1], P.Hamsagayathri[1]

[1]Department of Electronics & Communication Engineering, Bannari Amman Institute of Technology, Coimbatore-638 401, India.

[*]Author for correspondence E-mail: perarasi@bitsathy.ac.in

Abstract. An on chip Radio Frequency(RF) latent band sidestep channel is intended for different groups of frequencies dependent on a three post LC low pass get out and Direct Current (DC) barring capacitors that is prepared by means of Gallium Arsenide (GaAs) Monolithic Microwave consolidated Circuit innovation. The utilization of this model, the RF band skip channel with a tuneable focus recurrence and a legitimate data transfer capacity might be figured it out. A planar winding inductor required on this design and the format of the channel out is limited. The addition misfortune is improved and Substrate misfortunes might be limited. The recreation bring about the schematic and configuration of the band pass channel out that is executed by means of Agilent design software gadget. This sort of LC band pass might be actualized in RF control speaker to the front line report and radar machine for achieving the recurrence determination with an incredible advantage of chosen signal.

Keywords: Band pass filter; GaAs material; MMIC technique; Radio Frequency; LC.

1. Introduction- Radio Frequency

Radio repeat (RF) and microwave channels address a class of electronic channel, which is planned to deal with sign in the Mega Hertz (MHZ) to Gigahertz (GHZ) repeat ranges (medium to extremely high repeat). This repeat range is generally used to impart radio, TV remote correspondence (cell phones, Wi-Fi, etc.) and along these lines, most RF and microwave devices join a type of filtering on the sign that are transmitted and got. Such channels are ordinarily used as structure deters for duplexers and diplexers to join or separate different repeat gatherings.

1.1 RF Filter

In stylish, most RF and microwave channels are most generally comprised of one or more prominent coupled resonators and thus any age that might be utilized to cause resonators can likewise be utilized to make channels. The emptied excellent factor of the resonators being utilized will typically set the selectivity which the channel can achieve. The best resonator shape that might be utilized in RF and microwave channels is a LC tank circuit, for example, parallel or arrangement inductors and capacitors. Those have the advantage of being conservative. Those channels work with full frequencies and coupling coefficients of coupled resonators in a microwave filter.

1.2 *Types of RF Filter*

These are the 4 assortments of RF filter.

1.2.1 Band pass filter

A band-stop filter is a device that passes frequencies inside a specific assortment and rejects (weakens) frequencies open air that assortment. A truly flawless band-sidestep channel may have an absolutely level pass-band (for example without a favourable position/lessening during) and may totally constrict all frequencies open the pass-band as in Figure 1. In each transmitting and getting programs, band-skip channels have the top-quality transmission capacity for the mode and speed of correspondence being utilized, which boost the wide assortment of sign transmitters which can exist in a contraption, while limiting the obstruction or resistance among sign.

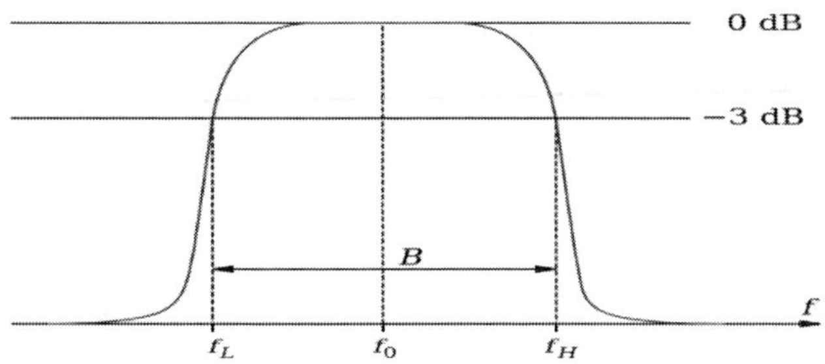

Figure 1. Importance switch function as opposed to frequency for a band-skip Filter.

1.2.2 Band Stop Filter

A band-stop channel or band-dismissal channel is one which passes most frequencies unaltered, anyway constricts those in a chose range to low levels as given in Figure 2. It's far something contrary to a band pass channel out. A notch circuit is a band counteract get out with a narrow stop band (high Q factor). By and large, the width of the stop band is 1 to 2 numerous years (that is, the absolute best recurrence lessened is 10 to 100 examples the least recurrence weakened). In any case, inside the sound band, an indent get out has high and low frequencies that might be handiest semitones separated.

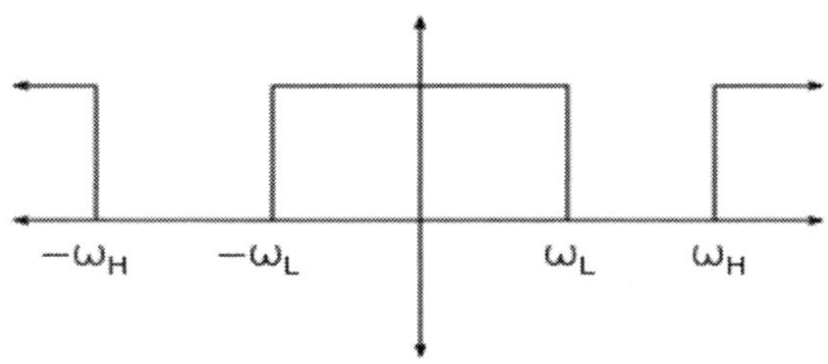

Figure 2. An ordinary best band-stop clear out, showing each wonderful and bad angular frequency

1.2.3 Low Pass Filter

Low-sidestep channel is a channel that passes low recurrence pointers and lessens (decreases the sufficiency of) markers with frequencies higher than the diminish off recurrence is given in Figure 3. The genuine measure of weakening for every recurrence changes relying upon exact channel format. Its miles from time to time known as an over the top cut channel out, or treble cut get out in sound bundles. A band pass filter is a mix of a low-pass and an over the top band. Low-pass channels exist in numerous select administration.

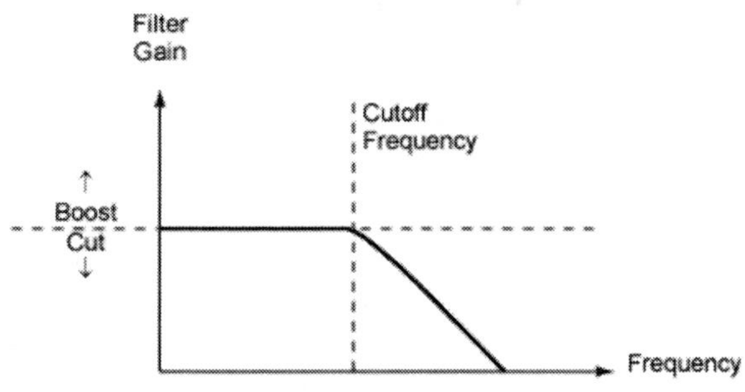

Figure 3. Low Pass Filter frequency response graph

IOP Conf. Series: Materials Science and Engineering **764** (2020) 012045 doi:10.1088/1757-899X/764/1/012045

1.2.4 High pass filter

A High pass is an electronic filter that passes over the top recurrence pointers anyway constricts (diminishes the abundance of) markers with frequencies and decline the upper level. The real measure of attenuation for each recurrence shifts from filter to filter is depicted in Figure 4. An extreme detour filter is commonly demonstrated as a direct time invariant framework. It's far at times known as a low profile channel or bass-cut get out. Extreme pass channels have numerous utilizations, which incorporates barring DC from hardware delicate to non-zero regular voltages of RF contraptions. They can likewise be utilized alongside a low-skip filter to make a band pass configuration.

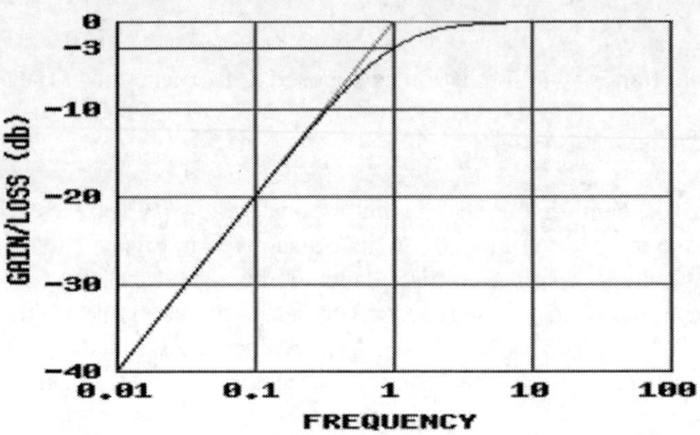

Figure 4. Normalized recurrence reaction of High Pass Filter

1.3 Band Pass Filter

Band pass filter can likewise be made through joining both low pass and over the top pass filter. A fantastic programming of a band pass is in sound sign preparing, where a particular assortment of frequencies of sound are required even as discarding the rest as depicted in Figure 5. Some other application is inside the determination of a chose sign from a scope of alarms in correspondence structures. A band skip channel out can be developed by falling an over the top pass RL channel out with a move off recurrence f_L and a Low pass RC channel with a move off recurrence f_H, to such an extent that $f_L < f_H$ .The lower lessen off recurrence is given as $f_L = \frac{R}{2\pi L}$. The higher cut-off recurrence is given as: $f_H = \frac{1}{2\pi RC}$. The Band Width of frequencies surpassed is given by means of: $BW = f_H - f_L$.

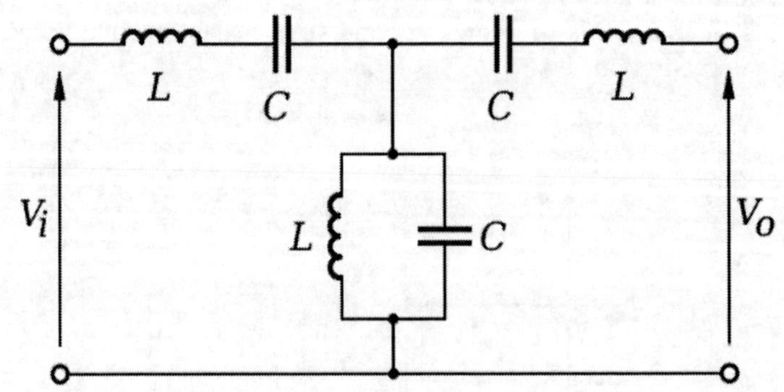

Figure 5. A schematic chart of a band pass filter

1.4 Microwave Monolithic Integrated Circuit (MMIC)

A Monolithic Microwave incorporated Circuit, or MMIC, is a sort of coordinated circuit apparatus that works at microwave frequencies of 300 MHz to 300 GHz. The capacities are as per the following

i. Microwave mixing,

ii. Energy enhancement,

iii. Low-commotion enhancement,

iv. High-recurrence exchanging.

Information sources and yields on MMIC devices are much of the time coordinated to a capacity impedance of 50 ohms, which makes them less confounded to utilize, in light of the fact that falling of MMICs does now not then require an outside coordinating network. MMICs are dimensionally little (from around 1 mm^2 to ten mm^2) and can be mass delivered, which has permitted the expansion of high-recurrence devices which incorporate phones. There are three materials in MMIC. They're Substrate fabric - GaAs, Conductor material - Al, Cu, Gold, Resistive fabric - Cr, NiCr, Ta, Ti. The MMIC materials are appropriate for territory and military programming, since they satisfy a need of brief temperature condition and serious vibration.

1.5 *Silicon*

Silicon has a high warm conductivity of 149 W·m−1·K−1, silicon conducts warmness pleasantly and as a final product isn't in every case habitually used to protect warm items. Silicon is widely utilized in semiconductor substances. Silicon keeps up preferable temperature over the germanium and because of the reality its neighbourhood oxide is easily developed in a heater. Silicon administrative work a higher semiconductor/dielectric interface than practically all other material mixes.

1.6 *Gallium Arsenide* (GaAs)

Gallium Arsenide (GaAs) is the compound component which incorporates gallium and arsenic. It is utilized inside the production of hardware which incorporates microwave recurrence included circuits, solid microwave included circuits, infrared gentle emanating diodes, laser diodes, sun cells and optical home windows.

1.6.1 *Why GaAs than Silicon?*

GaAs devices are gigantically cold-hearted toward warmth because of their more extensive band opening. Likewise, GaAs gadgets for the most part will in general have considerably less noise than silicon gadgets, especially at high frequencies. That is a final product of higher transporter portability's and abatement resistive instrument parasitic. Those homes recommend GaAs hardware in mobile phones, satellite correspondences, microwave factor-to-factor connections and better recurrence radar structures.

In view of its wide band gap, common GaAs is especially resistive. Mixed with the high dielectric consistent, this property makes GaAs a wonderful electrical substrate and as opposed to Silicon offers characteristic detachment among gadgets and circuits. This has made it a magnificent fabric for microwave and millimetre wave included circuits, MMICs, wherein exuberant and pivotal detached added substances can promptly be delivered on a single cut of GaAs.

IOP Conf. Series: Materials Science and Engineering **764** (2020) 012045 doi:10.1088/1757-899X/764/1/012045

1.7 *Tool Used*

Advanced Design System (ADS) is an advanced format mechanization programming machine created by method for Agilent EEs of EDA. Agilent plugs underpins each progression of the format technique schematic seize, design, recurrence and time region circuit recreation and electromagnetic order re-enactment allowing the designer to totally portray and improve a RF structure without changing over hardware.

Agilent EEs has given duplicates of the plugs programming project to the electric building offices at numerous colleges and a gigantic level of new alumni are knowledgeable about its utilization. Subsequently, the machine has watched wide notoriety in industry.

1.8 *Goal of the work*

The examination of RF on chip band pass channel dependent on GaAs solid microwave consolidated circuit. To format RF band pass channel out at 1-8 GHz dependent on utilizing three post LC low pass and two dc blocking capacitor. To acquire the design of band skip channel by methods for the use of promotions apparatus.

2. Existing System

Bands skip Filters plays out a fundamental position in RF and microwave incorporated circuits (ICs). In RF bundles, limit of the monetarily accessible RF band pass channels like ceramic filters or floor acoustic wave channels have genuine in general execution. Unfortunately, those channels are massive and off-chip discrete added substances, which is hard for the last scaling down and coordination of the framework.

The microwave design model of the designed channel with the director misfortunes is proposed principally dependent on the microwave network thought and checked through the EM reproduction. The utilization of the design form, the band pass with a tuneable focus recurrence and a proper transmission capacity might be discovered. In light of best utilizing an on-chip winding inductor with a littler inductance and four MIM capacitors in this plan.

2.1 Design and Analysis

Inactive band pass channels are by and large structured dependent on the channel out blend. Inside the method, the change of a model low-pass channel out to a band pass channel out requirements not best a monstrous assortment of the inductors but rather likewise a terrible parcel more inductance cost, consistently achieving many Nano Henry's, so the structure of the band sidestep filter with on-chip inductors is undermined.

2.1.1 Lossless Prototype

The change of a model low-pass to a band skip filter necessities now not most straightforward a gigantic scope of the inductors but rather moreover parcels more prominent inductance esteem so the plan of the band sidestep circuit with on-chip inductors is undermined. A small scale machined RF on-chip uninvolved band pass channel out at 1–8 GHz is proposed by means of using a 3-post LC low pass get out and dc-closing off capacitors as appeared in Figure 6.

IOP Conf. Series: Materials Science and Engineering **764** (2020) 012045 doi:10.1088/1757-899X/764/1/012045

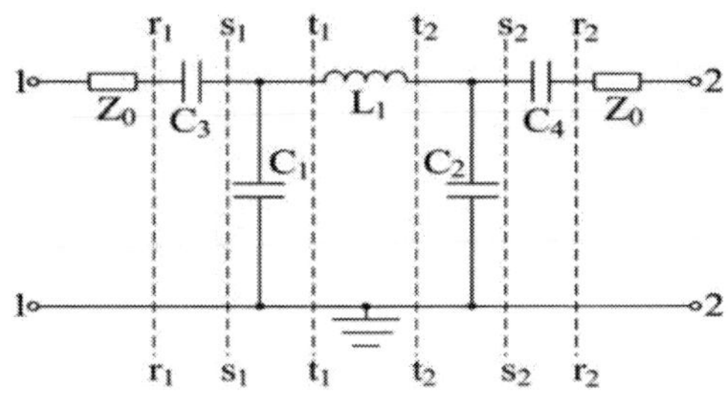

Figure 6. Simple π - arrange topology circuit of the band pass channel out.

Ports 1 and two or three are information and yield ports, Z_0 is the capacity impedance of the coplanar waveguide (CPW) transmission line, C_3 and C_4 are dc-barring capacitors and $C_1 - L_1 - C_2$ establishes a 3-shaft low skip channel. The equivalent circuit model of the Band pass channel is depicted by a lumped system and separated into seven components with the guide of the dashed line $r_1 - r_1$, $s_1 - s_1$, $t_1 - t_1$, $t_2 - t_2$, $s_2 - s_2$ and $r_2 - r_2$.With respect to the $r_1 - r_1$ side, underneath lossless circumstances, the ABCD framework standardized of the lumped equivalent circuit between port r_1 and port r_1 can take conveyance of as

$$\begin{bmatrix} A & B \\ C & D \end{bmatrix}_{1-r_1} = \begin{bmatrix} \cos\theta & j\sin\theta \\ j\sin\theta & \cos\theta \end{bmatrix} \tag{1}$$

Where θ is the electric length of the transmission line. For an easy analysis, the electrical period can be described as 0. Therefore, (1) can be simplified as

$$\begin{bmatrix} A & B \\ C & D \end{bmatrix}_{1-r_1} = \begin{bmatrix} 0 & 1 \\ 1 & 0 \end{bmatrix} \tag{2}$$

In accordance with the microwave thought, the ABCD grid standardized of the band jump get out between ports 1 and 2 might be acquired as

$$\begin{bmatrix} A & B \\ C & D \end{bmatrix}_{1-r_1} = \begin{bmatrix} A & B \\ C & D \end{bmatrix}_{t_2-s_2} \begin{bmatrix} A & B \\ C & D \end{bmatrix}_{s_2-r_2} \begin{bmatrix} A & B \\ C & D \end{bmatrix}_{r_2-2}$$

$$\begin{bmatrix} A & B \\ C & D \end{bmatrix}_{1-2} = \begin{bmatrix} 0 & 1 \\ 1 & 0 \end{bmatrix} \begin{bmatrix} 1 & \frac{1}{j\omega C_3 Z_0} \\ 0 & 1 \end{bmatrix} \begin{bmatrix} 1 & 0 \\ j\omega C_1 Z_0 & 1 \end{bmatrix} \begin{bmatrix} 1 & \frac{j\omega L_1}{Z_0} \\ 0 & 1 \end{bmatrix}$$

$$\begin{bmatrix} 1 & 0 \\ J\omega C_3 Z_0 & 1 \end{bmatrix} \begin{bmatrix} 1 & \frac{1}{J\omega C_2 Z_0} \\ 0 & 1 \end{bmatrix} \begin{bmatrix} 1 & 0 \\ 0 & 1 \end{bmatrix} \tag{3}$$

Correspondingly above condition might be disentangled and it could be communicated as

$$\begin{bmatrix} A & B \\ C & D \end{bmatrix}_{1-2} = \begin{bmatrix} \frac{C_3 - \omega^2 C_1^2 L_1 + C_1(2 - \omega^2 C_3 L_1)}{C_3} & \frac{j(C_1 + C_3)(-2 + \omega^2 C_1 L_1 + \omega^2 C_3 L_1}{\omega C_3^2 Z_0} \\ -j\omega C_1(-2 + \omega^2 C_1 L_1) Z_0 & \frac{C_3 - \omega^2 C_1^2 L_1 + C_1(2 - \omega^2 C_3 L_1)}{C_3} \end{bmatrix}$$

(4)

In which ω = 2πf, the precise recurrence and f is the recurrence. With the expectation to likewise rearrange the investigation in (3) so that $C_1 = C_2$ and $C_3 = C_4$, the ABCD lattice are standardized between ports 1 and two or three. Steady with the change relationship of the ABCD lattice and the dissipating framework, in light of (4), the S-parameters of the band sidestep get out among ports 1 and a couple of might be determined.

$$S_{11} = \left[\begin{matrix} (C_3(2-\omega^2 C_3 L_1) \\ -\omega^2 C_1^2 L_1 (1+\omega^2 C_3^2 Z_0^2) \\ +2C_1(1-\omega^2 C_3 L_1 + \omega^2 C_3^2 Z_0^2)) \end{matrix} \middle/ \begin{matrix} ((-jC_3 + C_1(-j+\omega C_3 Z_0)) \\ *(2j+\omega C_3(-j\omega L_1 - 2Z_0) \\ +\omega^2 C_1 L_1 (-j+\omega C_3 Z_0))) \end{matrix} \right]$$

$$S_{21} = (2j\omega C_3^2 Z_0)/((-jC_3 + C_1(-j+\omega C_3 Z_0))$$

$$* \left(2j + \omega C_3(-j\omega L_1 - 2Z_0) + \omega^2 C_1 L_1(-j+\omega C_3 Z_0)\right))$$

2.2 Dimension and Discussion

The capacity impedance Z_0 of the transmission line is 50 Ω. In order to uncover the structure rendition of the band pass channel out, the S-parameters of the hypothetical outcomes are demonstrated in Figure 7.The addition misfortune S_{21} of the band skip channel out for uncommon inductance L_1, unmistakable dc-blocking capacitance C_3 and C_4 and particular shunt capacitance C_1 and C_2 are appeared in Fig. 2.3. L_1 can be utilized to plan the tuneable focus recurrence of the band pass get out.

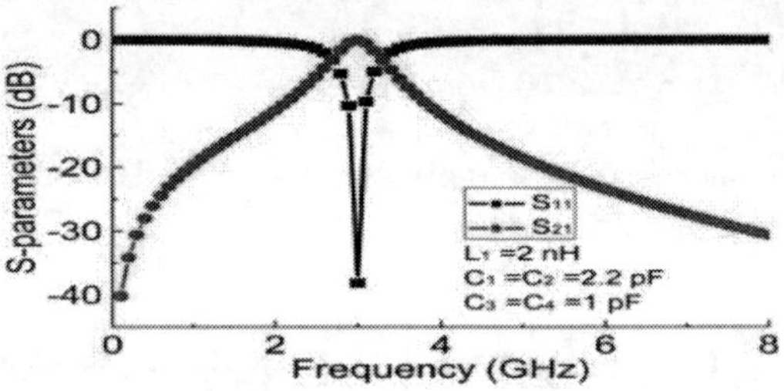

Figure 7. S-parameters of hypothetical outcome for band sidestep.

Indeed, even as C_3 and C_4 might be utilized to get a thin or broad data transfer capacity. C_1 and C_2 play an urgent position in planning the middle recurrence and the data transmission of the band pass channel and it can be discovered with the guide of microwave configuration model.

To have the option to affirm the precision of the design model with the conductor misfortunes, the hypothetical and recreated results are as thought about. The re-enactment of the band pass filter is done with the utilization of EM test system (HFSS).

The hypothetical impacts with the director misfortunes bring about the addition misfortune from −0.01 dB to −3.37 dB, which proposes that the microwave misfortune model can all the more likely foresee the general execution of the manufactured band pass channel out. In the re-enactment, the GaAs substrate parameter is prepared to be 0.006 in dielectric misfortune point of view and endless in resistivity. The S-parameters of the band skip filter can be planned and streamlined the use of format model with director misfortunes and therefore the inside recurrence and data transfer capacity of channel might be acquired.

The plan of RF on-chip LC detached band pass channels are provided with the goal that you can confine the substrate misfortunes. These band pass channels furnish the completely similarly invested usefulness with the GaAs MMIC time. Besides, the methodology of the microwave plan variant can be used in various RF contraptions. Estimations show that the band pass has come about inside the improvement of more noteworthy than 24% inclusion misfortune, with the effect at the middle recurrence of considerably less than 0.34%. The addition misfortune and the format size of the band pass channels have adaptable updates.

The table summarizes the theoretical, simulated, and measured effects of these band bypass filters.

Table.1 Precise of Theoretical, Simulated and measured consequences of the RF Band bypass filter out

	F0 (GHz)	IL(dB)	RL(dB)	Attenuation at 2 GHz (dB)	Attenuation at 4 GHz (dB)	-3dB Bandwidth (GHz)
Design Model	3	-3.2	-10.2	-11.3	-12.3	0.8
HFSS (h=2 m)3.25	3.25	-3.1	-10.6	-13.1	-9.9	0.84
HFSS(h=0.41 m) (No MS and PGS)	3.08	-3.5	-9.6	-12.7	-13.0	0.78
Measurement (No MS and PGS)	3.08	-7.4	-4.8	-17.3	-16.1	0.75
HFSS(MS and PGS)	3.05	-3.3	-10.1	-12.8	-11.9	0.78
Measurement(MS and PGS)	3.01	-5.6	-7.5	-15.4	-14.4	0.75

The format and creation of small scale machined RF on-chip LC latent band pass channels with MS, PGS and the back empty space are given inside the paper in order to limit the substrate misfortunes. Those band pass channels furnish the completely similar capacity with the GaAs MMIC period.

3. Proposed System

A filter has been planned at a recurrence scope of one-eight GHz and 1-12GHz. The channel had been planned and completed in the promotions gadget. The yields of these channels in schematic view were acquired. The structure of those channel is inside the state of LC blend circuit. The centre recurrence of those channels has been determined. The final product proposes the diagram between S-parameter vs. recurrence.

3.1 EM simulation

A mechanical three-D EM field solver, Ansoft's High Frequency Structure Simulator (HFSS), transformed into used to investigate the RF by and large execution of the suspended inductor contraptions. The product program instrument is based at the limited component technique and has been recommended to have exorbitant exactness in displaying customary on-chip IC inductors. The outcomes of substrate resistivity, air hole thickness, conductor material, and thickness on the inductor's part had been examined. Inside the EM reproduction, the styles of the wind and winding inductor plans recently portrayed were developed. Re-enacted Q components of the suspended winding inductor at the silicon substrates of different resistivity.

The graphical buyer interface of Ansoft's HFSS. The picked materials have been then appointed to the relating objects inside the styles. Recreation constraints had been characterized and ports had been

IOP Conf. Series: Materials Science and Engineering **764** (2020) 012045 doi:10.1088/1757-899X/764/1/012045

doled out to every finish of the inductor contraptions. The floor air ship is on the base part of the substrate. The recurrence range become one of a kind as from 0.2 to 15 GHz with a stage size of 0.1 GHz all through the circle assessment, metals inside the form were coincided mechanically and the ports were energized with a voltage limit. The EM assessment motor of Ansoft's HFSS processed the EM zone the use of the limited component procedure at every recurrence breadth factor.

After the investigation transformed into completed, the dispersing parameters (S-parameters) of the inductor model had been removed from the field arrangement. The dissipating parameters have been then changed into impedance parameters and induction parameters. The thing of the inductor at every recurrence factor was determined from its element parameter or S-parameter.

3.2 Layout and Analysis

The investigation and the design worried inside the channel is talked about. The inside recurrence is determined by utilizing the utilization of this framework

$$f_c = \sqrt{f_b \cdot f_a} \tag{5}$$

To locate the bypass band frequency this method is used.

$$f_p = f_b - f_a \tag{6}$$

Wheref_b is the upper cut off frequency

$f_a$is the lower reduce off frequency

3.3 At 1-8 GHz

By way of the usage of this formula the centre frequency and the skip band frequency is $f_0 = 2.82$ GHz and $f_p = 7$GHz respectively. This middle recurrence and the detour band recurrence is utilized to compute the estimations of inductors (L) and capacitors(C) inside the circuit. The impedance expense is $Z_0 = 50\Omega$ and the angular frequency is $\frac{1}{2\pi f}$. The tabular column shows the calculation of the LC combinations within the circuit.

The normalized price (g_k) may be calculated by using the following method$Z_0 = 1\Omega\,;\ \omega = 1\ rad/sec$

$$g_k = 2sin\frac{((2k-1)\pi)}{2n} \tag{7}$$

Wherein, k= role of the element.

n= number of detail.

Table. 2 Calculation of the L and C for the circuit

Circuit Region	$Z_0 = 1\Omega$, $\omega = 1\ rad/sec$	$Z_0 = 50\Omega$, $\omega = \frac{1}{2\pi f}$	$Z_0 = 50\Omega$, $f_p = 7GHz$
$g_1 or\ C_1$	1.618F	$\frac{1.618}{50}$	0.28pF
$g_2 or C_2$	1.618F	$\frac{1.618}{50}$	0.73pF
$g_3 or L_3$	2H	2*50	2.27nH
$g_4 or C_4$	1.618F	$\frac{1.618}{50}$	0.73pF
$g_5 or C_5$	0.618F	$\frac{1.618}{50}$	0.28pF

The very last capacitance and the inductance of the band skip filter out is given inside the below desk.

Table.3 Capacitance and Inductance values for the Band bypass filter

Capacitance	Inductance
C_1 =0.28Pf	L_1 =11.3nH
C_2 =0.73pF	L_2 =4.3nH
C_3 =1.4pF	L_3 =2.27nH
C_4 =0.73pF	L_4 =4.3nH
C_5 =0.28pF	L_5 =11.3nH

This filter out designed at 1-eight GHz the usage of ads tool is shown in Figure 8 and Figure 9., wherein the circuit shows the schematic view of the band pass filter out.

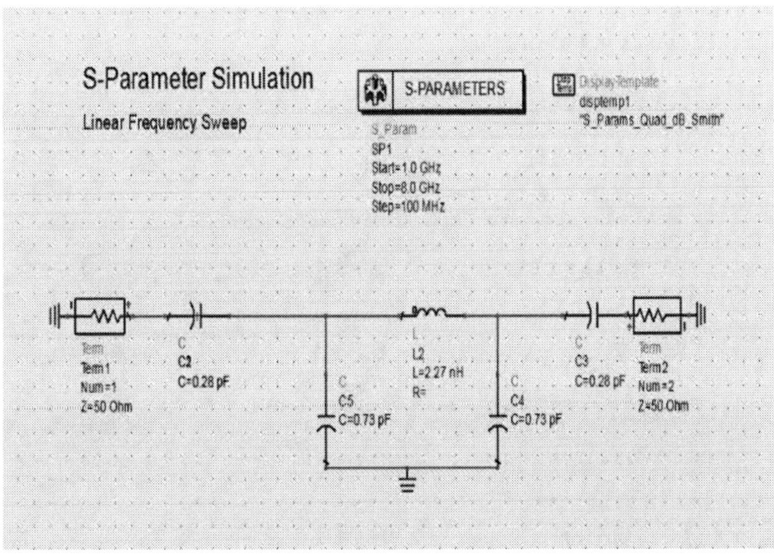

Figure 8. Schematic view of the band skip filter

IOP Conf. Series: Materials Science and Engineering **764** (2020) 012045 doi:10.1088/1757-899X/764/1/012045

From this circuit the schematic is exchanged by method for the LC pair total and the circuit yield is tried through changing the capacitance and inductance esteems correspondingly. Different diagram is procured and these charts are checked and as looked at for the better focus recurrence.

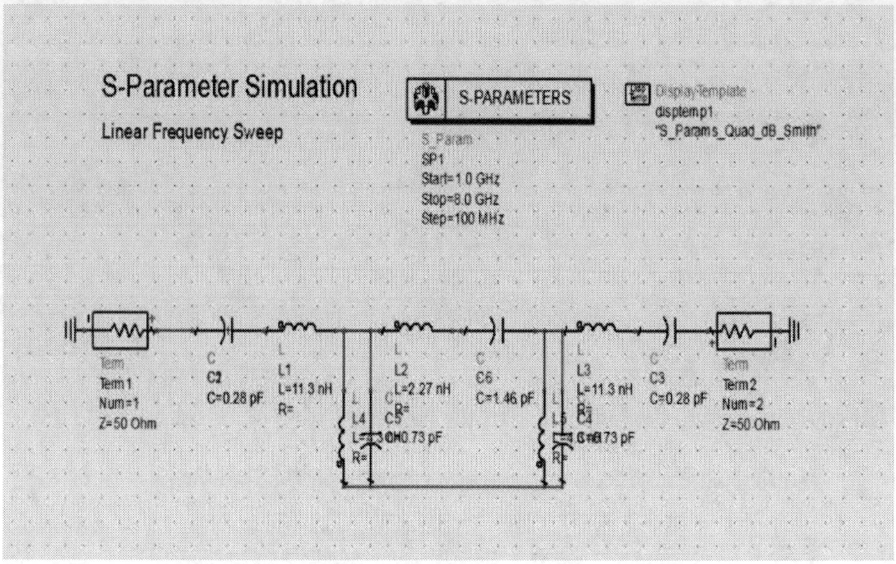

Figure 9. Schematic view of the LC pair band bypass filter out

3.4 At 1-12 GHz

With the guide of the use of this equation the middle recurrence and the detour band recurrence is f_o= 3.4GHz and f_p=11 GHz separately. Those middle recurrence and the detour band recurrence is utilized to figure the estimations of inductors (L) and capacitors(C) inside the circuit. The impedance charge is $Z_0 = 50\Omega$ and the angular frequency is $\frac{1}{2\pi f}$ the unthinkable section recommends the computation of the LC mixes in the circuit.

Table.4 Calculation of the L and C fee for the circuit

Circuit Region	$Z_0 = 1\Omega$ $\omega = 1\, rad/sec$	$Z_0 = 500\Omega$ $\omega = \frac{1}{2\pi f}$	$Z_0 = 50\Omega$ $f_p = 11GHz$
g_1 or C_1	0.618F	$\frac{0.618}{50}$	0.17pF
g_2 or C_2	0.618F	$\frac{0.618}{50}$	0.46pF
g_3 or L_3	2H	2*50	1.44nH
g_4 or C_4	0.618F	$\frac{0.618}{50}$	0.46pF
g_5 or C_5	0.618F	$\frac{0.618}{50}$	0.17pF

The very last inductance and capacitance fee is listed inside the desk under.

Table.5 L and C fee for the circuit

Capacitance	Inductance
C_1 =0.17pF	L_1 =12.4nH
C_2 =0.46pF	L_2 =4.5nH
C_3 =1.46pF	L_3 =1.44nH
C_4 =0.46pF	L_4 =4.5nH
C_5 =0.17pF	L_5 =12.4nH

The beneath diagram Figure 10. shows the clear out designed at 1-12 GHz using commercials tool. It suggests the schematic view of the designed filter.

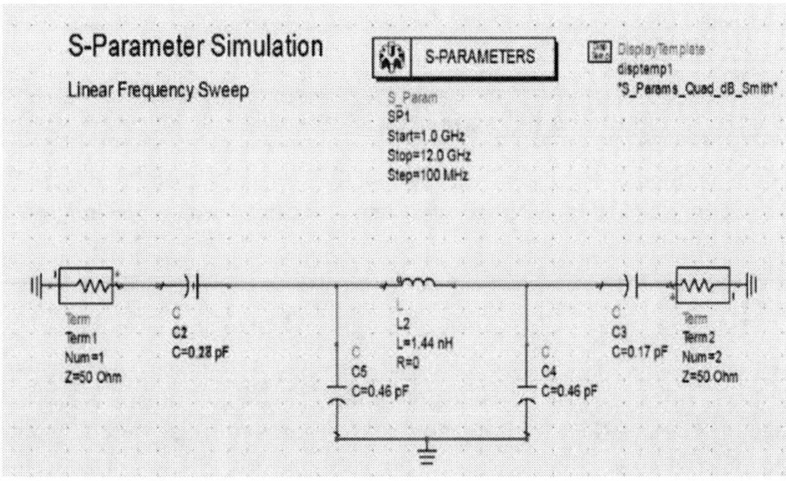

Figure 10. Schematic view of the band pass clear out at 1-12GHz

From the above circuit the format might be connected by means of the blend of inductors and capacitors and the results can be gotten and demonstrated. This circuit is actualized inside the promotions gadget and mimicked in schematic view as in Figure 11. This LC pair circuit helps in getting the inside recurrence tons higher.

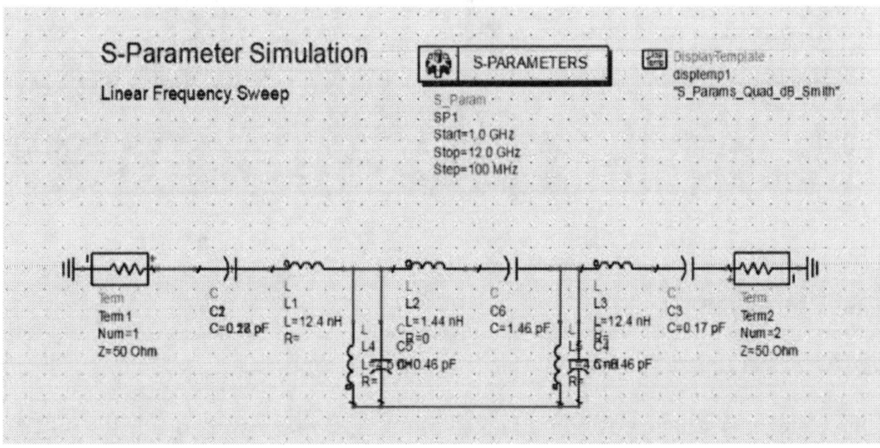

Figure 11. Schematic view of the LC paired 1-12 GHz band skip filter out

3.5 Substrate Materials

RF on-chip winding MMIC inductors are manufactured utilizing the GaAs MMIC strategy shows a creation of the MMIC inductor. The procedure starts on three-inch GaAs wafers with a thickness of around 625 μ m.

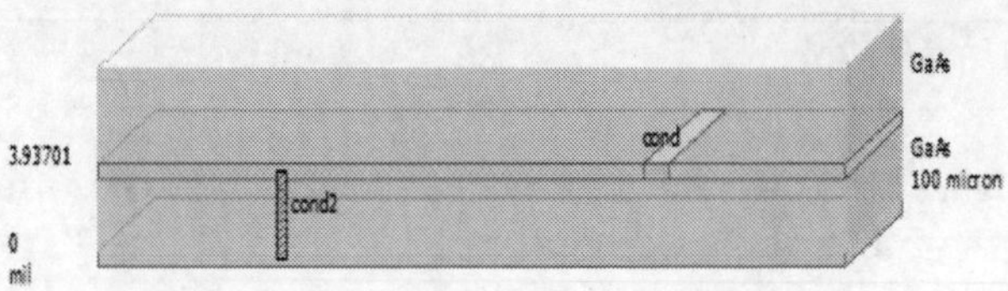

Figure 12. Substrate material worried in the layout process

The above outline in Figure 12. offers the perspective on the substrate material worried in the organization format in which the GaAs material has a width of a hundred micron and a dielectric material is included.

3.6 Layout Design

The 1-eight GHz band skip channel is structured inside the design see wherein its miles recreated inside the EM re-enactment. In this organization Gallium Arsenide substrate fabric is utilized. The thickness and the dielectric texture is chosen and with regards to it the yield is gained. The diagram gained is between the Discrete frequencies vs. Filted (AFS or direct). The organization configuration got is appeared in the underneath outline in Figure 13.

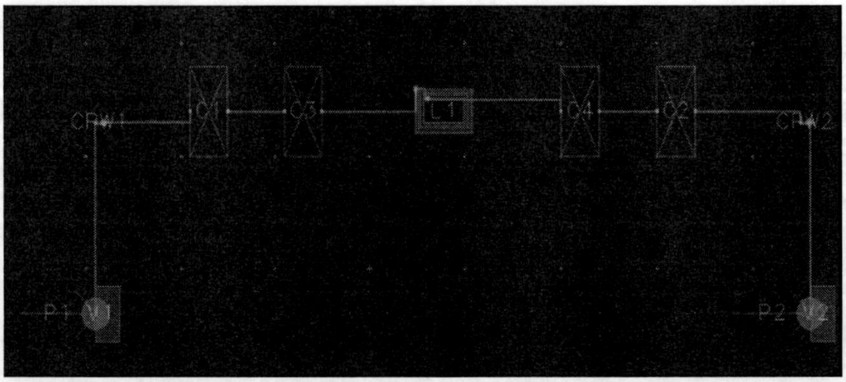

Figure 13. Layout View of the 1-8GHz Band skip filter

4. Results and Discussions

4.1 S Parameter at 1-8 GHz

The band skip channel is structured and mimicked through EM recreation inside the plugs gadget. The middle recurrence got from the hypothetical expense is tried from the reasonable expense. The chart is drawn between the S parameter and recurrence. The underneath chart in Figure 14 is drawn for the basic π network topology with a 1-eight GHz band pass channel out.

IOP Conf. Series: Materials Science and Engineering **764** (2020) 012045 doi:10.1088/1757-899X/764/1/012045

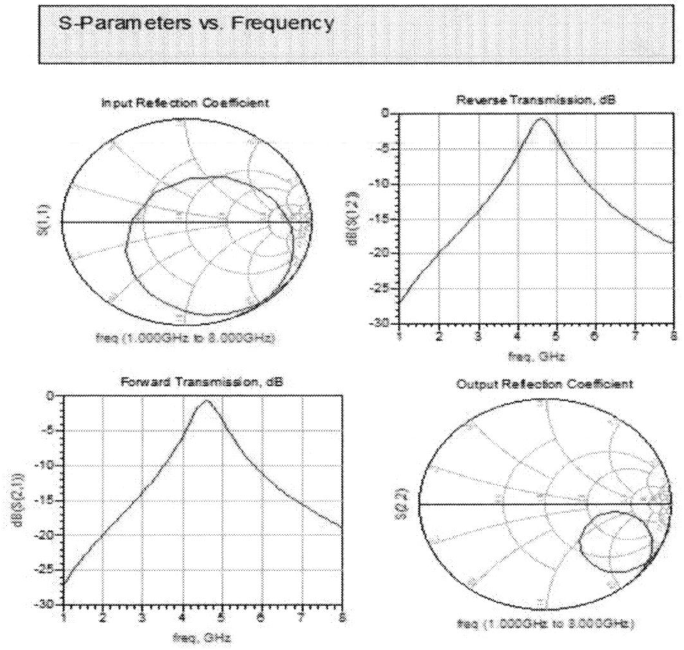

Figure 14. End result of 1-eight GHz π community topology

At that point the band sidestep channel is planned with inductance and capacitance total and re-enacted inside the EM reproduction. The chart proposes a few varieties inside the middle recurrence as contrasted and the π network topology. This chart is in like manner obtained between S-parameter vs. recurrence. in the beneath diagram the band pass area is gotten and the inside recurrence matches with the hypothetical recurrence.

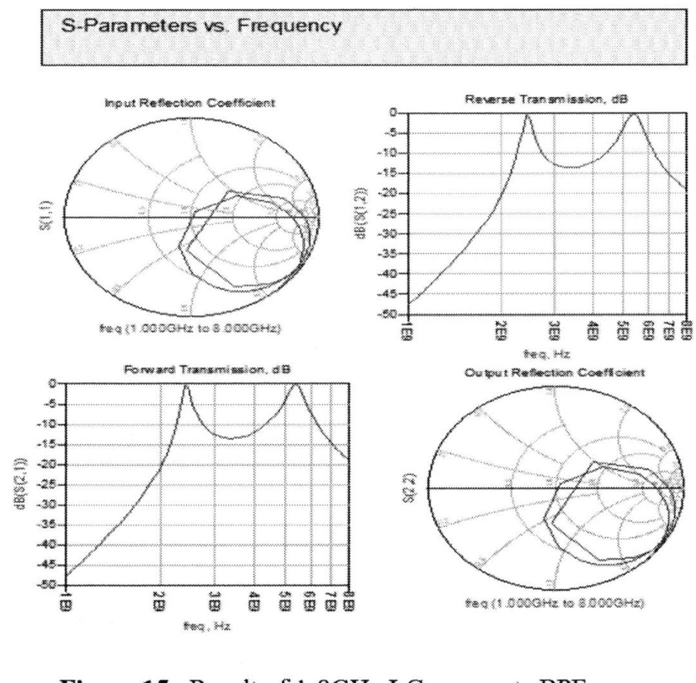

Figure 15. Result of 1-8GHz LC aggregate BPF

The above end result Figure 15 suggests the result of the 1-eight GHz BPF in schematic view within the ads device.

IOP Conf. Series: Materials Science and Engineering **764** (2020) 012045 doi:10.1088/1757-899X/764/1/012045

4.2 Substrate Material

RF on-chip winding MMIC inductors are manufactured the utilization of the GaAs MMIC strategy shows a creation buoy of the MMIC inductor. Because of the GaAs substrate with an over the top resistivity, these dielectric misfortunes are littler than the director misfortunes after the schematic view the band pass channel is planned inside the configuration. In the configuration see the GaAs substrate fabric is utilized and it's far appointed as appeared in Figure 16.

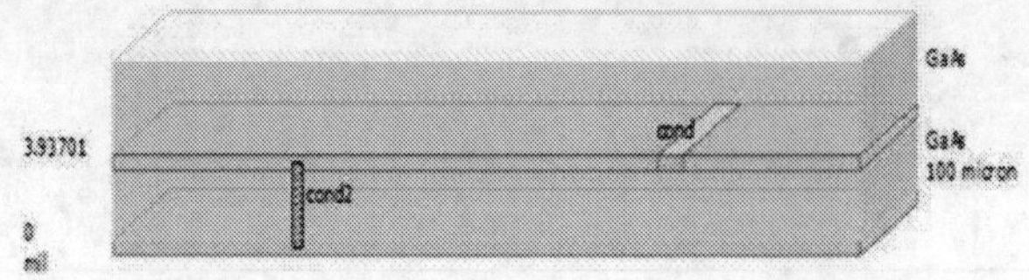

Figure 16. Substrate fabric used in the BPF

4.3 Layout Design

At 1–eight GHz, various parasitic outcomes and a few misfortune components notably affect band sidestep channels, specifically on-chip winding inductors to decrease the impact of substrate electric and attractive misfortunes brought about by the inductor, the miniaturized scale machined RF on-chip detached band pass get out with MS and PGS is proposed. The 1-8 GHz band pass channel is structured inside the design see in which its miles re-enacted in the EM recreation. The results of the configuration is gained and approved for better outcomes the use of the plugs apparatus. The beneath diagram of Figure 17 & Figure 18 is for discrete frequencies vs. channel out (AFS or Linear).

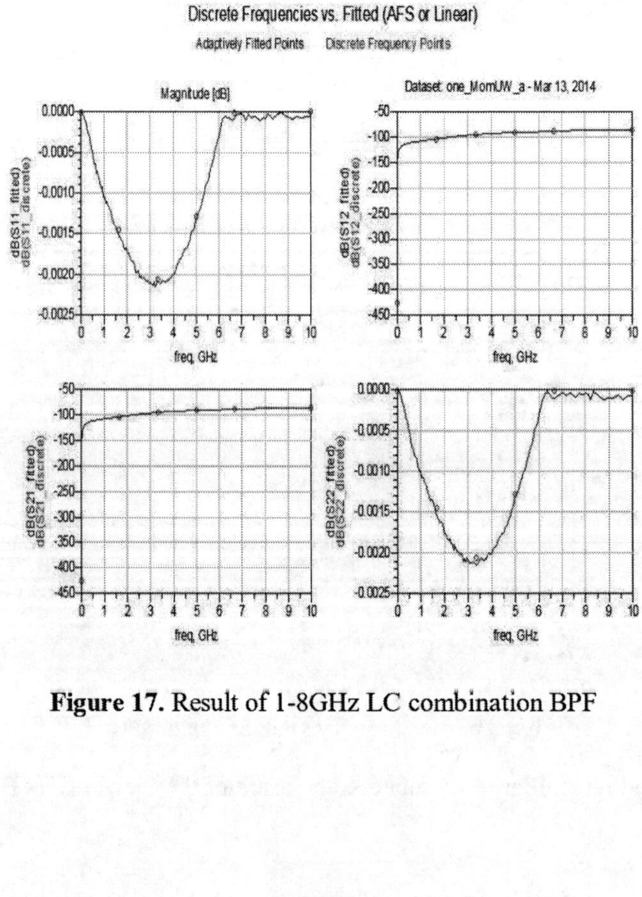

Figure 17. Result of 1-8GHz LC combination BPF

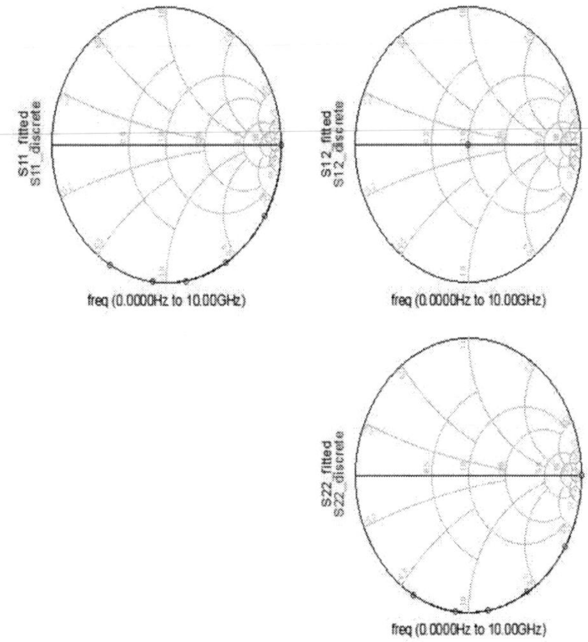

Figure 18. Result of 1-8GHz BPF in layout view

5. Conclusion

The method of the microwave style model is utilized in totally different RF gadgets. These small machined RF on-chip LC passive band skip filters could also be applied conjointly with RF electricity amplifiers to the modern non-public despatch system and also the radio location system for accomplishing the frequency selection, with a decent advantage of the chosen sign.

Reference

[1] L. Gu , X. X. Li, Concave-suspended high-Q solenoid inductors with an RFIC-compatible bulk-micromachining technology, *IEEE Transactions on Electron Devices*, 2007, 54, 882–885.

[2] L. H. Guo, Q. X. Zhang, G. Q. Lo, N. Balasubramanian, and D.-L.Kwong, High performance inductors on plastic substrate, *IEEE Electron Device Letters*, 2005, 26, 619–621.

[3] M. Kang, J. Gil, H. Shin, A simple parameter extraction method of spiral on-chip inductors, *IEEE Transactions on Electron Devices*,2005,52, 1976–1981.

[4] Y.-S. Lin, W.-C.Ku, C.-H. Wang, C. H. Chen, Wideband coplanar waveguide band pass filters with good stop band rejection, *IEEE Microwave Wireless Component Letters*, 2004, 14, 422–424.

[5] C.-M. Nam, Y.-S. Kwon, High-performance planar inductor on thick oxidized porous silicon (OPS) substrate, *IEEE Microwave Guided Wave Letters*, 1997, 7, 236–238.

[6] J.-M. Yook, K.-M.Kim, Y.-S. Kwon, Suspended spiral inductor and band-pass filter on thick anodized aluminium oxide, *IEEE Microwave Wireless Component Letters*, 2009, 19, 620–622.

[7] C. P. Yue , S. S. Wong, Physical modelling of spiral inductors on silicon, *IEEE Transactions on Electron Devices*, 2000, 47,560–568.

IOP Conf. Series: Materials Science and Engineering **764** (2020) 012045 doi:10.1088/1757-899X/764/1/012045

[8] C. P. Yue, S. S.Wong, On-chip spiral inductors with patterned ground shields for Si-based RF ICs, *IEEE Journal on Solid-State Circuits*, 1998, 33, 743–745.

[9] Z. Q. Zhang, X. P. Liao, Micro machined GaAs MMIC-based spiral inductors with metal shores and patterned ground shields, *IEEE Sensors*, 2012, 12, 1853–1860.

[10] N. B. Zhang, Z. L. Deng, C. Shu, and H. S. Wang, Design and analysis of a tuneable band pass filter employing RF MEMS capacitors, *IEEE Electron Device Letters*, 2011,32, 1460–1462.

[11] Nikhil Agrawal ,Anil Kumar ,Varun Bajaj ,G. K. Singh, Design of Band pass and Band stop Infinite Impulse Response Filters Using Fractional Derivative, *IEEE Transactions on Industrial Electronics*, 2019, 66 ,1285 – 1295.

[12] Dimitri Psychogiou,Roberto Gómez-Garcia, Reflection less Adaptive RF Filters: Band pass, Band stop, and Cascade Designs, *IEEE Transactions on Microwave Theory and Techniques*, 2017,65,4593 – 4605.

[13]Amir Eghbali, Håkan Johansson, On Efficient Design of High-Order Filters With Applications to Filter Banks and Trans multiplexers With Large Number of Channels, *IEEE Transactions on Signal Processing*, 2014,62, 1198 – 1209.

[14] Qian Liu ,Li Peng ,Yong Kang ,Shiying Tang ,Deliang Wu ,Yu Qi, A Novel Design and Optimization Method of an *LCL* Filter for a Shunt Active Power Filter, *IEEE Transactions on Industrial Electronics*, 2014,61, 4000 – 4010.

[15] Liang-Feng Qiu,Lin-Sheng Wu ,Bing Xie,Wen-Yan Yin ,Mao, Substrate Integrated Waveguide Filter With Flat Pass band Based on Complex Couplings, *IEEE Microwave and Wireless Components Letters*, 2018,28, 494 – 496.

[16] Dakotah J. Simpson ,Dimitra Psychogiou, Coupling Matrix-Based Design of Fully Reconfigurable Differential/Balanced RF Filters, *IEEE Microwave and Wireless Components Letters*, 2018, 28,888 – 890.

[17] Hee-Ran Ahn,Sangwook Nam, Design Method for Butter–Cheby Band pass Filters With Even Number of Resonators, *IEEE Transactions on Microwave Theory and Techniques*, 2012,60, 1549 – 1559.

[18] Sumit J. Darak, Vinod A. Prasad ,Edmund M.-K. Lai, Efficient Implementation of Reconfigurable Warped Digital Filters with Variable Low-Pass, High-Pass, Band pass, and Band stop Responses, *IEEE Transactions on Very Large Scale Integration (VLSI) Systems*, 2013, 21, 1165 – 1169.

[19] Boyoung Lee ,Seunggoo Nam ,Juseop Lee, Optimization-Free Design Equations for Narrowband Equal-Division Filtering Power Divider With Pre-Specified Filtering Response and Wideband Isolation, *IEEE Transactions on Circuits and Systems I: Regular Papers*,2019,66, 2496 – 2507.

[20] Thomas R. Jones ,Mojgan Daneshmand, Miniaturized Slotted Band pass Filter Design Using a Ridged Half-Mode Substrate Integrated Waveguide, *IEEE Microwave and Wireless Components Letters*, 2016,26, 334 – 336.

[21] M. Naimul Hasan ,ShahrokhSaeedi,Qun Jane Gu, Hjalti H. Sigmarsson,Xiaoguang Liu, Design Methodology of N-Path Filters With Adjustable Frequency, Bandwidth, and Filter Shape, *IEEE Transactions on Microwave Theory and Techniques*, 2018,66, 2775 – 2790.

[22] Bingo Wing-Kuen Ling ,Charlotte Yuk-Fan Ho,Kok-Lay Teo,Wan-Chi Siu ,Jiang zhong Cao , Qingyun Dai, Optimal Design of Cosine Modulated Non uniform Linear Phase FIR Filter Bank via Both Stretching and Shifting Frequency Response of Single Prototype Filter, *IEEE Transactions on Signal Processing*,2014,62, 2517 – 2530.

[23] Runqi Zhang ,Sha Luo ,Lei Zhu ,Li Yang , Synthesis and Design of Harmonic-Suppressed Wideband Band pass Filters With Mixed Modified Richard's Transformations, *IEEE Microwave and Wireless Components Letters*, 2017,27, 455 – 457.

[24] JinXu ,Zhi-Yu Chen ,Qi-Hang Cai, Design of Miniaturized Dual-Band Low-Pass–Band pass and Band pass Filters, *IEEE Transactions on Components, Packaging and Manufacturing Technology*, 2018, 8, 132 – 139.

[25] Runqi Zhang ,Lei Zhu, Synthesis Design of a Wideband Band pass Filter With Inductively Coupled Short-Circuited Multi-Mode Resonator, *IEEE Microwave and Wireless Components Letters*, 2012, 22,509 – 511.

[26] Shahrokh Saeedi, Juseop Lee ,Hjalti H. Sigmarsson, Tuneable, High-Q, Substrate-Integrated, Evanescent-Mode Cavity Bandpass-Bandstop Filter Cascade, *IEEE Microwave and Wireless Components Letters*, 2016, 26,240 – 242.

IOP Conf. Series: Materials Science and Engineering **764** (2020) 012046 doi:10.1088/1757-899X/764/1/012046

Experimental Investigation on PCC for Workability within Partial Replacement of M-Sand by Ecosand

N Sathyakumar[1*], M Harisbaskar[2], S Gowthaman[2] and K Murugananthan [2]

[1]Assistant Professor Department of Civil Engineering, Bannari Amman Institute of Technology, Coimbatore-638 401, India.
[2]PG Scholar Department of Civil Engineering, Bannari Amman Institute of Technology, Coimbatore-638 401, India.

*Author for correspondence E-mail: baskarm47@gmail.com

Abstract. In whole nation, concrete production events are enchanting abode on vast scale. Natural sand (river sand) is unique of the main requirements of concrete and it is pretty demand and exclusive. Manufactured sand (M-sand) is a standby of river sand for concrete assembly which is produced from rigid granite pebble by devastating. Eco-sand is fine particles which is outgrowth from cement trade. Effective use of this waste material should be done for cost efficiency. The present study is about the compressive strength of concrete block with limited surrogating of M-sand by Eco-sand. The conventional river sand is replaced with M sand and partial amount of silica sand Replacement of M-sand by Eco-sand at various percentages 0%,6%,12% 18%,24%&30% in PCC.It is verified that the finest spare percentage of M-sand by Eco-sand and the properties of fresh concrete and hard concrete in PCC with eco-sand. The comparison of economic advantages of m sand and eco sand. The main objective of this project is to reduce the air pollution and eco-friendly. The various proportions is carried out and tested.

1. Introduction

Bond concrete has been a mainstream development material on the planet and has fulfilled practically all the necessity of a decent building material. Concrete is characterized as composite blend of bond, sand, and rock in positive extents with water. The functionality for position and quality improvement with age rely on the extent of the constituent materials and their joined activity. Concrete is ordinarily delivered utilizing alluvial waterway rock and sands. These materials are richly accessible and are generally economical to process and consequently they are broadly utilized. Besides they produce great cement because of their physical properties, for example, shape degree. The common procedure of arrangement of these stores happened a large number of years. These stores are over misused. Subsequently there is a shortage of sand. There is overabundance sand mining bringing about consumption of groundwater, making ecological issues. Mining is denied by certain state governments for the accompanying reasons. Diving the sand from riverbed in overabundance amount is dangerous to condition. The profound pits delved in the riverbed influence the ground water level. Disintegration of close by land is likewise because of overabundance sand lifting. Concrete is a regular building substantial that is collected of cement, collective (usually a coarse cumulative made of gravel or crushed shakes and a fine cumulative such as sand) water and chemical admixtures. Thus true to its name, the constituent materials of concrete grow together in a condensed form. Truth is told concrete is utilized more than some other man-made material on the planet. Starting at 2011, around 9 cubic kilometres of cement made every year which is more than one cubic meter for each individual on earth.

2. Materials

2.1. Eco-sand

Eco sand are extremely fine particles, a side-effect from bond fabricate which can be utilized to expand productivity in concrete. Its small scale filling impact lessens pores in cements and gives better dampness resistivity and along these lines sturdiness. It has more reliable reviewing than many extricated totals. Powerful use for squander material and in this manner savvy and executes just as normally happening sand. The utilization of eco sand as opposed to extricated or dug normal sand will support fashioners and contractual workers address issues of manageability.

Eco sand is finely powdered crystalline silica which can supplant up to a changing level of regular sand utilization in cement and mortars. Its miniaturized scale filling impact diminishes pores in cements and gives better dampness resistivity and hence sturdiness. It has more predictable evaluating than many extricated totals. Viable use for squander material and along these lines practical and executes just as normally happening sand. The utilization of eco sand as opposed to removed or dug normal sand will support originators and contractual workers address issues of maintainability. The present examination is checking the functionality of Solid Square utilizing eco sand as fine total. The eco sand has different focal points, for example, vitality proficient, fireproof, decrease of dead burden, earth amicable, tough, light weight, low support, and low development cost.

2.2. Fine Aggregate

The Mass-produced sand (M-Sand) was used for the concrete. It is a substitute of river sand for concrete structure. Mass-produced sand is produced from solid granite stone by over whelming. The chemical composition of M-Sand is Si, Al, Ca, Na, K, Fe, etc. The specific gravity of M-sand was resulted as 2.6.

2.3.Coarse Aggregates

Coarse Aggregates are the maximum quarried supplies in the world. They are a constituent of combined ingredients such as concrete and asphalt concrete. Here 20mm aggregates are used. The specific gravity of aggregate was resulted as 2.6

3. Literature Review

M Indira April 2017 Increases the compressive quality and elasticity in a higher request of 15% substitution by eco sand increments compressive quality by 24.02% and split rigidity was expanded at 58.14 %. Be that as it may, in mortar, 10% substitution mortar shows 10.45% expansion in compressive quality

Selvakumaret.alApril 2016Eco sand reduces consumption of natural sand by up to 50% and hence reduces ecological footprint.

B.Durg et.alMarch 2016 the optimum replacement percentage arrived at compressive strength was 60% and split tensile strength it was 40% at 28 days. It was found that the compressive strength increases by 32% on 60% replacement and tensile strength increases by 13% on 40% replacement at 28 days. The particle size of silica sand by magnifying it to different ranges. They found that crystalline in nature it can be used as a filler material.

IOP Conf. Series: Materials Science and Engineering **764** (2020) 012046 doi:10.1088/1757-899X/764/1/012046

M. prabu et.alJune 2015. Quality of cement by fractional substitution (0%, 10%, 20%, 30% and 40%), ideal qualities are By GGBS - 30%, By Eco sand-20%. Eco sand supplanting with fine total with 20% gives ideal outcome however after that the quality got gradually diminishes. The level of functionality of cement improved with expansion of eco sand in concrete up to 20% swap level for M20 grade concrete.

4. Mix Design

Mix proportion is the one of the reason which decides the characteristics and properties of construction components. M 25 grade was taken for the experiment work.

Grade of concrete - M25
Size of aggregate - 20mm
Use of cement - PPC -A Grade
Relative density of cement - 3.15
Relative density of C.A - 2.74
Relative density of F.A - 2.63
Relative density of Eco-sand- 2.46
Maximum w/c - 0.45

4.1. Mix Proportion

Mix proportion -1.0: 1.43: 2.91
Wt of cement - 438.13 kg/m3
Wt of sand (M-sand) - 627.74 kg/m3
Wt of aggregate - 1276.74 kg/m3

5. Workability Test

5.1 Slump Cone Test

The solid droop test measures (functionality) the consistency of new cement before it sets. It is performed to check the functionality of newly made cement and in this way the simplicity with which solid streams. It can likewise be utilized as a marker of an inappropriately blended group.

This is a well-known because of the effortlessness of device utilized and straightforward method. The droop test is utilized to guarantee consistency for various heaps of cement under field conditions.

The cone is loaded up with new cement in three phases. Each time, each layer is packed multiple times with a 2 ft. (600 mm) long slug nosed metal pole. Toward the come to an end of the third stage, the solid is hit off flush with the highest point of the form. The form is deliberately lifted upwards, so as not to upset the solid cone. Thus the droop is named as obvious droop, shear droop or breakdown droop. A breakdown droop means that the blend is excessively wet.

5.2 Compacting Factor Test

The proportion of the mass of halfway compressed cement to the mass of the solid when completely compacted in a similar shape. To make a decision the compaction factor of cement with low, medium and high functionality according to IS 1199-1959.

The solid example is set on the upper container up to overflow, at that point open the snare entryway stream the solid stream uninhibitedly into lower container and afterward rehash the progression to permit to fall into the chamber. The overabundance concrete staying over the top degree of the chamber at that point cut off it and afterward gauged. This is known as weight of halfway compacted cement.

6. Results And Discussion

6.1 Slump Cone Test

% Replacement	Slump (mm)
0	72
6	74
12	79
18	82
24	85
30	91

Table 6.1 Slump Cone Test

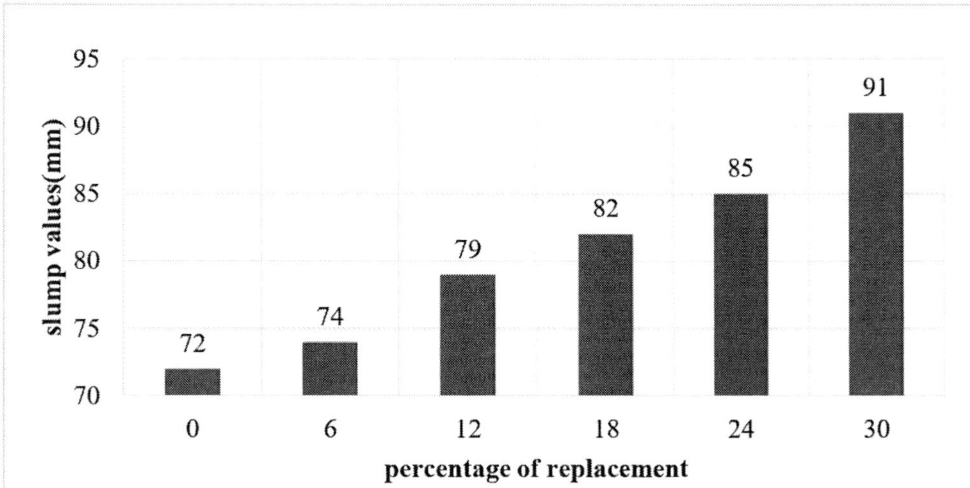

Figure 6.1 Slump Cone Test

The slump cone test shows that gradually increases at 2.78%, 9.72%, 13.89%, 18.05% and 26.38% as compared with conventional mix for the replacement of 0%, 6%, 12%, 18%, 24% and 30% in term of mm respectively

6.2 Compaction Factor Test

% Replacement	Compaction Factor
0	0.85
6	0.86
12	0.90
18	0.91
24	0.92
30	0.94

Table 6.2 Compaction Factor Test

IOP Conf. Series: Materials Science and Engineering **764** (2020) 012046 doi:10.1088/1757-899X/764/1/012046

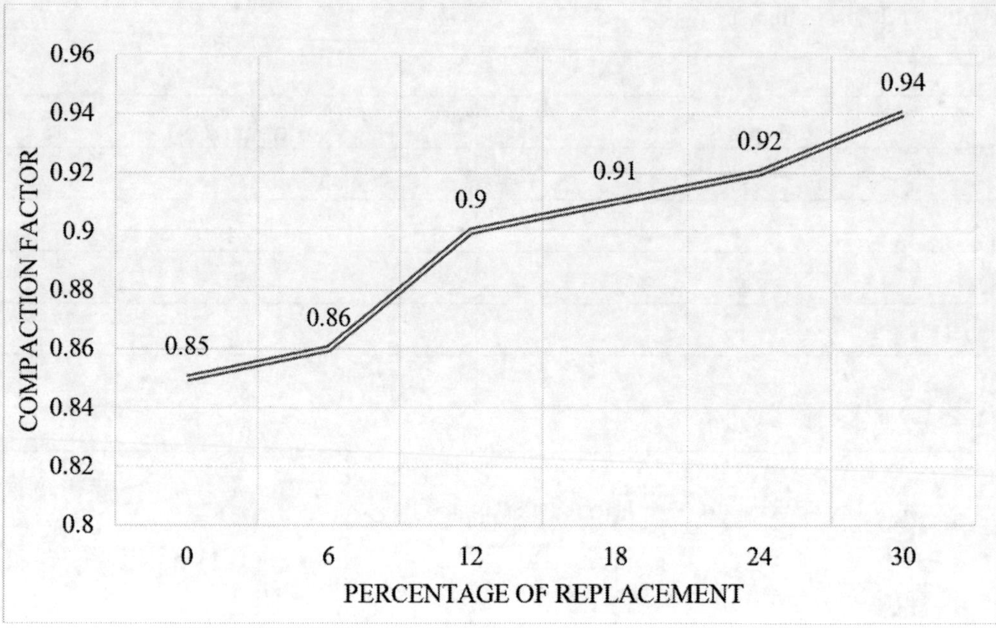

Figure 6.2 Compaction Factor Test

The compaction factor test shows that gradually increases at 1.17%, 5.88%, 7.05%, 8.23% and 10.58% as compared with conventional mix for the replacement of 0%, 6%, 12%, 18%, 24% and 30% respectively.

7. Conclusion

After conducting all the tests it has been detected that,

- The slump cone test of concrete is improved by adding Eco-sand in concrete.
- The compaction factor test of concrete is augmented by adding Eco-sand in concrete.

References

[1]. R Lakshmi Priyanka"Use of M-Sand as a Replacement of Cement and Fine Aggregate"SSRG International Journal of Civil Engineering (SSRG-IJCE) – volume 4 Issue 5 – May 2017.

[2]. M Indira "study on replacement for cement and fine aggregates using eco-sand" International Journal of Civil Engineering and Technology, volume 8, Issue 4, April 2017.

[3]. Biju Mathew "Study on Strength of Concrete by Partial Replacement of Fine Aggregate with M-Sand and Laterite with Super plasticizers" International Journal of Engineering Trends and Technology (IJETT) – Volume 38 Number 8- August 2016.

[4]. Selvakumar. "Experimental Investigation of Eco Sand in Concrete"International Journal of Recent Research and Applied Studies Received 8th April 2016.

[5]. B.Durg "Experimental Study on Various Effects of Partial Replacement of Fine Aggregate with Silica Sand in Cement Concrete and Cement Mortar" International Journal of Engineering Trends and Technology (IJETT) – Volume 33 Number 5- March 2016.

IOP Conf. Series: Materials Science and Engineering **764** (2020) 012046 doi:10.1088/1757-899X/764/1/012046

[6]. M. prabu " Influence of GGBS and Eco-sand in Green Concrete", International Journal of Innovative Research in Science, Engineering and Technology, volume 4, Issue 6,June 2015.

[7]. Vishnumanohar.A "Performance of normal concrete with eco-sand as fine aggregate" International Journal of Engineering Science Invention, volume 3, Issue 5, May 2014.

[8]. Priyanka A "Effect of replacement of natural sand by manufactured sand on the properties of cement mortar" International Journal of Civil And Structural Engineering Volume 3.No 3, 2013

IOP Conf. Series: Materials Science and Engineering **764** (2020) 012047 doi:10.1088/1757-899X/764/1/012047

Parametric study and optimization of regression model in WEDM using genetic algorithm

G Daniel[1], Girishkumar[2], Ali-Ul-Hasan-Rizvi[1] and P Mathan Kumar[3,*]
[1]Al-Falah University, Dhauj, Faridabad, India.
[2]Delhi Technological University, Delhi, India.
[3]Department of Mechanical Engineering, Bannari Amman Institute of Technology, Sathyamangalam.
[*]Author for correspondence E-mail: mathankumarp@bitsathy.ac.in

Abstract. Wire EDM is an unconventional machining processused for producing complicated profile on a hard material which is difficult in conventional cutting process. In WEDM heat affected zone is very minimum when compare to other unconventional cutting process. This distinct advantage made the WEDM to use in modern precision industries. In WEDM many process parameter which decides the quality of the machined component. In this study pulse on time, pulse off time, flushing pressure, wire tension, servo voltage and wire feed rate were selected as a process parameters. The effect of each parameter on material removal rate (MRR) was analyzed. Mathematical model was developed for material removal rate using regression analysis. The developed mathematical model is useful in predicting the response in the given range of input process parameter. Genetic algorithm was used to optimize the developed model. The optimized model would be the technical chart for the given material for the selected input parameter. The obtained results were experimentally verified.

1. Introduction

Wire EDM is an electro-thermal machining process particularly used for machining the hard materials with complex geometric shapes. Melting and vaporization is the material removal method in WEDM [1]. S. S. Mahapatra and Amar Patnaik, has developed a mathematical model to relate input process parameter and responses like MRR, Surface finish (SF) and kerf [2]. H.Singh et al. selected the input process parameters like pulse on time, gap voltage, peak current, pulse off time, wire tension and wire feed to analyse the MRR. They found that MRR increase with increase in the peak current and pulse on time [3]. Nihat Tosun et al analysed the effect of machining parameters on the MRR in wire EDM operations. The experiment was conducted by varying open circuit voltage, pulse duration, flushing pressure and wire speed.Machining parameters settings were determined by taguchi experimental design method [4]. Bijaya Bijeta Nayaket et al. have machined Inconel718 material with selected process parameter in EDM. Taguchi technique was used to conduct experiment. Pulse duration, part thickness, taper angle, wire speed, discharge current and wire tension were selected as a process parameter. Artificial neural network (ANN) model was developed to relate the process parameter and output responses [5].

J.Udaya Prakash, et al. have optimized the process parameters in WEDM for Aluminium alloy (A413) boron carbide hybrid composites material. The selected process parameter are gap voltage, pulse off time, wire feed, pulse on time, and percentage reinforcement. The observed responsesare material removal rate and surface roughness. Taguchi's L27 orthogonal array was used for the experimentation under different combination of parameters [6]. Amar Patnaik et al. used zinc coated copper as electrode with the selected input process parameters like discharge current, dielectric flow rate, pulse frequency, pulse duration, wire speed and wire tension,. By using selected factors maximization of material removal rate (MRR) and minimization of SR was done in WEDM process using taguchi method. Taguchi method was an effective technique to predict input process parameters for optimal process performance [7]. Genetic algorithm was an effective optimization technique used for WEDM machining process [8]. In this paper taguchi technique is used for modelling and GA is employed to optimize the MRR during WEDM.

2. Experimentation

The experiment is performed on Electra Spring cut 734 four axis Wire-EDM. The experiment is designed using Taguchi method. Totally 27 experiments conducted at various level of input parameter. As six input parameters namely current, pulse on, pulse off time, Servo voltage wire feed and fluid pressure are chosen for the experiments. Input parameters and their range is shown in table 1.

Table 1. Input variable and their levels.

S.No	Symbol	Input process parameter	Levels		
			I	II	III
1	A	Pulse on Time- Ton (μs)	110	115	120
2	B	Pulse off time T-off (μs)	40	45	50
3	C	Flushing Pressure (kgf/cm2)- Fp	8	10	12
4	D	Wire Tension (kgf)-Wt	550	750	950
5	E	Servo Voltage (volt)- Sv	15	20	25
6	F	Wire feed rate (m/min)- Wf	6	8	10

The input process parameters range is selected based on the trail run which conducted on the machine. Some of the parameter are kept constant due to limitation on the machine. Molybdenum wire electrode shown in figure 1 is used for to machine the work piece. It is reusable type wire.

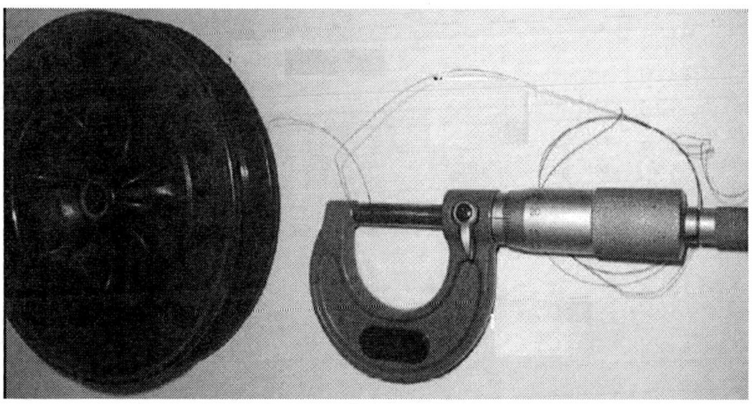

Figure 1. Molybdenum wire electrode.

IOP Conf. Series: Materials Science and Engineering **764** (2020) 012047 doi:10.1088/1757-899X/764/1/012047

The fixed parameter for this study is shown in the table 2. Material removal rate is selected response process parameter and it is calculated by weighing the workpiece before and after machining with respect to time. Calculated MRR for the experiments are shown in table 3.

Table 2. Fixed parameters.

S.No	Fixed parameter	Set value
1	Wire material	Molybdenum wire of diameter of 0.25 mm
2	Peak current	230 Amps
3	Pulse in peak voltage	2
4	Servo feed setting	250

Table 3. Experiments using L_{27} Orthogonal array.

Run	Ton	Toff	Fp	Wt	Sv	Wf	MRR
1	110	40	8	550	15	6	6.5826
2	110	40	8	550	20	8	6.096
3	110	40	8	550	25	10	5.7664
4	110	45	10	750	15	6	6.0711
5	110	45	10	750	20	8	5.7778
6	110	45	10	750	25	10	5.4864
7	110	50	12	950	15	6	5.6018
8	110	50	12	950	20	8	5.4668
9	110	50	12	950	25	10	4.978
10	115	40	10	950	15	6	8.3965
11	115	40	10	950	20	8	7.748
12	115	40	10	950	25	10	7.2169
13	115	45	12	550	15	6	7.029
14	115	45	12	550	20	8	6.6949
15	115	45	12	550	25	10	6.338
16	115	50	8	750	15	6	5.9778
17	115	50	8	750	20	8	5.777
18	115	50	8	750	25	10	5.4474
19	120	40	12	750	15	6	8.5164
20	120	40	12	750	20	8	8.1123
21	120	40	12	750	25	10	7.9431
22	120	45	8	950	15	6	8.1194
23	120	45	8	950	20	8	7.8864
24	120	45	8	950	25	10	7.8596
25	120	50	10	550	15	6	7.6905
26	120	50	10	550	20	8	7.1651
27	120	50	10	550	25	10	6.9987

2.1. Regression model

There are two different types of models are been used in regression analysis. One is linear regression and another one is multiple regressions. In Linear regression only one independent variable to predict the outcome of Y. In multiple regression two or more independent variables is available to predict the outcome. Mathematical representation of each type of regression is given in the form equation in (1)&(2)

$$Y = C_o + C_1X_1 + C_2X_2 + \cdots + C_nX_n \qquad -(1)$$

Where X_1 X_2....... X_n- attributes of the data set
C_0, C_1 ..C_n - coefficients
Y -Response.

$$Y = C_o + C_1X_1 + C_2X_2 + C_3X_1^2 + C_4X_2^2 + C_5X_1 \times X_2 \qquad -(2)$$

Where X_1, X_2 - attributes of the data set,
C_0 and C_1- coefficients
Y- response

2.2. MRR *Regression Equation*

To compute the classification accuracy three fold cross validation is used. The observed datasets are randomly divided into three sets. Two sets of observed data are used to train LSE-MRLM and to develop the model. Third set of data is taken for testing the developed model. The instances in each class in each set are randomly picked up. The each simulation is performed for 10 trials. The developed MRR equation (3) is shown below

MRR = -2.1323 + 0.224568 Ton - 0.401748 Toff - 1.05715 Fp + 0.00286929 Wt - 0.0661178 Sv + 0.0276469 Toff*Fp - 0.000206035 Fp*Wt-(3)

3. Genetic algorithm

The genetic algorithm is a technique for solving both constrained and unconstrained optimization problems. It is based on natural selection, the process that drives biological evolution.

The objective of this current research work is to maximize the MRR

MRR = f(Ton_Toff_ Fp_Wt_Sv_ Wf)

Subject to constraints
$110 \leq Ton \leq 120$
$40 \leq Toff \leq 50$
$8 \leq Fp \leq 12$
$550 \leq Wt \leq 950$
$15 \leq Sv \leq 25$
$6 \leq Wf \leq 10$

Where Ton, Fp, Toff, Wt, Wf and Sv are pulse on time, fluid pressure, pulse off time, wire tension, wire feed and servo voltage respectively. Parameters of genetic algorithm and their selected value shown in table 4. The input parameters limitation range assigned based on experimental setup. f denotes relation between MRR and six input parameters. For every iteration of Genetic algorithms, MRR values are predicted by regression models for a specified population size. The Genetic algorithm randomly generated a real valued population of possible input process parameter chromosomes. A new population for the subsequent generation is produced based on old population using different operator

IOP Conf. Series: Materials Science and Engineering **764** (2020) 012047 doi:10.1088/1757-899X/764/1/012047

functions. Optimization was carried out using MATLAB software. Obtained optimized value for selected rage of process parameter is shown in table 5.

Table 4. Genetic Algorithm parameter and their values.

Number of generations	500
population size	20
crossover rate	0.8
crossover mechanism	Two point
mutation rate	0.01

4. Confirmation Experiment

In order to prove the output which is obtained from genetic algorithm confirmation experiments is conducted. The specific set of input parameter is given by the genetic algorithm is experimentally verified. The experiment is repeated for three times to obtain an average output of the performance.

Table 5. Optimum process parameter.

Process parameter						MRR		Error
Ton	Toff	Fp	Wt	Sv	Wf	Predicted	Experimental	
120	40	12	750	20	8	8.256	8.426	2.03%

5. Conclusions

In this research work mathematical model for MRR was developed using regression analysis for the given range of input process parameter. The following conclusion were made from the study,

- The developed MRRregression mathematical model results are close agreement with experimental values.
- For material removal rate, Pulse current is the most significant factor and followed by pulse current.
- Optimum process parameter was derived from genetic algorithm process. The obtained optimum parameters are Ton of 120, Toff of 40, Fp of 12, Wt of 750, Sv of 20 and Wf of 8.

References

[1] K.H. Ho, S.T. Newman, S. Rahimifard and R. Allen, "State of art in wire electric discharge machining (WEDM)", International Journal of Machine Tools and Manufacture, Vol 44, pp.1247-1259, 2004.

[2] S. Silambarasan and G. Prabhakaran, "Optimization of WEDM process parameter using genetic algorithm, International Journal of Mechanical and Production Engineering Research and Development, vol 8(8), pp. 426-430, 2009.

[3] H. Singh and A. Singh, "Effect of Pulse On/Pulse Off Time On Machining Of AISI D3 Die Steel Using Copper And Brass Electrode In EDM", International Journal of Engineering and Science, vol.1(9), pp.19-22, 2012.

[4] C.Tosun, Cajun and H. Pihtili, "The effect of cutting parameters on wire crater sizes in wire EDM", International Journal of Advanced Manufacturing Technology. vol. 21(11), pp. 857–

IOP Conf. Series: Materials Science and Engineering **764** (2020) 012047 doi:10.1088/1757-899X/764/1/012047

865, 2003.

[5] B.B. Nayak and S.S. Mahapatra, "Optimization of WEDM process parameters using deep cryo-treated Inconel 718 as work materia", Engineering Science and Technology an International Journal, vol. 19(1), pp.161-170, 2016.

[6] J.UdayaPrakash, T. V. Moorthy and J. Milton Peter, "Experimental Investigations on Machinability of Aluminium Alloy (A413)/Flyash/B4C Hybrid Composites Using Wire EDM" "Procedia Engineering", vol. 64, pp.1344–1353,2013.

[7] S. S. Mahapatra and Amar Patnaik, "Optimization of wire electrical discharge machining (WEDM) process parameters using Taguchi method", International Journal of Advanced Manufacturing Technology, vol 34(9-10)., pp.2006.

[8] P. Shandilya and A.Tiwari, "Artificial Neural Network Modeling and Optimization using Genetic Algorithm of Machining Process",Journal of Automation and Control Engineering, vol. 2(4), pp.348-352, 2014.

IOP Conf. Series: Materials Science and Engineering **764** (2020) 012048 doi:10.1088/1757-899X/764/1/012048

Characterization and Performance Analysis of Graphite and Molybdenum disulphide based Hybrid Composite Materials

Anandha Moorthy Appusamy*, Prakash Eswaran, Madheswaran Subramani, Santhanakumar Sadaiappan and Karthik Shanmugavel

Department of Mechanical Engineering, Bannari Amman Institute of Technology, Sathyamangalam-638401.

*Author for correspondence E-mail: anandhamoorthya@bitsathy.ac.in

Abstract. An attempt has been performed to observe the influence of heat treatment on composite specimen by dry sliding wear and hardness testing. The particulate reinforcements embedded with AA2218 Aluminium metallic element alloy with 4% constant weight proportion of Graphite, MoS_2, with a variable amount of hard reinforcement particulates. Availability of fly ash is abounded and it has to be utilized in any other way. Instead of carbides and nitrides, introducing fly ash as reinforcement satisfies the structural and thermal properties. 5%, 10% & 15% of fly ash added to the composite by liquid scientific discipline route. To get better mechanical properties of composite specimens are treated with ASTM T6 standard. Performance of heat-treated hybrid composite was investigated with a pin on disc with a constant load 2 kg sliding at 300 rpm. Hardness of the specimen at temperature was conjointly measured before and once the heat treatment is completed. The results were compared with base alloy and untreated composite specimens. Heat treated metal exhibits high hardness and wear resistance. This investigation reveals that AA2218 Aluminium matrix composites with high hardness and wear resistance will replace the traditional material employed in automobile elements to get improved performance and longer life.

Keywords- AA2218 Aluminium alloy, Graphite, MoS2 & Heat treatment

1. Introduction

The world is trying to find the utmost doable optimisation in each field in recent times. With no exception in engineering, engineers are trying to find making the acute best from the most effective[1]. Throughout these developments, in direct or opposite, the surroundings are affected. The composites strengthen with biodegradable material has been one in all the foremost modernizations within the field of material development in the past few decades. Metallic element alloys embedded with ash particles exhibit greater mechanical properties to reinforced metallic element and therefore hybrid materials are the possible way to achieve desired properties. Particle-reinforced Metal Matrix Composites (MMCs) containing set on AA2218 have agreeable attention within the past few decades due to their improvised wear resistance, improved hardness, dimensional stability, and enhanced strength compared to unreinforced metallic element alloys[2]. Though they found a potential application in automobile and aerospace the applying base of those particulate MMCs is admissible by their low cost.

In recent times, inexpensive metallic element alloy MMCs strengthened with ash particles. Ash is the devastate consequence of coal combustion, has been built to act a substitute for typical particulate MMCs in many applications to broaden the applying basis of this category of MMCs. The addition of ash into metallic element MMCs could be an added initiative that lowers the disposal value of ash, will increase energy savings by dropping the number of metallic elements made, and creates healthier surroundings. Fly ash could be a used as reinforcement which might be found profusely as solid waste derivative throughout coal incineration in power plants. It's additionally terribly cheap. It's being employed in forged and shaped metallic element alloys to create aluminium-alloy–fly ash (ALFA) composites. Hence, ash bolstered composites area unit a lot of probably to beat the value barrier for wide unfold applications. The stirring arrangement improves the distribution and wettability of composites[3]. Mechanical behaviour of Al-fly ash composites is investigated. Once the investigation of ash composites it's been discerned that this material is appropriate for automobile. The general objective of this study is to develop and characterize the fabricated one distinctive from the normal metal or similar varieties of metal alloy. Once finishing the experimental work, it was ensured that the composites have a tendency to found increased in physical properties. These composites are used for enormous and advanced applications with low cost[4]. Besides, this sort of composites squares the property and setting friendly than traditional metal alloy.

2. Preparation of MMC using Stir Casting Method

2.1 Matrix Material:

The alloys are widely utilized as matrices in aluminium are being engrossed on the A356, 2xxx and 6xxx series alloys[5]. Though very little experimentation have been depict on the 2xxx sequence materials embedded with solid lubricant and ash particles, much fewer concentration has been paid to the 2xxx Al alloy matrix composites, in which it gives the utmost potency of all marketable Aluminium alloy materials and used for structural application. The main intention of this learning is to explore the wear, friction and hardness of Fly ash and Graphite / Molybdenum disulphide reinforced aluminium matrix composite. Here AA2218 has been selected as structural material to get desired shape. The main constituent material of AA2218 is copper, second metal is magnesium, which is predominantly added to amplify the interfacial bonding between reinforcements and matrix material. Contents of aluminium AA2218 is tabulated in Table 2.1.

Table 2.1. Composition (wt. %) of Aluminium 2218

Cu	**Ni**	**Mg**	**Si**	**Fe**	**Ti**	**Al**
3.870	1.900	1.470	0.510	0.160	0.020	Remaining

2.2 Reinforcement Material:

Fly ash is embedded with matrix to improve mechanical properties. The main purpose of adding Graphite and Molybdenum disulphide is to provide lubrication effect at the sliding interface[6]. Constituent particle size of fly ash varies from submicron to 500 µm. 2.1 to 2.6 of Specific gravity, whereas bulk density varies between 1 and 1.8 g/cm^3. Density of Graphite is lies between 2.09 – 2.23 g/cm^3.

Table 2.2. Reinforcements Properties

Reinforcements	**Density(g/cm^3)**	**Melting Temperature (k)**
Fly ash	1.6	1173
MoS_2	5.06	1458

3. Composite Preparation and Characterization

3.1 Processing of Composites using Stir Casting Method:

Coal fired furnace slag powder of fly ash is strengthened with AA2218 by liquid metal processing methodology[7]. Metal stir method was used to yield the required specimens. The structural material AAl2218 was primarily raised the temperature higher than its solidifying temperature and consequently the required quantities of embedded particles are weighed accurately and fed into the chamber. Blending of metals was carried at 650 rpm about fifty seconds turn over the interface between the matrix alloy and then the particles promote wetting and consequently the particles were homogeneously distributed. The liquid metal in the furnace is poured in a cast iron mould and solidifies to get circular bar samples. Consequently three unrelated compositions were prepared as mentioned above through metal stir casting method. The stir casting setup used for the study and board for dominant the stirrer speed and therefore the temperature is portrayed in Figure 1.

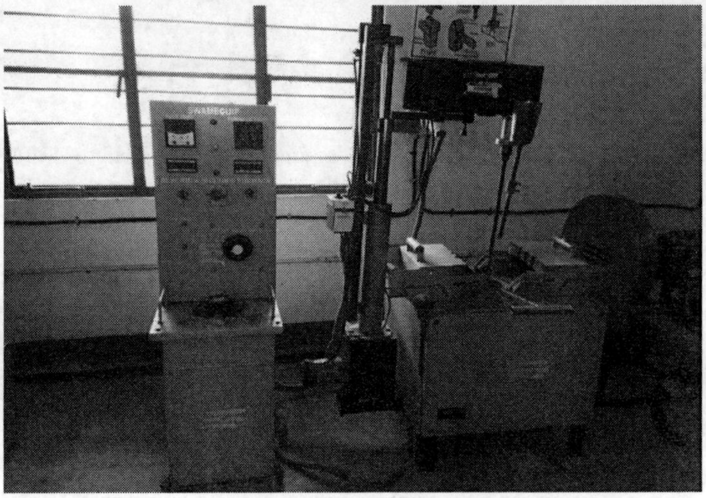

Figure 1. Stir casting set up

The two set of specimens were prepared with ash particles as hard particles and Graphite and MoS_2 as second phase reinforcements. Developed specimen details are tabulated below.

Table 3.1. Hardness of Specimens

Specimen	Matrix	Reinforcement		Hardness(BHN)
1	85%	15% of fly ash	5% MoS_2	42
2	85%	15% of fly ash	5% Graphite	58

3.2 Characterization of Composites by T6:

Coal fired furnace slag powder of fly ash is strengthened with AA2218 material by metal stir casting technique[8]. The developed composite specimen is heated to 180^{0}C for 6 & 8 hours subsequently cooled by air, brain solution and water to allow for precipitation hardening. This will enables the mixture of aluminium copper crystalline structures, the size of which will dictate the strength. The effect of quenching with different agents and aging hours has been evaluated through harness and wear test[9].

Table 3.2. Characterization of Specimens

Specimen No.	Composition	Aging hours	Quenching Medium
1	15% of fly ash + 5% Graphite	6	Air
2	15% of fly ash + 5% Graphite	6	Brain Solution
3	15% of fly ash + 5% Graphite	6	Water
4	15% of fly ash + 5% Graphite	8	Air
5	15% of fly ash + 5% Graphite	8	Brain Solution
6	15% of fly ash + 5% Graphite	8	Water
7	15% of fly ash + 5% MoS_2	6	Air
8	15% of fly ash + 5% MoS_2	6	Brain Solution
9	15% of fly ash + 5% MoS_2	6	Water
10	15% of fly ash + 5% MoS_2	8	Air
11	15% of fly ash + 5% MoS_2	8	Brain Solution
12	15% of fly ash + 5% MoS_2	8	Water
13	AA2218	8	-

Prepared specimens are displayed in figure number 2.

Figure 2. Specimens

IOP Conf. Series: Materials Science and Engineering **764** (2020) 012048 doi:10.1088/1757-899X/764/1/012048

4. Test & Result Analysis

4.1 Hardness test:

Hardness is the quantities of how resistant solid matter is to numerous kinds of permanent shape variation when a compressive load is applied or deformation. The hardness tester is shown in figure 3.

Figure 3. Hardness Tester

The Brinell hardness of different composite material has a great impression on its mechanical properties[10]. In our experimental work specimen, it has been observed that the hardness of different work pieces is improved due to an effect of heat treatment.

Table 4.1. Brinell Hardness of AA2218 with 5% MoS_2

Air		Brine solution		Water	
6	8	6	8	6	8
42	43	46	48	52	53

Table 4.2. Brinell Hardness of AA2218 with 5% Graphite

Air		Brine solution		Water	
6	8	6	8	6	8
45	47	55	57	50	53

There is no much effect on heat treatment of base alloy moreover the hardness values are lesser than embedded materials. For this reason matrix alloy result is not included for comparison. Brinell hardness of the heat treated specimens is list in the above table 4.1 and 4.2. The results are plotted as graph in figure number 4 and 5.

IOP Conf. Series: Materials Science and Engineering **764** (2020) 012048 doi:10.1088/1757-899X/764/1/012048

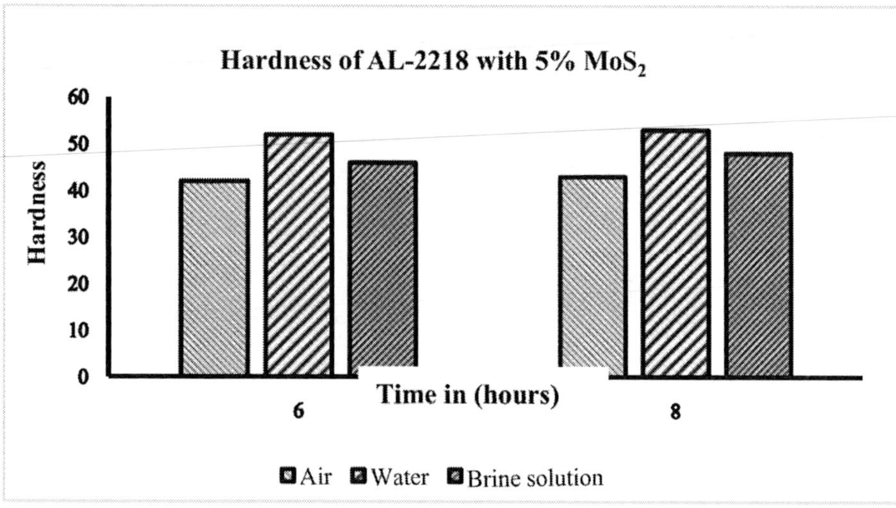

Figure 4. Hardness of MoS_2 based Specimens

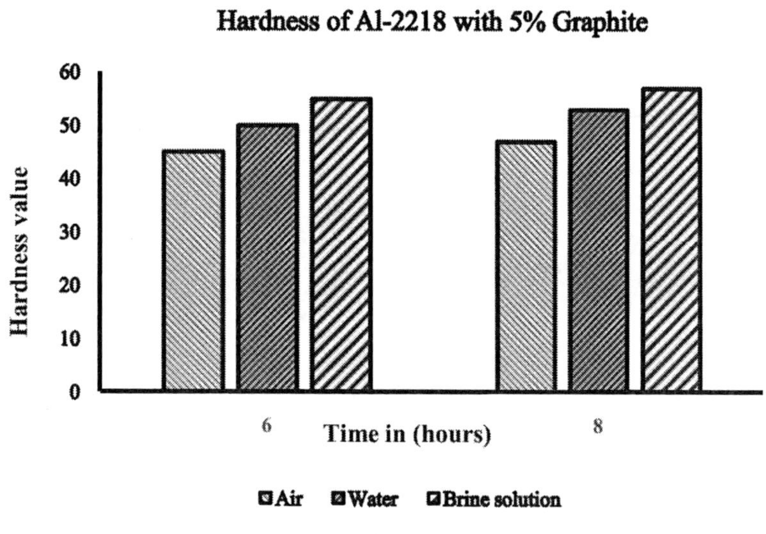

Figure 5. Hardness of Graphite based Specimens

From the test results, it was observed that the graphite specimen aging to 8 hours and quenched with brain solution gives higher hardness value than other specimen. Because of salt present in the brain solution enhance the heat transfer capacity. This has been effect on rapid cooling of specimens resulted with higher hardness.

4.2 Friction and Wear test:

The pin on disc is a device used to govern tribological properties of coating these properties comprised frictional coefficient and wear rate the coated specimen is placed on a rotating disc spring at an adaptable angular speed. A ball held by vertically reciprocating pin is constrained against the sample for a stable amount of full rotations of the disc. The tangential force and frictional coefficient are measured and the volume of removed materials can be measured by other Techniques.

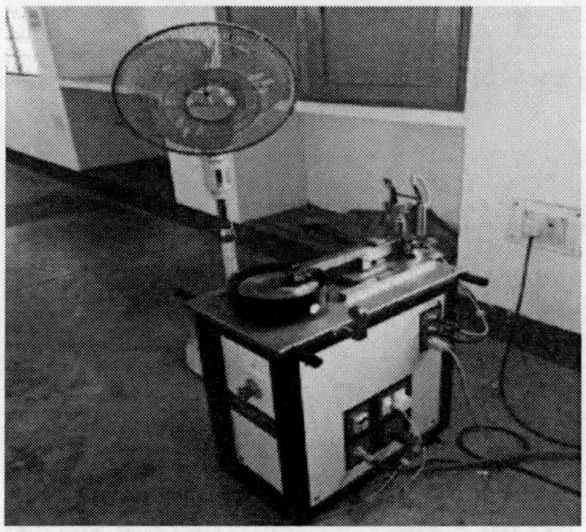

Figure 6. Pin on Disc wear test apparatus

Table 4.3. Wear of Al-2218 with 5% MoS_2

Wear rate (10^{-6} mm^3/m)					
Air		Brine solution		Water	
6	8	6	8	6	8
391	343	290	340	284	247

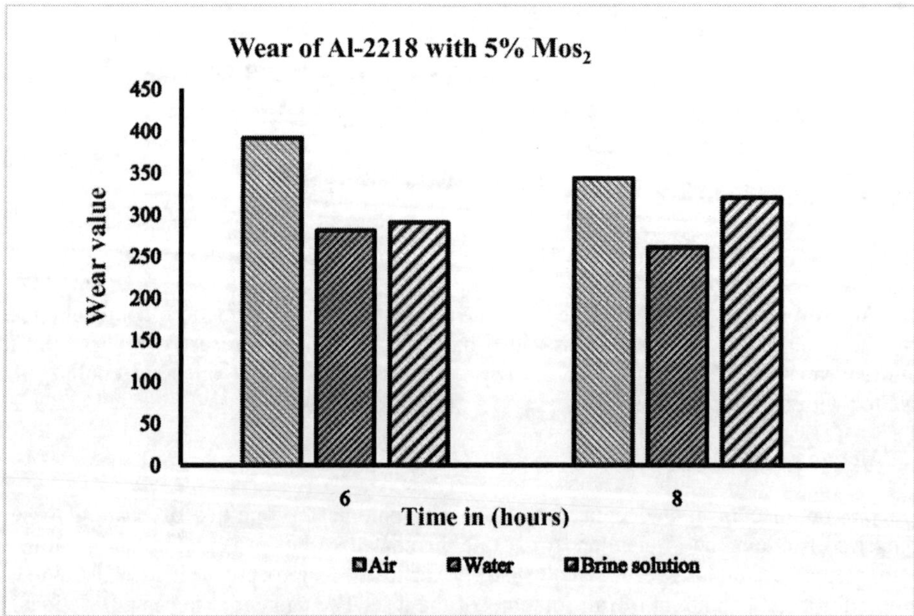

Figure 7. Wear of Al-2218 with 5% MoS_2

Table 4.4. Wear of Al-2218 with 5% Graphite

Wear rate (10^{-6} mm^3/m)					
Air		**Brine solution**		**Water**	
6	8	6	8	6	8
328	359	253	243	354	309

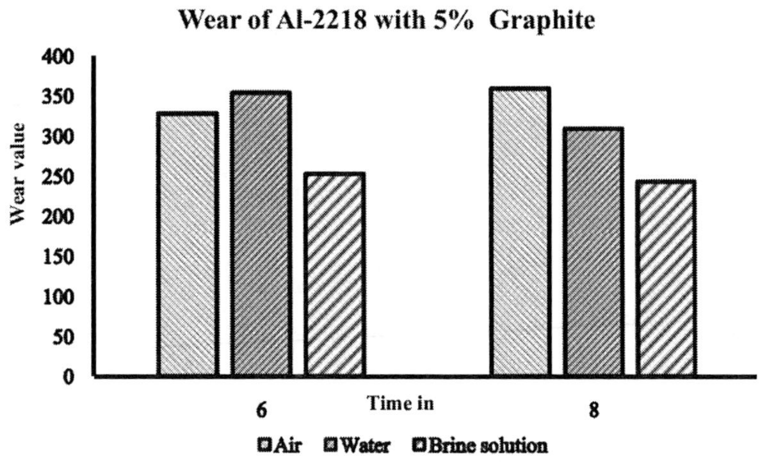

Figure 8. Wear of Al-2218 with 5% Graphite

Volume loss of metal is evaluated through dry sliding wear test. The figure number 7 & 8 shows the wear test results of the developed specimens[11]. Wear analysis gives inverse result with hardness value. Increase in hardness correspondingly reduces the wear loss of metal.

Conclusion

This research work is to explore the useful properties of a Stir casting processed Al-based composite materials reinforced with particles reinforcement. Stir casting method is used to make the composites could yield a uniform distribution of the reinforcement. Mainly physical, mechanical properties of Al matrix composite improved. Graphite reinforced composite specimen aging to 8 hours and quenched with brain solution exhibits higher hardness and improved wear loss. Thus, the fabricated reinforcement composites are used in manufacturing sectors where further strength is needed.

References

[1] A. Ahmed, A.J. Neely and K.Shankar, "Effect of Ceramic Reinforcements on the Mechanical Behaviour of 7xxx series Aluminium Matrix Composites", Vol. 7, ACAM December 2007.

[2] Salman, B Gabbitas, J Li and D Zhang, "Tribological properties of thermally sprayed TiAl-Al_2O_3 composite coating", Materials Science and Engineering, vol. 2, 2009.

IOP Conf. Series: Materials Science and Engineering **764** (2020) 012048 doi:10.1088/1757-899X/764/1/012048

[3] Rajesh Kumar Bhushan and Sudhir Kumar, "Influence of SiC Particles Distribution and Their Weight Percentage on 7075 Al Alloy". Journals of materials engineering and technology, Vol. 4, March 2011.

[4] Arun Kumar M. B and R. P. Swamy, "Evaluation Of Mechanical Properties of Al6061, Flyash And E-Glass Fiber Reinforced Hybrid Metal Matrix Composites", ARPN Journal of Engineering and Applied Sciences, Vol. 6, May 2011.

[5] Kala, H., Mer, K.K.S. and Kumar, S. A Review on Mechanical and Tribological Behaviors of Stir Cast Aluminium Matrix Composites. Procedia Materials Science, vol. 6, pp.1951-1960, 2014.

[6] L Grabowski, A Baier, M Majzner and M Sobek. Composite materials molding simulation for purpose of automotive industry. IOP Conference Series: Materials Science and Engineering, Volume 145, Engineering of Manufacturing Processes, pp.1-10, 2016.

[7] Selvam, M.D. and Sivaram, N.M., Optimal Parameter Design by Taguchi Method for Mechanical Properties of Al6061 Hybrid Composite Reinforced with Fly Ash/Graphite/Copper. International Journal of ChemTech Research. Vol. 10(13), pp.128-137, 2017.

[8] Selvam, M.D., Senthil, P. and Sivaram, N.M. Parametric optimisation for surface roughness of AISI 4340 steel during turning under near dry machining condition. International Journal of Machining and Machinability of Materials, vol. 19(6), pp.554-569., 2017

[9] Mastariyanto Perdana, Prastiawan and Syafrul Hadi. Mechanical Properties of Composite Waste Material Based Styrofoam, Baggase and Eggshell Powder for Application of Drone Frames. IOP Conference Series: Earth and Environmental Science, Volume 97, conference 1, pp.1-7, 2017.

[10] Balakumar, S., Selvam, M.D. and Nelson, A.J.R., Wear and Friction Characteristics of Aluminium Matrix Composites Reinforced with Flyash/Cu/Gr Particles. International Journal of ChemTech Research. vol. 11(01), pp.121-133, 2018.

[11] Anandha Moorthy A, Natarajan N, Palani PK and Manojkumar M, "Microstructure and Dry Sliding Wear Behavior of AA2218 Based Self-Lubricating Metal Matrix Composites", International Review of Mechanical Engineering (I.RE.M.E.), Vol. 7, N. 6 ISSN 1970 – 8734, 2013.

IOP Conf. Series: Materials Science and Engineering **764** (2020) 012049 doi:10.1088/1757-899X/764/1/012049

Enhancing the Tribological Properties of Talc and Boron Nitride based Composites by T6 - Heat Treatment

Anandha Moorthy Appusamy*, Prakash Eswaran, Madheswaran Subramani, Santhanakumar Sadaiappan and Karthik Shanmugavel

Department of Mechanical Engineering, Bannari Amman Institute of Technology, Sathyamangalam-638401.

*Author for correspondence E-mail: anandhamoorthya@bitsathy.ac.in

Abstract. Aluminium matrix composites (AMC) strengthened with hard particles has given rise to a potential material particularly for wear resistant and weight basic applications. The effect of heat treatment of aluminium alloys sliding wear nature are observed under varying load and speed. Irrespective of materials heat treatment improves hardness of the material. The particulate reinforcements embedded with Al2218 Aluminium metallic element alloy with 4% constant weight proportion of Talc, Boron Nitride and a variable amount of hard reinforcement particulates. By introducing fly ash as reinforcement satisfies the structural and thermal properties. 5%, 10% & 15% of fly ash added to the composite by metal stir casting technique. The specimens are treated with ASTM T6 standard. Performance of heat-treated matrix composite was investigated with a pin on disc apparatus through a constant load 2 kg and with a desirable rpm. Hardness of the specimen at temperature were measured before and once the heat treatment. Heat treated metal exhibits high hardness and wear resistance. This investigation reveals that 2218 Aluminium matrix composites with high hardness and wear resistance will replace the material employed in Automobiles like brake drum, cylinder block, connecting rod, piston etc for higher performance and longer life.

Keywords- Aluminium (Al2218), ASTM T6, Boron Nitride, Hardness and Talc

1. Introduction

The two significant parameters that are Light weight and Eco-friendliness in automobile field are most common genuine natural issues these days. [1]Aluminium alloys are remarkable materials because of their high explicit strength and stiffness for the use in automobile fields as segments of internal combustion engines e.g. liner, bearing, cylinder and piston. But their applications are turned down due to poor wear resistance. It is well known that this aluminium matrix composite have predominant mechanical properties, particularly high wear resistance than other composites. Besides, some investigators have studied the friction and wear properties of the tailor made materials comprising a combination of soft and hard reinforcements. Particle-reinforced as compound element in metal matrix composites (MMCs) containing ash and Al_2O_3 have established nice consideration within the past few years. Because of their reduced coefficient of thermal expansion, enhanced wear resistance, high elastic modulus and improved strength related to unreinforced chemical element alloys. [2]Though they have found possible applications in weight-critical components in automobile, aerospace region and defence systems, the applying base of those particulate MMCs is increased by their performance.

Recently, low-cost chemical element alloy MMCs embedded with ash, a waste by-product of coal combustion, has been considered to act an additionally for typical particulate MMCs in numerous applications to widen the applying basis of this class of MMCs. [3]The addition of ash into tailor made

MMCs may well be one more imitativeness that lowers the disposal worth of ash, could increase energy savings by dropping the amount of chemical element created, and makes healthier surroundings. Aluminium ash composite is metal matrix dispersion strengthened composite inside that soft and ductile chemical element matrix is strengthened by heavy and brittle ash particles. It's being used in cast and formed chemical element alloys to form aluminium alloy and fly ash based composites. Hence, ash bolstered composites materials are widen their unfold applications in the result of improved mechanical properties. [4]The stirring arrangement improves the distribution and wettability. The overall approach of this study is to develop and characterize by heat treatment of unique composites from the traditional metal or similar types of metal alloy. This composite squares the usage in advanced applications with low price.

2. Experimental Procedure

2.1 Matrix Material:

So far most of the alloys that are used as matrices in aluminium being targeted on the A356, 2xxx and 6xxx series alloys. Although only a few studies are based on the 2xxx series alloys strengthened with ash and solid lubricating particles, abundantly less attention has been paid to the 2xxx Al alloy matrix composites, within which it provides the best strength of all industrial Al alloys and used for structural application. the most important objective of this entire research is to analyse the friction, wear and fracture of ash and solid lubricant strengthened metal matrix composite. [5]Here Al 2218 was used as matrix material. The main component is copper. The second is atomic number 12 which is magnesium that is preponderantly value-added to extend the moistening between matrix and reinforcement to improve interfacial effect. Composition of metal AA2218 tabulated in Table 1.

Table 1. Composition (wt %) of Aluminium 2218

Copper	Nickel	Magnesium	Silicon	Iron	Titanium	Aluminum
3.87	1.90	1.47	0.51	0.16	0.02	Bal

2.2 Heat treatment:

The constituent alloy material and reinforcement embedded composites are heat treated through an electrically powered Muffle furnace. [6]There are several stages of treatment are undergone during the heat treatment of the selected material (Al2218) are: (i) Solution treatment: the aluminium alloy or matrix composite are heated for 6 hrs with a temperature of 520° C till the solute elements are mostly been dissolved in Al solid solution, (ii) Quenching: the solution treated material is cooled rapidly in water to prevent the indefinite quantity of the solute elements and to get a saturated solid solution and (iii) Artificial aging: In this process hardening is done by reheating the solute alloy to a temperature of 200° C for 6 and 8 hours to achieve the desired properties.

2.3 Reinforcement Material:

Fly ash is incorporated as main particulates to strengthen the composites, Talc and Boron Nitride are the additional reinforcement materials used to reduce the friction and wear.

Table 2. Properties of Reinforcement Particles

Reinforcements	Density (g/cm^3)	Melting Temperature (°k)
Fly ash	1.6	1409.15
Talc	2.7	1773.15
BN	2.1	3246.15

IOP Conf. Series: Materials Science and Engineering **764** (2020) 012049 doi:10.1088/1757-899X/764/1/012049

2.4 Processing of Composites using Stir Casting Method:
[7]Stir casting is a liquid state method used for the fabrication of composite materials, in which a dispersed phase is mixed with a molten matrix metal by means of mechanical stirring. Stir casting is the simplest and most cost-effective method of liquid state fabrication. Thermal power plant ash powder and selected lubricant particles are reinforced with Al2218 by stir casting methodology. Liquid metallurgy technique was used to manufacture the hybrid composite specimens. The matrix alloy Al2218 was primarily superheated over its melting temperature and thus the essential quantities are evaluated exactly and fed into the chamber. Stirring was carried at 650 rpm relating to fifty seconds till the interface among the particle and thus the matrix promotes moistening and thus the particles were consistently distributed. The metal is coagulated during a forged iron mould to induce flat plate samples. so, all three are in totally different compositions were solid as declared over by stir casting methodology. The stir casting setup used for study and board for dominant the stirrer speed and thus the temperature is delineating in Figure 1.

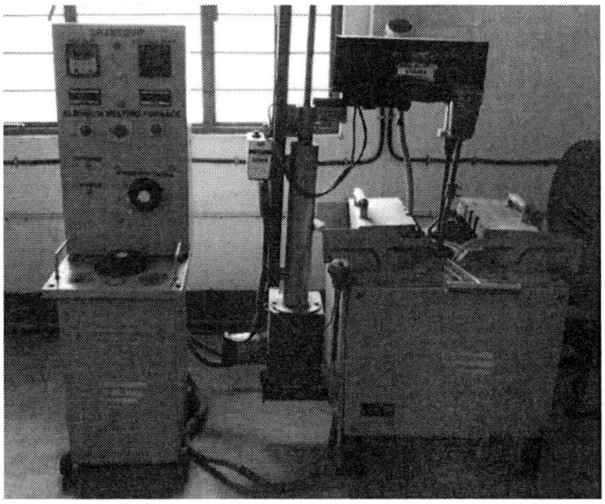

Figure 1. Stir casting set up

3. Investigation & Result Analysis

3.1 Brinell indentation test:
Hardness is that quantity of resistant force of solid matter. The hardness of specimens is evaluated through hardness tester with Brinell hardness number. [8]Penetration of intender is depends on the resistivity force created by the substance embedded with base alloy. Weak bonding between the atoms is results with low hardness value. Results of hardness have also been used to evaluate the proper mixing and uniform distribution of particulates.

To improve the adhesiveness between the particulates and base alloy is the result of wettability. [9]This effect has been improved by the addition of Mg. In the present investigational work specimen has to be tested for three times and the average value is taken for evaluation because indenter meets the soft reinforcement particulates, the impression is much larger than the hard particle approach.

This test experimentation has to be carried out on both base alloy and heat treated metals. Heat treatment of metals above re-crystallization temperature results with refinement of grain structure. Ultimately, hardness of the metal improves drastically with respect to the cooling rate. Faster cooling gives higher hardness. Figure number 2 shows the hardness tester, which is calibrated properly as per the ASTM standard and verified with master gauge.

IOP Conf. Series: Materials Science and Engineering **764** (2020) 012049 doi:10.1088/1757-899X/764/1/012049

Figure 2. Rockwell and Brinell hardness Tester

3.2 Result of indentation:

Brinell indentation tests were executed on developed composite and base alloy specimens. The BHN of the test specimen were deliberated with Brinell hardness measuring scheme with 10 mm diameter ball intender by a compressive force of 500 kg. The confinement time of indentation is 30 seconds. Three tests were taken on all samples to eradicate opportunity of isolation and average assessment has to be considered. Table number 3 shows the experimental results of Boron Nitride bases specimens.

Table 3. Hardness of AL-2218 with 5% Boron Nitride

Brinell Hardness of Al-2218 with 5% BN					
Air		**Water**		**Brine solution**	
6	8	6	8	6	8
77	79	78	80	83	85

Figure 3 shows the variation of hardness value of specimen results with heat treatment process. Quenching with brain solution exhibits high hardness value because of salt contamination in the solution improves the heat transfer capacity.

Table No. 4 Hardness of Al-2218 with 5% Talc

Brinell Hardness of Al-2218 with 5% Talc					
Air		**Water**		**Brine solution**	
6	8	6	8	6	8
49	51	51	52	54	55

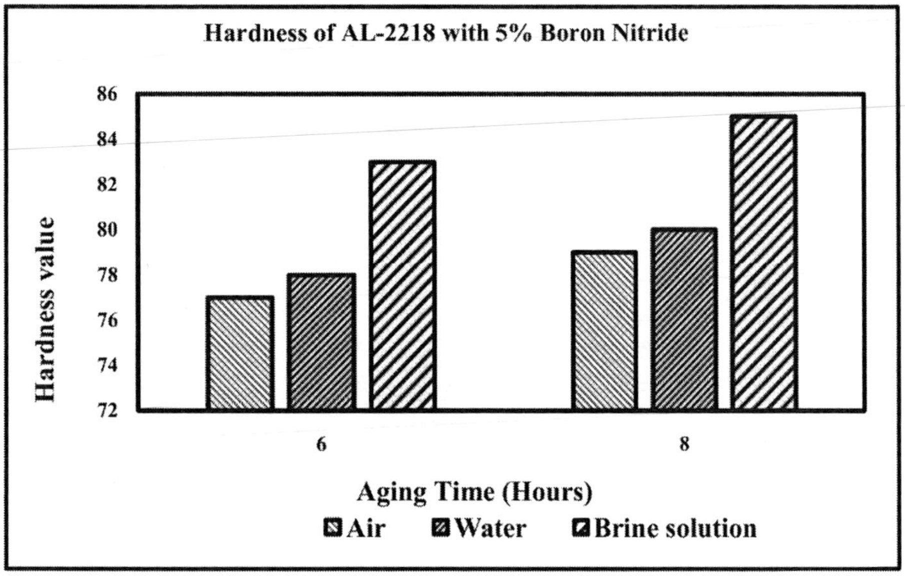

Figure 3. Hardness Test Result Graph

Al 2218 embedded with 5% Talc and 15% Fly ash particulate based composite specimens exhibits same results as that of BN. The results are tabulated in Table number 4 and the correlation between the aging and quenching medium is significantly explains in figure number 4.

It is evident from figure number 3 & 4 that the quenching reaches the highest hardness value than base alloy and untreated metals. In heat treatment process quenching is the fastest cooling method than annealing, tempering and normalizing. Quenching of specimen with different medium leads to the improved hardness, as to be expected. This has to be possible by the development of massive quantity of Martensite in the composite structure, commencing from austenite structure.

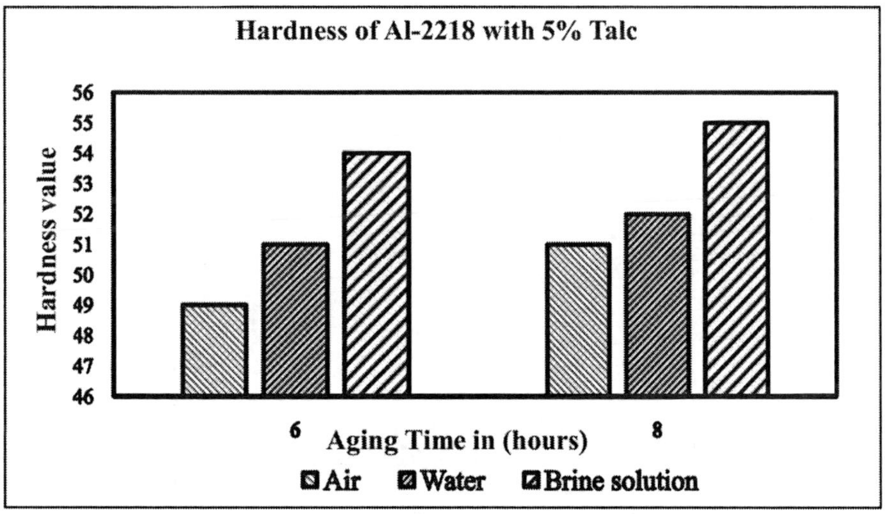

Figure 4. Hardness of Al-2218 with 5% Talc

IOP Conf. Series: Materials Science and Engineering **764** (2020) 012049 doi:10.1088/1757-899X/764/1/012049

3.3 Friction and Wear test

[10]The pin on disc is an equipment to test the tribological properties of circular specimens these properties comprised resistance to the sliding wear rate and coefficient of friction. Arrangement of wear tester is shown in figure number 5. Developed specimen is placed on a rotating disc in radial direction to the rotation. Rotation of disc is controlled by drive system and control unit. Radial load is applied on the pin and the opposing frictional force opposite to the direction of rotation is measured as frictional force by LVDT. The tangential force and resistance force are measured and therefore the volume of removed materials is often measured by weighing the specimen.

Figure 5. Wear Apparatus – Pin on Disc

Table 5. Wear loss of Al 2218 with 5% Boron Nitride

Wear rate (10^{-6} mm^3/m)					
Air		Water		Brine solution	
6	8	6	8	6	8
575	530	436	411	430	375

[11]Progressive loss of metal due to rubbing together with other metal is known as wear. The relative motion is caused by rotating the disc and radial load on pin. Life and performance of the engineering components is limiting by its wear resistance, whether the machine or mechanism is as large as heavy apparatus, or as tiny as a small mechanical, electrical and electronic device.

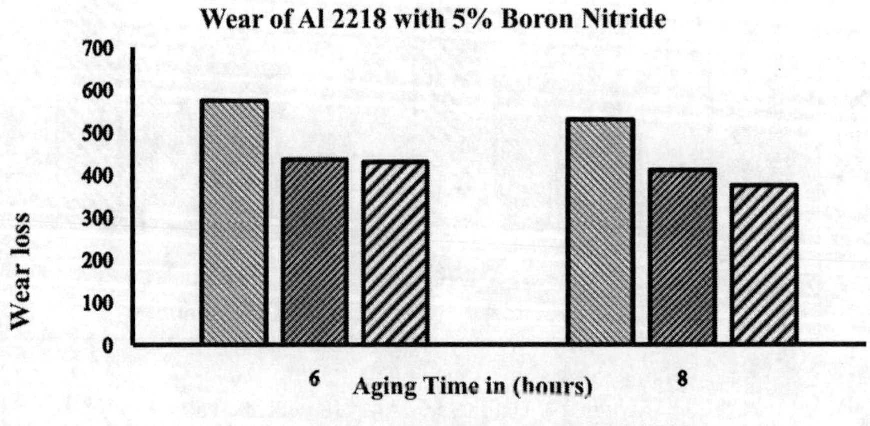

Figure 6. Wear loss of Al 2218 with 5% Boron Nitride

The figure number 6 and 7 shows the correlation between harness and effect of aging hours in terms of mass loss of metal. It was clearly identified that the improved hardness ruined to reduce the mass loss of metal with dry sliding wear

Table number 6 and figure number 7 be evidence of volume loss[12] of the heat treated specimens with three different quenching medium. Mass loss of every composite specimen is improved with rising aging times. The dry sliding wear of specimens shows that the volume loss has been low while quench with brain solution. At the same time increase in aging time also have significant effect on hardness and wear loss of metal.

Table 6. Wear loss of Al 2218 with 5% Talc

Wear rate (10^{-6} mm^3/m)					
Air		**Water**		**Brine solution**	
6	8	6	8	6	8
426	486	305	289	241	276

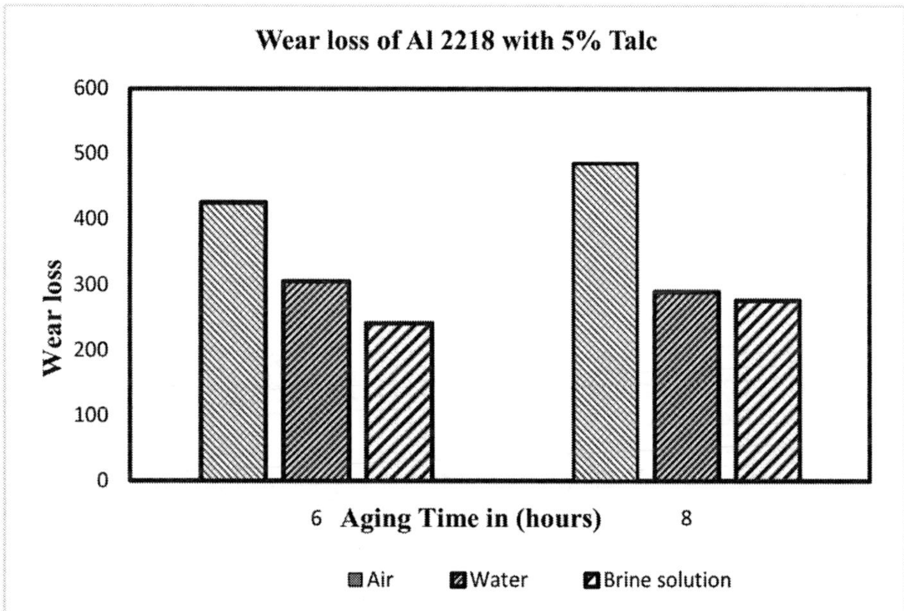

Figure 7. Wear of Al 2218 with 5% Talc

4. Conclusions

In this research, the effects of heat treatment of composite specimens are experimented. Based on the experimental results,

- This analysis work is to explore the helpful properties of a heat-treated Stir casting processed Al-based composite materials bolstered with particles reinforcement.
- Stir casting methodology is employed to synthesize the composites might produce the uniform spreading of the reinforcement.
- Primarily physical, mechanical properties of Al2218 matrix composite have also been analysed.
- With the evident of experimental results, increase in aging time has improves the hardness of metal.
- Quenching with brain solution results with lesser mass loss. Eight hours aging with brain solution quenching exhibits improved performance.

References:

[1] Ahmed, A, et al, "Effect of Ceramic Reinforcements on the Mechanical Behaviour of 7xxx series Aluminium Matrix Composites",vol. 7, ACAM, 2007.

[2] Salman, B Gabbitas, J Li and D Zhang, "Tribological properties of thermally sprayed TiAl-Al_2O_3 composite coating", Materials Science and Engineering, vol. 2, 2009.

[3] Rajesh Kumar Bhushan and Sudhir Kumar, "Influence of SiC Particles Distribution and Their Weight Percentage on 7075 Al Alloy". Journals of materials engineering and technology, vol. 4, 2011.

[4] Arun Kumar M. B and R. P. Swamy, "Evaluation Of Mechanical Properties of Al6061, Flyash And E-Glass Fiber Reinforced Hybrid Metal Matrix Composites", ARPN Journal of Engineering and Applied Sciences, vol. 6, 2011.

[5] Kala, H., Mer, K.K.S. and Kumar, S. A Review on Mechanical and Tribological Behaviors of Stir Cast Aluminium Matrix Composites. Procedia Materials Science, vol. 6, pp.1951-1960, 2014.

[6] Selvam, M.D. and Sivaram, N.M., Optimal Parameter Design by Taguchi Method for Mechanical Properties of Al6061 Hybrid Composite Reinforced with Fly Ash/Graphite/Copper. International Journal of ChemTech Research.vol. 10(13), pp.128-137, 2017.

[7] Selvam, M.D., Senthil, P. and Sivaram, N.M. Parametric optimisation for surface roughness of AISI 4340 steel during turning under near dry machining condition. International Journal of Machining and Machinability of Materials, vol. 19(6), pp.554-569, 2017.

[8] Balakumar, S., Selvam, M.D. and Nelson, A.J.R., Wear and Friction Characteristics of Aluminium Matrix Composites Reinforced with Flyash/Cu/Gr Particles. International Journal of ChemTech Research. vol. 11(01), pp.121-133, 2018.

[9] Mahdavi, S. and Akhlaghi, F. Fabrication and characteristics of Al6061/SiC/Gr hybrid composites processed by in situ powder metallurgy method. Journal of Composite Materials, vol. 47(4), pp.437-447, 2013.

[10] Palash Biswas et al, Effect of heat treatment on microstructure behavior and hardness of EN 8 steel, IOP Conf. Series: Materials Science and Engineering 377, pp. 1-6, 2018.

[11] Yamashita, Y et al, Microstructural development of a heavily neutron-irradiated ODS ferritic steel (MA957) at elevated temperature, Journal of Nuclear Material, 367–370, pp. 202-207, 2007.

[12] Anandha Moorthy A, Natarajan N, Palani PK and Suresh M, "Study on Tribological Characteristics of Self-lubricating AA2218-Fly ash-White graphite Composites" International Journal of Engineering and Technology (IJET), ISSN : 0975-4024, vol. 5, No. 5, 2013.

IOP Conf. Series: Materials Science and Engineering **764** (2020) 012050 doi:10.1088/1757-899X/764/1/012050

Experimental Study on Thermal Behaviour of Double Pipe Helical Flow Heat Exchanger

R Karthikeyan[1], P Thangavel[2], R Naveen[3], P Selvakumar[4], S Astra[5]
[1] Assistant Professor, Department of Mechanical Engineering, Kongu Engineering College, Erode
[2] Associate Professor, Department of Mechanical Engineering, Kongu Engineering College, Erode
[3] Assistant Professor, Department of Aeronautical, Bannari Amman Institute of Technology, Erode
[4] Associate Professor, Department of Mechanical Engineering, Kongu Engineering College, Erode
[5] UG Scholar, Department of Aeronautical, Bannari Amman Institute of Technology, Erode

*Author for correspondence E-mail: aeror.naveen@gmail.com

Abstract. The warmth among the fluids is transferred by a tool named warmth exchanger. Now a days, various types of heat exchangers existed with different configurations, in each configuration different types of flow of fluids are followed. In this paper single helical tube made of copper is arranged inside the shell like structure and helical baffle. Double pipe helical flow warmth exchanger which is used to transfer warmth or otherwise known as heat is considered as main configuration. The main advantage of the pipe-in-pipe helical flow is more contact on uppermost area for transferring heat as compared with the everyday based pipe-in-pipe exchanger. The performance of this model is analyzed with water in both inlets in same directions and then opposite directions. The inlet, outlet temperatures and flow costs of the both fluids are measured. From the acquired effects, the effectiveness of double helical drift warmth exchanger is sort of 20% extra than the normal double pipe warmness exchanger and it gives warm water output temperature almost 4^0C lower than the regular warmthexchanger.

1. Introduction

Timothy J. Rennie [11] (2005) carried out a numerical study on pipe-in-pipe helical exchanger. It is structured for laminar type flow and for two different diameters. The warmth transfer coefficient was calculated for parallel and counter flows. Comparison in inner tubes were made with the help of dimensionless number named Nusselt number, various simulations were conducted. In inner annular place the heat resistance was observed. Nusselt range in the annulus became interrelated with a changed quantity of dean. Ashkan Alimoradi [3] (2017) presented energy analysis for forced convection in helical tube warmth exchanger in which water acts as the working substance. The experiment suggests that efficacy of the warmth exchanger decreases linearly and it also indicates that 50% of boom inside the diameter of the coil resulting in 10.7% decrease within the effectiveness and 50% boom in the coil height outcomes 8.7% lower in efficacy. Also, increase in 50% of flow rate leads to decrease in effectiveness by 12.5%. Usman et al., [12] (2015) provided a overview approximately principal paintings executed on helical baffles in helical tube warmth exchanger and mentioned the foremost elements affecting the overall performance of warmth exchanger like shell as well as tube with the assessment between segmented and helical baffles. The experimental results concluded that the segmented baffles are less advantageous and the overall performance of the both warmness exchanger is stepped forward via usage of discontinuous helical baffles with low baffle spacing and folded baffle in place of plainbaffle.

Weinjing et al., [13] (2015) described the outcome of baffle space in shell as well as helical tube warmth exchanger. In this they investigated the bodily models with baffles of different form, i.e. Tri, qua and sextant quarter. The leakage of fluid in triangular region is evidently decreased and the streamline seems plenty towards an idle flow which is spiral. Among those, sextant helical baffle heat exchanger suggests the better warmness transfer performance than that of trisection and quad helical baffle heat exchanger. Insertion of sextant helical baffle in the place of tri helical baffle reduces the strain drop on the shell. Askhan Alimoradi [2] (2017) presented a study about the efficacy of both warmth exchangers and the outcome of geometrical parameters and operational parameters. Analysis becomes completed on constant state situation with the running fluid of water for both aspects. Based on the result two correlations were advanced and is used to find the thermal efficacy for an extensive range of mass flow ratio, the parameters mentioned, and product of dimensionless number like Reynolds numbers. It is determined that, the efficacy is 12.6% less than the efficacy of warmness exchanger in which the flow is parallel. Ghorbani et al., [9] (2010) Studied the blended convection warmness transfer in helical coiled exchanger for finding various parameters. The effects indicated that the ratio of mass flow rate is powerful on the axial temperature profiles. Also, NTU relations of the test have been as compared with the usual counter flow members. Jundika et al., [8] (2016) the heat transfer performance is discussed and flow which is laminar in a coiled tube of various sections including round, ellipse and square relative to the instant tube of comparable sections are also discussed. The outcome of overall geometry, the temperature on the wall, Reynolds quantity and heat as well as cold mode has been mentioned. The outcomes indicated that coiled tube has more warmness transfer than the instant tube. The outcomes also indicated that coiled tube with squared section supply extra warmness transfer rate than others, however, it requires better stress. But, the coiled tube with round section suggests mild warmness transfer with a low-pressure drop; as a consequence, it is going to be suitable for many applications. Simin Wang et al., [10] (2017) discussed the parameters of helical perspective and overlapped attitude of shell and tube heat exchanger with helical baffles for the thermal conduct analysis. The most efficacy results indicated that the coefficient of heat transfer will increase averagely by means of 14.1% which affords steering for business layout of warmth exchanger with helical baffles. Davood. P and Kamiar Zamzamian [5] (2017) Studied the overall performance of both tubes warmness exchanger in which the helical tube acts as turbulator. The experiment was performed in two modes (with turbulator and without turbulator). The shell facet and the tube facet fluids are warm water and bloodless water in both modes respectively. The cloth for turbulator may be identified as the material of the helical tube. The effects show that universal heat transfer rate has accelerated inside the case wherein turbulator was attached.

In present layout of a pipe-in-pipe warmth exchanger, location of touch between the two fluids is small; the effectiveness of those heat exchangers is very low when as compared to other warmness exchangers. In those locations, helical tube double pipe heat exchangers can be effective than the immediately tube or U-tube warmth exchanger. The fundamental goal of this work is to design double-helical go with the flow warmness exchanger and experimental assessment of the warmth transfer characteristics for each parallel flow and counter flow configurations. Schematic view of double pipe warmness exchanger is proven in Figure 1 for counter driftconfiguration.

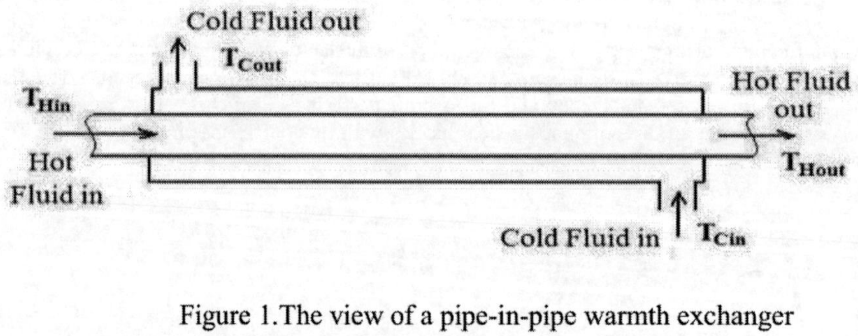

Figure 1.The view of a pipe-in-pipe warmth exchanger

IOP Conf. Series: Materials Science and Engineering **764** (2020) 012050 doi:10.1088/1757-899X/764/1/012050

2. Proposed Heat Exchanger

Helical tube with helical baffle configuration is shown in Figure 2. It contains inner tube in helical manner which is made of copper. An outer tube helical baffle is used between the coils of inner helical tube. The tube with baffle arrangement is covered with a cylindrical arrangement. Thus, the outer shell and baffles are act as outer tube of double pipe warmth exchanger. The working substance used in pipe-in-pipe flow heat exchanger is water. Dimensions of warmth exchanger components are given below in Table 1.

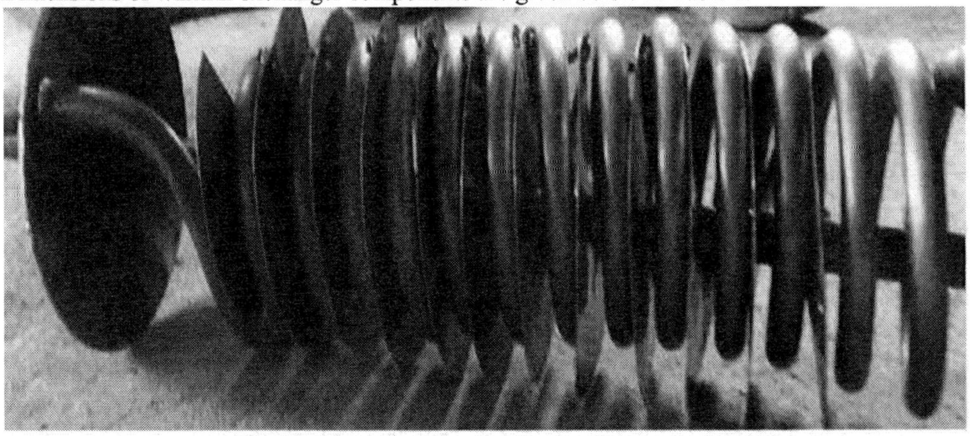

Figure 2.Helical tube with helical baffle configuration

Table 1. Dimensions of proposed heat exchanger components

s.no	Inner conduit	Values (mm)
1	Outer diameter	25.4
2	Inner diameter	23.9
3	Thickness	1.5
	Helical coil	
1	Diameter	90
2	Pitch	25
3	Length	500
	Helical baffle	
1	Thickness	1
2	Diameter	90
3	Pitch	25
	Shell side	
1	Outer diameter	101
2	Inner diameter	90
3	Thickness	5
4	Length	500

3. Experimental Procedure

The various components like double helical flow heat exchanger, heater, pipelines, flow meters, valves and thermocouples are assembled and the experimental study has been completed. Initially, all the components are connected with valves to make the configurations in parallel and counter flow. Power supply has applied to heating coil and the water is heated. The heating process has applied for 30 minutes initially. The various parameters like inlet temperature of hot water, inlet temperature of cold water, outlet temperature of hot water, and outlet temperature of cold water are measured in experimental arrangement shown in figure 3.

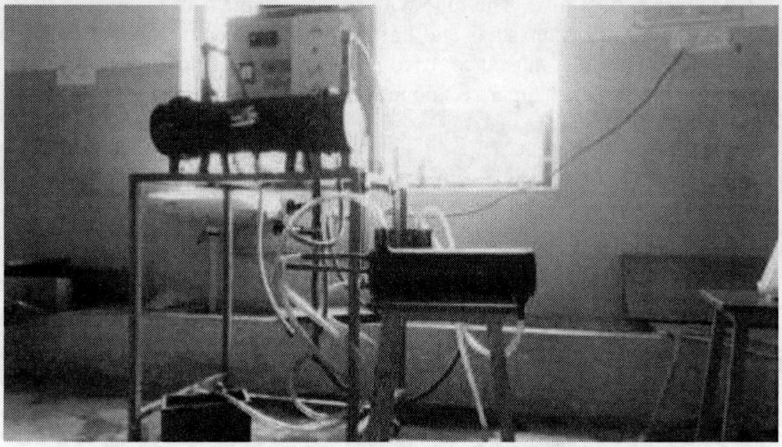

Figure 3. Experimental arrangement

The various properties of hot and cold fluids are tabulated in Table2.

Table 2. Properties of hot and coldwater

Fluids/Properties	Cold water	Hot water
Specific heat Capacity(Cp) in J/kg°C	4178	4170.2
Viscosity (μ) in Ns/m2	0.000757	0.000630
Density (ρ) in kg/m3	994.622	1000
Thermal conductivity (k) in W/mK	0.6206	0.6335
Prandtl Number (Pr)	5.00	4.00

4. Results andDiscussions

The observations are made at the proposed double-helical flow warmth exchanger and the outcome of mass flow charge on effectiveness; average warmness transfer of coefficient, average warmth transfer, and warm water outlet temperature are analyzed. The overall performance of double helical glide warmness exchanger is compared with the performance of base double pipe heat exchanger. In this observation, the impact of mass flow rate of hot water is studied. For that exceptional set of readings were excited by diverse viable combinations of the mass flux of warm water and mass flow rate of cold water. There are nine mixtures are made with three one of a kind mass flow rates of bloodless water including 50 cc/s, 70 cc/s, and ninety cc/s.

4.1 Outcome of Mass flow rate onEffectiveness

The impact mass flux at the efficacy of parallel and counter flow configurations is proven in Figure 4. It indicates that the counter flow is simplest than the parallel flow in the heat exchanger. It is located that the maximum efficacy attained for counter and parallel flow preparations are 0.88 and 0.75 respectively.

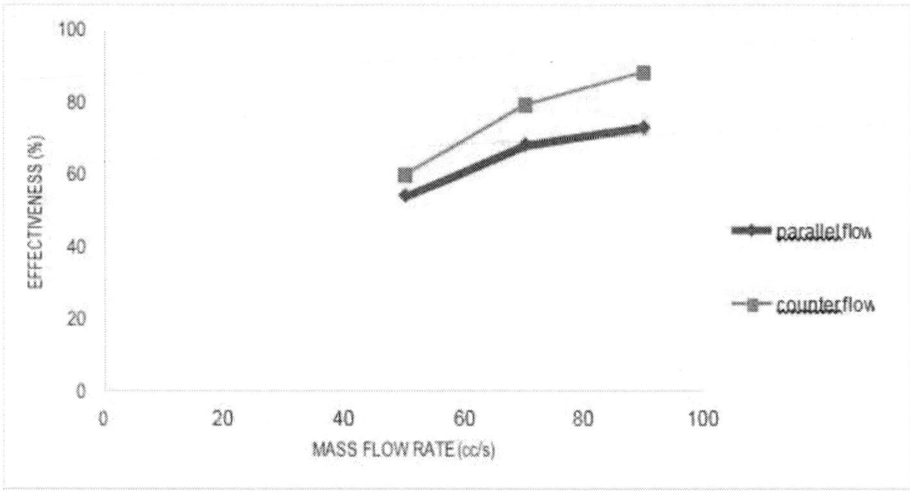

Figure 4. The impact mass flux at the efficacy of parallel and counter flow configurations

The variations of the mass flux of warm water at the effectiveness of the 3 distinctive flow quotes of cold fluid are proven in Figure 5. The mass with the flow rate of warm water will increase the efficacy of double-helical go. It is also shown that the maximum effectiveness is carried out while the mass flow rate of warm water is excessive with the low cold-water mass flow rate. The contact period for heat transfer is expanded in the above-referred to the situation of fluid flow in the double-helical go with the flow heat exchanger. The decreasing effectiveness is executed while the mass flux of warm water is minimum.

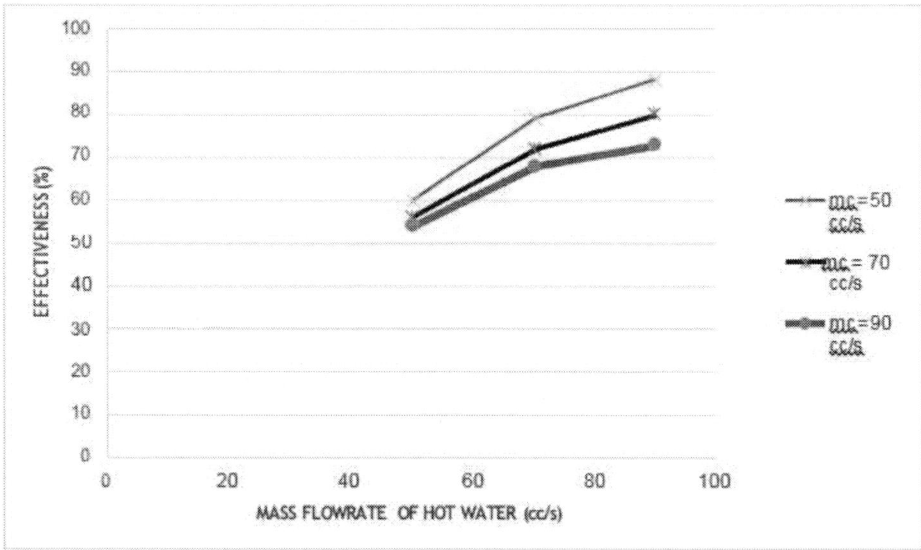

Figure 5.The variations of the mass flux of warm water at the efficacy of the 3 distinctive flow

IOP Conf. Series: Materials Science and Engineering **764** (2020) 012050 doi:10.1088/1757-899X/764/1/012050

4.2 Outcome of Mass Flow Rate on overall Heat TransferCoefficient

The outcome of mass flow rate on the overall warmth transfer coefficient of parallel and counter flow configurations is shown in Figure 6. and indicates the warmth coefficient transfer is better within the counter flow than the parallel flow. Hence, the extra heat transfer rate is attained for counter flow warmth exchanger. The variations of mass go with the flow rate of warm water on the overall warmness transfer coefficient three distinctive float costs of bloodless fluid are shown in Figure 7. The normal warmness transfer coefficient of the double-helical flow warmness exchanger will increase with the increase in mass go with the mass flux of the hot water and at steady mass flux of cold water for the set of three readings. The impact of mass go with the flow rate on average warmness transfer coefficient is equal to the effect of mass go with the flow rate on the effectiveness of the pipe-in-pipe helical flow warmth exchanger. The maximum value of ordinary heat transfer coefficient is achieved when the mass flow rate of hot water is better than the mass flow rate of coldwater.

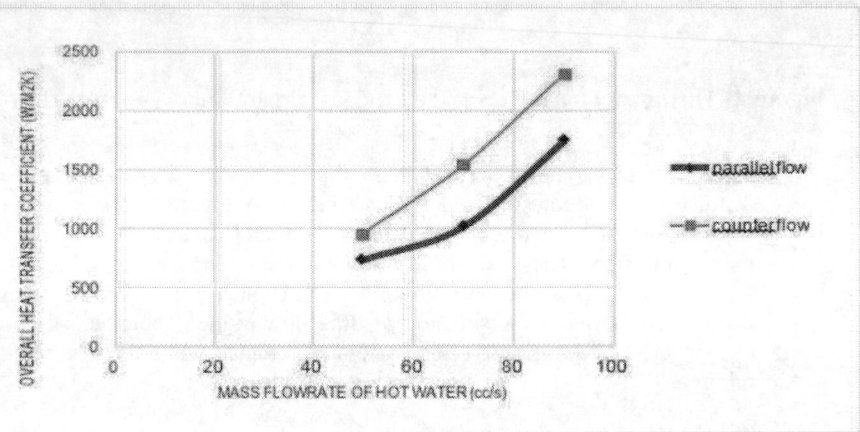

Figure 6. The effect mass flux on the overall warmth transfer coefficient of parallel and counter flow configurations

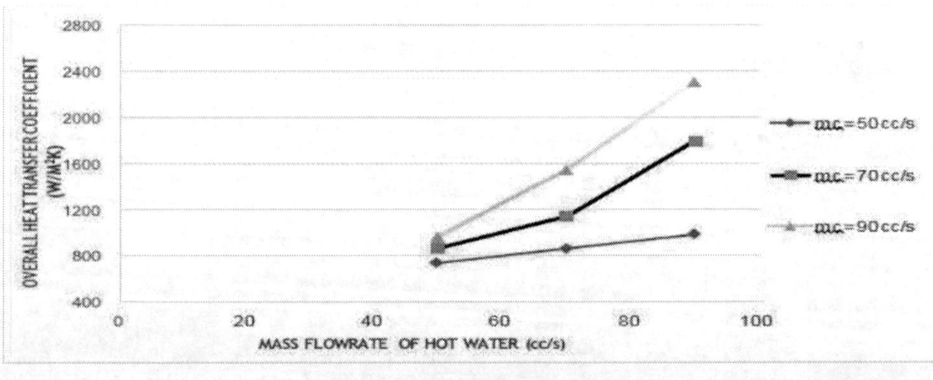

Figure 7. The variations of mass flux of heated water on the effectiveness for the three different flow rates of cold fluid

4.3 Outcome of Mass Flux on Heat TransferRate

The outcome mass moves with the flow rate on the warmth transfer of parallel and counter flow configurations are shown in Figure 8 and propose the warmth transfer is better inside the counter waft than the parallel flow. Hence, the extra warmth switch fee is attained for counter drift warm temperature exchanger. The versions of the mass flux of heat water at the warmth switch for three precise flow rate of cold fluid are validated in Figure 9. The warm temperature transfer of the double-helical flow warm temperature exchanger increases with the boom in the mass flux of the brand new water and at a constant mass flux of cold water for the set of three readings. The maximum price of heat transfer is completed even as the mass go with the flow rate of warm water is higher than the mass glide price of cold water. The not unusual warmth sporting functionality is increasing gradually with the growth in mass go along with the flow rate of hot water and the mass go with the flow exchanger will boom with growth inside the mass float rate of the recent water and at a constant mass flux of bloodless water for the set of three readings. The effect of mass float rate on the regular warmness switch coefficient is the same as the effect of mass float charge on the effectiveness of the pipe-in-pipe flow warmness exchanger. The most rate of normal warmth transfer coefficient is executed even as the mass flux of warm water is better than the mass flux of cold water. The maximum cost of a regular warmth switch coefficient is accomplished even as the mass go with the flow rate of warm water is higher than the mass flux of coldwater.

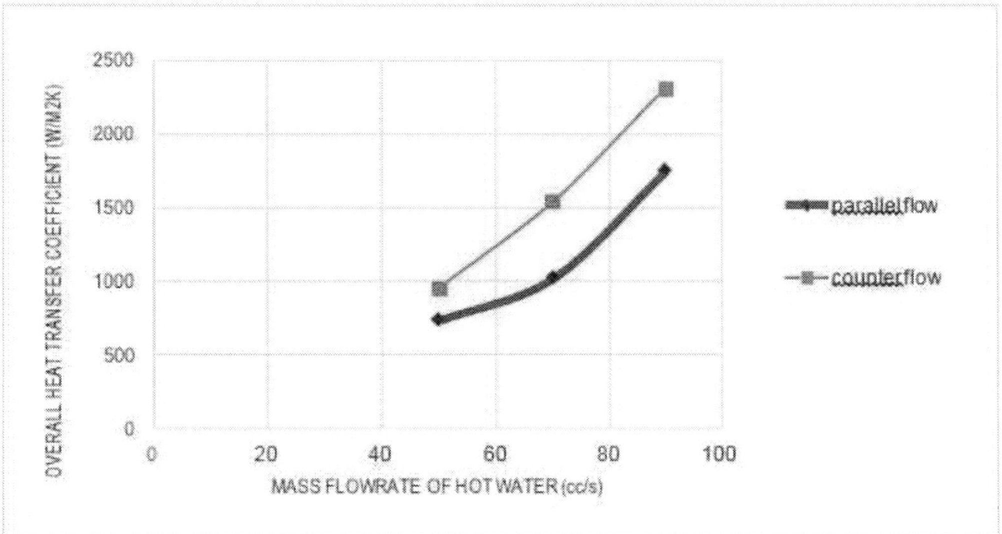

Figure 8.The outcome mass flow rate on the overall heat transfer coefficient of parallel and counter flow configurations

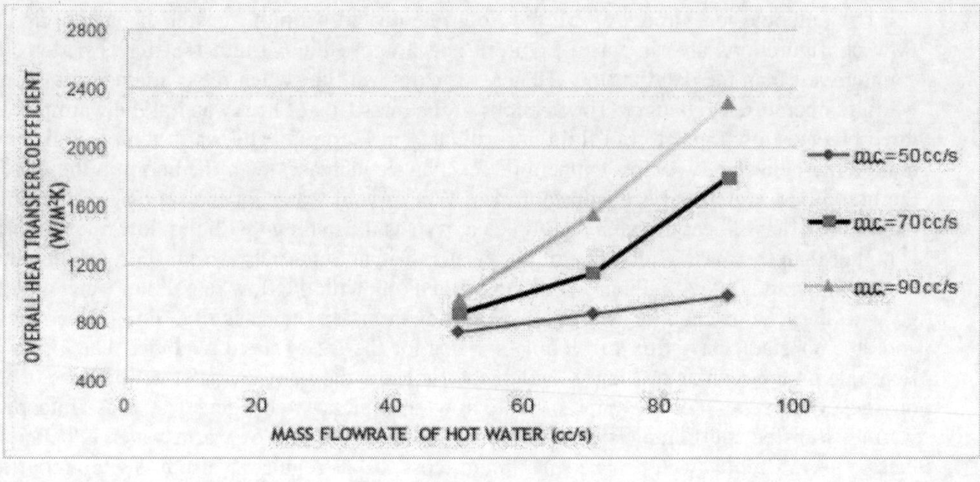

Figure 9. The variations of mass flux of hot water on the effectiveness for the three different flow rates of cold fluid

4.4 *Outcome of Mass Flux on Heat TransferRate*

The impact mass flux on the warmth transfer of parallel flow and as well as counter flow configurations is proven in Figure 10 and shows the warmth transfer is better inside the counter flow. Hence, the greater warmness transfer rate is attained for counter go with the flow heat exchanger. The versions of the mass flux of heated water on the warmth transfer for three one of kind flow rates of cold fluid is proven in Figure 11. The heat switch of the double-helical flow warmness exchanger will increase with growth in the mass flux of the hot water and at steady mass go with the flow rate of cold water for the set of three readings. The maximum price of heat transfer is accomplished while the mass flux of warm water is higher than the mass flow rate of cold water. The average warmth wearing capability is will increase step by step with the increase in the mass flux of warm water and the mass flux of bloodless water stored steady for the set of threereadings

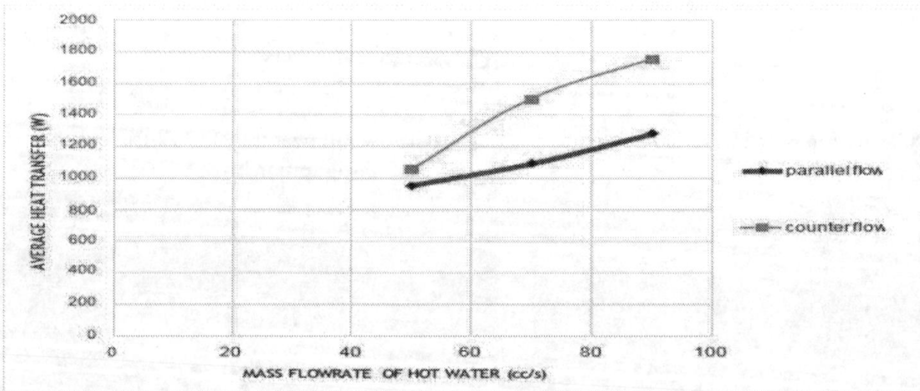

Figure 10.The outcome mass flux on the heat transfer of parallel and counter flow configurations

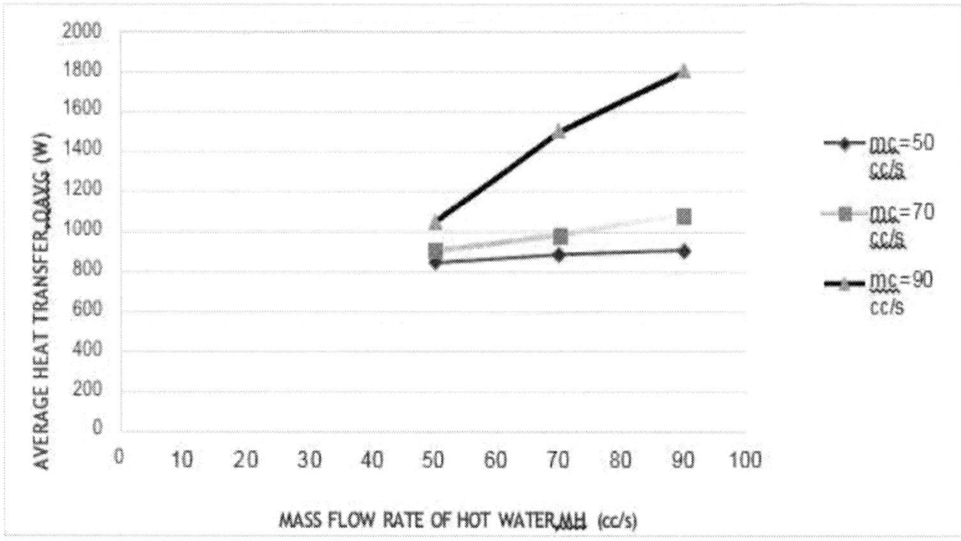

Figure 11.The variations of mass flux of hot water on the heat transfer for the three different flow rates of cold fluid

4.5 Effect of Mass Flowrate on Outlet Temperature and HeatExchanger

The impact mass flow rate on the opening temperature of parallel and counter flow configurations is shown in Figure 12 and shows the outlet temperature is lower in the counter flow than the parallel flow. Hence, the greater warmth transfer rate is attained for counter-flow heat exchanger.

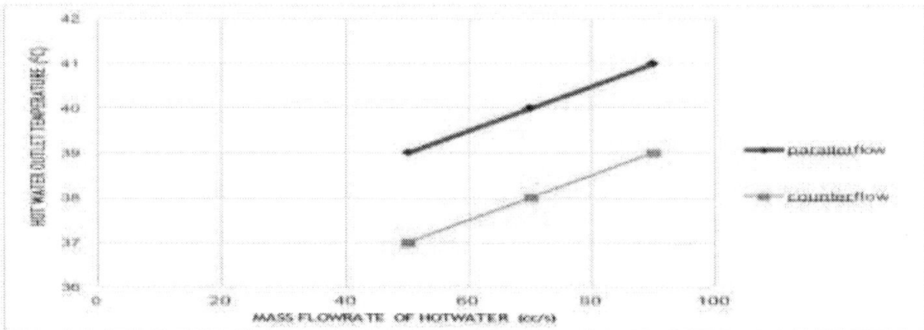

Figure 12. The impact mass flux on the outlet temperature of parallel and counter flow configurations

Double pipe warmth exchanger without the helical tube is taken into consideration because the base warmth exchanger for this is taken and comparison has been carried out. Figure 13 shows an assessment of the effectiveness of base warmth exchanger with the double-helical flow heat exchanger. The end result suggests that better effectiveness is performed for the double-helical flow heat exchanger. Better overall performance may be received by the double-helical tube warmth exchanger because of extra warmness transfer place

IOP Conf. Series: Materials Science and Engineering **764** (2020) 012050 doi:10.1088/1757-899X/764/1/012050

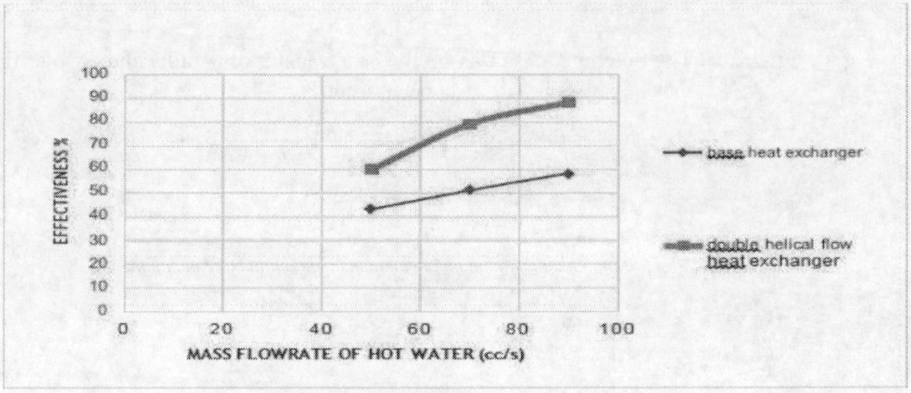

Figure 13.Comparison of effectiveness for base heat exchanger and double helical flow heat exchanger

5. Conclusion

Double helical flow warmness exchanger has been fabricated and assembled with base warmth exchanger. Various observations are cited and the overall performance evaluation of parameters like effectiveness, general warmness transfer coefficient, warmness transfer, and outlet temperatures are performed. Comparing to the above components, the counter-flow heat exchanger holds better overall performance. The efficacy of pipe-in-pipe helical flow warmth exchanger is sort of 20% more than the double pipe heat exchanger. The outlet temperature of heated water is sort of 4^0C much less than the normal base warmness exchanger. The effects show that the overall performance of the double-helical go with the flow heat exchanger is greater than the performance of the base warmth exchanger and may be implemented for industrial applications

References

[1] Amithkumar.M, (2015) "Design and Thermal Evaluation of Helical Coil Heat Exchanger", International Journal of research in Engineering and Technology (04)416-423.

[2] Askhan Alimoradi, (2017) "study of thermal effectiveness and its relation with NTU in helically coiled tube heat exchangers", Case Studies in Thermal Engineering (9)100-107.

[3] Askhan Alimoradi, (2017) "Investigation of Excergy effectiveness in Shell and Helically Coiled Heat Exchanger", Case Studies in Thermal Engineering (10)1-8.

[4] C. Dong, Y.P. Chen, J.F. Wu, (2014) "Influence of baffle configurations on flow and heat transfer characteristics of trisection helical baffle heat exchangers", Energy Conversation Management (88)251–258.

[5] Davood Panahi and Kamiar Zamzamian, (2017) "Heat transfer enhancement of shell and coiled tube heat exchanger utilizing helical wire turbulator", Applied Thermal Engineering (115) 607-615.

[6] D. Kral, (1996) "Helical baffles in shell-and-tube heat exchangers", Part I: Experimental verification, Heat Transfer Eng. 17 (1)93–101.

[7] Dravid, A. N., Smith., K. A., Merrill, E.A., and Brian, P.L.T., (1971) "Effect of secondary fluid motion on laminar flow heat transfer in helically coiled tubes", AICE Journal, Vol.17(5):1114-1122.

IOP Conf. Series: Materials Science and Engineering **764** (2020) 012050 doi:10.1088/1757-899X/764/1/012050

[8] Jundika C. Kurnia, Agus P. Sasmito, Tariq Samim, ArunS.Mujumdhar, (2016) "Numerical investigation of heat transfer and entropy generation of laminar flow in helical tubes with various cross sections", Applied Thermal Engineering (102)849-860.

[9] N. Ghorbani, H. Taherian, M. Gorji, H. Mirgolbabaei, (2010) "An experimental study of thermal performance of shell-and-coil heat exchangers", International Communication in Heat and Mass Transfer (37)775–781.

[10] Simin Wang, Juan Xiao, Jiarui Wang, Guanping Jian, Jian Wen, ZoaxioZang, (2017) "Configuration optimization of shell-and-tube heat exchangers with helical baffles using multi-objective genetic algorithm based on fluid-structure interaction", International communications in Heat and Mass Transfer (85)62-69.

[11] Timothy J. Rennie, (2005) "Numerical Studies of Double Pipe Helical Exchanger", Applied Thermal Engineering (26)1266-1273.

[12] Usman Salahuddin, Muhammad Bilal, Haider Ejaz, (2015) "A Review of Advancements made in Helical Baffles used in Shell and Tube Heat Exchanger", International Communications in Heat and Mass Transfer (67)104-108.

[13] Weinjing. J, H.Z. Yang, S.M. Wang, S.F. Xu, Y.L. Xue, H.F.Tuo, (2015) "Numerical investigation on baffle configuration improvement of the heat exchanger with helical baffles", Energy Conversation Management (89) 438–448.

IOP Conf. Series: Materials Science and Engineering **764** (2020) 012051 doi:10.1088/1757-899X/764/1/012051

Modelization of elevator control system

RS Karrthik[1], S Raghavendra Prabhu[2], K K Megavarthini[3], K Nandakumar[4], B Mathesh[5]

[1, 2, 3, 4, 5] Department of Mechatronics Engineering, Bannari Amman Institute of Technology, India.

*Author for correspondence E-mail: karrthik@bitsathy.ac.in

Abstract. Elevator is a transport device that is used to move people or goods in vertical direction. In this paper programmable logic based elevator control system is constructed to simulate as an actual lift in real life. Here, a new method for traction based elevator system is introduced which doesn't need a counterweight. By this method friction along the traction sheave of the elevator can be improved. To reduce friction along the rails bearings are used in the hoisting mechanism. Servo motor (Mitsubishi's HJ-KS43J) is used as the prime mover for its accurate positioning capability. SIEMENS S7 200 SMART PLC was used for controlling the entire system. PLC is interfaced with hooters, floor switches, indicators and sensors for effective functioning of the elevator system.

Keywords: Elevator, PLC, Counterweight, Friction, Servo, traction.

1. Introduction

Traditional equipments and heavy machines are constantly developing with a goal in order to make them easy to operate along with safety features. Elevator is one of such machine which is used daily. Elevators are used in various environments, based on their hoisting mechanism they are classified as hydraulic elevator, Traction elevator, climbing elevators and Pneumatic elevators.

As the life standards of people increases it is necessary to increase the comfort and safety in elevators. This project describes the study of programmable logic based elevator control system which incorporates a new method for traction based elevator system. Replacing traditional control of elevator system with PLC's not only reduces the space but also makes a way for future upgradation as these systems are reprogrammable and can be programmed easily for future requirements. PLC's are less vulnerable to environmental effects such as dust, moisture and heating effects, this rugged nature enables them to fit into any surroundings [1]. The purpose behind all elevators is the same, it is attaching a lifting system to a compartment built into a hoist way. But elevators get much more complicated once you introduce different kinds of lifting mechanisms [2].

Hydraulic elevators are typically used in low-rise applications (2-8 story buildings) to move up and down via a piston that supports the elevator cab from the bottom. The maximum travelling speed of these hoist systems is about 200 feet per minute, with the maximum travelling height lying anywhere between 20-60 feet, dependent upon the type of hydraulic system used. There are numerous types of hydraulic systems including: holed, hole-less, telescopic, non-telescopic, and roped hydraulic elevator systems.

Climbing elevators have their own electric or combustion engines mounted to them. These types of elevators used in construction areas and other work-zones.

Pneumatic elevators move their passenger cabs up and down by controlling air pressure within the elevator's chamber. Vacuum pumps or turbines control the flow of air in the chamber in order to transport the passenger to and from their respective floor. This type of application could be found in residential homes where a pit and hoist-way would not be feasible.

Traction elevators use ropes that pass over a wheel attached to a motor that sits above the elevator shaft. These elevators are used for mid- to high-rise applications due to their increased speeds and maximum travelling heights. There are three types of traction elevators used today: geared, gear-less, and machine-room-less.

Traction elevators are the most used ones. Traction elevators are lifted by rope and motor arrangement, they are used for mid and high rise applications which require high travel speed. As discussed earlier a new method for traction based elevator system is prototyped and analysed under various conditions.

2. Experimental Design and Analysis

The elevator control system prototype is fabricated using sheet metal. This prototype consists of five floors including ground floor. Sheet metal made of mild steel with thickness about 1.6 mm is used for fabrication. The prototype was designed using SOLID WORKS 2018 software and the rail through the elevator car slides was made with galvanized iron.

2.1 CAD design

Initially part files were created for individual parts and finally they are assembled together to form a complete prototype. The front and back layout of design modules were represented in the figure 2.1.1.

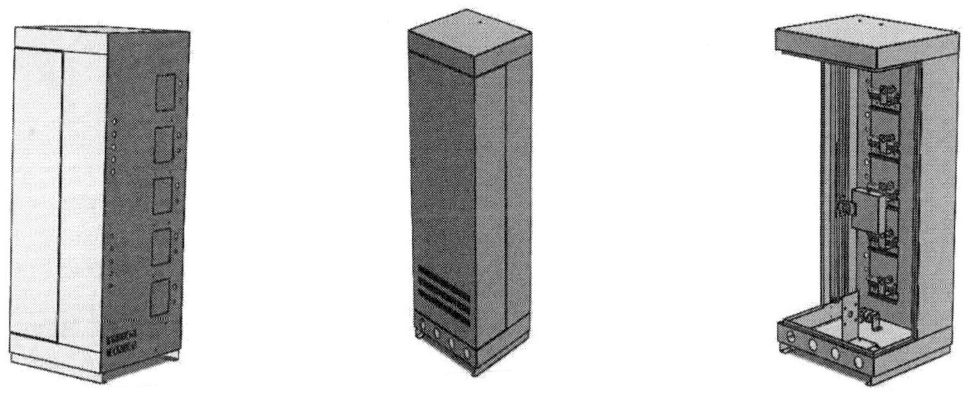

A) Front layout B) Back layout C) Inner View

Figure 2.1.1 Solid works design layout

The needed sub assemblies like pulley, flange to hold the motor, rack and pinion assembly for door opening and closing, rail and elevator car which makes the sliding portion etc.., were designed and fabricated. A series of pulleys were used which provides the required tension along the elevator cable. The elevator cable is looped back to the opposite end of elevator car using pulleys. The friction along the traction sheave of the elevator can be improved and the grip between the traction sheave and the hoisting ropes are improved. This makes it possible to reduce the load on motor [3]. Rack and pinion arrangement was made for door opening and closing mechanism used in elevator control system. Design was made and the same is converted into binary STL file and fabricated through 3D printing.

IOP Conf. Series: Materials Science and Engineering **764** (2020) 012051 doi:10.1088/1757-899X/764/1/012051

Figure 2.1.2 represents the designed pulley, pinion and sub assembly of rack and pinion arrangement. Mitsubishi's HJ-KS43J servo is a flange mounted motor, so separate flange is designed for mounting the

servo.

A) Pulley assemble B) Pinion gear C) Rack and pinion assembly

Figure 2.1.2 Design and layout of sub assemblies

2.2 Fabrication

Once the design was completed the same was fabricated using rolled sheet of thickness about 1.6 mm. Initially the rolled sheet was cut for required dimensions. Then, the required holes and other specifications were carried out using laser cutting process. After this bending was carried out, these operations are shown in the below Figure 2.2.1.

A) Cutting process B) Specified sheet C) Bending process

Figure 2.2.1 Initial fabrication process

Then the assembly is given powder coating to avoid rusting. The fabricated front and back assembly of the proposed prototype is shown in the below Figure 2.2.2.

A) Front layout B) Inner layout C) Back layout

Figure 2.2.2 Completed fabrication of outer layout

Bearings were provided with the elevator car to reduce friction among the railways. This sub assemblies were shown in the below figure. Since four pulleys (including motor mounted pulley) is required for the proposed design they are fabricated. Sliding doors for opening and closing of the doors were also made by rolled sheet of thickness 1.6 mm for the designed specifications. Motor holding flange is made of rolled sheet of thickness 2 mm. 3D printing was carried out to make the rack and pinion assembly. In 3D printing fill of about 10 % is used in the binary STL file for fabricating the rack and pinion. Figure 2.2.3 represents the above mentioned assemblies.

PULLEY ELEVATOR CAR GUIDE WAY WITH SENSOR CLAMP

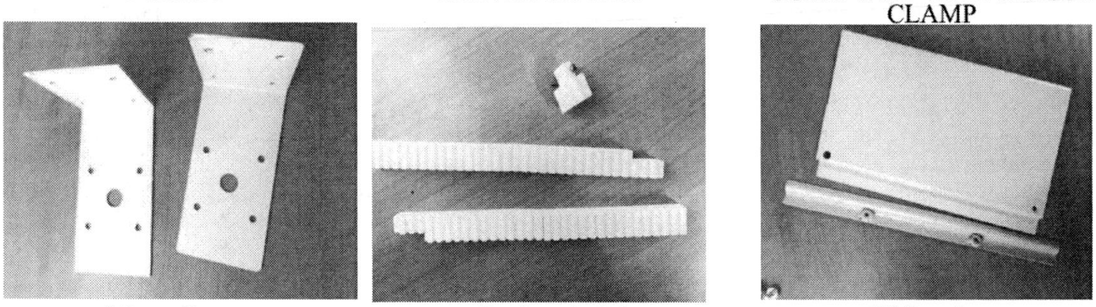

Motor flange Rack and pinion Sliding door

Figure 2.2.3 Fabricated sub assemblies

The complete fabricated prototype is about a volume of 950*425*375 mm. This prototype weighs about 27.03 kg. Figure 2.2.4 represents the completed layout along with its sub assemblies.

IOP Conf. Series: Materials Science and Engineering **764** (2020) 012051 doi:10.1088/1757-899X/764/1/012051

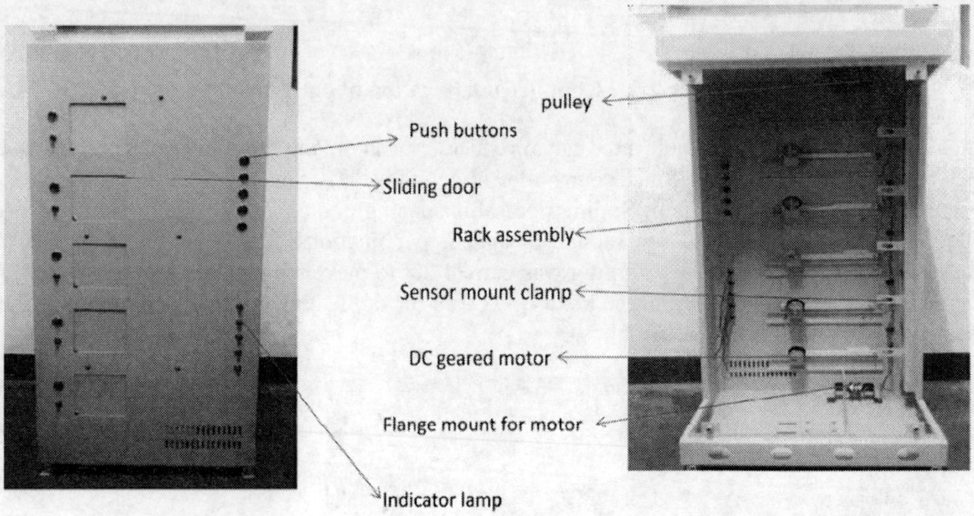

Figure 2.2.4 Completed prototype front and back layout

2.3 Methodology and Analysis

For traction based elevator system a new method is proposed which has mechanical advantage over the former one. Pulley based mechanism is used which reduces the input effort [4]. The pulley generally has a wheel and axle which is designed for supporting movement and change of direction. Here a continuous wire rope is to transmit a tension around four pulleys to lift the load. One end of the rope is connected directly with the load then transferred around anchor pulleys and return backs to the load. Here 4 pulleys are used with a wire rope wounded between them so that the force and weight are equally distributed between them there by reducing the input effort. Also, this method can reduce the space needed for counterweight arrangement. Assumption based calculation is made below for better understanding. Figure 2.3.1 gives basic illustration of proposed method.

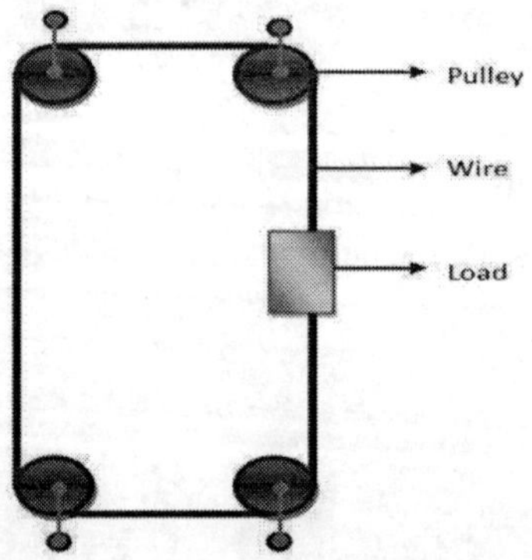

Figure 2.3.1 Basic illustration of proposed method

Here the load to be carried is 10 Kg (assumed) = 10 * 9.81 = 98 N But by using these pulleys here, 3T = 98 N (3 pulleys are considered one is acting as prime mover) T =32.66M = T/g = 32.66/9.8 M = 3.33 Kg.

Therefore by using 4 pulleys the amount of input power given to do a particular task is only 33.33% of the power given usually using counterweight.

Since a new method was proposed it is necessary to analyse them to find how the system behaves for various conditions. The proposed system doesn't contain counterweight. So, it is mandatory to compare the analysis of both the systems. Design was made for both the system using SOLIDWORKS software and analysis is carried out by the same software. Design was made for original model with counterweight and for the derived model without counterweight using pulley system. The design layout for both the models was given in the figure 2.3.2.

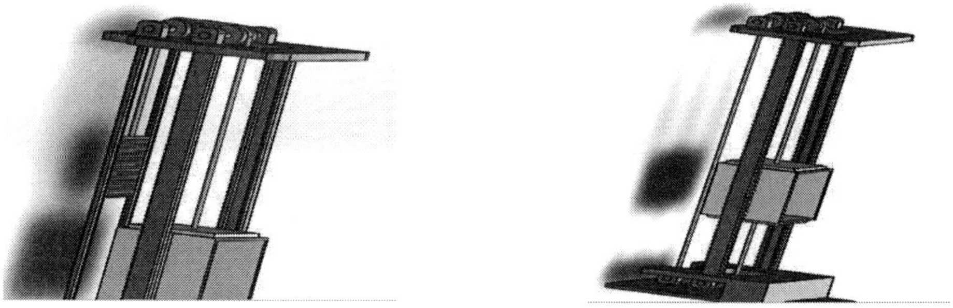

A) Original model B) Proposed layout

Figure 2.3.2 Original and proposed design layout

Initially the original model with counterweight was analyzed for various conditions and loads followed by the proposed model the strain diagram for various fixtures and loads for both the analysis is shown in figure 2.3.3 for comparison. Observations and inferences were made form this study.

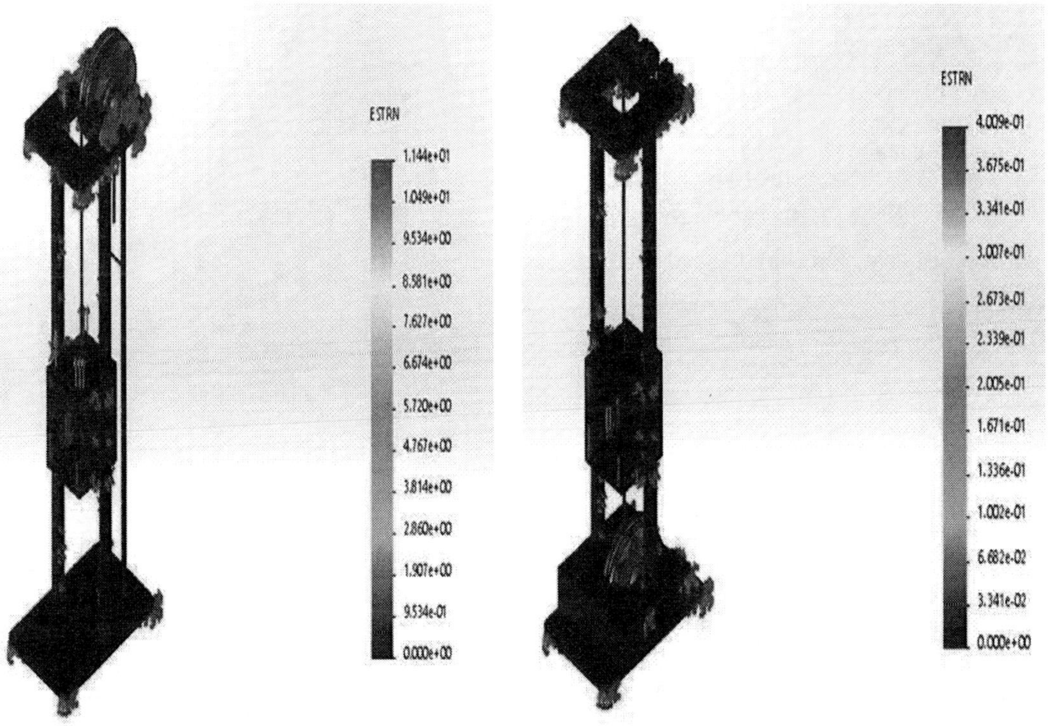

A) Original model B) Derived model

Figure 2.3.3 Strain response of original and derived model

IOP Conf. Series: Materials Science and Engineering **764** (2020) 012051 doi:10.1088/1757-899X/764/1/012051

Comparing both the analysis original model and the derived one, it is found that the maximum value of stress at the pulleys is almost the same hence it is obvious to use the proposed system in place of existing model. For larger weights in hoisting mechanism counterweight helps in balancing the load hence, for heavy load hoisting mechanism elevator with counterweight is an effective one [5]. For short and mid-range traction based elevator system we can use the proposed system of traction elevators.

3. PLC

3.1 PLC Interface and Programming

The Programmable Logic Controller is a digital computer used in several industries and machines for automation of Industrial electromechanical processes. The main advantage of PLC is that it is modular, so various input and output modules can be readily interfaced based upon the need [6]. The processing unit in the PLC executes the user control instructions contained in the user program. The program stored is Non-volatile so even if the power is lost the program remains uninterrupted. It is a real time system used to produce output within limited time as a response to changes in input conditions [7].

The S7-200 SMART has two different types of CPU modules. One is standard type and another one is compact type. Standard type CPUs are expandable with I/O expansion modules and signal boards. Compact type CPUs are fixed one which has non expandable with I/O expansion modules and signal boards [8]. It is equipped with dedicated Siemens processor chip with basic instruction execution time is up to 0.15 micro second, it has high performance and fast process. It has Ethernet interconnectivity which is economic and convenient for communication. Table 3.1 gives the specifications for SIMATICS S7 200 in detail.

Table 3.1 Simatics s7 200 specifications

CPU Model	CPU ST20 DC/DC/DC
Dimension	90*100*81
Weight	320g
Power consumption	20W
Input Voltage	24 V DC
On board digital I/O	12 Input point/8 Output Point
I/O module extension	6 extension module
High speed counters	4
Pulse output	2 of 100 KHz
Number of ports	1 Ethernet port/ 1serial (RS485)/ 1additional serial (optional RS232/ 485 signal board) port
Signal board extension	Max 1 signal board extension
Memory	Micro SDHC card (optional)
Data transmission rate	Ethernet 10/100 Mb/s

3.2 PLC Interface

The programming software is used in the project to develop ladder logic program to control plc system for elevator control. The address locations for the elevator control system components interfaced with plc are given in table 3.2.

Table 3.2 Memory address

S.No	Input address		Output address		Memory address	
1	Ground floor switch	I0.0	Pulley up motor signal	Q0.0	Ground floor memory	M0.0
2	Floor 1 limit switch	I0.1	Pulley down motor signal	Q0.1	Floor 1 memory location	M0.1
3	Floor 2 limit switch	I0.2	Motor stops signal	Q0.2	Floor 2 memory location	M0.2
4	Floor 3 limit switch	I0.3	Ground floor light indicator	Q0.3	Floor 3 memory location	M0.3
5	Floor 4 limit switch	I0.4	Floor 1 light indicator	Q0.4	Floor 4 memory location	M0.5
6	Ground floor call	I0.5	Floor 2 light indicator	Q0.5	Ground floor last visit	M0.6
7	Floor 1 call switch	I0.6	Floor 3 light indicator	Q0.6	Floor 1 last visit memory	M0.7
8	Floor 2 call switch	I0.7	Floor 4 light indicator	Q0.7	Floor 2 last visit memory	M1.4

9	Floor 3 call switch	I1.0	Ground floor car indicator	Q1.0	Floor 3 last visit memory	M1.0
10	Floor 4 call switch	I1.1	Floor 1 car indicator	Q1.1	Floor 4 last visit memory	M1.1
11	Ground floor call	I1.2	Floor 2 car indicator	Q1.2	Ground floor visit	M1.2
12	Floor 1 car switch	I1.3	Floor 3 car indicator	Q1.3	Floor 1 visit memory	M1.3
13	Floor 2 car switch	I1.4	Floor 4 car indicator	Q1.4	Floor 2 visit memory	M1.4
14	Floor 3 car switch	I1.5			Floor 3 visit memory	M1.5
15	Floor 4 car switch	I1.6			Floor 4 visit memory	M1.6

4. Conclusion

The experimental model is verified by functionality of the proposed control and results were verified under various conditions. The proposed method can be used for short and mid range traction based elevator systems. Based on the arrangement of diverting pulley the elevator car is balanced, this made us to build a system without counterweight. This prototype serves as an appropriate model demonstration in laboratory.

4.1 Future enhancements

Various studies can be made to understand the behavior for numerous load conditions and how to implement the system in lower space constraints. Extended safety features like cushioning mechanism at the bottom of the elevator chamber can be implemented in real time elevators for increasing the viable chances of escaping the mishap. SCADA interface of the elevator system provides remote monitoring and control in constrained situations and also this enables data logging facilities. This system can be enhanced by using automatic door opening mechanisms, command reorganization and extended functionalities.

5. References

1. Gurmeet Singh, Anshul Agarwal, R.K. Jarial, Vineeta Agarwal, Mithun Mondal (2013) "PLC controlled elevator system," Students Conference on Engineering and Systems (SCES), *IEEE conferences,* **DOI:** 10.1109/SCES.2013.6547517.
2. Songkran Kantawong, Ronnahrit Kanchanawiboon (2017) "Multipath tall Building Elevator Module for flexible Transport System with PLC based on Cloud Control," International Electrical Engineering Congress (iEECON), **DOI:** 10.1109/IEECON.2017.8075900.
3. Dhara G. Rangani, Nikunj V. Tahilramani (2017) "Automation based elevator control system,"presented at 3rd International Conference on Applied and Theoretical Computing and Communication Technology (iCATccT), *IEEE conference*, **DOI:** 10.1109/ICATCCT.2017.8389116.
4. https://patents.google.com/patent/US8235179.
5. Peng Wang (2011) "A control system Design for elevator based on PLC" presented at Fourth International Symposium on Computational Intelligence and Design, *IEEE conferences,* Volume: 1, pages: 71-74.

IOP Conf. Series: Materials Science and Engineering **764** (2020) 012051 doi:10.1088/1757-899X/764/1/012051

6. Pierre Bernard, Ioan-Dragoş Deaconu, Sergiu Valentin Popescu, Constantin Ghiţa, Anca-Simona Deaconu, Aurel-Ionuţ Chirilă (2015) "PLC controlled elevator drive system," 9th International Symposium on Advanced Topics in Electrical Engineering (ATEE), *IEEE conferences,* **DOI:** 10.1109/ATEE.2015.7133775.
7. Norbert Ferencik, Jan Jadlovsky, Michal Kopcik, Iveta Zolotova (2017) "PI control of laboratory model elevator via ladder logic in PLC," published in IEEE 15th International Symposium on Applied Machine Intelligence and Informatics (SAMI), *IEEE conferences,* Pages: 000383 – 000386.
8. Kamalu, U.A, Ogunleke, F.A (2018) "Implementation of a Four-Floor Programmable Logic Controlled Elevator System", International Journal of Scientific & Engineering Research Volume 9, Issue 6,ISSN 2229- 5518.
9. Xiaoling Yang, Xiaoling Yang, Qunxiong Zhu, Hong Xu (2008) "Design and Practice of an Elevator Control System Based on PLC,"published in Workshop on Power Electronics and Intelligent Transportation System. **DOI:** 10.1109/PEITS.2008.44
10. William Bolton "Programmable Logic Controllers," 6th Edition, 3rd March 2015.

IOP Conf. Series: Materials Science and Engineering **764** (2020) 012052 doi:10.1088/1757-899X/764/1/012052

Optimization of machining parameters in Electric Discharge Machining of Magnesium and Aluminium by Taguchi Technique

Tajdeen A[1*], SakthivelMuruganE[1], Wasim khan M[2], Praveen Kumar S[3], PraveenKumar M[3], Vidhyaa Prakash S[3]

Assistant Professor[1], Research Scholar[2], U.G Students[3]
Department of mechanical Engineering
[1, 3] BannariAmman Institute of Technology, Sathyamangalam,Erode,Tamilnadu, India
[2]Anna University, Chennai, Tamilnadu, India.
*Corresponding Author Email: **tajdeena@bitsathy.ac.in**

Abstract. Aluminium, Magnesium and its alloys are preferred wheneverweight reduction is needed without compromising its strength. Both alloys have been used in numerous fields like in aerospace, defence field, automobile wheeling parts and also, they have been used in Medical field because of their biodegradability. In this study, Statistical investigation on surface roughness in Electric discharge machining of aluminium and Magnesium using Taguchi optimization Technique has been reported. Electric discharge machining was carried out under the condition of Current (5, 10, 15 A), Voltage (50, 55, 60 V), Pulse On time (10, 20, 30 µs) and constant flushing pressure of 0.50 Kg/sq.cm. Twenty-seven Experiments are carried out based on L_{27} orthogonal array. After machining the surface roughness (Ra,Rq,Rz) of the machined surface is measured using SURFTESTSJ-210. Optimizations of the results were done by Taguchi technique and the optimal parameters for the machining operation of aluminium and magnesium are determined. It is found that for Ra, Rq and Rz the most influencing factor is Current followed by the pulse on time, pressure and voltage.

1. Introduction

Aluminium and its alloys have been used in various applications because of their properties light weight, corrosion resistance. Strength to weight ratio is comparable high when compared to any other alloys. They also have very good machining properties which make to applicable for the various applications [1]. Magnesium is the lightest and low density material which is used in constructional applications. The weight of the Magnesium compound is 35% lower than that of Al and shows predominant vibration characteristics. Mg is a preferable material for alloy especially with the Aluminium increasing both corrosion resistance and hardness. Mainly because of this property the Mg-alloys have been used in the aircraft applications especially in aircraft fuselages [2]. Electrical discharge machining (EDM) process is used for machining the Al and Mg alloys. Because an Al and Mg alloy requires high accuracy for machining, this can be achieved by the EDM process [3].

Nine EDM experiments of three levels with four parameters were conducted on the AZ31 magnesium alloy to detect its optimal machining parameters by using Taguchi method. Parameters considered are Pulse on time, pulse off time, peak current and voltage. It is found that out of these parameters Pulse on time is the most influencing factor. Optimum result is obtained when the peak current is 47A, voltage is 80 and Pulse on time is16µs. The cracks and craters formation were formed on the machined

IOP Conf. Series: Materials Science and Engineering **764** (2020) 012052 doi:10.1088/1757-899X/764/1/012052

surface even though the use of EDM which has very good machining properties. Thereby in this process Powder mixed EDM is used to avoid the craters and cracks formation during the process [4].In biodegradable Magnesium using EDM, because they have a various uses in medical field especially in the bone replacement. Magnesium alloys are degraded inside the human body leaving without any trace and they have the high biocompatibility and very good structural stability. An attempt made to analyze the micro and macro surface properties using the EDX and SEM analysis. Thereby the continuous corrosion of the magnesium leads to decrease its Viability and observed that toxicity of machined magnesium occurs which can be avoided by using of optimized process parameters [5].

In recent days, normal and cryogenic cooled EDM process was carried out with different machining parameters such as discharge current, pulse on time and gap voltage. The response parameters are material removal rate, electrode wear rate, surface roughness and also they taken the micro, macro surface using the SEM and EDX analysis. The results obtained were the electrode wear rates are reduced by 18% when cooled by the cryogenics. And also the use of the liquid nitrogen gives a better EDMed surface region avoiding the cracks and craters that formed during the machining process [6].

Both C- EDM and PM- EDM process were carried out in nine experiments with same setting parameters. The optimized surface roughness obtained from PD-EDM is reduced to 44% from the C-EDM. Also the Powder Mixed – EDM is proven to reduce the surface roughness from the machining process. In order to reduce the corrosion of the machined magnesium alloy by the use of the PD-EDM process [7].

The machining parameters of Mg alloy and Al alloys were optimized using various single and multiple parameters optimization techniques. The parameters considered are pulse on time, pulse off time, and peak current. Analysis based on the ANOVA found out that the pulse off time is least affecting factor and the pulse on time and the peak current are the most affecting factor [8, 9,10].

In EDM process in Mg alloys, accuracy of the holes were examined by giving the various inputs and calculate the outputs. From the result it is came to know that the surface roughness and the taper majorly depends on the pulse on time and servo speed of the input given to the EDM during the process [11].

In our present work, optimize the machining parameters in aluminium and magnesium during EDM process.

2. Materials and Methods

Magnesium is a combination of silvery white, light- weight metal. Unlike other alloys it doesn't require oxygen free environment to store them because they are naturally protected by their thin layer of oxide.

Aluminium alloys have very good physical properties like high hardness, low density, high specific strength, good casting ability and also having very good machining capability.

2.1. Experimental setup

The selected material is cut into require dimension based on the working process shown in fig 1. Then the material is machined through the selected machine. Experiments were conducted using an EDM machine mode 550×300 with EDM oil as dielectric medium.

Figure1. Workpiece material

IOP Conf. Series: Materials Science and Engineering **764** (2020) 012052 doi:10.1088/1757-899X/764/1/012052

2.2. Experimental Procedure

The 12 mm diameter copper tool is selected as tool material. Copper tool is connected with positive power supply terminal and work table connected with negative power supply terminal as shown in fig 2. Three most influencing process parameters of EDMsuch as current (A), pulse on time (Ton) and Gap voltage (V) were taken as input parameters with three different levels. Surface Roughness (Ra, Rz andRq) were chosen as response parameter. The dielectric fluid is circulated at pressure 0.5Kg/cm2 at maintaining the constant pulse off time 10 and 18 duty cycle per min.

Figure 2. Electric discharge machining

2.3. Experiment Work

By setting these provided parameters on the EDM machine is operated. Set the first parameter as given in the Table 1. Those are called Process Parameters. After setting those parameters, place the tool on the required position for the given time. After doing this, change the parameters and do the process once again. Repeat the process for 27 times. Then measure the surface roughness.

2.4 Measurement of the Surface Roughness

Once the process is completed, then the work pieces were subjected to the Surface Roughness measurement. The surface roughness should be measured for all the work pieces on the machined surface. Surface roughness is measured using the MITUTOYO Surface roughness tester SJ-210 shown in fig 3. First the SJ210 is set into calibration mode. Then the probe of SJ210 is fixed on the machined surface of the work piece, then start the tester the probe will move traverse and back. After that the reading will be shown in the monitor attached to the probe detector. Thus the three trails are taken and their average is tabulated.

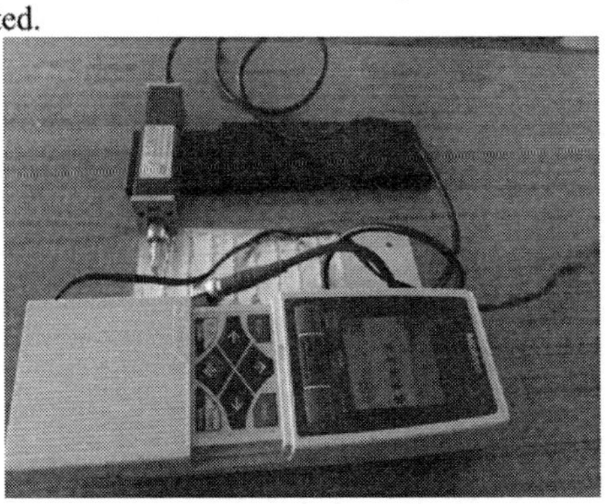

Figure 3.surftest SJ-210

Table 1. Experimental values of Surface Roughness

SI.No	Input parameters			Response Parameters					
	Current (A)	Voltage (V)	Pulse on time (µs)	Surface Roughness Aluminium			Surface Roughness Magnesium		
				Ra	Rq	Rz	Ra	Rq	Rz
1.	5	50	10	3.771	4.592	18.053	1.718	2.065	7.305
2.	5	50	20	3.881	4.674	17.473	1.414	1.706	5.634
3.	5	50	30	4.202	4.961	18.113	1.593	1.921	6.459
4.	5	55	10	3.290	4.145	16.052	1.620	1.942	6.849
5.	5	55	20	4.250	5.156	20.089	1.788	2.134	7.397
6.	5	55	30	3.618	4.489	16.804	1.612	1.898	6.643
7.	5	60	10	3.641	4.420	17.472	1.851	2.206	7.590
8.	5	60	20	3.733	4.547	17.447	1.569	1.866	6.368
9.	5	60	30	4.375	5.210	19.454	1.536	1.796	6.121
10.	10	50	10	4.111	4.911	18.684	1.710	2.077	7.122
11.	10	50	20	4.607	5.590	21.258	1.726	2.073	7.365
12.	10	50	30	4.490	5.377	19.239	1.515	1.906	6.608
13.	10	55	10	3.704	4.516	18.015	1.651	1.983	7.055
14.	10	55	20	4.479	5.307	19.137	1.864	2.168	7.023
15.	10	55	30	4.492	5.347	19.714	1.402	1.681	5.952
16.	10	60	10	4.297	5.180	19.464	1.488	1.808	6.295
17.	10	60	20	4.583	5.455	19.770	1.811	2.528	7.380
18.	10	60	30	4.107	5.030	18.754	1.305	1.541	5.696
19.	15	50	10	4.539	5.536	20.787	1.413	1.759	5.780
20.	15	50	20	4.814	5.734	21.163	1.199	1.480	5.194
21.	15	50	30	4.327	5.195	19.164	1.424	1.743	6.034
22.	15	55	10	4.278	5.056	18.525	1.599	1.895	6.785
23.	15	55	20	4.925	5.897	21.816	1.588	1.974	6.857
24.	15	55	30	5.642	6.749	23.599	1.103	1.393	4.814
25.	15	60	10	4.033	4.964	18.331	1.398	1.686	5.649
26.	15	60	20	5.491	6.480	23.154	1.533	1.874	6.797
27.	15	60	30	5.533	6.699	24.497	1.127	1.342	4.636

3. Result and Discussion

After completing the experiments readings were noted. With the help of the readings S/N ratio is calculated with the help of Taguchi Method. In this method the four parameters are fixed and they are called Factors. The four factors are Voltage, current, Pulse on time and pressure. They are noted as A, B, C and D. Then the levels are calculated. After it the average levels are calculated and find their difference. Difference values are ranked at last.

3.1 Calculationof S/Nratio

$\Lambda = S/N = -10\times\log(\Sigma(Y2))$ (Lower the better value)---------- (1)

By using this formula Ra, Rq, and Rz is calculated, once completing it, then the level1 , level2 and the level 3 for the factors A,B,C and D is calculated with the help of the S/N ratio for each and every parameters is calculated using formula 1 and shown in table 2. After this the average and the rank of the respective factors are calculated.

Table 2. calculated S/N ratios for Aluminium and Magnesium

Input parameters				S/N Ratio					
SI.NO	Current (A)	Voltage (V)	Pulse on time (μs)	Surface Roughness – Aluminium			Surface Roughness – Magnesium		
				Ra	Rq	Rz	Ra	Rq	Rz
1.	5	50	10	-11.529	-13.240	-25.131	-4.702	-6.300	-17.273
2.	5	50	20	-11.779	-13.394	-24.848	-3.007	-4.641	-15.016
3.	5	50	30	-12.470	-13.911	-25.160	-4.043	-5.672	-16.203
4.	5	55	10	-10.343	-12.350	-24.111	-4.192	-5.765	-16.713
5.	5	55	20	-12.568	-14.246	-26.059	-5.049	-6.584	-17.381
6.	5	55	30	-11.169	-13.042	-24.508	-4.147	-5.567	-16.447
7.	5	60	10	-11.224	-12.909	-24.847	-5.348	-6.871	-17.605
8.	5	60	20	-11.440	-13.155	-24.834	-3.911	-5.420	-16.080
9.	5	60	30	-12.819	-14.337	-25.780	-3.728	-5.088	-15.737
10.	10	50	10	-12.280	-13.823	-25.429	-4.658	-6.347	-17.052
11.	10	50	20	-13.269	-14.949	-26.550	-4.741	-6.331	-17.343
12.	10	50	30	-13.044	-14.611	-25.684	-3.606	-5.601	-16.401
13.	10	55	10	-11.374	-13.094	-25.113	-4.353	-5.948	-16.970
14.	10	55	20	-13.023	-14.497	-25.637	-5.407	-6.720	-16.930
15.	10	55	30	-13.048	-14.562	-25.896	-2.933	-4.511	-15.494
16.	10	60	10	-12.663	-14.286	-25.785	-3.450	-5.146	-15.980
17.	10	60	20	-13.224	-14.736	-25.920	-5.157	-8.056	-17.361
18.	10	60	30	-12.271	-14.032	-25.462	-2.310	-3.756	-15.111
19.	15	50	10	-13.139	-14.863	-26.356	-3.001	-4.905	-15.238
20.	15	50	20	-13.650	-15.169	-26.511	-1.579	-3.405	-14.310
21.	15	50	30	-12.723	-14.312	-25.650	-3.072	-4.816	-15.613
22.	15	55	10	-12.624	-14.076	-25.355	-4.079	-5.552	-16.631
23.	15	55	20	-13.849	-15.413	-26.775	-4.015	-5.905	-16.723
24.	15	55	30	-15.029	-16.584	-27.458	-0.854	-2.881	-13.650
25.	15	60	10	-12.113	-13.917	-25.264	-2.908	-4.535	-15.039
26.	15	60	20	-14.794	-16.232	-27.293	-3.709	-5.457	-16.646
27.	15	60	30	-14.859	-16.520	-27.782	-1.036	-2.557	-13.322

Figure4 shows the main effects plot for S/N ratio for aluminium. It is observed thatbetter surface finish can be obtained at lower current(5A), medium Voltage (55V) and lower pulse on time (10μs).

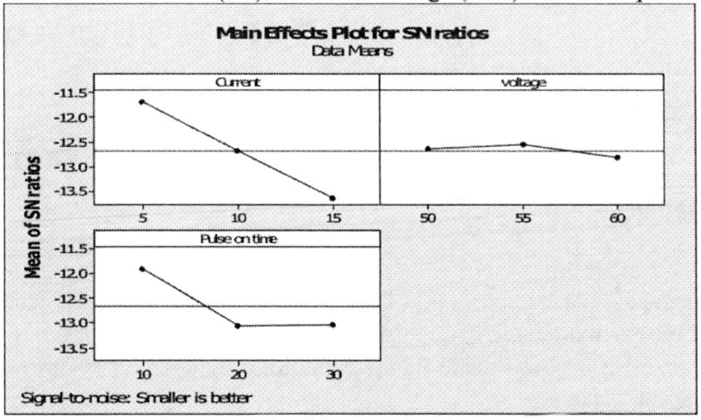

Figure 4. Main effects plot for S/N ratio for Aluminium (Ra,Rq and Rz)

Figure 5 shows the main effects plot for S/N ratio for aluminium. It is observed thatbetter surface finish can be obtained at higher current(15 A), high voltage (60 V) and high pulse on time (30μs).

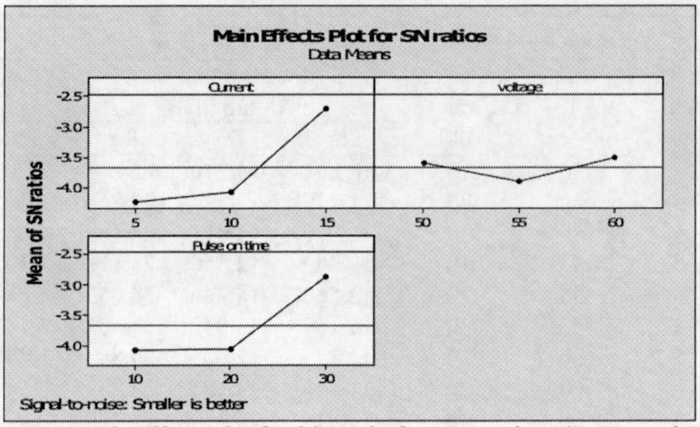

Figure 5. Main effects plot for S/N ratio for Magnesium (Ra,Rq and Rz)

Table 3 and 4shows the Average and the rank value of the Ra, Rq and Rz. After the completion of finding levels for the Factors, then the average of the respective factors is determined. Then the difference between the maximum and the minimum is calculated for the Levels. After the values of the differences is ranked according to the descending type.

Table 3. Influencing factor and rank for Magnesium

	Influencing factor and rank of Ra			Influencing factor and rank of Rq			Influencing factor and rank of Rz		
Factor	**A**	**B**	**C**	**A**	**B**	**C**	**A**	**B**	**C**
Level 1	-4.24	-3.6	-4.07	-5.77	-5.34	-5.71	-16.50	-16.05	-16.50
Level 2	-4.07	-3.89	-4.06	-5.82	-5.49	-5.84	-16.52	-16.33	-16.42
Level 3	-2.69	-3.5	-2.85	-4.45	-5.21	-4.49	-15.24	-15.88	-15.33
Difference	1.54	0.38	1.21	1.38	0.28	1.34	1.27	0.45	1.17
Delta	0.49	0.12	0.38	0.46	0.09	44.66	0.44	0.15	0.40
rank	1	3	2	1	3	2	1	3	2

Table 4. Influencing factor and rank for Aluminium

	Influencing factor and rank of Ra			Influencing factor and rank of Rq			Influencing factor and rank of Rz		
Factor	A	B	C	A	B	C	A	B	C
Level 1	-11.70	-12.65	-11.92	-13.40	-14.25	-13.62	-25.03	-25.70	-25.27
Level 2	-12.70	-12.56	-13.07	-14.29	-14.21	-14.64	-25.72	-25.66	-26.05
Level 3	-13.60	-12.82	-13.05	-15.23	-14.46	-14.66	-26.49	-25.89	-25.93
Difference	1.94	0.26	1.15	1.83	0.25	1.04	1.46	0.23	0.78
Delta	0.58	0.077	0.34	0.59	0.08	0.33	0.59	0.09	0.32
rank	1	3	2	1	3	2	1	3	2

4. Conclusion

AftercompletingthecalculationsbyusingtheTAGUCHIMethodtheresults are tabulated and the optimal parameters for the machining operation ofMagnesiumare determined.

- It is found that for Ra, Rq and Rz the most influencing factor is Current followed by the pulse on time and voltage.
- It can be proved with the help of the tables 3 and 4 calculated by the Taguchi Method and with the ranks determined by the method.
- The optimal result for Ra, Rq and Rz obtained, when theparameters are 15A, 60 V and 30 µs.

For Aluminium

- Itisfoundthatfor R_a, Rqand$R_z$the mostinfluencingfactor isCurrent followedbythe pulseontime,pressure andvoltage.
- It can be proved with the help of the tables 3 and 4 calculated byTaguchi Method and with the ranks determined by the method.
- The optimal result for Ra, Rq and Rz obtained, when the parameters are 5 A, 55 V and 10 μs.

5. References

[1] PulkitGarga, AnbeshJamwal, Devendra Kumara, Kishor Kumar Sadasivuni,ChaudheryMustansar Hussain, Pallav Gupta., "ArticleAdvance research progresses in aluminium matrixcomposites: manufacturing & applications" *Journal of material research and technology*. Volume **74**, pp1-16, 2019.

[2] J.P. Weiler, "A review of magnesium die-castings for closure applications" *Journal of Magnesium and Alloys*, Volume **7**, pp297-304, 2019.

[3] L. Selvarajana, J. Rajavel, V. Prabakaran ,B.Sivakumar and G.Jeeva, "A Review Paper on EDM Parameter of Composite material and Industrial Demand Material Machining" *Materials Today: Proceedings*, Volume **5**, pp5506-5513, 2018.

[4] M.A. Razak, A.M. Abdul-Rani, T.V.V.L.N. Rao, S.R. Pedapati and S. Kamal., "Electrical Discharge Machining on Biodegradable AZ31 Magnesium Alloy using Taguchi method" *4th International Conference on Process Engineering and Advanced Materials*, volume **148** , pp.916–922, 2016.

[5] F. Klocke, M. Schwade, A. Klink, D. Veselovac, A.Kopp.," Influence of Electro Discharge Machining of Biodegradable Magnesium on the Biocompatibility" *TheFirstCIRP Conference on Bio manufacturing*, volume **5**,pp.88-93, 2013.

[6] S. Vinoth Kumar, M. Pradeep Kumar.," Machining process parameter and surface integrity in conventional EDM and cryogenic EDM of Al–SiCpMMC". *Journal of Manufacturing Processes*, volume **2**, pp 216-220, 2015.

[7] A.M. Abdul- rani, M.ARazak, G. Littlefair, I.GibsonA.M.Nanimina., "Improving EDM process on AZ 31 Magnesium Alloy towards Sustainable Biodegradable Implant Manufacturing" *International conference on Sustainable Materials Processing*. Volume **7**, pp504–509, 2016.

[8] Sohil Parsana ,NishilRadadia a MohakSheth , NisargSheth , VimalSavsani , N. Eswara Prasad , T. Ramprabhu., "Machining parameter optimization for EDM machining of Mg–RE–Zn–Zr alloy using multi-objective Passing Vehicle Search algorithm" *Archives of civil and mechanical engineering*, volume **8**,pp799-817,2018.

[9] K. Ponappa, S. Aravindan, P. V. Rao , J. Ram Kumar , M. Gupta.," The effect of process parameters on machining of magnesium nano alumina composites through EDM" *International Journal of Advanced Manufacturing Technology*.Volume **46**, pp1035–1042, 2010.

[10] A. Pramanik, A. K. Basak , M. N. Islam , G. Littlefair.,"Electrical discharge machining of 6061 aluminium alloy" *Transaction of Nonferrous Metals Society of China*. Volume **25**, pp2866-2874, 2015.

[11] PujariSrinivasaRao, KoonaRamjii,Beela.,"Experimental Investigation and optimization of Wire EDM parameters for Surface Roughness, MRR and White layer in Machining of Aluminium Alloys"*International Journal of Advanced Manufacturing Technology*. Volume **5**, pp 2197-2206, 2014.

IOP Conf. Series: Materials Science and Engineering **764** (2020) 012053 doi:10.1088/1757-899X/764/1/012053

Experimental Investigation on Production of Jatropha Methyl Ester

Arun Kumar Rajamanickam[1], Uvaraja V C[2], Selvamuthukumaran D[2] and Shanmugasundaram A[3]

[1]Department of Mechanical Engineering, Sri Krishna College of Engineering and Technology, Coimbatore-641 008, India.
[2]Department of Mechanical Engineering, Bannari Amman Institute of Technology, Coimbatore-638 401, India.
[3]Department of Mechanical Engineering, Sri Shakthi Institute of Engineering and Technology, Coimbatore-641 062, India.

arunkumarr@skcet.ac.in

Abstract. At present transportation and agricultural machinery are heavily dependent on petroleum fuels. In current scenario few nations only produce the petroleum fuel that leads to variability in the cost and demand in supply of fuels. Due to this problems it is necessary to find an alternate source for the mineral oil. Based on many research work biogas, ethanol, methanol and vegetable oils are best alternate fuel for mineral oil. The vegetable oils are consider as an alternate to mineral oils since they are environmentally friendly and biodegradable. Compare to mineral oils vegetable oils has high viscosity, low friction coefficient, low volatile, high flash and fire point. In this work it shows that Jatropha oil Methyl Ester (JME) based biofuel is an alternative to fossil fuels. JME can be used as an alternate to fossil fuel and also due to its viscosity index and lubricity properties it has the potential to use as lubricant. The Ester is formed by transesterification process and the viscosity, flash and fire point tests are conducted to analyse the properties of JME over mineral oil. Further, the maximum yield of JME is obtained by varying the proportions of KOH (catalyst), methanol (alcohol) and stirring speed by L27 Orthogonal Array based on Taguchi method.

Keywords: jatropha oil, transesterification, diesel, lubricant

1. Introduction

Current situation indicates that the conventional mineral oils used in these days will not be used for longer period due to price and uncertainty in supply of fuels. It leads to focus the research on finding alternate fuel for mineral oil by taking various vegetable oil. As an initiation for this, the sunflower vegetable oil is being tested for its performance in two stroke engines and to attain a better result out of it.

Transesterification process is one the best method for conversion of the methyl ester from vegetable oil. In this research work Jatropha curcas oil is used to prepare the methyl ester. Among the non-edible oil source of India Jatropha curcas is one of the prime source, catalyst presence make the reaction of fatty acid to alcohol in transesterification process the chemical composition of jatropha are as follows: 6.2-6.5% of moisture, 18-20% of protein, 38-40% of fat, 17-20% of carbohydrates, 15.5-17% of fiber and 5-6% of ash. The presence of fatty acid develop the saponification reaction in vegetable oil while the process of transesterification.

Jatropha curcas seeds contains around 25-30 percentage of oil content. Jatropha oil contain 79% of unsaturated fatty acids and 21% of saturated fatty acids. In cosmetic industry soap and candles are manufactured from jatropha viscous oil, it's an alternate for the commercial fuels available in the market. It explore the great alternate for fossil fuel.

Development of the nation is depends on the growth of energy sector, but consumption of present conventional fuel may exhaust shortly due to the high rate of consumption. So the worldwide courtesy is moving towards the usage of bio degradable fuel. Renewable energy resources are used to prepare the biofuels. In this research work it was also found that the transesterification process is affected by the following parameters reaction temperature, oil temperature and purity of reactants. Comparing sodium hydroxide (NaOH) catalyst the use of potassium hydroxide (KOH) catalyst increase the ester yield during transesterification process.

Glycerine will be separated from vegetable oil to prepare the mono-alkyl ester and the ester is user to produce bio fuels and lubricants. The ester floats on top and the glycerine descends to the bottom and glycerine can drain off in this work. Reactants purity, temperature of reaction of acid, mixture of liquids, type of catalyst are some of the parameter which improves the esterification process.

1.1. Jatropha curcas

There are plenty of variety is available in jatropha among all of these jatropha curcas is the best oil to extract the methyl ester. In rural areas jatropha oil plays the vital role for cooking and lighting needs. Energy extracted in jatropha oil is equal to the energy extracted in diesel. Jatropha oil directly used in diesel engine by added to diesel fuel or it can be converted as a bio diesel fuel by transesterification process. While using a jatropha oil directly in diesel engine it create the technical issue in the engine that major problem needs to be overcome. Same time preparing bio diesel from jatropha curcas is quite expensive compare to the preparation of diesel fuel.

2. Production Process

2.1. Transesterification

Transesterification is the process of chemical reaction of oil with alcohol in the presence of catalyst. Hydrochloric acid is used as an alcohol and sodium hydroxide or potassium hydroxide is mainly used as a catalyst for the transesterification process. Ester and glycerin are the product produced by the transesterification process. Glycerine is co product produced during the transesterification process.

2.2. Separation

Purification process is done to separate the biodiesel from the glycerine after completing the transesterification process.

2.3 Investigational Procedure

2.3.1 Neutralization

In nature nearly 14 to 20 percentage of free fatty acid will presence in the vegetable oil, the fatty acid must remove from the vegetable oil before taking into the conversion process of vegetable oil to biodiesel. Fatty acid presence nearly 14% in jatropha curcas it unsuitable for biofuel production in industry. For neutralization of jatropha oil, the dehydrated oil is disturbed with 4 % of hydrochloric

IOP Conf. Series: Materials Science and Engineering **764** (2020) 012053 doi:10.1088/1757-899X/764/1/012053

acid solution about thirty minutes then 0.82g of potassium hydroxide (KOH) was added per 100 ml of jatropha oil to counterbalance the fatty acid.

$$RCOOH + NaOH \_ RCOONa + H_2O$$

Figure 1. Experimental setup

Figure 2. Neutralization

2.3.2 Ester Production

In this present work biodiesel is produced from jatropha curcas by transesterification process. Batch reactor is used to carry out the transesterification-ion reaction process. During the transesterification process, to eliminate the moisture content in the jatropha oil, 250 ml of oil is heated and stirred vigorously at 700C in round bottom flask. High purity methanol is used in the transesterification process having a density of 0.79 g/cm3. Based on weight percent ratio for 250ml oil 2.5 grams of KOH is liquefied in methanol, in separate vessel and stirred vigorously then it was added to round bottom flask while stirring the mixture continuously. The mixture was maintained in an atmospheric pressure at 700C for 60 minutes. At the end of the transesterification process the ester and glycerine mixture is allowed to settle under gravity for 24 hours in separating funnel. The product made during the process are jatropha methyl ester and glycerine. The upper layer consists of methyl ester and the bottom layer consists of glycerine, excess alcohol, unreacted oil and catalyst. The vaporisation of water and alcohol gives 82-86 % pure glycerine, which can be sold as crude glycerine. After removing glycerine from separating funnel unreacted alcohol will be removed from jatropha methyl ester by washing the jatropha methyl ester with hot distilled water. Then the ester was allowed to settle under gravity for 24 hours. The separated biodiesel is taken for characterization.

Figure 3. Separation of ester and glycerine

3. Factorial Design

Factorial design was framed to indicate the most influencing factors in jatropha crucas transesterification process. Three level and three factors were selected and L27 orthogonal array is used to produce the required number of samples. Reaction temperature was kept constant and reaction speed, catalyst and methanol are consider as varying parameter.

4. Ester Characterization

Transesterification process reduce the viscosity value from 57 to 4.7 centistokes, which is equivalent to viscosity of biodiesel as per ASTM standard. Transportation, handling and storage of biodiesel depends on the flash and fire points. Jatropha methyl ester flash and fire point was found to be 195°C and 210°C. Work output per unit volume is high in biofuel compared to diesel due to higher mass of fuel per unit volume in vegetable oil. After transesterification process flash point of jatropha oil will be reduced, it shows that it safe to handle and improvement in volatile characteristics. Usage of vegetable oil in diesel engine is a major issue due to its high viscosity. Fuel will be considered as low temperature performance of fuel based on its pour and cloud point criterion. For example pour point of diesel is –4°C and the cloud point for Diesel is 4°C which is very low compare to methyl ester so it can performs adequately even in cold climatic conditions. If the methyl ester has higher pour point it cannot be used in diesel engine during winter season. If pour point of biofuel is greater than the ambient temperature it will lose the flow characteristics and the presence of wax will affect the fuel supply and filter. Due to this injector will not inject the fuel during working cycle of engine. Fuels which are has above 65°C as flash point are considered as safe fuels.

5. Results and Discussion

During the transesterification process mixing of methoxide with KOH is much faster compare to mixing with NaOH. NaOH is cheaper and easy to get but it takes long time compare to KOH. KOH dissolves in the methanol more easily than NaOH and it will be ready to use in ten minutes. To avoid the interferes of water with the lye catalyst it is vital to remove the presence of water content in the oil. While storing the oil it may react with ecological elements and atmospheric oxygen may lead to reduce the quality of oil by increasing the free fatty acid. It is essential to neutralize the oil to avoid the formation of free fatty acid content in the oil. Based on factorial design the required number of samples were prepared and the following results are obtained.

IOP Conf. Series: Materials Science and Engineering **764** (2020) 012053 doi:10.1088/1757-899X/764/1/012053

Table 1. Experimental result.

Combination No	Combination			Ester Yield			
	Methanol (ml)	KOH (Grams)	Speed (rpm)	In Wt. (Grams)	In Wt. (%)	In (ml)	In ml (%)
1	90	0.871	1000	203.08	93.26	231	92.4
2	100	1.306	1000	166.47	76.45	187	74.8
3	90	1.088	1100	137.7	63.24	158	63.2
4	90	1.306	1200	118.37	54.36	137	54.8
5	100	0.871	1100	215.66	99.04	246	98.4
6	100	1.088	1200	210.54	96.69	239	95.6
7	110	0.871	1200	207.63	95.35	235	94.0
8	110	1.088	1000	197.89	90.88	214	85.6
9	110	1.306	1100	209.93	96.41	237	94.8
10	90	1.088	1000	153.10	70.31	174	69.6
11	90	1.088	1100	165.55	76.03	186	74.4
12	90	1.088	1200	162.08	74.43	181	72.4
13	100	0.871	1000	134.77	61.89	153	61.2
14	100	0.871	1200	166.77	76.58	188	75.2
15	100	1.088	1000	186.80	85.79	213	85.2
16	100	1.088	1100	179.17	82.28	202	80.8
17	100	1.306	1100	175.04	80.39	189	75.6
18	100	1.306	1200	172.22	79.09	177	70.8
19	90	1.306	1000	164.78	75.67	182	72.8
20	90	1.306	1100	172.20	79.08	195	78
21	90	1.306	1200	178.34	81.90	204	81.6
22	110	0.871	1000	180.26	82.78	210	84
23	110	0.871	1100	184.02	84.51	215	86
24	110	1.088	1100	181.33	83.27	211	84.4
25	110	1.088	1200	194.94	89.52	228	91.2
26	110	1.306	1000	187.6	86.15	213	85.2
27	110	1.306	1200	161.01	73.94	184	73.6

6. Conclusion

In the present research, conforms that jatropha oil may use to produce the biofuel and it can be used as an alternate to mineral fuel. The investigational outcome conforms that alkaline catalyzed transesterification is a hopeful area of research for the large scale production of biofuel. Influence of constant reaction temperature and different level of reaction speed, reactant ratio and catalyst concentration on the biodiesel yield were analyzed. The result indicates that 100ml of Methanol to oil, 0.871g KOH catalyst, and reaction temperature of 70°C and reaction time of 60 minutes are the best combination to produce the methyl ester from the jatropha oil.

References

[1] Antolın G, Tinaut F.V, Briceno Y, Castano V, Perez C, Ramırez A I 2002, "Optimisation of biodiesel production by sunflower oil transesterification", Bioresource Technology, Vol.83, 111–114.

[2] Antony Raja S, .Robinson smart D.S, and Lindon Robert Lee C 2011, "Biodiesel Production from Jatropha Oil and Its Characterization". Research Journal of Chemical Sciences, Vol.1 pp-81-87 April.

[3] Balamurugan K, Kanagasabapathy N and Mayilsamy K 2010, "Studies on soya bean oil based lubricant for diesel engines", Journal of Scientific & Industrial Research, Vol.69, pp.794- 797.

[4] Bang-Quan He, Shi-Jin Shuai, Jian-Xin Wang, Hong He 2003, "The Effect of Ethanol Blended Diesel Fuels on Emissions from A Diesel Engine" Atmospheric Environment Vol.37, pp 4965–4971.

[5] Devendra Vashist 2011, "A Comparative Study of Castor and Jatropha oil Source and its Methyl Ester Test on the Diesel Engine", International Journal of Engineering Science and Technology, Vol. 3 No.6, pp- 4765-4773.

[6] Ejilah I.R., Asere A.A., Adisa, A. B. and Ejila, A 2010, "The effect of diesel fuel-Jatropha curcas oil methyl ester blend on the performance of a variable speed compression ignition engine", AJAE 1(3):80-85.

[7] Foidl N, Foidl G, Sanchez M, Mittelbach M, Hackel S 1996, "Jatropha Curcas L. As a Source for the Production of Biofuel in Nicaragua", Bioresource technology 77-82.

[8] Hawash S,.El Diwani G, Abdel Kader E 2011, "Optimization of Biodiesel Production from Jatropha Oil By Heterogeneous Base Catalysed Transesterification" International Journal of Engineering Science and Technology Vol. 3 No. 6, pp-5242-5251.

[9] Nakpong P and Wootthikanokkhan S 2010 "Optimization of biodiesel production from Jatropha curcas L. oil via alkali-catalyzed methanolysis" Materials Science and engineering, pp-105-109.

[10] Niraj S. Topare, Shruti G. Chopade, Sunita j. Raut,V. C. Renge, Satish V. Khedkar and S. L. Bhagat 2011 "Biodiesel Production From Jatropha Curcas Oil", Int. J. Chem. Sci.: Vol.9(4), 1607-1612.

[11] Noor Hafizah Arbain and Jumat Salimon 2011 "The Effects of Various Acid Catalyst on the Esterification of Jatropha Curcas Oil based Trimethylolpropane Ester as Biolubricant Base Stock" E-Journal of Chemistry, Vol.8 (S1), S33-S40.

[12] Pramanik K 2003, "Properties and Use of Jatropha Curcas oil and Diesel Fuel Blends in Compression Ignition Engine". Published in Renewable Energy, Vol.28 Issue 2, pp-239- 248.

IOP Conf. Series: Materials Science and Engineering **764** (2020) 012053 doi:10.1088/1757-899X/764/1/012053

[13] Vijittra Chalatlon, Murari Mohon Roy, Animesh Dutta and Sivanappan Kumar 2011, "Jatropha oil Production and an Experimental Investigation of its use as an Alternative Fuel in a DI Diesel Engine". Journal of Petroleum Technology and Alternative Fuels Vol.2(5), pp. 76-85.

[14] Xin Deng, Zhen Fang, Yun-hu Liu, Chang-Liu Yu 2011, "Production of Biodiesel From Jatropha oil Catalyzed by Nanosized Solid Basic Catalyst" Energy 36, pp. 777-784.

[15] Arun Kumar Rajamanickam, Lokesh.M, Ramaswamy.K, Sudharson.G.K 2018, "A Review - Production of Methyl Ester from Vegetable Oil" International Journal of Mechanical Engineering and Technology, Vol.9, pp. 749 – 7 54.

IOP Conf. Series: Materials Science and Engineering **764** (2020) 012054 doi:10.1088/1757-899X/764/1/012054

Non-linear Stochastic Fractional Neutral-type Differential Systems with State Delay : A Controllability Interpretation

B.Sundaravadivoo *

Department of Mathematics, Alagappa University, Karaikudi-630 004, India.

Abstract

Aforementioned manuscript, stochastic neutral fractional systems with bounded operator having state delay is explored. We use a Grammian matrix defined by Laplace transform to get the solution representation for the considered model. Controllability results are obtained through the application of the Banach contraction principle. Finally, a numerical example shows the accuracy of the controllability parameters obtained.

Keywords. *Delays; controllability; implicit derivative; integro differential equations.*

1 Introduction

The design of stochastic models to explain the functions of intrinsic noise, uncertainty of natural phenomenon, extrinsic noise and variability in the system has been of rapid interest in recent years. Accordingly, the theory play a consistent elucidation of various perceived circumstances in the real world phenomena, one can refer for more details [1, 3, 4, 11, 12]. Concurrently, non-integer order differential equations has grow universally in the furthest three decades adequate to its capability to represent the complex occurrence systematically by reproducing the non-local relations in space and time. Especially, in many engineering and scientific disciplines fractional differential equation play a significant role to develop the mathematical modeling such as biological sciences, chemical sciences, physical sciences, medical sciences, industries etc.

Furthermore, in numerous medical and physical phenomena the involvement of delay features arises in the fractional differential equation. The aroused delay may be with the state-dependent one or in the non-constant delay. Basically, delay differential equations are solved by using numerical methods, asymptotic solutions and graphical tools.

On the other hand, controllability play a key role in the qualitative reaction of a dynamical system. The theory was started in the sixties of nineteenth century based on the mathematical description of the dynamical system. Generally, it can have the ability to control a dynamical system from the initial position to the desired final position with the assistance of some set of admissible controls. Hence the combination of fractional-order derivatives and as well as integrals in the theory of controllability leads to better results than the integer-order derivatives. In most of the dynamical system, the whole state of the dynamical system is not affected by the control, but it acts as a part of it. Furthermore, very often in real industrial processes it is possible to observe only a certain part of the complete state of the dynamical system. Therefore, it is very important to determine whether or not control of the complete state of the dynamical system is possible,see for references [2, 5, 6, 7, 8, 9].

The main contributions of this paper is exposed as follows:

- In the existing literature, there is no work that describes the controllability criteria of neutral stochastic fractional differential equations with delay.

*Corresponding author is B.Sundaravadivoo with e-mail: sundaravadivoon@gmail.com.

- Here controllability results were proved using algebraic approach which is very effective and easier to calculate the results.
- An example is provided to illustrate the existing theory in the available source of literature.

The contour of the manuscript is classified in this way. Section 2, provides preliminaries and solution representations. Section 3, describes the controllability theorems. Section 4, contributes an example to embellish the efficacy and utility of controllability. Eventually, epilogues are worn in Section 5.

2 Preliminaries

Notations: Let (Y, U) denotes real separable Hilbert spaces and L represents bounded linear operator such that $L : Y \to U$. Suppose that a filtered probability space $(\Omega, \mathcal{F}, \mathcal{F}_s,\ s \geq 0, P)$ contains probability measure P on Ω, G represents the expectation corresponding to the measure P. Consider the following;

(i) $X = \mathbb{L}_2(\Omega, \mathcal{F}_s, X)$, denotes the Hilbert space of all $\mathcal{F}_t$-measurable quadratically integrable random variables in terms of efficacy in X.

(ii) $\mathcal{H}_2 \subseteq PC(J, \mathbb{L}_2(\mathcal{F}, X))$ comprised of all $\mathcal{F}_t$-measurable procedures involving values in X and equipped by $\parallel \phi \parallel^2_{\mathcal{H}_2} = \sup_{s\in J} E \parallel \phi \parallel^2$.

(iii) $U_{ad} = \mathcal{L}_2^{\mathcal{F}}(J, U)$, represents the Hilbert space of all quadratically integrable and $\mathcal{F}_t$-measurable processes involving values in U

Definition 2.1 [1] Let $\alpha > 0$, $f(t) : (0, \infty) \to \mathbb{R}$ the Caputo fractional derivative is given by

$$({}^C D^{\alpha}_{0+} f)(t) = \frac{1}{\Gamma(n-\alpha)} \int_0^t (t-s)^{n-\alpha-1} f^{(n)}(s) ds$$

Definition 2.2 [1] The Mittag-Leffler function in two parameters is given as

$$E_{\alpha,\beta}(z) = \sum_{k=0}^{\infty} \frac{z^k}{\Gamma(k\alpha+\beta)}, \quad for\ \alpha, \beta > 0$$

Lemma 2.3 [16] Let $\mathbb{C}$ be complex plane, for any $\alpha > 0, \beta > 0$ and $\mathcal{A} \in \mathbb{C}^{n\times n}$ then

$$L[t^{\beta-1} E_{\alpha,\beta}(\mathcal{A}t^{\alpha})] = s^{\alpha-\beta}(s^{\alpha}I - \mathcal{A})^{-1}, \Re(s) > \parallel \mathcal{A} \parallel^{\frac{1}{\alpha}}$$

In general,

$$\begin{aligned} E_{\alpha,1}(\lambda z^{\alpha}) &= E_{\alpha}(\lambda z^{\alpha}) \\ {}^C D^{\alpha} E_{\alpha}(\lambda t^{\alpha}) &= \lambda E_{\alpha}(\lambda t^{\alpha}) \\ \text{and } L[E_{\alpha}(\pm a t^{\alpha})](s) &= \frac{s^{\alpha-1}}{s^{\alpha} \pm a} \end{aligned}$$

Assumption (A1): The operator $\mathcal{A} \in L(Y)$ such that $\parallel \mathcal{A} \parallel^2 < \frac{(2\alpha-1)(\Gamma(\alpha))^2}{T^{2\alpha}}$

Evaluate the linear neutral fractional delay differential structure as follows.

$$\begin{aligned} {}^C D^{\alpha}(y(s) - q(s, y(s))) &= Ny(s) + Ky(s-\tau) + Bu(s) + f(s) + \sigma(s, y(s-\delta))\frac{dw(s)}{ds}, \quad s \in J = [0, T] \\ y(s) &= \mu(s), s \in [-\delta, 0] \end{aligned} \tag{1}$$

The solution of system (1) is by employing (A1), we get

$$\begin{aligned} y(s) &= \mu - q(0,\mu(0)) + q(s,y(s)) + \int_0^s (s-t)^{\alpha-1} E_{\alpha,\alpha}(N(s-t)^{\alpha}) \\ &\quad [N\mu(0) - Nq(0,\mu(0)) + Nq(s,y(s)) + Ky(t-\delta) + Bu(t) + f(t)]dt \\ &\quad + \int_0^s (s-t)^{\alpha-1} E_{\alpha,\alpha}(N(s-t)^{\alpha})\sigma(t,y(t-\delta))dw(t) \end{aligned} \tag{2}$$

Definition 2.4 The set $x(s) = \{y(s), u_s\}$ is the complete state of equation(1).

Definition 2.5 Equation (1) is completely controllable on J if for every $y_1 \in X$ there exists $u \in U_{ad}$ such that $y(T) = y_1$. The operator $L_T : U_{ad} \to X$ is defined by

$$L_T u = \int_0^T (T-s)^{\alpha-1} E_{\alpha,\alpha}(N(T-s)^{\alpha})Bu(s)ds.$$

$$(L_T^* y) = (T-s)^{\alpha-1} B^* E_{\alpha,\alpha}(N^*(T-s)^{\alpha})G\{y|\mathcal{F}_s\}$$

Definition 2.6 $W_T : Y \to Y$ is defined by

$$W_T y = \int_0^T (T-t)^{2\alpha-2} E_{\alpha,\alpha}(N(T-t)^{\alpha})BB^* E_{\alpha,\alpha}(N^*(T-t)^{\alpha})G\{y|\mathcal{F}_s\}ds$$

where $*$ represents the adjoint operator and $W_(T)$ represents the controllability Grammian operator.

Lemma 2.7 The operator $W_T \in L_T L_T^* \in L(Y)$ is well defined and bounded for all $\frac{1}{2} < \alpha \leq 1$.

Proof. For $u \in U_{ad}$ and $s \in J$ and employing Holders inequality and (A1)we get

$$\begin{aligned} G \parallel L_s u \parallel_y^2 &= G \parallel \int_0^s (s-t)^{\alpha-1} E_{\alpha,\alpha}(N(s-t)^{\alpha})Bu(t)dt \parallel^2 \\ &= \sum_{k=0}^{\infty} G \parallel \int_0^s N^k \frac{(s-t)^{\alpha k+\alpha-1}}{\Gamma(\alpha k+\alpha)} Bu(t)dt \parallel^2 \\ &\leq sum_{k=0}^{\infty} G((\int_0^s \frac{(s-t)^{2\alpha k+2\alpha-2}}{\Gamma(\alpha k+\alpha)} dt)^{\frac{1}{2}} (\parallel N \parallel^{2k} \parallel u(t) \parallel^2 \parallel B \parallel^2)^{\frac{1}{2}})^2 \\ &\leq \parallel u \parallel_{U_{ad}}^2 T^{2\alpha-1} k \sum_{k=0}^{\infty} \frac{(\Gamma(\alpha))^{2k}}{(2\alpha k+2\alpha-1)(2\alpha-1)^k} \end{aligned}$$

where $k = \sup_{0 \leq s \leq T} \parallel B \parallel^2$. If $\alpha \in (\frac{1}{2}, 1]$ the last series converges. Hence L_s is bounded for any $s \in J$.

3 Controllability of Neutral Stochastic Fractional Differential Systems with Delay

Evaluate the nonlinear neutral stochastic fractional structure with delay,

$$\begin{aligned} {}^C D^{\alpha}(y(s) - q(s,y(s))) &= Ny(s) + Ky(s-\delta) + Bu(s) + f(s,y(s),y(s-\delta), \int_0^s p(s,t,y(t),y(t-\delta))dt)) \\ &\quad + \sigma(s,y(s),y(s-\delta), \int_0^s h(s,t,y(t),y(t-\delta))dt)\frac{dw(s)}{ds}, \quad s \in J = [0,T] \\ y(s) &= \mu(s), \quad s \in [-\delta, 0] \end{aligned} \tag{3}$$

IOP Conf. Series: Materials Science and Engineering **764** (2020) 012054 doi:10.1088/1757-899X/764/1/012054

where $N : Y \to Y$ and $K : Y \to Y$ are bounded linear operators, $B : U \to Y$ is a bounded linear operator, $q : J \times Y \to Y$ is continuous function and q is continuously differentiable, δ is a positive constant, $y \in \mathbb{R}^n$ is the state variable, $u \in \mathbb{R}^m$ is the control input, $\frac{1}{2} < \alpha \leq 1$ and the non linear function $f : J \times Y \times Y \times Y \to Y$, $\sigma : J \times Y \times Y \times Y \to L_2^0$, $p : J \times J \times Y \times Y \to Y$ and $h : J \times J \times Y \times Y \to Y$ are continuous. $W(s)$ is a K-valued Wiener process with positive symmetric trace class covariance operator.

Assumption (A2): Assume that there exists constants $M_i > 0, i = 1, 2, 3, 4$ such that

$$\begin{aligned}
\| f(t, y_1, y_1^{'}, x_1) - f(t, y_2, y_2^{'}, x_2) \|^2 &\leq M_1(\| y_1 - y_2 \|^2 + \| y_1^{'} - y_2^{'} \|^2 + \| x_1 - x_2 \|^2) \\
\| \sigma(t, y_1, y_1^{'}, x_1) - \sigma(t, y_2, y_2^{'}, x_2) \|^2 &\leq M_2(\| y_1 - y_2 \|^2 + \| y_1^{'} - y_2^{'} \|^2 + \| x_1 - x_2 \|^2) \\
\| [\int_0^s p(t, s, y_1, y_1^{'}) - \int_0^s p(t, s, y_2, y_2^{'})]ds \|^2 &\leq M_3(\| y_1 - y_2 \|^2 + \| y_1^{'} - y_2^{'} \|^2) \\
\| [\int_0^s h(t, s, y_1, y_1^{'}) - \int_0^s h(t, s, y_2, y_2^{'})]ds \|^2 &\leq M_4(\| y_1 - y_2 \|^2 + \| y_1^{'} - y_2^{'} \|^2) \\
\| g(s, y_1) - g(s, y_2) \| &\leq M_5 \| y_1 - y_2 \|^2
\end{aligned}$$

for all $y_1, y_2, x_1, x_2 \in Y$.

$$Let \ \ M_6 = \sup_{t \in J} \| \sigma(t, 0, 0, 0) \|, M_7 = \sup_{t \in J} \| \int_0^t r(t, s, 0, 0) \|,$$

$$M_8 = \sup_{t \in J} \int_0^t \| h(t, s, 0, 0) \|, M_9 = \sup_{t \in J} \| f(t, 0, 0, 0) \|, M_{10} = \sup_{t \in J} \| g(t, 0) \|$$

For each fixed $z \in \mathcal{H}_2$, equation(3) becomes

$$\begin{aligned}
{}^C D^\alpha(y(s) - q(s, y(s))) &= Ny(s) + Ky(s - \delta) + Bu(s) + f(s, z(s), z(s - \delta), \int_0^s p(s, t, z(t), z(t - \delta))dt)) \\
&\quad + \sigma(s, z(s), z(s - \delta), \int_0^s h(s, t, z(t), z(t - \delta))dt)\frac{dw(s)}{ds}, \quad s \in J = [0, T] \\
y(s) &= \mu(s), \quad s \in [-\delta, 0]
\end{aligned} \tag{4}$$

The solution of (4) is written by

$$\begin{aligned}
y(s) &= \mu(0) - q(0, \mu(0)) + q(s, y(s)) + \int_0^s (s - t)^{\alpha - 1} E_{\alpha,\alpha}(N(s - t)^\alpha)[N\mu(0) - Nq(0, \mu(0)) \\
&\quad + Nq(t, y(t)) + Ky(t - \delta) + Bu(t)]dt + \int_0^s (s - t)^{\alpha - 1} E_{\alpha,\alpha}(N(T - t)^\alpha) \\
&\quad \times f(t, z(t), z(t - \delta), \int_0^t p(t, r, z(r), z(r - \delta))dr)dt + \int_0^s (s - t)^{\alpha - 1} E_{\alpha,\alpha}(N(T - t)^\alpha) \\
&\quad \times \sigma(t, z(t), z(t - \delta), \int_0^t h(t, r, z(r), z(r - \delta))dr)dw(t)
\end{aligned} \tag{5}$$

$$Let \ \ N_1 = \sup_{0 \leq s \leq T} \| E_{\alpha,\alpha}(Ns^\alpha) \|^2 \ and \ \ N_2 = \sup_{0 \leq s \leq T} \| E_{\alpha,\alpha}(N(s - t)^\alpha) \|^2 .$$

Assumption (A3): Let $\rho_1 = \frac{8T^{2\alpha}}{2\alpha - 1} N_2(M_2 T^{-1} + M_2 M_4 + M_1 M_3 T + M_9)$, *where* $0 \leq \rho_1 < 1$.

IOP Conf. Series: Materials Science and Engineering **764** (2020) 012054 doi:10.1088/1757-899X/764/1/012054

Theorem 3.1 Assumptions $(A1)-(A3)$ are fulfilled and if equation(4)is controllable, then equation(3)is controllable.

Proof. Let $y_1 \in X$ be arbitrary. Define the operator ψ on $\mathcal{H}_2$ by

$$\begin{aligned}\psi y(s) &= \mu(0) - q(0,\mu(0)) + q(s,y(s)) + \int_0^s (s-t)^{\alpha-1}E_{\alpha,\alpha}(N(s-t)^\alpha)[N\mu(0) - Nq(0,\mu(0))\\ &\quad + Nq(t,y(t)) + Ky(t-\delta) + Bu(t)]dt + \int_0^s (s-t)^{\alpha-1}E_{\alpha,\alpha}(N(s-t)^\alpha)\\ &\quad \times f(t,z(t),z(t-\delta),\int_0^t p(t,r,z(r),z(r-\delta))dr)dt + \int_0^s (s-t)^{\alpha-1}E_{\alpha,\alpha}(N(s-t)^\alpha)\\ &\quad \times\sigma(t,z(t),z(t-\delta),\int_0^t h(t,r,z(r),z(r-\delta))dr)dw(t) \qquad (6)\end{aligned}$$

W_T is invertible(see [10]).
Let

$$\begin{aligned}u(s) &= (T-s)^{\alpha-1}B^*E_{\alpha,\alpha}(N^*(T-s)^\alpha)G\{W_T{}^{-1}(y_1 - \mu(0) - q(0,\mu(0)) + q(s,y(s))\\ &\quad + \int_0^T (T-t)^{\alpha-1}E_{\alpha,\alpha}(N(T-t)^\alpha)[N\mu(0) - Nq(0,\mu(0)) + Nq(t,y(t)) + Ky(t-\delta)]dt\\ &\quad + \int_0^T (T-t)^{\alpha-1}E_{\alpha,\alpha}(N(T-t)^\alpha)f(t,z(t),z(t-\delta),\int_0^t p(t,r,z(r),z(r-\delta))dr)dt\\ &\quad + \int_0^T (T-t)^{\alpha-1}E_{\alpha,\alpha}(N(T-t)^\alpha)\sigma(t,z(t),z(t-\delta),\int_0^t h(t,r,z(r),z(r-\delta))dr)dw(t))|\mathcal{F}_s\}\end{aligned}$$

clearly,$\psi(y(T)) = y_1$. To prove that ψ maps $\mathcal{H}_2$ into itself.

$$\begin{aligned}\sup_{s\in J} G\parallel u(s)\parallel^2 &= \sup_{s\in J}\parallel (T-s)^{\alpha-1}B^*E_{\alpha,\alpha}(N^*(T-s)^\alpha)G\{W_T{}^{-1}(y_1 - \mu(0) - q(0,\mu(0)) + q(s,y(s))\\ &\quad + \int_0^T (T-t)^{\alpha-1}E_{\alpha,\alpha}(N(T-t)^\alpha)[N\mu(0) - Nq(0,\mu(0)) + Nq(t,y(t))\\ &\quad + Ky(t-\delta)]dt + \int_0^T (T-t)^{\alpha-1}E_{\alpha,\alpha}(N(T-t)^\alpha)f(t,z(t),z(t-\delta),\\ &\quad \times \int_0^t p(t,r,z(r),z(r-\delta))dr)dt + \int_0^T (T-t)^{\alpha-1}E_{\alpha,\alpha}(N(T-t)^\alpha)\\ &\quad \times\sigma(t,z(t),z(t-\delta),\int_0^t h(t,r,z(r),z(r-\delta))dr)dw(t))\}\parallel^2\\ &\le \parallel L_T^*\parallel^2\parallel W_T{}^{-1}\parallel^2 [G\parallel y_1\parallel^2 + G\parallel\phi(0)\parallel^2 + G\parallel q(0,\mu(0))\parallel^2 + G\parallel q(s,y(s))\parallel^2]\\ &\quad + N_2[G\parallel N\mu(0)\parallel^2 + G\parallel Nq(0,\mu(0))\parallel^2 + G\parallel Nq(t,y(t))\parallel^2 + G\parallel Ky(t-\delta)\parallel^2]\\ &\quad + N_2\int_0^T (T-t)^{\alpha-1}dt\int_0^T (T-t)^{\alpha-1}E\parallel f(t,z(t),z(t-\delta),\\ &\quad \times\int_0^t p(t,r,z(r),z(r-\delta))dr\parallel^2 dt + N_2\int_0^T (T-t)^{\alpha-1}dt\int_0^T (T-t)^{\alpha-1}\\ &\quad \times G\parallel\sigma(t,z(t),z(t-\delta),\int_0^t h(t,r,z(r),z(r-\delta))dr\parallel^2 ds\end{aligned}$$

$$\begin{aligned}\sup_{s\in J} G\parallel u(s)\parallel^2 &\le \parallel L_T^*\parallel^2\parallel W_T{}^{-1}\parallel^2 [G\parallel y_1\parallel^2 + G\parallel\mu(0)\parallel^2 + G\parallel q(0,\mu(0))\parallel^2\\ &\quad + M_9\parallel y(t)\parallel^2 + M_{10}] + N_1 M\frac{T^{2\alpha}}{2\alpha-1} + N_1 M'\frac{T^{2\alpha-1}}{2\alpha-1} = K_1 < \infty.\end{aligned}$$

where

$$M = (M_1M_3 + M_1)\sup_{s\in J} G \parallel y(t) \parallel^2 + (M_1M_3 + M_1)\sup_{s\in J} G \parallel y(s-\delta) \parallel^2 + M_1M_9 + M_1M_7 < \infty.$$
$$and M' = (M_2M_4 + M_2)\sup_{s\in J} G \parallel y(s) \parallel^2 + (M_2M_4 + M_2)\sup_{s\in J} G \parallel y(s-\delta) \parallel^2 + M_2M_8 + M_1M_6 < \infty.$$

using the assumptions ,we get

$$\sup_{s\in J} G \parallel \mu y(s) \parallel^2 \le 4G \parallel \mu(0) \parallel^2 + 4G \parallel y_1 \parallel^2 + 4M_{10} + M_9 sup_{s\in J} G \parallel y(s) \parallel^2 + 4N_2G \parallel Ky(t-\delta) \parallel^2 + 4N_2K_1 \parallel B \parallel^2 \frac{T^{2\alpha}}{2\alpha-1} + 4N_2M\frac{T^{2\alpha}}{2\alpha-1} + 4N_2M'\frac{T^{2\alpha}}{2\alpha-1}$$

Hence ψ maps $\mathcal{H}_2$ into itself. Let $y_1, y_2 \in \mathcal{H}_2$, we have

$$\begin{aligned}
\sup_{s\in J} G \parallel \psi y_1(s) - \psi y_2(s) \parallel^2 &= \sup_{s\in J} G \parallel \int_0^s (s-t)^{\alpha-1} E_{\alpha,\alpha}(n(s-t)^\alpha) B L_T^* W_T{}^{-1} \{ \int_0^T (T-\theta)^{\alpha-1} \\
&\times E_{\alpha,\alpha}(N(T-\theta)^\alpha)[f(\theta, y_1(\theta), y_1(\theta-\delta), \int_0^\theta p(\theta, r, y_1(r), y_1(r-\delta))dr) \\
&- f(\theta, y_2(\theta), y_2(\theta-\delta), \int_0^\theta p(\theta, r, y_2(r), y_2(r-\delta))dr)]d\theta \\
&+ \int_0^T (T-\theta)^{\alpha-1} E_{\alpha,\alpha}(N(T-\theta)^\alpha)[\sigma(\theta, y_1(\theta), y_1(\theta-\delta), \\
&\times \int_0^\theta h(\theta, r, y_1(r), y_1(r-\delta))dr) - \sigma(\theta, y_2(\theta), y_2(\theta-\delta), \\
&\times \int_0^\theta h(\theta, r, y_2(r), y_2(r-\delta))dr)] dW\theta\} dt + \int_0^s (s-t)^{\alpha-1} \\
&\times E_{\alpha,\alpha}(N(s-t)^\alpha)[ky_1(t-\delta) - ky_2(t-\delta) + g(t, y_1(t)) \\
&- g(t, y_2(t))]dt + \int_0^s (s-t)^{\alpha-1} E_{\alpha,\alpha}(N(s-t)^\alpha)[f(t, y_1(t), y_1(t-\delta), \\
&\times \int_0^t p(t, r, y_1(r), y_1(r-\delta))dr)]dt - \int_0^s (s-t)^{\alpha-1} E_{\alpha,\alpha}(N(s-t)^\alpha) \\
&\times f(t, y_2(t), y_2(t-\delta), \int_0^t p(t, r, y_2(r), y_2(r-\delta))dr)]dt \\
&+ \int_0^s (s-t)^{\alpha-1} E_{\alpha,\alpha}(N(s-t)^\alpha)[\sigma(t, y_1(t), y_1(t-\delta), \\
&\times \int_0^t h(t, r, y_1(r), y_1(r-\delta))dr) - \sigma(t, y_2(t), y_2(t-\delta), \\
&\times \int_0^t h(t, r, y_2(r), y_2(r-\delta))dr)]dW(t) \parallel^2 \\
&\le \frac{8T^{2\alpha}}{2\alpha-1} N_2(M_2T^{-1} + M_2M_4 + M_1M_3T + M_9)[\sup_{s\in J} G \parallel y_1(s) - y_2(s) \parallel^2 \\
&\quad + \sup_{s\in J} G \parallel y_1(s-\delta) - y_2(s-\delta) \parallel^2] \\
&\le \rho_1 \parallel y_1 - y_2 \parallel^2_{\mathcal{H}_2}
\end{aligned}$$

By (A3) ψ is a contraction mapping. So,there exists a unique fixed point $y \in \mathcal{H}_2$ for ψ. Any fixed point of ψ satisfies $y(T) = y_1$ for arbitrary $y_1 \in Y$. Hence equation (3) is controllable on J.

4 Controllability of Neutral Stochastic Fractional Systems with delay Implicit Derivative

Consider the following non-linear stochastic fractional system with delay,

$$\begin{aligned} {}^C D^{\alpha}(y(s) - q(s, y(s))) &= Ny(s) + Ky(s-\delta) + Bu(s) + f(s, y(s), y(s-\delta), {}^C D^{\alpha} y(s), {}^C D^{\alpha} y(s-\delta)) \\ &\quad + \sigma(s, y(s), y(s-\delta))\frac{dw(s)}{ds}, \quad s \in J = [0, T] \\ y(s) &= \mu(s), \quad s \in [-\delta, 0] \end{aligned} \tag{7}$$

where $N : Y \to Y$ and $K : Y \to Y$ are bounded linear operators, $B : U \to Y$ is a bounded linear operator, $q : J \times Y \to Y$ is continuous function and q is continuously differentiable, $u \in \mathbb{R}^m$ is the control input, $\frac{1}{2} < \alpha \leq 1$ and the non linear function $f : J \times Y \times Y \times Y \times Y \to Y, \sigma : J \times Y \times Y \to L_2^0$, are continuous.

Assumption (A4): Assume that there exists constants $R_i > 0, i = 1, 2$ such that

$$\begin{aligned} \| f(s, y_1, y_2, y_3, y_4) - f(s, z_1, z_2, z_3, z_4) \|^2 &\leq R_1 \sum_{i=1}^{4} (\| y_i - z_i \|^2) \\ \| \sigma(s, y_1, y_2) - \sigma(s, z_1, z_2) \|^2 &\leq R_2 \sum_{i=1}^{2} (\| y_i - z_i \|^2) \end{aligned}$$

for any $s \in J$ and $y_i \in Y$, $z_i \in Y (i = 1, 2, 3, 4)$.

$$Let \quad R_3 = \sup_{s \in J} \| f(s, 0, 0, 0) \|, \quad R_4 = \sup_{s \in J} \| \sigma(s, 0) \| .$$

For each fixed $z \in \mathcal{H}_2^{\alpha}$, consider the corresponding linear system of (7) as

$$\begin{aligned} {}^C D^{\alpha}(y(s) - q(s, y(s))) &= Ny(s) + Ky(s-\delta) + Bu(s) + f(s, z(s), z(s-\delta), {}^C D^{\alpha} z(s), {}^C D^{\alpha} z(s-\delta)) \\ &\quad + \sigma(s, z(s), z(s-\delta))\frac{dw(s)}{ds}, s \in J = [0, T] \\ y(s) &= \mu(s), s \in [-\delta, 0] \end{aligned} \tag{8}$$

The solution of (8) is written by

$$\begin{aligned} y(s) &= \mu(0) - q(0, \mu(0)) + q(s, y(s)) + \int_0^s (s-t)^{\alpha-1} E_{\alpha,\alpha}(N(s-t)^{\alpha})[N\mu(0) - Nq(0, \mu(0)) \\ &\quad + Nq(t, y(t)) + Ky(t-\delta) + Bu(t)] + \int_0^s (s-t)^{\alpha-1} E_{\alpha,\alpha}(N(s-t)^{\alpha}) \\ &\quad \times f(t, z(t), z(t-\delta), {}^C D^{\alpha} z(s), {}^C D^{\alpha} z(s-\delta))dt \\ &\quad + \int_0^s (s-t)^{\alpha-1} E_{\alpha,\alpha}(N(s-t)^{\alpha})\sigma(t, z(t), z(t-\delta))dw(t) \end{aligned} \tag{9}$$

Assumption (A5): Let $\rho_2 = max\{\frac{8T^{2\alpha}}{2\alpha-1} N_2(R_2 T^{-1} + R_1), \frac{8T^{2\alpha}}{2\alpha-1} N_2 R_1\} where \quad 0 \leq \rho < 1.$

Theorem 4.1 Assumptions $(A1) - (A3)$ are fulfilled and if equation (7)is controllable. Then equation(8)is controllable.

IOP Conf. Series: Materials Science and Engineering **764** (2020) 012054 doi:10.1088/1757-899X/764/1/012054

Proof. Let $y_1 \in X$ be arbitrary. Define the operator ψ on $\mathcal{H}_2$ by

$$\begin{aligned}\psi y(s) &= \mu(0) - q(0,\mu(0)) + q(s,y(s)) + \int_0^s (s-t)^{\alpha-1}E_{\alpha,\alpha}(N(s-t)^\alpha)[N\mu(0) - Nq(0,\mu(0)) \\ &\quad +Nq(t,y(t)) + Ky(t-\delta) + Bu(t)]dt + \int_0^s (s-t)^{\alpha-1}E_{\alpha,\alpha}(N(s-t)^\alpha) \\ &\quad \times f(t,z(t),z(t-\delta),{}^C D^\alpha z(s),{}^C D^\alpha z(s-\delta))dt \\ &\quad + \int_0^s (s-t)^{\alpha-1}E_{\alpha,\alpha}(N(s-t)^\alpha)\sigma(t,z(t),z(t-\delta))dw(t) \qquad (10)\end{aligned}$$

So W_T is invertible (see[10]).
Define

$$\begin{aligned}u(s) &= (T-s)^{\alpha-1}B^*E_{\alpha,\alpha}(N^*(T-s)^\alpha)G\{(W_T{}^{-1}(y_1 - \mu(0) - q(0,\mu(0)) + q(s,y(s)) \\ &\quad + \int_0^T (T-t)^{\alpha-1}E_{\alpha,\alpha}(N(T-t)^\alpha)[N\mu(0) - Nq(0,\mu(0)) + Nq(t,y(t)) + Ky(t-\delta)]dt \\ &\quad + \int_0^T (T-t)^{\alpha-1}E_{\alpha,\alpha}(N(T-t)^\alpha)f(t,z(t),z(t-\delta),{}^C D^\alpha z(s),{}^C D^\alpha z(s-\delta))dt \\ &\quad + \int_0^T (T-t)^{\alpha-1}E_{\alpha,\alpha}(N(T-t)^\alpha)\sigma(t,z(t),z(t-\delta))dw(t))|\mathcal{F}_s\}\end{aligned}$$

clearly,$\psi(y(T)) = y_1$. To prove that ψ maps $\mathcal{H}_2$ into itself.

$$\begin{aligned}\sup_{s\in J} G \parallel u(s) \parallel^2 &= \sup_{s\in J} \parallel (T-s)^{\alpha-1}B^*E_{\alpha,\alpha}(n^*(T-s)^\alpha)G\{W_T{}^{-1}(y_1 - \mu(0) - q(0,\mu(0)) + q(s,y(s)) \\ &\quad + \int_0^T (T-t)^{\alpha-1}E_{\alpha,\alpha}(N(T-t)^\alpha)[N\mu(0) - Nq(0,\mu(0)) + Nq(t,y(t)) \\ &\quad +Ky(t-\delta)]dt + \int_0^T (T-t)^{\alpha-1}E_{\alpha,\alpha}(N(T-t)^\alpha) \\ &\quad \times f(t,z(t),z(t-\delta),{}^C D^\alpha z(s),{}^C D^\alpha z(s-\delta))dt \\ &\quad + \int_0^T (T-t)^{\alpha-1}E_{\alpha,\alpha}(N(T-t)^\alpha)\sigma(t,z(t),z(t-\delta))dw(t))\} \parallel^2 \\ &\le \parallel L_T^* \parallel^2 \parallel W_T{}^{-1} \parallel^2 [G \parallel y_1 \parallel^2 + G \parallel \mu(0) \parallel^2 + G \parallel q(0,\mu(0)) \parallel^2 + G \parallel q(s,y(s)) \parallel^2] \\ &\quad +N_2[G \parallel N\mu(0) \parallel^2 + G \parallel Nq(0,\mu(0)) \parallel^2 + G \parallel Nq(t,y(t)) \parallel^2 \\ &\quad +G \parallel Ky(t-\delta) \parallel^2] + N_2 \int_0^T (T-t)^{\alpha-1}dt \int_0^T (T-t)^{\alpha-1} \\ &\quad \times G \parallel f(t,z(t),z(t-\delta),{}^C D^\alpha z(s),{}^C D^\alpha z(s-\delta) \parallel^2 dt \\ &\quad +N_2 \int_0^T (T-t)^{\alpha-1}dt \int_0^T (T-t)^{\alpha-1}E \parallel \sigma(t,z(t),z(t-\delta) \parallel^2 dt \\ \sup_{s\in J} G \parallel u(s) \parallel^2 &\le \parallel L_T^* \parallel^2 \parallel W_T{}^{-1} \parallel^2 [G \parallel y_1 \parallel^2 + G \parallel \mu(0) \parallel^2 + G \parallel q(0,\mu(0)) \parallel^2 \\ &\quad +M_9 \parallel y(t) \parallel^2 + M_{10}] + N_1 M \frac{T^{2\alpha}}{2\alpha - 1} + N_1 M' \frac{T^{2\alpha-1}}{2\alpha-1} = K_2 < \infty.\end{aligned}$$

where

$$\begin{aligned}M &= R_1 \sup_{s\in J} C \parallel y(s) \parallel^2 + G \parallel y(s-\delta) \parallel^2 + G \parallel {}^C D^\alpha z(s) \parallel^2 + G \parallel {}^C D^\alpha z(s-\delta) \parallel^2 + R_3 < \infty. \\ \textit{and } M' &= R_2 \sup_{s\in J} E \parallel y(s) \parallel^2 + R_4 < \infty.\end{aligned}$$

IOP Conf. Series: Materials Science and Engineering **764** (2020) 012054 doi:10.1088/1757-899X/764/1/012054

using the assumptions ,we get

$$\begin{aligned}\sup_{s\in J} G \parallel \phi y(s) \parallel^2 &\leq 4G \parallel \mu(0) \parallel^2 + 4G \parallel y_1 \parallel^2 + 4M_{10} + M_9 sup_{s\in J} G \parallel x(s) \parallel^2 + 4N_2 G \parallel Ky(t-\delta) \parallel^2 \\ &+4N_2K_2 \parallel B \parallel^2 \frac{T^{2\alpha}}{2\alpha-1} + 4N_2 M \frac{T^{2\alpha}}{2\alpha-1} + 4N_2 M' \frac{T^{2\alpha}}{2\alpha-1}\end{aligned}$$

Hence ψ maps $\mathcal{H}_2$ into itself. Let $y_1, y_2 \in \mathcal{H}_2$, we have

$$\begin{aligned}\sup_{s\in J} G \parallel \psi y_1(s) - \psi y_2(s) \parallel^2 &= \sup_{s\in J} G \parallel \int_0^s (s-t)^{\alpha-1} E_{\alpha,\alpha}(N(s-t)^\alpha) BL_T^* W_T{}^{-1} \{ \int_0^T (T-\theta)^{\alpha-1} \\ &\times E_{\alpha,\alpha}(N(T-\theta)^\alpha)[f(\theta, y_1(\theta), y_1(\theta-\tau), {}^C D^\alpha y_1(\theta), {}^C D^\alpha y_1(\theta-\delta)) \\ &-f(\theta, y_2(\theta), y_2(\theta-\delta), {}^C D^\alpha y_2(\theta), {}^C D^\alpha y_2(\theta-\delta))]d\theta \\ &+ \int_0^T (T-\theta)^{\alpha-1} E_{\alpha,\alpha}(N(T-\theta)^\alpha)[\sigma(\theta, y_1(\theta), y_1(\theta-\delta), \\ &-\sigma(\theta, y_2(\theta), y_2(\theta-\delta))]dW\theta\}dt + \int_0^s (s-t)^{\alpha-1} E_{\alpha,\alpha}(N(s-t)^\alpha) \\ &\times [ky_1(t-\delta) - ky_2(t-\delta) + q(t, y_1(t)) - g(t, y_2(t))]dt \\ &+ \int_0^s (s-t)^{\alpha-1} E_{\alpha,\alpha}(N(s-t)^\alpha)[f(t, y_1(t), y_1(t-\delta), \\ &\times {}^C D^\alpha y_1(t), {}^C D^\alpha y_1(t-\delta))dt - \int_0^s (s-t)^{\alpha-1} E_{\alpha,\alpha}(N(s-t)^\alpha) \\ &\times f(t, y_2(t), y_2(t-\delta)]dt, {}^C D^\alpha y_1(t), {}^C D^\alpha y_1(t-\delta))dt \\ &+ \int_0^s (s-t)^{\alpha-1} E_{\alpha,\alpha}(N(s-t)^\alpha)[\sigma(t, y_1(t), y_1(t-\delta))dr) \\ &-\sigma(t, y_2(t), y_2(t-\delta)]dW(t) \parallel^2 \\ &\leq \frac{8T^{2\alpha}}{2\alpha-1} N_2(R_2T^{-1} + R_1)[\sup_{s\in J} G \parallel y_1(s) - y_2(s) \parallel^2 \\ &+ \sup_{s\in J} G \parallel y_1(s-\delta) - y_2(s-\delta) \parallel^2] \\ &+ \frac{8T^{2\alpha}}{2\alpha-1} N_2 R_1 [\sup_{s\in J} G \parallel^C D^\alpha y_1(s) - ^C D^\alpha y_2(s) \parallel^2 \\ &+ \sup_{s\in J} G \parallel^C D^\alpha y_1(s-\delta) - ^C D^\alpha y_2(s-\delta) \parallel^2] \\ &\leq \rho_2 \parallel y_1 - y_2 \parallel^2_{\mathcal{H}_2}\end{aligned}$$

By (A4)ψ is a contraction mapping. So,there exists a unique fixed point $y \in \mathcal{H}_2$ for ψ. Any fixed point of ψ satisfies $y(T) = y_1$ for arbitrary $y_1 \in Y$. Hence the system (7)is controllable on J.

5 Example

Example 5.1 Consider the nonlinear stochastic neutral fractional system for $s \in [0, 1]$

$$\begin{aligned}{}^C D^{0.6}\Big(\Big(y(s) - \frac{e^{-s}|x(s)|}{(9+e^s)(1+|x(s)|)}\Big) &= \begin{pmatrix} 0.3 & 1.5 \\ 1.8 & -0.9 \end{pmatrix} y(s) + 1.5y(s-\frac{1}{3}) + \begin{pmatrix} In(cosh(y_1)) \\ \frac{tan^{-1}(y_2}{)} 30s \end{pmatrix} \frac{dw(s)}{ds} \\ y(s) &= 1, \ -1 \leq s \leq 0. \qquad (11)\end{aligned}$$

for $s \in J$ and $0 < \alpha < 1$. In matrix form

$$A = \begin{pmatrix} 0.3 & 1.5 \\ 1.8 & -0.9 \end{pmatrix},$$

IOP Conf. Series: Materials Science and Engineering **764** (2020) 012054 doi:10.1088/1757-899X/764/1/012054

Here

$$y(s) = \begin{pmatrix} y_1(s) \\ y_2(s) \end{pmatrix}, \sigma(s, y(s)) = \begin{pmatrix} In(cosh(y_1)) \\ \frac{tan^{-1}(y_2)}{30s} \end{pmatrix}, and \ \alpha = 0.6.$$

Our aim is to steer the solution from the initial point $x(0)$ to the final point $x(1)$. To show the nonlinear system (11)is controllable it is enough to check the hypotheses of Theorem (4.1) are satisfied.
The operator

$$W_T = \begin{pmatrix} 29.886 & 34.8387 \\ 34.8387 & 40.7205 \end{pmatrix}. \tag{12}$$

Now consider

$$\begin{aligned} < W_T y, y > &= 29.886 y_1^2 + 69.6774 y_1 y_2 + 40.7205 y_2^2 \\ &\geq \gamma(y_1^2 + y_2^2) \end{aligned}$$

where $0 < \gamma \leq 0.0009$. $\sigma(s, y(s))$ is Lipschitz continuous with $M_1 = \frac{1}{300}$. The value of $\rho = 0.7859 < 1$. Hence the system (11)is controllable.

6 Conclusion

This paper executes the idea of controllability criteria of neutral type stochastic fractional differential systems with delays. By using the controllability Grammian matrix and the Mittag-Leffler matrix function the solution representation were obtained and the controllability results for the addressed system is proved using Banach Contraction principle. Additionally a numerical example is provided to justify the obtained theoretical results.

References

[1] A.A.Kilbas, H.M.Srivastava, J.J.Trujillo, Theory and Applications of Fractional Differential Equations, Elsevier, *Amsterdam* 2006.

[2] S.A.Ammour, S.Djennoune, M.Bettayeb, A slidingmode control for linear fractional systems with input and state delays, *Communications in Nonlinear Science and Numerical Simulation* 14 (5) (2009) 2310-2318.

[3] B.Bonilla, M.Rivero, L.Rodriguez-Germa, J.J.Trujillo, Fractional differential equations as alternative models to nonlinear differential equations, *Applied Mathematics and Computation* 187 (2007) 79-88.

[4] K.Balachandran, S.Divya, L.Rodriguez-Germa, J.J.Trujillo, Relative controllability of nonlinear neutral fractional integro-differential systems with distributed delays in control, *Mathematical methods in the Applied Sciences* November 2014.

[5] I.Gyri, J.Wu, A neutral equation arising from compartmental systems with pipes, *Journal of Dynamics and Differential Equations* 3 (2) (1991) 289-311.

[6] M.Kamrani, Numerical Solution of Stochastic Fractional Differential Equations, *Numerical Algorithms* 68 (2015) 81-93.

[7] J.Klamka, Constrained Controllability of Semilinear Systems, *Nonlinear Analysis* 47 (2001) 2939-2949.

[8] J.Klamka, Controllability of Dynamical Systems: *A Survey of Bulletin of the Polish Academy of Sciences Technical Sciences* 61 (2013) 221-229.

[9] E.Kreyszig, Introductory Functional Analysis with Applications, *John Wiley and Sons Inc*, New York (1978).

[10] J.Klamka, Stochasstic Controllability and minimum energy control of systems with multiple delays in control, *Applied Mathematics and Computation* 206 (2008) 704-715.

[11] V.Lakshmikantham, A.S.Vatsala, Basic theory of fractional differential equations, *Nonlinear Analysis* 69 (2008) 2677-2682.

[12] R.Mabel Lizzy, K.Balachandran, M.Suvinthra, Controllability of nonlinear stochastic fractional systems with distributed delays in control, *Journal of Control and Decision*, 4 (3) (2017) 153-167.

[13] N.I.Mahmudov, Controllability of Linear Stochastic Systems in Hilbert Spaces, *Journal of Mathematical Analysis and Applications* 259 (2001) 64-82.

[14] N.I.Mahmudov, Controllability of Semilinear Stochastic Systems in Hilbert Spaces, *Journal of Mathematical Analysis and Applications* 288 (2003) 197-211.

[15] K.S.Miller, B.Ross, An Introduction to the Fractional Calculus and Fractional Differential Equations, *Wiley, New York*, 1993.

[16] I.Podlubny, Fractinal Differential Equations of Mathematics in Science and Engineering, Technical University of Kosice, Kosice, *Slovak Rebublic* (198) 1999.

[17] R.Sakthivel, Y.Ren, Approximate controllability of fractional differential equations with state-dependent delay, *Results in Mathematics* 63 (2013) 949-963.

[18] R.Sakthivel, S.Suganya, S.M.Anthoni, Approximate controllability of fractional stochastic evolution equations, *Computers and Mathematics with Applications* 63 (2012) 660668.

[19] X.Zhang, C.Zhu, C.Yuan, Approximate controllability of impulsive fractional stochastic differential equations with state-dependent delay, *Advances in Difference Equations* 2015 (91) (2015) DOI 10.1186/s13662-015-0412-z.

IOP Conf. Series: Materials Science and Engineering **764** (2020) 012055 doi:10.1088/1757-899X/764/1/012055

A Review on 3D Printed Sensors

S K Dhinesh[1] and K L Senthil Kumar[2]

[1&2] Department of Mechatronics, Bannari Amman Institute of Technology, Sathyamangalam, Erode

Abstract

Industry 4.0 is a revolution that has greater potential to connect machines to monitor and control them through data acquisition and analysis. It helps the industries in getting quality products at economic cost with flexibility and processability incorporated in it. 3D printing is a subset of Industry 4.0, which can support the digital manufacturing and sensors are devices, which can measure the physical parameters from the environment and play key role in implementing Industry 4.0. Thus, this paper aims to review the recent works carried out in 3D printed sensors. Here an overview of sensors and their fabrication methods practiced using 3D printing technology for various applications are discussed. From the review, it is found that there is major scope for fabricating the complete sensor with 3D printing.

Keywords: 3D printing, pressure sensor, force sensor, sensor materials and sensor fabrication

1. Introduction

In the digitization era, industry 4.0 is going to play a major role including Industrial Internet of things (IIOT). The current industrial revolution will uplift the automation industries especially the process of manufacturing by facilitating mass production with atomization and flexibility as given in figure 1. The major challenge faced by industry 4.0 includes security, capital, employment and privacy. This indicates in near future machines are provided with adequate intelligence to function itself to work with human beings. More effectively, to produce products of customized requirements.

Industry 4.0 concepts range from factory automation to agile production lines, smart manufacturing to retail inventory control and supply chain. Recently premier automobile industry, in its assembly section, tried interconnecting of systems with digital technologies using IOT based sensors and had revealed significant improvement in its manufacturing efficiency. Sensor is a device used to convert physical parameter into electrical signal that could help to monitor real world data, which can further used for failure prediction. With the development of micro-machinery and advances in micro-controller platforms, elaborate sensors are widely used in production, avionics, robotics and biomedical applications. Growth rate is estimated that is likely to impact sensor market growth to about 21.6 billion USD by next five years. Fabrication cost of sensors is the major challenge to which 3D printing technology has provided the flexible solution. Remarkable amount of work has been carried out on sensors fabricated using 3D printing machines for various applications.

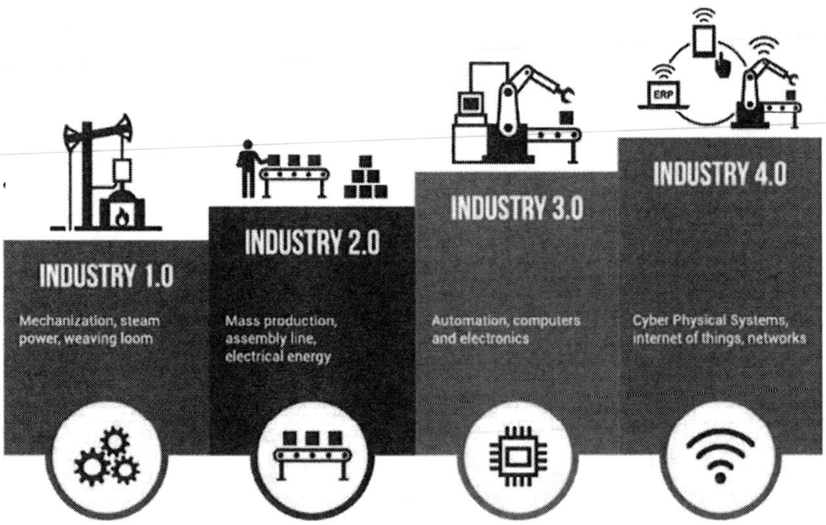

Figure 1. Growth of Industry 4.0

3D printing in another upcoming technology that builds the product layer by layer upon reading CAD file of the components. Variants such as Fused Deposition Modelling (FDM), Laminated Object Manufacturing (LOM) and Stereolithography (SLA) are available. The vital part of industry 4.0 is taken by 3D printing technologies as it could be easily integrated with other systems. The major factors of importance linked to 3D printing machines are quality, safety, environmental impact and printing speed. In this review, an overview of recent 3D-printed sensors is discussed with the pros and cons of various fabrication processes and their application in different field.

2. 3D Printed sensors

The recent development in the additive manufacturing helps in manufacturing different types of sensors for different applications. The researchers are working to fabricate a total 3D printed sensor in the near future because of its low manufacturing cost, easy availability and ease of customization for different applications. As the traditional sensor fabrication requires huge cost and time where the customization is very challenging, the era of 3D printing technology has made it feasible. The 3D printed sensors are printed using various 3D printing technologies such as FDM, lamination and material jetting.

3. Force Sensor

Force sensors find its applications in robotics, automobile, manufacturing and biomedical field. A new approach was done by Kesner et al (2011) in which the author has fabricated a 3D printed micro force sensor. The outer body of the sensor is 3D printed and the body has the internal space for the sensor components like transducer and cables. The fabricated sensor has many limitations such as overloading flexure deformation, thermal sensitivity and accuracy. In the same way, Butzer et al (2015) fabricated an additive manufacturing-based force sensor in which the transduction principle is based on fiber optic displacement with good accuracy, reproducibility and temporal stability.

The elastic probe, which is normally, fabricated using injection molding or milling is replaced using 3D printing. Which the author claims that the fabrication process is fast and economic despite of its certain limitations. Another 3D printed micro force sensor was produced (Qu et al (2016)), in which a load cell is fabricated using FDM and a cantilever structure using stereolithography (SLA) process. Stereolithography is a type of additive manufacturing technology in which a light source is used to solidify the chemical monomers to create structures. For both the force sensors, strain gauges were used because of its easy installation process and simple signal readout circuit. Both the sensors had shown good resolution capabilities with high linearity properties.

A flexible capacitive based force sensor, which works with the principle of parallel plate capacitor in which the dielectric medium is printed, using the 3D printing technology, was developed. The developed sensor was able to measure the sub newton force with cost effectiveness. This sensor also suffered from limitations of higher drift (Schouten et al (2017)).

Wasserfall et al (2017) tried a cost-effective force sensor in which all the mechanical parts have been developed using the 3D printer. The force sensor had a flexible cantilever structure where the displacement is measured using the optical proximity sensor. The developed sensors had good linear properties but the mechanical deformation of the 3D printed cantilever structure was expected to affect the sensitivity and accuracy of the force sensor. A stretchable force sensor with micro-pyramid array arrangements was developed by Shao et al (2018). The fabricated sensor had one flat upper layer and bottom layer with pyramid structure. The flexible layers were coated with poly (3,4-ethylenedioxythiophene) polystyrene sulfonate (PEDOT: PSS) a conductive polymer with good flexible property. The change in resistance value with respect to the applied force is used as output signal from sensor. However, lower accuracy with higher error value was reported as major limitation. Yang et al (2018) fabricated a hydrogel based flexible force sensor using 3D printer. The sensor was printed in a zigzag pattern so that the solidification of hydrogel will be uniform in nature. PDMS is used as a substrate. The developed printer has shown high sensitivity and repeatability.

Figure 2. 3D printed force sensor a) Capacitive force sensor b) TPU based flexible capacitive force sensor c) Modular force sensor embedded in robotic foot d) Micro-pyramid based flexible force sensor and e) Hydrogel-based force sensor

4. Pressure Sensor

Pressure sensor finds a variety of everyday applications either directly or indirectly. An attempt is made from last decade to reduce the fabrication cost of these sensors. Shi et al (2015) represents a flexure pressure sensor forged by means of 3D printed technology combined with the bio simulant artefact. The fabricated sensor consisted of three layers naming successively namely leather skin, ballistic gelatin and bottom one a sensor. The top two layers are categorized as bio simulant artifact. The rigid disk present at the bottom of the sensor acted as a mechanical support for the stable positioning of the sensor. The application resembled the motion of a wiper along a resistive mechanism to detect the strain caused by applied pressure.

Microlithography can also be successfully used by merging functionalities of multiple components in a single printed sensor (Lucklum et al (2016)). The vital parts of the sensor include fluidic connector, pressure chamber, membrane wall, contact pad area completely made using additive manufacturing. The unique arrangement and assembly of the sensor makes it quite suitable for applications involving bridge circuits. 3D printed pressure chamber has the capability to provide highly précised outputs and resistive readout being inbuilt in the sensor makes it useful for various analytical process.

Lathers et al (2017) presented their analysis on the piezo resistive pressure sensor employed in prosthetic feedback system. They used Polystyrene Ethylene Butylene Styrene (PEBS) material of low elastic modulus comprised with the carbon in its graphite state and found out that 3D printed sensors are comparatively good in terms of reliability and handling. It was also quoted that; this sensor is an improvisation model of the current prosthetic pressure sensor. An innovative approach in fabricating a robust pressure sensor in terms of aesthetics and reliability was tried by Ntagios et al (2018). Component aggregates being used in the sensor makes it the best manufacturing method at the first place and ability to be printed using additive manufacturing eliminates the wiring complexity. This device was tested with varying loading conditions and it is found that the capacitance of the sensor increase with the pressure by which a small change can be measured.

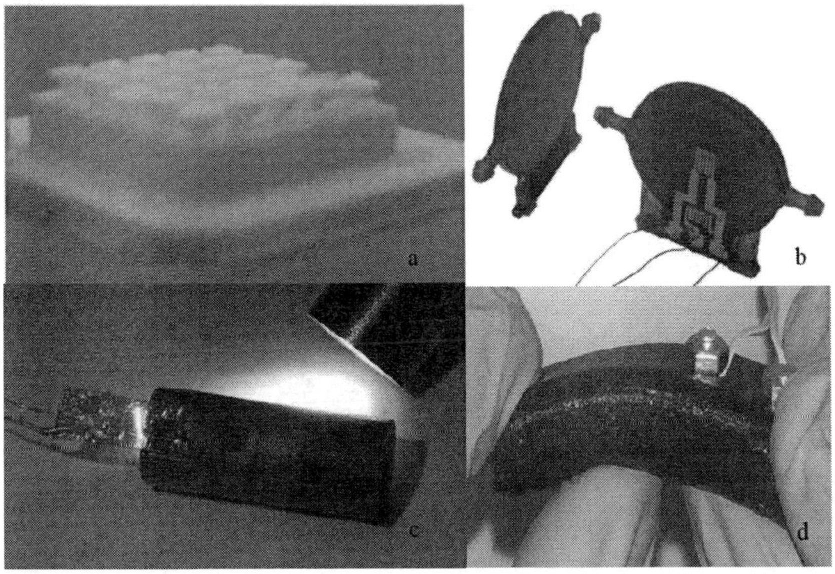

Figure 3. 3D printed pressure sensor **3a)** Structural pressure sensor; **3b)** Strain gauge pressure sensor; **3c)** Flexible pressure sensor; **3d)** Embedded pressure sensor

IOP Conf. Series: Materials Science and Engineering **764** (2020) 012055 doi:10.1088/1757-899X/764/1/012055

5. Other Sensors

Apart from pressure and force sensor, it is possible to fabricate different sensors such as accelerometer, angular sensor, acoustic, magnetic and optical sensors using this technology. Three types of capacitive sensors namely a single wire sensor (320 μm), large area meshed sensor, a completely encapsulated one are fabricated with bulk copper wire, and copper meshes by Shemelya C et al (2013). The developed sensor is able to differentiate salt water with distilled water. The other applications include biomedical sensing, human interface devices, electronic characterization and environmental sensing. Maldonado et al (2019) presented the fabrication of 3D printed humidity for sensor environmental condition monitoring. Process involves printed objects based on the copper (II) 1D coordination polymer whose sensing capacity depends on the temperature and the composition of the solution.

Sajid et al (2018) formulated a method to design and fabricate a linear resistive temperature sensor using FDM. Resistance of the material increases with increase in temperature. Thermal expansion alters the contact between the GNR thereby change in resistance is obtained at good stability. They claim that these sensors can be used with environmental conditions and weather monitoring systems, automated monitoring systems compatible at both wet and dry conditions and amphibious electronic gadgets.

A completely passive wireless soil moisture detection was suggested by Farooqui et al (2018) who prepared sensor that used a capacitive circuit. The capacitor had two metal electrodes with an air channel in between which when inserted in the soil acts as dielectric. Monitoring the soil moisture and texture of the agricultural lands in remote areas and plots distributed over a large area. Conducting hydrological studies at reduced time complexity are some of its applications recommended by the author.

Muth et al (2014) fabricated a stretchable strain sensor using 3D printed elastomer. A conductive ink is printed on an uncured elastomer substrate and it is capped with filler fluid, an array of strain gauges is fabricated and results are validated. Whereas Kim eta all (2017) used a composite dough for fabricating the ultra-stretchable strain sensor. The fabricated sensor has an ability to distinguish the actual body movement information and has the capability to detect even small strain values. Liu and Li (2017) used double network hydrogel to fabricate stretchable strain sensor and added an additional self-healing property.

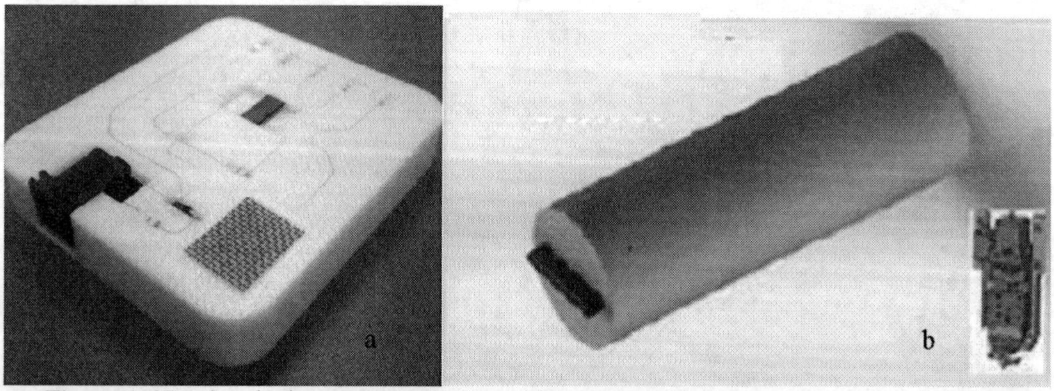

Figure 4. 3D printed other sensors; **4a)** Capacitive sensor; **4b)** Temperature sensor for robots

6. Materials and Fabrication

The material selection is the first step in fabricating the sensor. The important sensor characteristics such as sensitivity, accuracy is mainly depending on the material selection and the fabrication process. The sensor fabrication process is one of the time-consuming process, which requires very high technology including clean room technology to fabricate a simple sensor. The boom in additive manufacturing has eliminated the afore mentioned problems to a greater level.

Based on the sensor application the selection of material and fabrication method are done. It is possible to fabricate different sensors such as accelerometer, angular sensor, acoustic, magnetic and optical sensors using this technology. The materials include variety of thermoplastic polymers, photosensitive resins and some metal nano particles. The fabrication process depends upon the material selection and each material includes certain melting and curing temperature. Acrylonitrile Butadiene Styrene (ABS) and Polylactic Acid (PLA) are commercially available thermoplastic polymers. They are mostly preferred because of their easy fabrication and mechanical properties. Photo polymers are materials, which gets solidifies when exposed to ultraviolet lights. Researchers are working with different materials and fabrication process to reduce the printing time and cost.

7. Conclusion

Traditional sensor fabrication technique involves higher production cost, tedious fabrication steps with selected layout. The 3D printing of sensors can eliminate all the problems by incorporating new technology and materials. The demand for industry 4.0 and IOT increases with increase in sensor usage, variety of new technologies evolve out of it. The 3D printing of sensor increases the accuracy and helps in achieving customized molds which in turns reduces the overall cost factor. Development in multi-functional 3D-printing technology has the possibility to realize more sensors for future research and in biomedical and robotics applications.

References

[1] S. B. Kesner and R. D. Howe, "Design principles for rapid prototyping forces sensors using 3-D printing," *IEEE/ASME Trans. Mechatronics*, vol. 16, no. 5, pp. 866–870, 2011.

[2] T. Butzer, B. Vigaru, and R. Gassert, "Design and evaluation of a compact, integrated fMRI-compatible force sensor printed by additive manufacturing," *IEEE World Haptics Conf. WHC 2015*, pp. 158–164, 2015.

[3] J. Qu, Q. Wu, T. Clancy, and X. Liu, "Design and calibration of 3D-printed micro force sensors," *2016 Int. Conf. Manip. Autom. Robot. Small Scales, MARSS 2016*, no. Htm 140, 2016.

[4] M. Schouten, R. Sanders, and G. Krijnen, "3D printed flexible capacitive force sensor with a simple micro-controller based readout," *Proc. IEEE Sensors*, vol. 2017-December, no. 3, pp. 1–3, 2017.

[5] F. Wasserfall *et al.*, "June 26, 2017 10:3 WSPC - Proceedings Trim Size: 9in x 6in paper," vol. 3, 2017.

[6] Y. Shao, Y. Zhao, M. Liu, Q. Zhang, and C. Liu, "Flexible Force Sensor with Micro-Pyramid Arrays Based on 3D Printing," *Proc. IEEE Sensors*, vol. 2018-October, pp. 1–4, 2018.

[7] R. Yang *et al.*, "Transparent and flexible force sensor based on microextrusion 3D printing," *Micro Nano Lett.*, vol. 13, no. 10, pp. 1460–1464, 2018.

[8] I. S. Testing, "A 3D Printing Flexure Pressure Sensor for Robot Impact Safety Testing IEngineering Laboratory , National Institute of Standards and Technology , USA 2 Key Laboratory of Electronic Equipment Structure Design , Xidian University , China," pp. 2–7.

[9] I. S. Testing, "A 3D PRINTING FLEXURE PRESSURE SENSOR FOR ROBOT IMPACT SAFETY TESTING IEngineering Laboratory , National Institute of Standards and Technology , USA 2 Key Laboratory of Electronic Equipment Structure Design , Xidian University ,

IOP Conf. Series: Materials Science and Engineering **764** (2020) 012055 doi:10.1088/1757-899X/764/1/012055

China," pp. 2–7.

[10] F. Lucklum and G. Dumstorff, "3D printed pressure sensor with screen-printed resistive read-out," *Proc. IEEE Sensors*, pp. 5–7, 2017.

[11] C. Shemelya *et al.*, "3D printed capacitive sensors," *Proc. IEEE Sensors*, no. 3, pp. 0–3, 2013.

[12] J. Z. Gul, M. Sajid, and K. H. Choi, "3D printed highly flexible strain sensor based on TPU-graphene composite for feedback from high speed robotic applications," *J. Mater. Chem. C*, vol. 7, no. 16, pp. 4692–4701, 2019.

[13] M. F. Farooqui and A. A. Kishk, "Low-Cost 3D-Printed Wireless Soil Moisture Sensor," *Proc. IEEE Sensors*, vol. 2018-October, pp. 1–3, 2018.

[14] N. Maldonado *et al.*, "3D Printing of a Thermo- and Solvatochromic Composite Material Based on a Cu(II)–Thymine Coordination Polymer with Moisture Sensing Capabilities," *Adv. Funct. Mater.*, vol. 29, no. 15, pp. 1–11, 2019.

[15] S. Lathers, M. Mousa, and J. La Belle, "Additive Manufacturing Fused Filament Fabrication Three-Dimensional Printed Pressure Sensor for Prosthetics with Low Elastic Modulus and High Filler Ratio Filament Composites," *3D Print. Addit. Manuf.*, vol. 4, no. 1, pp. 30–40, 2017.

[16] M. Ntagios, W. T. Navaraj, and R. Dahiya, "3D Printed Phalanx Packaged with Embedded Pressure Sensor," *Proc. IEEE Sensors*, vol. 2018-Octob, pp. 1–4, 2018.

[17] M. Sajid, J. Z. Gul, S. W. Kim, H. B. Kim, K. H. Na, and K. H. Choi, "Development of 3D-printed embedded temperature sensor for both terrestrial and aquatic environmental monitoring robots," *3D Print. Addit. Manuf.*, vol. 5, no. 2, pp. 160–169, 2018.

[18] J. T. Muth *et al.*, "Embedded 3D printing of strain sensors within highly stretchable elastomers," *Adv. Mater.*, vol. 26, no. 36, pp. 6307–6312, 2014.

IOP Conf. Series: Materials Science and Engineering **764** (2020) 012056 doi:10.1088/1757-899X/764/1/012056

Characterization of Stainless Steel 410 L weld bead for Plasma Transferred Arc Hardfaced Valve Seat Rings

N Srimath[1*], P **Sathyabalan**[2]

[1]Department of Mechanical Engineering, Sri Ramakrishna Engineering College, Coimbatore-641 022, India.
[2]Department of Mechanical Engineering, Kumaraguru College of Technology, Coimbatore-641 049, India.
*Author for correspondence E-mail: srimath@srec.ac.in

Abstract. Wear brings about drastic fallout in metallic components during their application. It is established that hardfacing process has substantially matured over the recent times. To start with, Stainless steel 410 L is hardfaced over the face of low carbon steel IS-2062 valve seat rings by plasma transferred arc hardfaced process. The main and interaction effects of Plasma Transferred Arc hardfaced variables for stainless steel were obtained. These investigations were based on a design matrix of five factors, five level factorial method. The Regression analysis was tested for scrutinizing the adequacy. The percentage dilution was optimized. The conformity test was conducted and the optimized results were verified, with the percentage of error being calculated. The impact factors on the dry sliding wear resistance of Stainless Steel 410 L alloy was investigated under conditions leading to an obdurate metallic wear condition of the hard facing alloy. Mathematical model was developed relating the wear and the main factors, such as, normal load, Disc Speed, track radius. The Optimized wear model was obtained for minimizing the wear rate. Mathematical models relating wear testing parameters to wear and coefficient of friction were developed. The developed mathematical model can be used to assess the wear and the wear was found to increase.

Keywords- *Plasma Transferred Arc Hard facing, Friction, Wear*

1. Introduction

Fabrication or repeated rebuilding of composite wall section in materials by Plasma transferred arc hardfacing has significant advantages [11]. Substantial advancements are made in the hardfacing steam valve seat rings [14,8]. The Plasma transferred arc hardfacing process ensures a high degree of consistency permitting precise control of weld parameters [7]. Low dilution with crack free overlay necessitates the study of mathematical model, and the main and interaction effects of the various process parameters affecting the weld-bead dimensions, were factorial technique [2] was involved to reduce the cost and time. Automatic PTA hard facing was carried out by depositing stainless steel SS410L onto a carbon steel (ASTM A105) ring having inner diameter 76 millimeters, outer diameter 92 millimeters, thickness 15 millimeters and the observed data was used to develop the model. Weld-bead geometry of valve seat rings [9] were clad for prediction and thereby optimising dilution.

This research article discusses about the mathematical models used for predicting effects of Plasma transferred arc hardfaced variables on dilution and bead geometry from the experimental data obtained. The conformity tests were compared with the results of Artificial Neural Network by back propagation. Also, the wear behaviour of stainless steel 410 L alloy deposited on carbon steel IS-2062 ring are analysed. Sliding wear test using a pin on disc machine was conducted. Mathematical models relating to wear was developed using. The effect of wear testing parameters was analysed.

2. Methodology

2.1. Identification of process variables

The process variables identified for affecting the quality: Area denoted by A, travel speed denoted by S, powder feed rate denoted by F, oscillation frequency denoted by H and torch standoff denoted by N. The experiments were conducted by forming a single layer with DC electrode negative. Industrially pure argon was used at a constant flow rate for shielding at 12 $Lmin^{-1}$, for plasma at 3.5 $Lmin^{-1}$ and for powder feeding at 3 $Lmin^{-1}$.

By altering one of the parameters, trial experiments were conducted by keeping rest constant. Trial runs were conducted after examining the weld-bead for an even exterior when flaws. The coded factors for upper limit was taken +2, and the lower limit was taken as -2. The in-between range was calculated using Equation (I). The response function in lieu with the dimensions are expressed [3] as y = f (A, S, F, H, N) where y is the yield. The 2nd order polynomial regression equation for the 5 factors is articulated.

$$Xi = 2[2x-(Xmax - Xmin)]/ (Xmax - Xmin) \qquad (I)$$

Where, the required coded value of a variable is Xi extending from Xmax to Xmin ; with Xmax taken as the superior level and Xmin taken as the lower level of the variable.

2.2. Design matrix and Conducting experiments

A central composite rotatable factorial design matrix [5] consists of thirty-two sets of coded conditions and encompassing a half duplication of 24 with16 factorial design plus 6 centre points and 10 star points. An automatic Plasma Transferred Arc hardfacing system, designed and fabricated by M/s Omplas Systems, Coimbatore, India was used for experimentation. The surfacing system ranges for a table rotational speed of 10mms-1 and 750mm diameter with 0.05mm resolution. The design matrix experiments were conducted at random and thereby avoided errors in random. The deposition of Stainless steel SS410 L was completed over the un-preheated carbon steel IS-2062 valve seat ring.

2.3. Design matrix and Conducting experiments

A central composite rotatable factorial design matrix [5] consists of thirty-two sets of coded conditions and encompassing a half duplication of 24 with16 factorial design plus 6 centre points and 10 star points. An automatic Plasma Transferred Arc hardfacing system, designed and fabricated by M/s Omplas Systems, Coimbatore, India was used for experimentation. The surfacing system ranges for a table rotational speed of 10mms-1 and 750mm diameter with 0.05mm resolution. The design matrix experiments were conducted at random and thereby avoided errors in random. The deposition of Stainless steel SS410 L was completed over the un-preheated carbon steel IS-2062 valve seat ring.
The initial temperature is maintained constant throughout the simulation process. The speed of material deposition is equal to the travel speed of the nozzle. The extruded material can be modeled as a finite volume with imposed heat energy.

2.4. Responses Recorded

In-order to achieve a typical test specimen which is 10 mm wide, the hardfaced valve seat rings was cross-sectioned radially at their mid points. The bead geometry specimens were prepared by etching using 2% nital for the bead geometry measurements. An optical profile projector was used to trace the profiles of the weld-bead and the bead dimensions; viz, penetration P, reinforcement R and width W were measured. A digital planimeter was used to measure the weld metal forming and reinforcement thereby to calculate percentage dilution.

2.5. Assessing Adequacy of models

The Analysis of variance technique [5, 3, 10] was used to test the adequacies of models. For this, to assess the adequacy of models, it was checked if the calculated value of F-ratio of did not surpass the standard tabularized value for the preferred level of confidence (95%) and the same for R-squared value. The final mathematical models that predicted the effects of the altered parameters for the weld-bead dimensions are found to be adequate and are given in Table 1. The Confirmatory test experiments directed to determine the accuracy of the mathematical model using the same investigational research setup.

Table 1. Final Mathematical Models

S. No.	Mathematical Model
1.	Penetration, P=1.204+0.394A-0.135S-0.110F+0.048H+0.218N+0.150A2+ 0.116S2+ 0.094H2-0.268AS+0.132AN+0.137SF -0.336SN
2.	Reinforcement, R=4.179-0.202A+0.046S+0.289F-0.049H+0.083N-0.150S2-0.107N2 +0.160AF-0.238AH-0.213AN+0.255SF+0.319SN -0.228HN
3.	Dilution, D=5.990+1.215A-0.902S-0.569F-0.191H+0.189N+0.312F2+0.326H2 +0.444N2+1.077AN- 640SN+0.418HN
4.	Width,W=10.160+0.403A-0.086S+0.100F-0.173H+0.176N-0.318H2-0.414AF +0.444AH-0.765AN+0.239SF-0.245SH+0.261FH-0.363FN+0.650HN
5.	TotalArea,,TA=37.586+0.767A-0.556S+1.159F-0.309H-0.916N-1.374S2-1.23N2- 2.387AS-1.032AH- 3.009AN+2.281SF+1.748SN

3. Artificial Neural Network

In Artificial Neural Network, the output or the yield layer receives signals from the hidden layer and does calculations to give the yield signal, which represents the neural network response for the specific input vector [4, 13]. For predicting the weld-bead geometry of valve seat rings, efforts were done and the output layer was taken as the hidden layer [12, 16]. If the initial weights are too small the net input to the hidden or output unit would be close to zero, which also causes extremely slow learning [6]. Neural network inputs typically range from 0 to 1 (inclusive) and usually the output ranges from 0 to 1 continuously [1]. The Comparison of experimental and predicted output was calculated for the predicted dimensions of bead parameters using the test data from neural network.

4. Results and Discussion

For investigating the range of parameters and in order to predict the weld-bead geometry and dilution, the respective values were taken in the coded form and were substituted utilizing the mathematical model equations. The effects of process parameters on the bead geometry were plotted and computed using these models. The values of the control parameters were obtained by substituting the coded form values of the desired bead geometry. Figure 1 shows that the welding current A increases with an increase in penetration P and with a decrease in reinforcement R. With an increase in A, the total area TA increases. With an increase in A the Dilution D increases. Figure 2 shows that with an increase in travel speed S; the values of R, D, W and TA decreases. With any further increase in S, the value of P decreases with increase in S to an optimum value and then increases. Figure 3 shows that with an increase in F, the powder feed rate F increases with increase in the values of P, R, W and TA. And thus, Dilution D decreases. Figure 4 shows that while Oscillation frequency H increases with an increase in D and W and it has a very less effect on P and R. Figure 5 shows with an increase in torch standoff N, the total area TA increases and width W increases slightly while P and R have less significant effect on P and R. Dilution D decreases upto an optimum value and then increases with further increase in N. The mathematical model for the weld-bead by artificial neural network is predicted accurately. The Comparison of experimental and predicted data for dilution by artificial neural network is shown in figure 6.

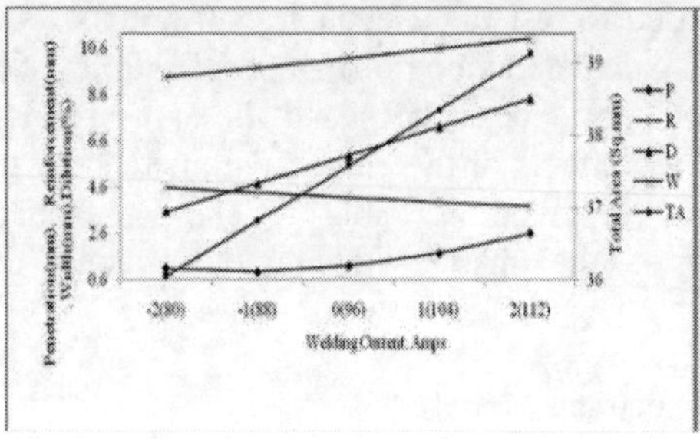

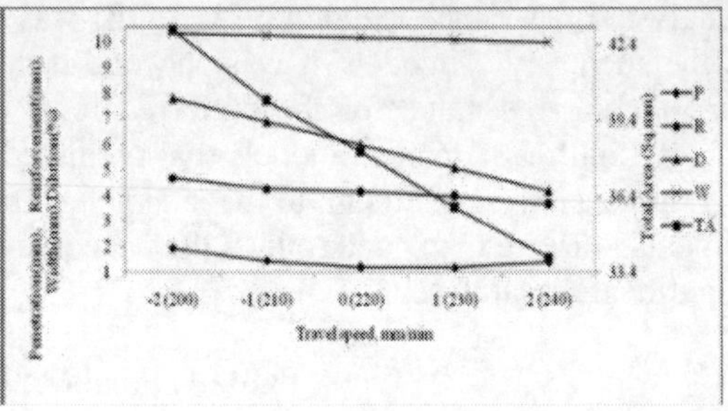

Figure 1. Direct effect of welding current on bead dimensions. **Figure 2.** Direct effect of travel speed on bead dimensions.

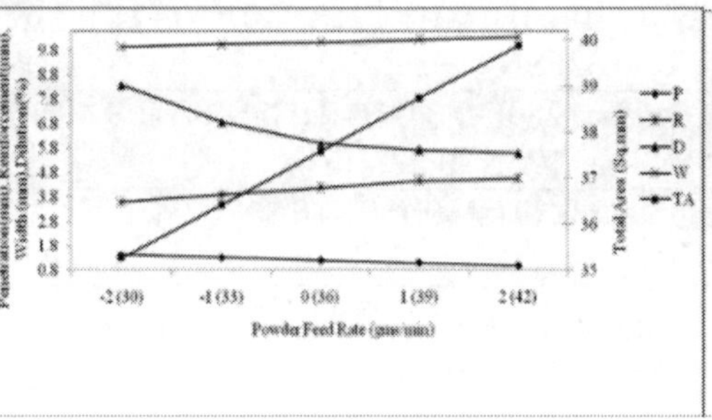

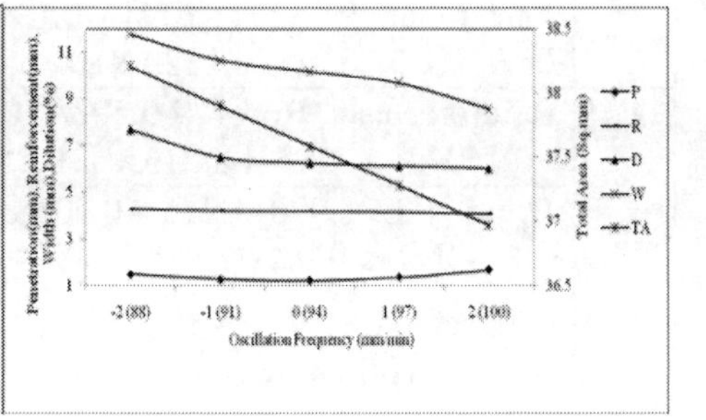

Figure 3. Direct effect of powder-feed-rate on bead dimensions. **Figure 4.** Direct effect of oscillation frequency on bead dimensions.

4.1. Wear Analysis

Wear testing pin specimens of size 10 x 10 x 10 mm3 were prepared from the stainless hard facing deposited at optimum dilution condition by machining. Dry sliding wear test was conducted at room temperature (30̊ C) with a Pin-on Disk type machine. A circular disk made up of C45 steel with a diameter of 200 mm and heat - treated to a hardness of HRC 60 prior to wear test was installed in the machine chuck to slide against the mating pin mounted in the holder. The disk was slid against a #1000 grade abrasive paper for 1 min to provide a sound surface contact between pin and disk. Lesser importance effects were found from sliding speed and effects of load on the wear behavior and on wear studies [15]. For conducting the experiments, Model TR 20, a computer controlled wear and friction monitor was employed.

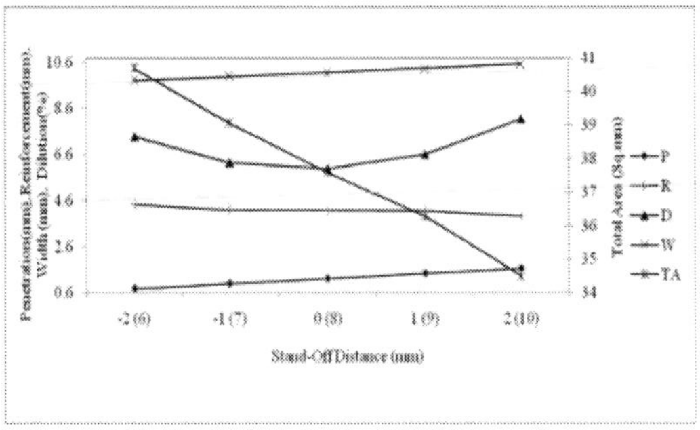

Figure 5. Direct Effect of Stand-Off Distance on Bead Dimensions

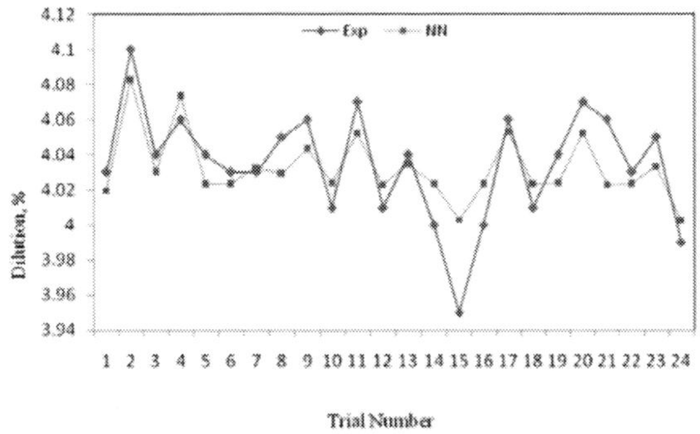

Figure 6. Comparison of experimental and predicted data for dilution

The wear and friction coefficient are recorded by the software. In order to avoid errors skulking, the experiments were conducted as per the design matrix. Coefficients of wear models were tested for their significance and the final mathematical models were constructed using the significant coefficients. The final models are given in Table 2.

Table 2. Final Mathematical Model of wear analysis

S. No.	Mathematical Model
1.	Wear = 37.55 – 1.64 NL - 1.64 TR + 1.86 DS + 4.82 NL 2 + 0.82 TR 2 + 3.12 DS 2 – 3.00 NL TR + 3.00 TR DS + 1.00 NL DS
2.	Coefficient of friction = 0.37 + 0.04 NL - 0.11 TR + 0.05 DS + 0.04 NL2 + 0.05 TR2 – 0.03NL TR + 0.04 DS TR + 0.01 NL DS

Using QA Six Sigma DOE package, the values of the coefficients were calculated by using linear regression analysis for finding out the responses. The Analysis of variance for testing adequacy of models, square multiple R-values and standard error of estimates are adequate. From the interaction effect of track radius and normal load on wear, it is found that normal load the wear increases gradually and reaches maximum as shown in Figure 7. At higher normal loads, the wear increases as the track radius increases. From the interaction effect of track radius and normal load on co-efficient of friction, it is evident that with the increase in track radius provided for hard facing, the coefficient of friction increases gradually. At higher normal loads, the coefficient of friction increases as the track radius T increases. From the interaction effect of disc speed and normal load on wear, shown in Figure 8, it is obvious that with the increase in normal load provided, the wear increases gradually and reaches a maximum. The increase in disc speed has a commendable impact on wear.

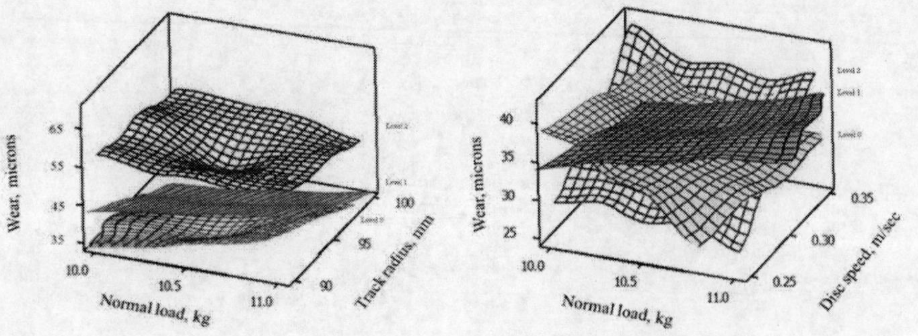

Figure 7. Response surface showing interaction effect of track radius and normal load on wear **Figure 8.** Response surface showing interaction effect of disc speed and normal load on wear

5. Conclusion

a) For developing mathematical models of desired quality for automated robotic surfacing, and a five level factorial technique program can be easily engaged. The dilution increases steadily, with an increase in welding current.

b) It can be observed from the scatter diagram for dilution in Figure 9, the neural network was more precise in predicting bead dimensions. Figure 10 shows the comparison of experimental and neural network predicted data of bead dimensions, indicating a very close proximity to each other.

c) For minimum number of experiments, the experimental model and predicted output of bead parameters from the neural network were compared. By performing the Conformity test, a small estimation error of 1.49% found out to for dilution reveals that dilution could be predicted accurately using the artificial neural network.

d) The QA Six Sigma DOE IV PC software package can be effectually employed for the predictions of wear behavior and finding the corresponding optimum process variables. The wear and coefficient of friction increased when the normal load and disc speed increased.

e) It was found that the average error for most of the models was less than 1%, thereby validating the model. on comparison of experimental and predicted output of wear. Also, it was found that the average error for the wear model was 1.148%

References

[1] Chang.W.S and Na.S.J, Prediction of laser spot weld shape by numerical analysis and neural network, Metallurgical and Materials Transactions, Vol.32, pp. 723-731, 2001.

[2] Cochran W.G. and Cox G.M., Experimental design, India, Asia Publishing House, pp.27-45,1992.

[3] Little R.J.A., DB Rubin D.B., Statistical analysis with missing data, New York, Longmount, pp15-27, 2014.

[4] Fausett, Fundamentals of Neural Networks, Prentice Hall PTR, pp.220-235, 1996.

[5] Grainer S., Engineering coatings- design and application, 1st edition; Cambridge, UK, Abington Publishing, pp 50- 68, 1989.

[6] Gunaraj V. and Murugan N., Prediction and Optimisation of weld-bead volume for submerged arc process, Welding Journal, Welding Research Supplement 79, (9-10), 331-337, 2000.

[7] Srimath N. and Murugan N., Development of mathematical models for prediction of weld-bead geometry in cladding mild steel valve seat rings by plasma transferred arc hardfacing , Procedia Engineering, 38, pp. 15-20, 2012.

[8] Kunrad H., Klimpke R.and Zilling H., VGB Kraftwerkstechnik, 74, (11), pp.795-802, 1994.

[9] Marimuthu K. and Murugan N., Prediction and optmisation of weld-bead geometry of Plasma transferred arc cladding valve seat rings, Surface Engineering, Vol. 19, No.2, pp.143-149, 2003.

[10] Murugan N. and Parmar R.S., Journal of Material Processing Technology 41, 381-398, 1994.

[11] Parmar R.S., Welding process and technology, New Delhi, Khanna publishers, pp.153-162, 1992.

[12] Philip.D. Wasserman, Neural Computing Theory and Practice, Van Nostrand Reinhold, pp 12-21, 1989.

[13] Rajasekaran.S and Vijialakshmi.G.A, Neural Networks, Genetic Algorithms and Fuzzy systems-Synthesis and applications, National conference on neural networks and fuzzy systems, Chennai, pp.298-309, 2003.

[14] Wu J. B. C. and Redman J.E., Welding Journal, 73, pp.63-68, 1994.

[15] Jerry Banks Discrete-Event System Simulation, Third Edition, Prentice-Hall of India Private Limited, pp. 16-24, 2004.

[16] Srimath N. and Murugan N., Prediction and optimization of weld-bead geometry of plasma transferred arc hard facing valve seat rings, European Journal for scientific research, 15, 2, pp. 285-298, 2011.

IOP Conf. Series: Materials Science and Engineering **764** (2020) 012057 doi:10.1088/1757-899X/764/1/012057

Comparison of seismic performance of reinforced concrete frame structure and composite frame structure using response spectrum analysis

Karthiga @ Shenbagam N[1*], Santhosh R[1] and Kannan V[2]
[1]Department of Civil Engineering, Bannari Amman Institute of Technology, Sathyamangalam, India.
[2]Department of Civil Engineering, Francis Xavier Engineering College, India.
[*]Author for correspondence E-mail: karthigashenbagamn@bitsathy.ac.in

Abstract. This research is mainly focused on the behaviour of the concrete and composite framed structures when such structural buildings are subjected to special loadings such as seismic loading. To evaluate the forces and failures in such building subjected to seismic forces linear dynamic statistical analysis method - Response spectrum analysis is carried out. This method results us with a complete response of time history at each and every particular joint by showing its displacements and member forces. In this an RC framed structure of M_{25} grade of concrete with G+7 storey and also a Composite structure of G+6 storey is modelled in ETABS as per Indian Standard Code 1893: 2002 Part(I). The building is situated in Himachal Pradesh with a seismic intensity of zone factor V. The concept of strong column weak beam is adopted here throughout the paper. Response spectrum analysis technique was used with the help of ETABS for both the composite and reinforced concrete structure. The results were explained in detail having the parameters in consideration with the base shear, storey deflection and storey drift are compared for each model.

1. Introduction

Low raise buildings are the types of building which prevail very common in India at the current situation, such buildings are constructed with the help of conventional reinforced concrete members that are found to be easier for finishing the structural work and is also found economical. But in the cities which are densely populated the only choice for construction is high-raise construction. High-raise buildings have the same conventional method of concrete construction that will be having many restrictions such as the span, depth of the member, stiffness etc., For such reasons Steel concrete composite structures can be adopted, Steel concrete composite structures are highly efficient than the conventional concrete structures from cost and performance factors such as increased stiffness, lesser deflection, longer span, less overall weight, reduction in the slenderness of column etc.

Since earthquakes are generally unpredictable natural disaster, in order to find the response of the building for such seismic loadings, we should confirm the serviceability of the structure to give proper warning before failure of the structure for the large scale of earthquakes and the structure must withstand for small scale of loadings. The performance based design of reinforced concrete and steel concrete composite buildings under seismic loading for various parameters are compared in this study and the results are given as follows.

2. Literature Survey

Vishwanatha S N and D S Sandeep Kumar investigated the seismic performance of concrete and composite buildings for different constraints of vertical irregularities in buildings and composite columns are installed in the structure to increase the strength and fire resistance of the column and also to resist lateral forces. Finally, they concluded from their study is that the maximum storey drift and displacement will increases as the vertical irregularities increases. Jingbo and Yangbing studied the behaviour of composite framed structure system under seismic loads are studied in detail. In this study four types of columns such as composite beam concrete filled square tubular column, Equivalent stiffness RC column and other two types are analyzed under response spectrum analysis. And from the analysis of those columns and frame they come to a conclusion that, in composite frames the maximum storey drift angle is reduced by 18% and in composite we can achieve greater span and height but in RC the columns must be enlarged to meet the desired bearing capacity.

3. Elastic Analysis of Systems

The analysis can be carried out in two methods one is by using the equivalent lateral force that is the static method and the other one is dynamic method which is response spectrum method. Both the methods will lead to the analysis procedures of the horizontal forces in the direction of the ground motion component. The main differences between both the methods are the distribution of the lateral forces throughout the height of the building and the magnitude. In this study we have analyzed the structure by using the dynamic method that is by using Response spectrum method.

3.1 Response Spectrum Analysis

It is the Linear Dynamic statistical method and this is also called as mode super position method or modal method. This is one of the type of linear analysis of structures. This method is based on the criteria that for definite forms of damping the response of each and every individual natural mode of vibration is calculated independently of the others and the results of the response are combined together to find the complete response of the structure. Each mode we analysed would reflect its own mode shape, natural frequency with its own modal damping. A complete modal analysis of the structure gives us the full history of the response in various parameters such as forces, deformations and displacements with a ground acceleration history. This method is mainly applicable for the analysis of the dynamic response of the structures, and for asymmetrical buildings and buildings with some irregularities such as vertical irregularity or discontinuity. Square root of the sum of squares method.

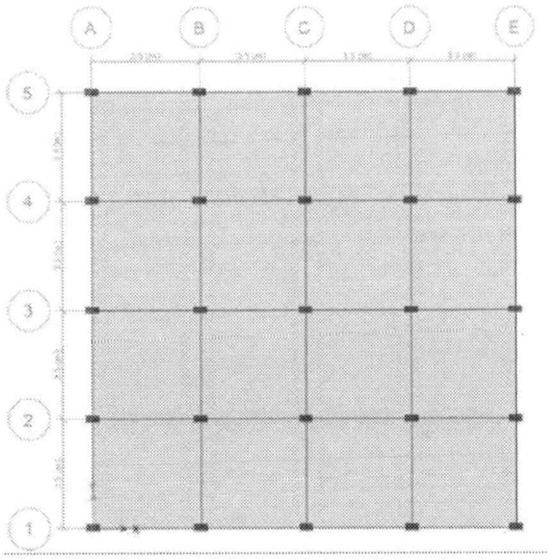

Figure 1. Plan of the structure.

IOP Conf. Series: Materials Science and Engineering **764** (2020) 012057 doi:10.1088/1757-899X/764/1/012057

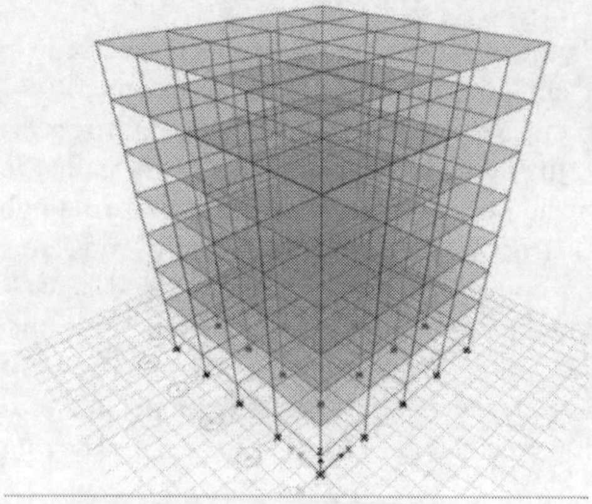

Figure 2. Frame model of structure in ETABS.

- Complete quadratic Combination method
- Absolute peak values are summed up

3.2. Description of the framed model

The building which is modelled in ETABS software are of the following dimensions.

No of storey	=G+6
Structure	= OMRF
f_{ck}	= 25N/mm^2.
f_y	= 415N/mm^2.
Spacing of frame	= 3.5m
Spacing of storey	= 3m
Type of Seismic analysis	= Response Spectrum analysis
Seismic zone	=V
Important factor	=1.5
Reduction factor	= 2 (medium)

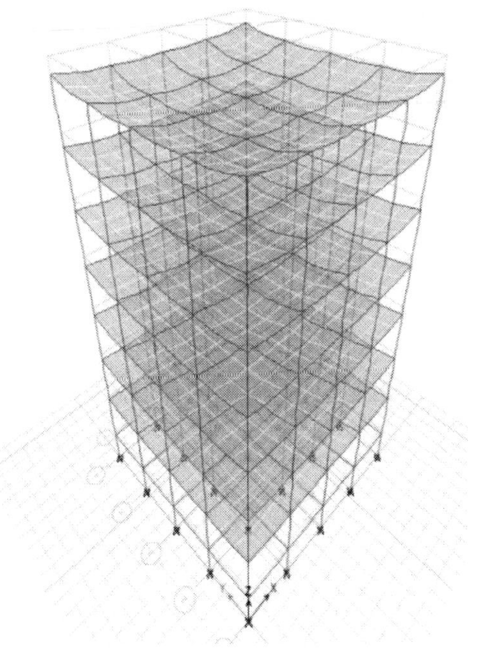

Figure 3. Deformed shape of RC structure.

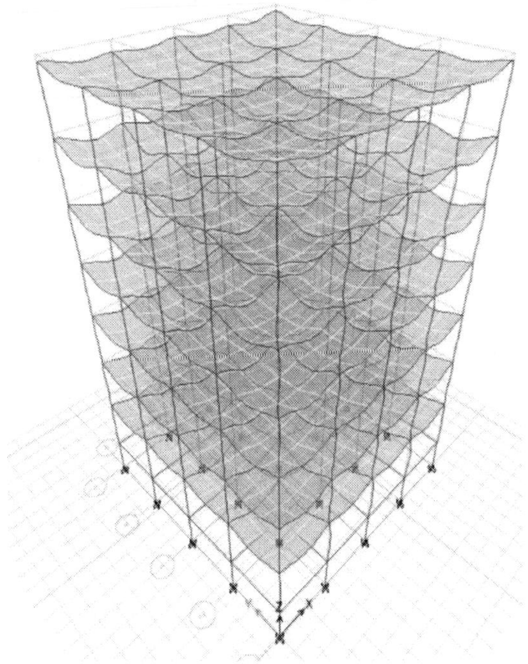

Figure 4. Deformed shape of composite structure.

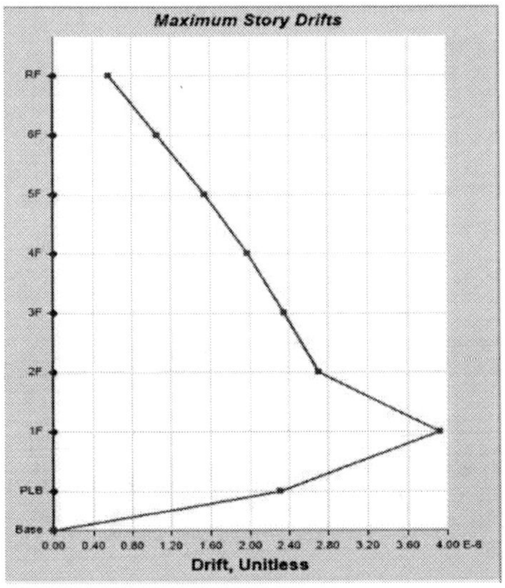

Figure 5. Deflection of RC structure.

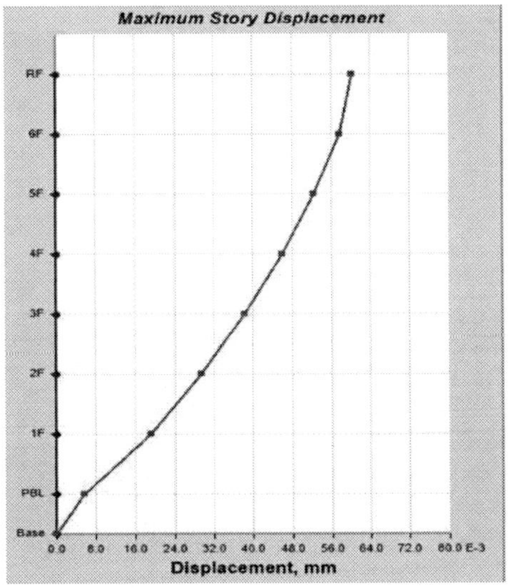

Figure 6. Deflection of composite structure.

IOP Conf. Series: Materials Science and Engineering **764** (2020) 012057 doi:10.1088/1757-899X/764/1/012057

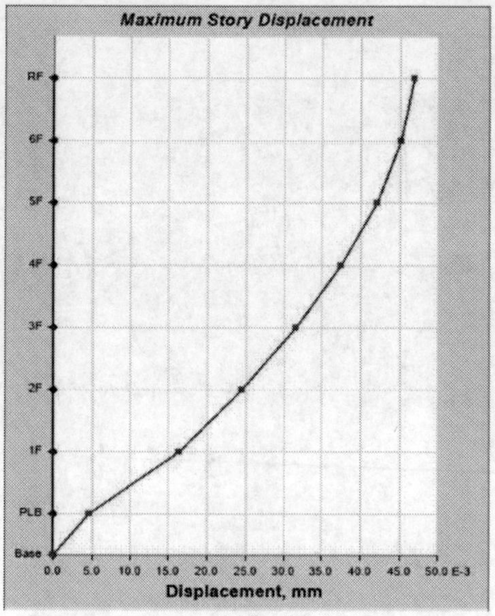

Figure 7. Story drift of RC structure.

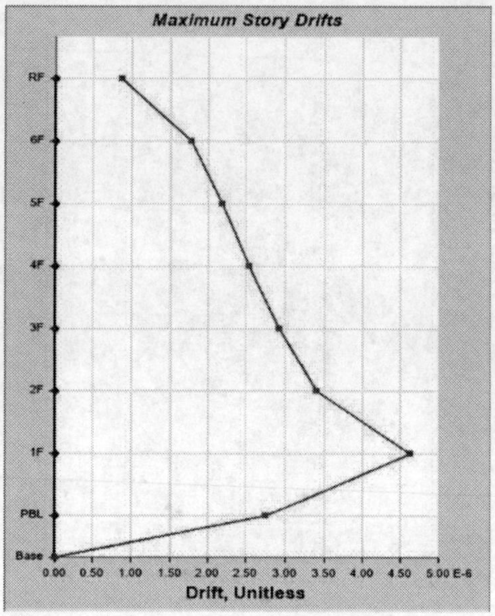

Figure 8. Story drift of composite structure.

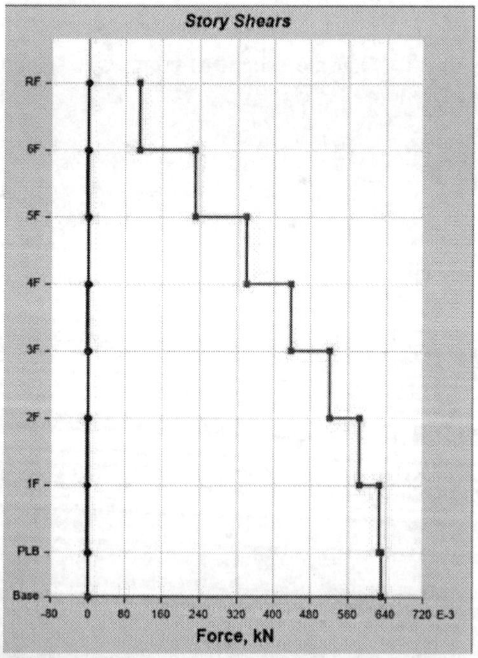

Figure 9. Base shear of RC structure.

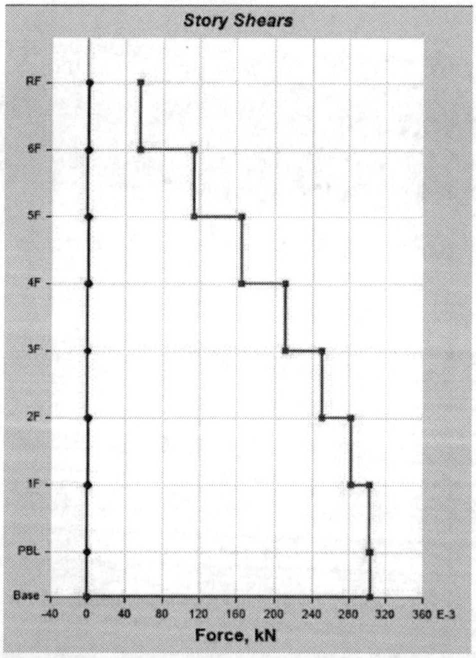

Figure 10. Base shear of composite structure.

4. Conclusion

On performing the seismic analysis of the RCC and composite G+6 Storey building under response spectrum method the following conclusions were obtained.

- The deflection of the composite structure is more when compared with the RC structure, Since the stiffness of the RC building is very high.

- The storey drift of the composite structure is found maximum when it is compared with the RC building. This is mainly because of the stiffness criteria. The RC structure has more stiffness so that it has minimum storey drift.
- When comparing the base shear parameter for both the composite and RC structure RC building has the maximum base shear. This is because of the self weight of the RC building.

References

[1] Vishwanatha S N, D S Sandeep Kumar, "Seismic Analysis Of Multistoreyed RCC And Composite Building Subjected To Vertical Irregularity", International Research Journal of Engineering and Technology, Vol.5(5), pp.3628-3633, 2018.

[2] LIU Jingbo, LIU Yangbing, "Seismic Behavior Analysis Of Steel-Concrete Composite Frame Structure Systems", The 14th World Conference on Earthquake Engineering, October 12-17, 2008, Beijing, China

[3] IS 11384:1985, "Code of Practice for Design of Composite Structure, Bureau of Indian Standards", New Delhi.

[4] IS 800: 2007, "Indian Standard Code of practice for General Construction of Steel in India, Bureau of Indian Standards", New Delhi.

[5] Earthquake resistant design of structures by S.K. Duggal.

[6] IS 456: 2000, "Code for practice of plain and reinforced concrete code of practice, Bureau of Indian Standards", New Delhi.

[7] IS 1893: 2002, "Code for earthquake resistant design of structures- general provisions for buildings, Part I, Bureau of Indian Standards", New Delhi.

IOP Conf. Series: Materials Science and Engineering **764** (2020) 012058 doi:10.1088/1757-899X/764/1/012058

Aerodynamic design and structural optimization of a wing for an Unmanned Aerial Vehicle (UAV)

K Raja Sekar*, M Ramesh, R Naveen, M S Prasath and D Vigneshmoorthy

Department of Aeronautical Engineering, Bannari Amman Institute of Technology, Sathyamangalam, Tamilnadu - 638401, India.

*Author for correspondence email id: rajasekark@bitsathy.ac.in

Abstract. This paper summarizes the process of structural design, analyses and optimization for weight reduction of a UAV wing. A semi-monocoque wing structure that consists of ribs, spars and skin are efficiently optimized for Weight minimization without reducing the Strength to weight ratio. Initially a solid wing surface is designed with the air foil co-ordinates of NACA 2412. The CFD Analysis is carried out to get pressure distribution on various arrangements of wing by changing angles of attack. From the CFD results the variation of coefficient of lift for different angle of attack is determined. The pressure fields obtained from CFD analysis is imposed on structural members for structural analysis. Optimization of this wing is done considering various cases until the required performance is obtained. The results are obtained for the optimized wing and this helps in obtaining reduced weight of the wing and increased strength to weight ratio. Results show that substantial improvement has been achieved in comparison to the standard model.

1. Introduction

Fixed wing Unmanned Aerial Vehicles (UAV) are aircrafts that operate without a human pilot onboard. They are controlled either remotely by a human operator or autonomously through onboard computers. These UAVs shown in the figure 1, fly by making use of the lift that is generated during the aircraft's forward motion and by the aerodynamic shape of the wing. They can be self-propelled, pure gliders or the mixture of two. Self-propelled fixed wing UAVs typically rely on forward thrust that is generated by an IC engine or electric motor powered propeller. Due to the heavier payload capabilities and long sustained flight with a minimum power makes them well suited to cover extended distances for missions such as mapping, surveillance and defense. They may also be better equipped to survive technical failure in the air, as many designs incorporate natural gliding capabilities when loss of propulsion occurs. Fixed wing UAV with a simpler structure and efficient aerodynamic design that provide the advantage of long range and endurance with high speeds. Only the wings structure of the UAV determines its efficiency in the sustained flight. The semi-monocoque wing consists of Spars, Ribs and Skin. The spar is the main load carrying member of the wing run through the tip to tip of a wing and ribs are gives the aerofoil shape to the wing placed in a frequent interval from the root to tip of the wing. The skin is very thin in nature and covers the internal structures also carry the shear loads.

Figure 1. Fixed wing UAV.

For increasing the payload capacity and optimum usage of construction materials structural optimization plays an important role [1-3]. The selection of suitable aerofoil for the high altitude and long endurance (HALE) UAVs are crucial in the UAV design process [4, 5] and to face challenges in the manufacturing and reducing the cost of materials and process the optimization provides guidelines for the designers [6-8]. The development of composite also provides the effective way for reducing the weight, increased structural efficiency, and provides reduction in cost of manufacturing and maintenance [9, 10]. From the Drag polar curves, at Mach number 0.7, confirm that the plan form of a wing strongly influences the pressure distribution and the drag coefficient with a lift coefficient ranging through – 0.25 and 0.25 the curved plan form wing develops the minimum of the drag coefficient thanks to reduction of both pressure drag and shock wave effects [11,12].

2. Problem statement

Unmanned Aerial Vehicle is considered as one of the promising research fields owing to its ability in carrying out critical operations. This development will abruptly increase in the next few decades. Lift is generated by the use of wing in a fixed wing UAV. The surface area is directly proportional to the lift generated by the wing. The weight of the entire UAV increases when long span wing is used. Assuming the total weight of UAV as 6 kg and 40% of the weight as structural weight, i.e2.4 kg is the structural weight of the aircraft. Assuming 50% of the structural weight for wing and the other 50% as fuselage, i.e 1.2 kg is the structural weight of the wings. With a help of structural optimization techniques how much weight can be reduced without reducing the strength with increase the payload capacity with long range and endurance.

3. Results and discussion

3.1. CADD modelling

The model is created by importing the co-ordinates of NACA 2412 in CATIA V5 software as indicated in the figure 2, figure 3 and figure 4. The type of wing model developed is tapered swept wing. The surface is extruded in the workbench by connecting the imported co-ordinates. The surface modelling is carried out for creating the surface. The rib at the root is modelled with 15 mm thickness and the rest of the ribs are of thickness 5 mm. The spar is welded with rib which runs from the leading to the trailing edge of the rib. There are 7 ribs that are positioned at appropriate spacing as listed in the table 1. The spar runs to the span of 1200 mm as the total length of the wing is 1200 mm. The chord length at root is 200 mm and tip is 50 mm. The wing length from root to mid measures about 400 mm and the length from mid to tip is about 800 mm. The sweep angle of the wing in the mid-section is taken as 12.4°.

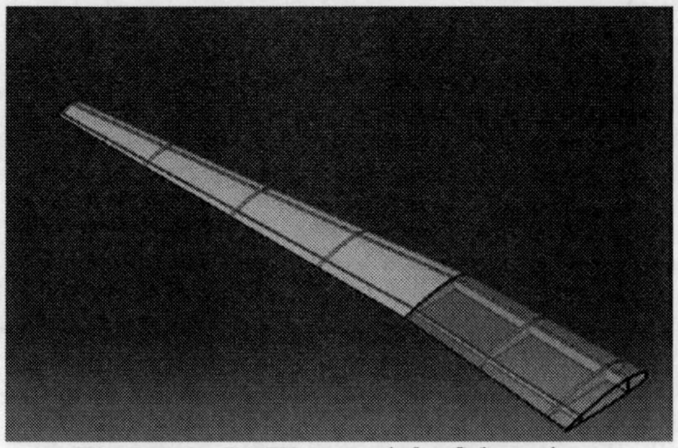

Figure 2. CATIA model of the wing.

Table 1. Dimensions of the model (mm).

Components	position	Thickness	Length	Chord length
Rib (1)	Root	15	-	200
Rib (3)	Mid	5	-	150
Rib (7)	Tip	5	-	50
Ribs (2,4,5,6)		5		
Spar (L)	Leading	5	1200	-
Spar (T)	Trailing	3	1200	-
Skin	-	2	1200	-

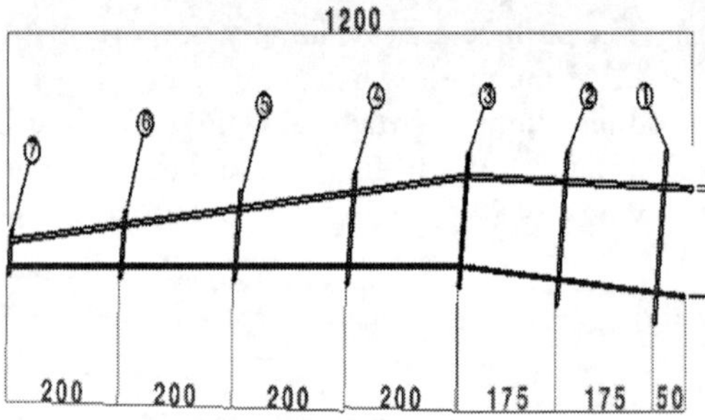

Figure 3. CADD model with dimensions in mm.

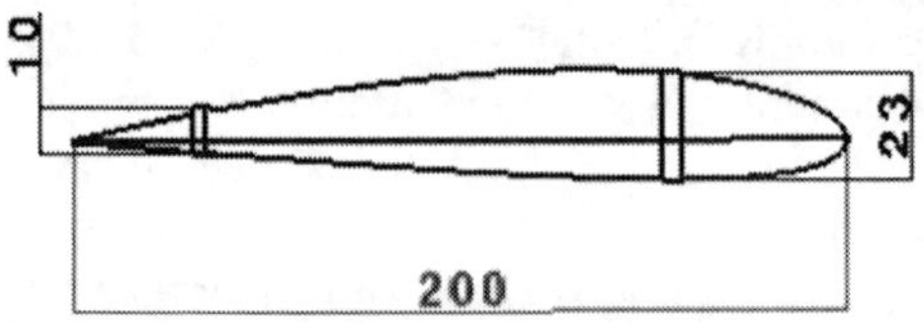

Figure 4. Root rib 2D drawing.

3.2. *CFD analysis*

The imported CADD model is subjected to mesh in the ANSYS (ICEM) software and analysis in the ANSYS (fluent) software. This analysis is carried out for different angles of attack and corresponding pressure distributions are taken out. The CADD model is imported into the ANSYS (ICEM 16.0). The wing surface alone is imported without the internal parts. After importing, the wing is sectioned and named as leading edge, trailing edge, wing tip and root to obtain fine meshing. A hemispherical control volume is created and the symmetry is created along the root of the wing in the geometry. The medium inside the far field is considered as air. Once the geometry is created, the model is now subjected to meshing.

3.3. *Meshing*

Tetrahedral meshing is done on the wing along with its control volume and symmetry. The obtained Mesh is now checked for its quality, skewness and orthogonality. Total number of elements after

IOP Conf. Series: Materials Science and Engineering **764** (2020) 012058 doi:10.1088/1757-899X/764/1/012058

meshing is 19, 05,495 and the total number of nodes is 3, 24,509. Prism layers are now created on the mesh to capture boundary layer with less numerical diffusion. The model is again checked for its quality, skewness and orthogonality. Total number of elements after the formation of prism layers is 25, 78, 240.The mesh used in the analysis is shown in the figure 5.

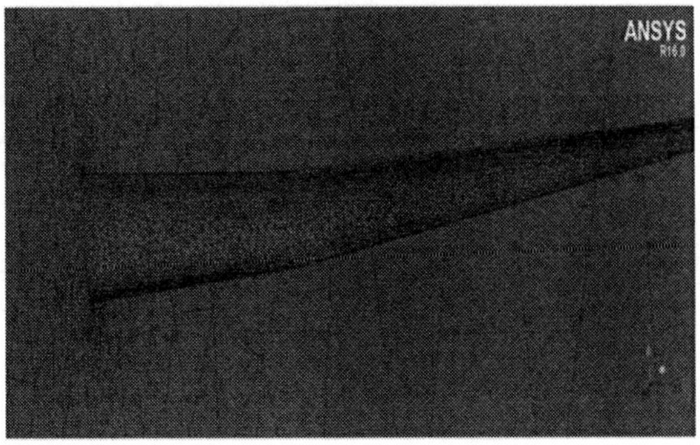

Figure 5.Mesh on wing.

3.4. Solver input and boundary conditions

The problem consists of flow around the airfoil at various angles of attack (-3, 0, 3, 6, 9, 12, 15, 18). The inputs and boundary conditions are presented in the table 2 below.

Table 2. Inputs given in CFD analysis.

Input	Value
solver	Pressure based
State	Steady state
Viscous model	Transition-k epsilon
Material	Air
Density	1.17664 kg/m^3
Mach number	0.2
Temperature	300 K
Enthalpy	4271.293 j/Kg
Pressure	101325 N/m^2
Reynolds number	6.38×10^6

3.5. CFD analysis and results

The mesh file from ANSYS is imported into the software FLUENT 16.0. The analysis at zero angle of attack is done applying appropriate boundary conditions. The pressure values are obtained and the values of lift and drag are calculated. The same procedure is repeated for other angles of attack and the corresponding values of lift and drag are calculated. The static pressure values obtained from the analysis for each angle of attack is exported as a NASTRAN file. The contours of pressure obtained for various angles of attack from CFD simulations are shown in the figure 6 and figure 7. According to the Bernoulli's principle the top surface will experience lower pressure when compared to the bottom surface. Hence, the wing is effectively pushed upward normal to the incoming flow stream.

IOP Conf. Series: Materials Science and Engineering **764** (2020) 012058 doi:10.1088/1757-899X/764/1/012058

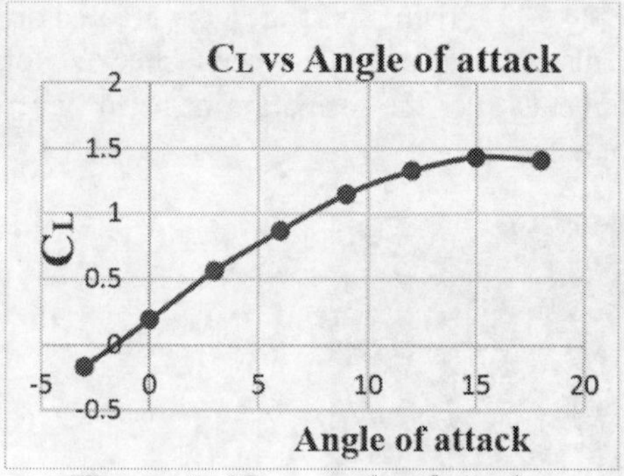

Figure 6. CL vs angle of attack.

Figure 7. Pressure contours at alpha 18°.

3.6. *Structural analysis*

The CADD model from CATIA is imported into MSC PATRAN. It is then subjected to meshing and the analysis is done from the results obtained from CFD. The CADD model of the wing along with the ribs and spars, which is saved as a STEP file is imported into a new database in the MSC PATRAN software. Total mass of the model is 2.27kg.

3.7. *Structural meshing*

The type of mesh created on the model is two-dimensional shell mesh. Initially, the ribs are meshed with quad and tri elements. Then, the spars are meshed using quad elements with respect to the mesh of the ribs to maintain the connectivity. Eventually, the skin is also meshed with quad elements maintaining the connectivity. Once the mesh is completed as shown in figure 8, equivalence is done and the FE model is verified.

Figure 8. Mesh on skin.

The pressure load at zero angle of attack is applied on the cantilever wing surface and analyzed for the structural stress and deformations are indicated in figure 9 and figure 10.

IOP Conf. Series: Materials Science and Engineering **764** (2020) 012058 doi:10.1088/1757-899X/764/1/012058

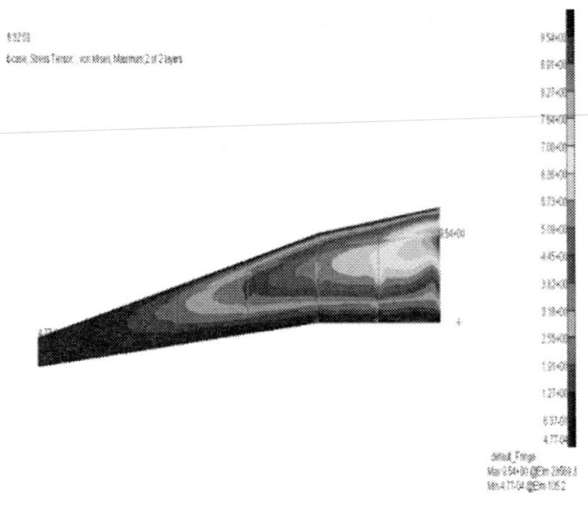

Figure 9. Von-mises stress on surface (α=0) in MPa.

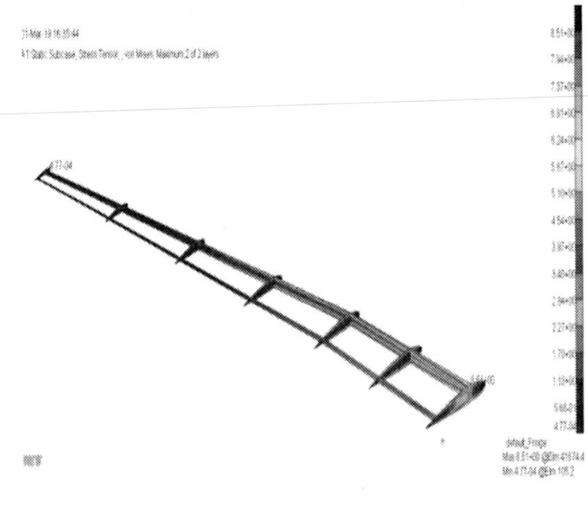

Figure 10. Von-mises stress on spars and ribs (α=0) in MPa.

4. Optimization

In order to increase the performance and reduce the weight of the wing, optimization is carried out. Series of optimization for angle of attack 0° is done to obtain better results. Here, manual optimization is carried out.

4.1. Optimization-1

Standard model after the analysis is taken into account. For optimization, the thickness of the skin is reduced to 1mm and analysis is carried out. The results are mentioned in the table 3 and result contours are displayed in the figure 11.

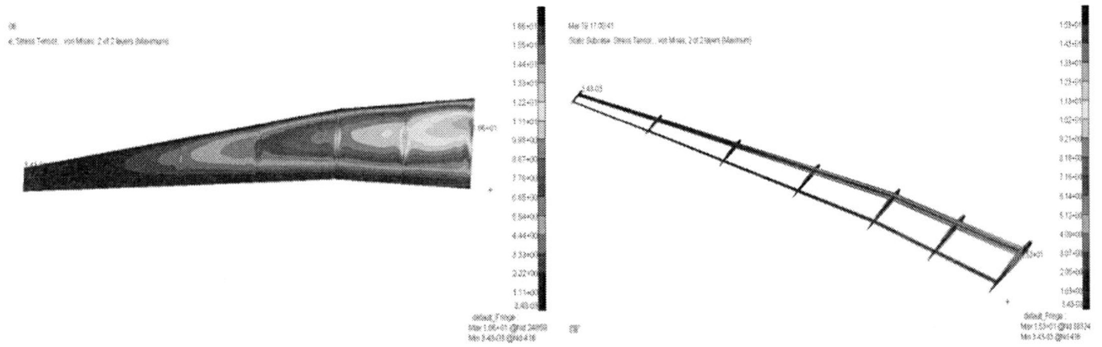

Figure 11. Von-mises stress on surface and spars and ribs (opt-1) in MPa.

4.2. Optimization-2

Mass of the ribs is reduced by creating cut-outs and analysis is carried out for angle of attack 0°. Optimized structure and resulting stress distributions are shown in the figure 12 and figure 13 below.

IOP Conf. Series: Materials Science and Engineering **764** (2020) 012058 doi:10.1088/1757-899X/764/1/012058

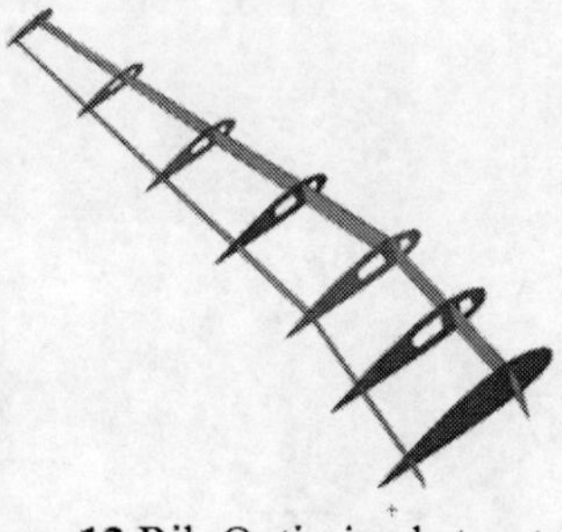

Figure 12.Rib Optimized structure.

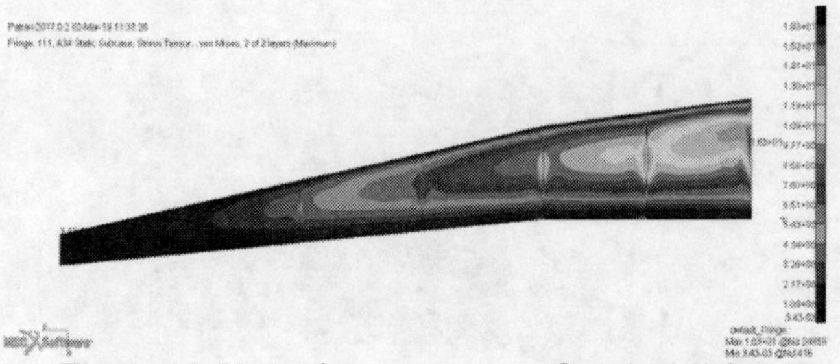

Figure 13. Von-mises stress on surface (opt-2) in MPa.

4.3. Optimization-3

Further optimization is done by reducing the number of ribs from 7 to 5. Analysis is carried out and results are obtained. The optimized structure is shown in the figure 14.

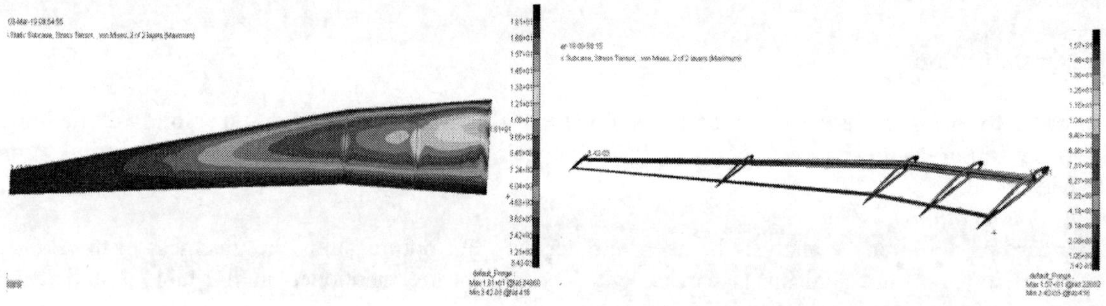

Figure 14.Von-mises stress on surface and spars and ribs (opt-3) in MPa.

4.4. Material Optimization-1

To reduce the weight of the wing further optimization is done. Now the material of the wing is changed from aluminium to composites. Composites are anisotropic material whose properties are directional dependent. For the case of laminate composite, each lamina is a continuum and behaves as a linear elastic material. Composites are used more in the aerospace industry due to its less weight and high strength. The composite material that we chose for the optimization process is Carbon epoxy, which is extremely strong and light according to the symmetric laminate is created in such a way that it has 8layers of lamina in an orientation of [0/90/45/-45] with a thickness of about 0.15mm. After the change in material of the wing structure, analysis is carried out. The material properties of the composite are listed in the table 3. The corresponding contour results are shown in figure 15.

Table 3. Properties of carbon-epoxy.

Young's modulus	E_{11}	70GPa
	E_{22}	25GPa
Poisson's ratio	N	0.1
Shear modulus	G_{12}	5 GPa
Density	P	1.6 gm/cm³

IOP Conf. Series: Materials Science and Engineering **764** (2020) 012058 doi:10.1088/1757-899X/764/1/012058

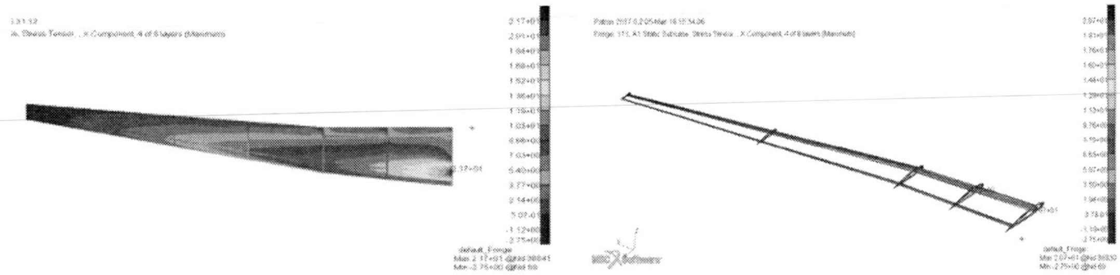

Figure 15. Layer stress obtained on surface and stress on spars and ribs (opt-4) in MPa.

4.5. Material Optimization-2

Further optimization is done to reduce weight by applying 4 layers of carbon-epoxy lamina on the skin and 8 layers of lamina on the ribs and spars. Analysis is carried out and the results are obtained as indicated in the figure 16 and figure 17.

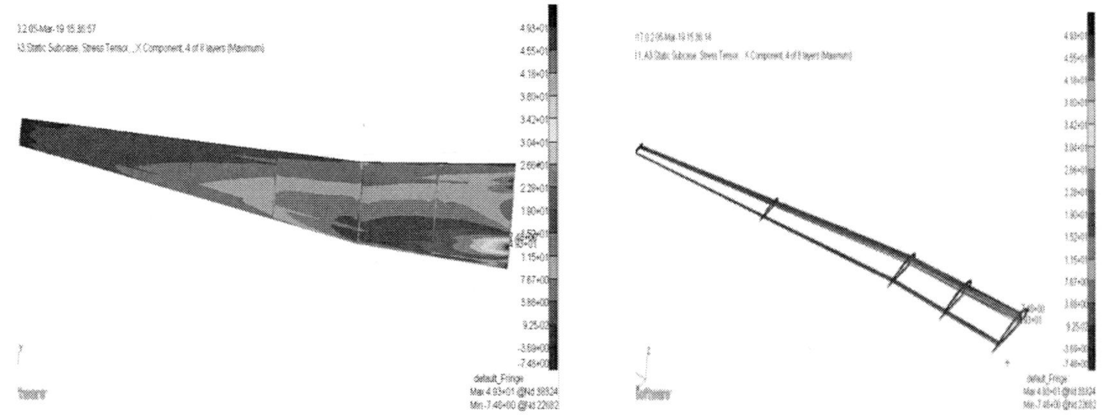

Figure 16. Layer stress obtained on surface and Von-mises stress on spars and ribs(opt-5) in MPa.

Figure 17. Deformation of the wing (opt-5) in mm.

Table 4. Optimization results.

Properties	**Structural optimization -1**	**Structural optimization -1**	**Structural optimization -1**	**Materials optimization - 1**	**Materials optimization - 2**
Mass (Kg)	1.46 (36% reduced from standard model)	1.42 (38% reduced from standard model)	1.35 (41% reduced from standard model)	0.64 (72% reduced from standard model)	0.35 (85% reduced from standard model)
Stress (MPa)	16	16.3	18.1	21.7	49.3
Deformation (mm)	10	10	10.9	17.5	32

5. Conclusion

From the above optimization process results indicated in the table 4 above, 40% of the mass is reduced in the optimized wing and the strength to weight ratio of the standard model is 7.47 whereas the optimized wing has strength to weight ratio of 31.8.The optimized wing is found to have more strength to weight ratio in comparison with the standard model. The factor of safety is an important parameter in the design of aerospace structures. From the above analysis we found that, for the optimized structure of the wing with the assumed dimensions the structure is safe even if the loads are six times more than the load that is applied, considering the factor of safety as 6.

References

[1] Balabanov V O, Haftka R T, "Topology optimization of transport wing internal structure", Journal of Aircraft, Vol.33(1), pp.232-233, 1996.

[2] Grihon S, Mahe M, "Structural and multidisciplinary optimization Applied to Aircraft Design", 3rd World Congress of Structural and Multidisciplinary Optimization, Buffalo, NY, May 17-21, 1999.

[3] De Faria, A R , Hansen J S, "On Buckling Optimization under Uncertain Loading Combinations", Structural and Multidisciplinary Optimization Journal, Vol. 21(4), pp.272-282, 2001.

[4] Alsahlani A , T Rahulan, "Aerofoil Design for Unmanned High-Altitude Aft-swept Flying-Wings", Journal of Aerospace Technology and Management, Vol.7(3), pp.335-345, 2017.

[5] Chiarelli M R, Cagnoni M, Ciabattari M, De Biasio M, Massai A, "Curved Planform Wings with a High Aspect Ratio for Aircrafts Operating in the Transonic Regime", Pub. No. WO/2010/150212, World Intellectual Property Organization, 2010.

[6] Hinrichsen J, Bautista C, "The challenge of reducing both airframe weight and manufacturing cost", Air & Space Europe, Vol.3(3-4), pp.119-121. 2001.

IOP Conf. Series: Materials Science and Engineering **764** (2020) 012058 doi:10.1088/1757-899X/764/1/012058

[7] Starnes J H , Haftka R T, "Preliminary design of composite wings for buckling, strength, and displacement constraints", Journal of Aircraft, Vol.16(8), pp.564-570, 1978.

[8] Mohamed Hamdan A, Nithiyakalyani S, "Design and Structural Analysis of the Ribs and Spars of Swept Back Wing", International Journal of Emerging Technology and Advanced Engineering, Vol.4(12), pp.208-213, 2014.

[9] Kong H Park Y Kim, K Kang, "Structural design on wing of a small-scale wig vehicle with carbon/epoxy and foam sandwich composite structure", 16th International conference on composite materials, 2007.

[10] Milan Kumar Pal, "Wing Design Optimization for a Subsonic Unmanned Aerial Vehicle", M.Tech Thesis, IIT Kharagpur, 2008.

[11] A C Marta, P V Gamboa "Long endurance electric UAV for civilian surveillance missions", In 29th Congress of the International Council of the Aeronautical Sciences, St. Petersburg, Russia, 2014.

[12] Nguyen Minh Triet, Nguyen Ngoc Viet, Pham Manh Thang, "Aerodynamic Analysis of Aircraft Wing", VNU Journal of Science: Mathematics – Physics, Vol. 31(2), pp.68-75, 2015.

AUTHOR INDEX